LIGHT IN BIOLOGY AND MEDICINE, Volume 2

LIGHT IN BIOLOGY AND MEDICINE
Volume 2

Edited by

Ron H. Douglas
The City University
London, United Kingdom

Johan Moan
The Norwegian Radium Hospital
Oslo, Norway

and

Györgyi Rontó
Semmelweis Medical University
Budapest, Hungary

PLENUM PRESS • NEW YORK AND LONDON

Library of Congress Cataloging in Publication Data

(Revised for vol. 2)
European Society for Photobiology. Congress (2nd: 1987: Padua, Italy)
   Light in biology and medicine.
   "Proceedings of the Second Congress of the European Society for Photobiology,
held September 6–10, 1987, Padua, Italy"—T.p. verso.
   Includes bibliographies and index.
   1. Photobiology—Congresses. 2. Phototherapy—Congresses. 3. Light—
Physiological effect—Congresses. I. Douglas, Ron H. II. Moan, Johan. III. Dall'
Acqua, F. IV. Title.
QH515.E97   1988   574.19′153   89-113434

ISBN 978-1-4684-5993-7   ISBN 978-1-4684-5991-3 (eBook)
DOI 10.1007/978-1-4684-5991-3

Proceedings of the Third Congress of the European Society for Photobiology,
held August 27–September 2, 1989, in Budapest, Hungary

© 1991 Plenum Press, New York
Softcover reprint of the hardcover 1st edition 1991
A Division of Plenum Publishing Corporation
233 Spring Street, New York, N.Y. 10013

PREFACE

This is the third book chronicling the scientific activities of the European Society for Photobiology (ESP). It contains 56 chapters, written by authors from 16 countries, based on presentations at the <u>3rd Congress of the European Society for Photobiology</u> held in Budapest, Hungary on the 27th August - 2nd September 1989.

The science of photobiology, which can simply be defined as the study of the effects of light on living matter, covers so many subject areas that no single book can hope to do justice to them all. This multidisciplinary nature of photobiology is reflected by the material covered in this volume, which contains chapters on such diverse themes as motile photoresponses in bacteria, cancer therapy and photosynthesis. Interestingly, the emphasis placed on various subject areas differs quite markedly from the preceding volume ('Light in Biology & Medicine, volume 1, eds. R. Douglas, J. Moan & F. Dall'Acqua, Plenum Press, 1988). It is hoped that by highlighting different areas of photobiology these and future publications emanating from the ESP will, in time, produce a comprehensive record of photobiological research, not only in Europe but throughout the world.

Unlike many conference proceedings all the chapters con-tained within this book have been subjected to rigorous peer review and several potential contributions were rejected during the editing process. Furthermore, most manuscripts underwent extensive editing to try and produce chapters of a uniform format and standard. The editors would therefore like to express their gratitude to the many referees who gave considerable amounts of their time to achieve these ends and especially to Ms Annalese George without whose administrative and secretarial skills this volume would have been much the poorer.

CONTENTS

PHOTOSYNTHESIS

PLANT LUMINESCENCE

PSORALEN PHOTOSENSITIZATION

PHOTODYNAMIC THERAPY

PHOTOCARCINOGENESIS

LIGHT SENSITIVE SKIN DISEASES

CHROMOPHORE - MATRIX INTERACTION

SPECTROSCOPY

DNA DAMAGE AND REPAIR

LASER UV INACTIVATION

PHOTOSENSITIVITY IN MICROORGANISMS

PHOTOSENSITIVE PIGMENTS

PHOTOSYNTHETIC REACTION CENTERS:
COMMON FEATURES AND SPECIFIC PROPERTIES

Paul Mathis

Departement de Biologie, CEN Saclay
91191 Gif-sur-Yvette, France

INTRODUCTION

Photosynthesis is well known as the photobiological
process which permits some classes of organisms to utilize the
energy of light for their growth. Such a definition would
include those bacteria which use bacteriorhodopsin as a light
energy transducer. Usually, however, the name photosynthesis
refers to organisms which use chlorins for the absorption and
the transduction of light energy. The entire process is
highly complex and it includes reactions which go from the
absorption of light, at the femtosecond time scale, to the
metabolism and organism development, at a time scale of
seconds or days. The basic conversion of light energy into
chemical energy, however, is performed in membrane structures
named reaction centers which are together complex and simple,
and have been revealed as very interesting objects for basic
studies on light energy conversion, not mentioning their key
role in the biological realm.

Reaction centers are complex: in cases where they have
been isolated, they have been shown to be composed of several
(3 to ≈10) polypeptides, of pigments (chlorin derivatives:
chlorophyll, bacteriochlorophyll, etc., and carotenoids) and
of redox centers involved in electron transfer (some of the
chlorin derivatives and other classes of molecules: quinones,
hemes, iron-sulfur centers, Mn atoms, etc.). The complexity
is also associated to taxonomy: purple and green bacteria
possess only one class of reaction center, with some
differences in the chemical composition, whereas plants and
cyanobacteria possess two classes of reaction center (named
PS-I and PS-II), the properties of which are remarkably
similar when one goes from the ancient procaryots to the
modern eucaryotic plants.

These complexities should not let one forget that
reaction centers perform a simple task, which is to transfer
electrons across a membrane, from a terminal electron donor

Light in Biology and Medicine. Volume 2 Edited by R.H. Douglas *et al.*
Plenum Press, New York, 1991

1

(cytochrome c in purple bacteria; plastocyanin in PS-I; H_2O in PS-II) to a terminal electron acceptor which is a quinone or an iron-sulfur center. Electron transfer is probably the most simple chemical reaction, and accordingly the mode of functioning of reaction centers can be studied in great detail. It now appears that, in spite of variations among the photoautotrophic species, all reaction centers share a common basic structure and essential functional features. I shall try to analyze some of them and outline a few of the key questions which are now raised.

REACTION CENTER OF PURPLE BACTERIA

The reaction center of purple bacteria has been intensely analyzed. Some of its properties will be briefly recalled (Michel et al., 1986; Feher et al. 1989):

- the reaction center is composed of three polypeptides, named L, M, H, which hold four bacteriochlorophyll molecules, two bacteriopheophytins, two quinones and one non-heme Fe^{2+} atom. A cytochrome is more or less tightly bound to the other polypeptides;

- upon absorption of light, a dimer of BChl serves as a trap for energy and as a primary electron donor (named P or special pair). Electron transfer is very fast: in about 2ps an electron arrives on a BPheo molecule, which then reduces a quinone in about 200ps. These first steps take place even at 4K at a rate which is not slower than at room temperature; the initial step is even faster at low temperatures;

- the atomic structure of the RC of several bacteria has been determined by X-ray cristallography at a resolution of 2.3 - 2.9 Å. These achievements have been recognized as an outstanding contribution to the biochemistry of proteins. They reveal that sub-units L and M are intimately associated. The entire structure has an approximate two-fold symmetry axis which runs from the bacteriochlorophyll dimer to the Fe^{2+} atom;

- in general, the crystallographic structure fits nicely with what had been concluded, hypothesized or guessed on the basis of previous spectroscopic work. In particular the dimeric nature of the primary donor P has been confirmed. In several respects, however, the atomic structure is rather surprising; the symmetric organisation had not been predicted (the chlorin pigments are organized in two branches, as shown in Fig. 2; only one branch is photochemically active; the other is inactive, for reasons which are not fully understood (Michel-Beyerle et al., 1988)), and the position of a bacteriochlorophyll molecule between the primary donor and bacteriopheophytin, on the active branch as well as on the inactive branch, had not been predicted either. These two bacteriochlorophylls have been named "voyeur" by Clayton. This name implies that they have the "function" of informing the experimentalist on the redox state of P or of the bacteriopheophytins. They have also been named "accessory" bacteriochlorophylls, as if they had no essential function. These two qualifications are probably unfortunate. Perhaps they should simply be named "monomeric" bacteriochlorophylls, to differentiate them from the dimeric bacterichlorophylls of the special pair P. In the case of <u>Rps viridis</u>, the structure of the tetrahemic cytochrome c was also a surprising feature;

- detailed studies have been aimed at understanding the primary events of electron transfer. Excitation energy is conveyed from the antenna pigments to the special pair P, which has a low energy level permitting its function as an of energy trap. The location of a monomeric bateriochlorophyll between P and bacteriopheophytin is a puzzle for experimentalists and theoreticians (Marcus, 1988). Indeed, picosecond absoption spectroscopy has revealed that the decay of the singlet excited state P^* corresponds nearly to the reduction of BPheo. The reduction of BChl is not or hardly detectable (Breton et al., 1988). Two major reaction schemes are presently favored (the times are rough approximations):

- a two-step process:
P^* BChl BPheo ->0.2ps-> P^+ BChl$^-$ BPheo-> 2ps-> P^+BChl$^-$BPheo$^-$

- a superchange mechanism mixing locally excited states and charge transfer states including P, BChl and BPheo.

The properties of the excited state of P have recently been intensely studied. Photochemical hole-burning experiments, Stark effect measurements and electric field-induced fluorescence anisotropy measurements all agree in attributing to P^* a substantial charge transfer character (with detailed properties which cannot be discussed here). The initial events of photo-induced electron can thus be visualised as a progressive, very rapid electron transfer, permitting a hole and an electron to become widely spatially separated in an extremely short time (the steps are perhaps not really successive steps):

$(BChl - BChl)^* -> (BChl^+ - BChl^-) -> (BChl - BChl)^+$
$BChl^- -> (Bchl - BChl)^+ BChl - Bpheo^-$

The progressive charge separation would thus start within the special pair P, and this might be the very reason for its dimeric nature (this structure also induces excitonic interaction in the excited state and generates a low energy absorption band which permits P to be an efficient trap for energy).

Can we use the knowledge gained on bacterial reaction centers to improve our understanding of the structure of reaction centers (PS-I and PS-II) from oxygenic organisms?

STRUCTURE AND FUNCTION OF PS-I AND PS-II REACTION CENTERS

These reaction centers have properties which are widely different from the purple bacterial ones (Mathis and Rutherford, 1987). In PS-II, the primary donor P has a very oxidizing redox potential (couple P/P^+) which allows it to oxidize an enzyme system that oxidizes water and evolves O_2; this system is part of the RC and it includes a tyrosine residue and a cluster of Mn atoms as well-established redox species. The properties of this enzyme system are still poorly understood; they are one of the important research issues in present biochemistry (Rutherford, 1989). The case of PS-I is nearly symmetrical: it has low-potential electron acceptors, including several iron-sulfur centers, which permit it to reduce low-potential ferredoxins (Golbeck, 1987; Lagoutte and Mathis, 1989). At the protein level, both PS-II and PS-I are highly complex since both of them contain about ten polypeptides. This complexity is also reflected at the pigment level: the PS-I RC contains about 40 chlorophyll molecules. The atomic structure of neither PS-I or PS-II has been determined, although crystals of PS-I with reasonable diffracting quality have been obtained in several laboratories.

However, it can be firmly hypothesized that all kinds of reaction centers share basic common properties at the level of their "core", i.e. the part of them which realizes the primary photochemical steps. Let us mention a few of these properties:

- the acceptor sides of PS-II and of purple bacteria have structural, chemical and functional properties which are nearly identical. In both of them the first stable acceptor is a pheophytin, followed by a bound quinone Q_A (which operates as a one-electron carrier) and then a secondary quinone Q_B (which is a two-electron carrier and leaves the RC after its full reduction to the quinol state). An iron atom is located between Q_A and Q_B.

- the core polypeptides of the RC of purple bacteria (L and M) and of PS-II (polypeptides named D_1 and D_2) have highly significant sequence homologies (Trebst, 1986; Michel and Deisenhofer, 1988).

- in PS-I and PS-II, as in purple bacteria, the RC core is made of two homologous polypeptides, a property which leads one to propose symmetric structural models. Also, there are arguments for proposing that the primary donor P is a dimer of chlorophyll molecules (although this dimeric structure is still debated).

- in all reaction centers, the redox centers involved in primary photochemistry are chlorin derivatives.

- in all RC's, the photochemical step takes place at the ps time scale, the primary reactions are nearly temperature-independant and they include a succession of a few steps which permit a charge separation across the membrane (Fig. 1).

- in all RC's also, a very particular mechanism leads to triplet state formation when normal forward electron transfer is blocked. This is named the radical-pair mechanism. The primary radical-pair (P^+, I^-, where I is the first well-defined electron acceptor) is formed initially in the singlet state, since it is born out of

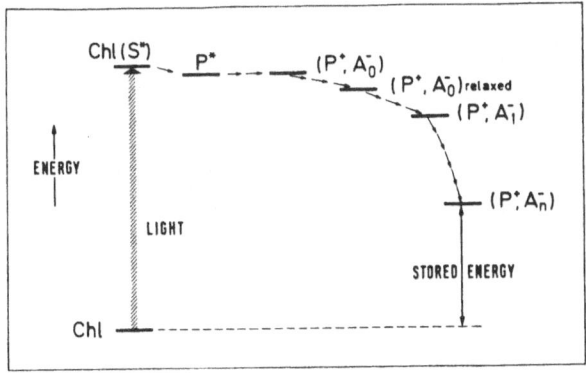

Fig. 1   Schematic presentation of the energy level
         of the successive states attained by the
         photosynthetic apparatus following
         absorption of light. Chl: pigment; P=
         primary donor ("special pair"); A_0, A_1, A_n:
         electron acceptors.

the singlet state P^*. I^- is normally reoxidized by
subnanosecond electron transfer to the secondary
acceptor. When this reaction is rendered impossible by
some artificial treatment, the radical pair (P^+, I^-)
lasts for a time long enough to enter into its triplet
state. The radical pair eventually decays by charge
recombination populating the triplet state of P. In all
kinds of RC's, this process leads to two specific
experimental features: an EPR spectrum of ^3P which has an
unusual polarization, and an effect of weak magnetic
fields (0-100G) on the triplet population.

A GENERAL SCHEME FOR REACTION CENTERS

     The large number of features which are common to all
types of reaction centers makes it possible to hypothesize a
common scheme for their core, i.e. the part of them which
realizes the very primary steps (Fig. 2). A pair of two
homologous polypeptides hold the primary donor, which has a
dimeric structure. The RC contains two "branches", only one
of which is active in electron transfer. The chemical nature
of the species constituting the active and the inactive
branches is indicated in Fig. 2. This scheme should be
considered as a working hypothesis since several of its
features are disputable or even totally hypothetical. In
particular the species named ("Chl")o has been evidenced only
in purple bacteria, and its eventual evidencing in PS-I or PS-
II will certainly have to await their crystal structure.

     If the model is correct, its nearly symmetric structure
is rather puzzling, specially since one half of that structure
is largely inactive. The large homology of the two
polypeptides probably originates in a gene duplication. The
close association of the two hydrophobic polypeptides may be
an important factor in the stability of the RC , but the
dilemna awaits for a precise interpretation.

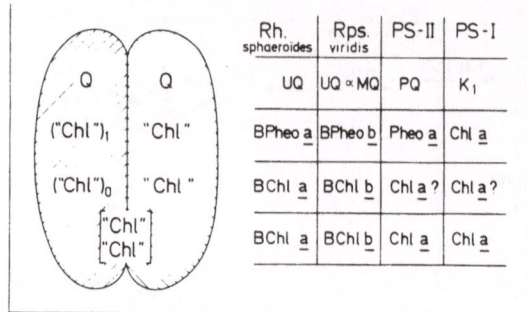

	Rh. sphaeroides	Rps. viridis	PS-II	PS-I
Q Q	UQ	UQ ∝ MQ	PQ	K_1
("Chl")$_1$ "Chl"	BPheo a	BPheo b	Pheo a	Chl a
("Chl")$_0$ "Chl"	BChl a	BChl b	Chl a ?	Chl a ?
"Chl" "Chl"	BChl a	BChl b	Chl a	Chl a

Fig. 2   Hypothetical structure of the core of all
photosynthetic reaction centers. "Chl"
stands for a chlorine derivative and Q for
a quinone. The round objects are
polypeptides. The redox centers are:
bacteriochlorophylls a and b (BChl a and b),
chlorophyll a, bacteriopheophytins a and b
(BPheo a and b), pheophytin a, ubiquinone,
menaquinone, plastoquinone and
phylloquinone (K_1).

It is worth recalling that the quantum yield of all RC's
in nearly unity: each absorbed photon leads to charge
separation (all measurements give a values between 80 and
100%). The energetic yield, however, appears to vary in a
large extent, as shown in Table 1. The stored energy is
evaluated from the difference in mid-point potentials of the
stable electron acceptors and donors. Even if this evaluation
is not rigourous, it is clear that the PS-I RC is more
efficient than PS-II, which is more efficient than purple
bacteria. These features show that all RC's are not optimum
energy converters.

A FEW QUESTIONS ABOUT RC STRUCTURE AND FUNCTION

a) What are the roles of the RC apoprotein?

The most obvious role is to act as a scaffold to maintain
pigments and redox centers in a favorable geometry for energy
and electron transfer, favoring forward reactions and
rendering less probable the wasteful electron-hole
recombinations. Our knowledge of the scaffolding has still to
progress considerably; if X-ray crystallography provides a
good picture of the positioning of pigments, redox centers and
amino-acid residues, the details are not precise enough to
define the modes of binding. Vibrational spectroscopy (IR-TF
and resonance Raman) is a good tool in this respect.

Table 1   Energetic yield of photoconversion in
reaction centers (in % of energy of the
lowest energy active photons)

Rh. sphaeroides	21	Cyt c to Q_B
PS-II	34	H_2O to Q_B
PS-I	51	Plastocyanin to Fe-S_A

Expanding on this static role, the apoprotein is certainly important in facilitating electron transfer. This role is the source of major debates concerning two of its aspects: i) the coupling of electron transfer to vibrational modes which permit to relax the excess of energy in going from one state to another. The reorganization energy is an important parameter in all theories of electron transfer, but its experimental significance is difficult to assess (Robert and Lutz, 1988). ii) the protein medium located between redox centers cannot be considered as isotropic and inert. Some channels for electrons can be facilitated, especially by aromatic amino-acids, by a superexchange mechanism. This facilitation has not yet been demonstrated, but serious presumptions are based on experimental observations following site-directed mutagenesis.

Amino-acid residues may also be directly involved in electron transfer. This has been shown for two tyrosine residues in PS-II, which extend toward the donor side of the symmetric arrangement shown in Fig.2: Tyr Z donates an electron to the special pair and then oxidizes the Mn cluster, whereas Tyr D seems to be only weakly active.

Finally, the apoprotein may be involved in proton transfer. Reaction centers essentially realize a light-induced electron transfer, but protons are involved in two types of phenomena (not considering the later stages of the photosynthetic process where H^+ fluxes become more important): in the full reduction of Q_B to the quinol state, a process which requires two electrons and two H^+, and in the oxidation of water, where four H^+ accompany the evolution of one O_2 molecule. A specific role of the polypeptides in H^+ transfer has recently been assigned on the basis of two types of approaches: site-directed mutagenesis and measurement of the net H^+ flux associated with water oxidation.

b) Is it possible to engineer reaction centers and to build artificial models?

Chemical and genetic engineering are useful tools for understanding structure-function relationships in reaction centers. As examples one may cite the systematic replacement of ubiquinones at the Q_A and Q_B sites in Rh. sphaeroides RC's (Gunner et al., 1986), or the replacement of amino-acids in the vinicity of the special bacteriochlorophyll pair, in Rh. capsulatus. These approaches should certainly develop in the future.

The synthesis of artificial reaction centers is much more difficult and progresses more slowly (Gust and Moore, 1989). This type of work has two basic goals. Firstly, it should permit the quantitative evaluation of the physical mechanisms involved in the primary electron transfer steps. Several groups of physical chemists are actively working on these theoretical aspects; most experimental data can be approximately understood, but in a way that is not sure enough for making reliable predictions. There is no doubt that well-defined artificial models would be helpful in this respect. Secondly, these models may be revealed to be important for building, in the long term, artificial systems for the conversion of solar energy by photochemical devices.

c) What insures the stability of RC's against photodamage?

Reaction centers are sites where an exceptional concentration of excitation energy takes place. They are coupled to an antenna system of typically 200 chlorophyll molecules and the ensemble is permanently exposed to sunlight. A rate of 100 excitations per second is quite typical. Chlorin derivatives are very efficient in forming radical species or singlet oxygen, and the stability of the photosynthetic apparatus requires exceptionally efficient protective mechanisms. I cannot examine here the details of the biological situations, but a few general mechanisms can be listed:

i   light-harvesting complexes contain both chlorophylls and carotenoids, in close proximity. Thus a nanosecond triplet-triplet energy transfer conveys to carotenoids eventual triplet states (Fig. 3). The lifetime of the chlorophyll singlet excited state is rather short because of an efficient transfer of singlet energy toward RC's. However the actual lifetime of about 100ps is enough to give a triplet yield of 1 - 5% which, in the absence of carotenoids, would be enough to generate dramatic photodynamic effects under sun light.

ii   Once singlet excitation energy arrives at the reaction center, electron transfer is very fast, so that practically no triplet states should be populated under normal conditions. There are, however, a number of abnormal cases which occur under physiological conditions, for example when reaction centers are not yet fully built, or when a second photon arrives at the RC when the secondary acceptor is still reduced (for statistical reasons, especially under strong sun light). In those cases, the triplet state ^3P is populated; it will be deactivated very quickly by energy transfer to the carotenoid molecule(s) of the RC.

iii   The redox centers are held with some rigidity in the protein matrix, which decreases the radical reactions. These have been shown to take place, however, specially in PS-II where very oxidizing radicals are created. This

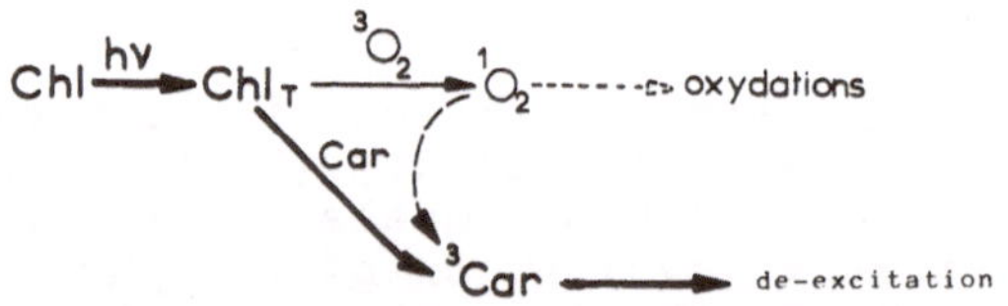

$$Chl \xrightarrow{h\nu} Chl_T \xrightarrow{^3O_2} {}^1O_2 \text{------} \leadsto \text{oxydations}$$

Car

$$^3Car \longrightarrow \text{de-excitation}$$

Fig. 3   A scheme showing how carotenoids act primarily by quenching triplet excited states formed in chlorophyll molecules, and much less importantly by quenching singlet oxygen (scheme adapted from Mathis, 1970).

may be the reason why PS-II is provided with a repair
mechanism, which is far from being understood, and which
compensates for the destructive processes named
"photoinhibition".

REFERENCES

Breton, J., Martin, J.L., Fleming, G.R., and Lambry, J.C.,
    1988, "Low-temperature femtosecond spectroscopy of the
    initial step of electron transfer in reaction centers
    from photosynthetic purple bacteria", Biochemistry.,
    27:8276.
Feher, G., Allen, J.P., Okamura, M.Y., and Rees, D.C., 1989,
    "Structure and function of bacterial photosynthetic
    reaction centres", Nature., 339:111.
Golbeck, J.H., 1987, "Structure, function and organization of
    the Photosystem I reaction center complex", Biochim.
    Biophys. Acta., 895:167-204.
Gunner, M.R., Robertson, D.E., and Dutton, P.L., 1986,
    "Kinetic studies on the reaction center protein from
    Rhodopseudomonas sphaeroides: the temperature and free
    energy dependence of electron transfer between various
    quinones in the Q_A site and the oxidized
    bacteriochlorophyll dimer", J. Phys. Chem. 90:3783.
Gust, D., and Moore, A., 1989, "Mimicking photosynthesis",
    Science., 244:35.
Lagoutte, B., and Mathis, P., 1989, "The photosystem I
    reaction center : structure and photochemistry",
    Photochem. Photobiol., 49:833.
Marcus, R.A., 1988, "An internal consistency test and its
    implications for the initial steps in bacterial
    photosynthesis", Chem. Phys. Lett., 146:13
Mathis, P., 1970, "Transitory states of carotenoids", Ph. D.
    Thesis, Paris-Orsay.
Mathis, P., and Rutherford, A.W., 1987, "The primary reactions
    of photosystems I and II of algae and higher plants",
    Photosynthesis. (J.Amesz,ed.), pp 63-96
Michel, H., Epp, O., and Deisenhofer, J., 1986, "Pigment-
    protein interactions in the photosynthetic reaction
    centre from Rhodopseudomonas viridis", EMBO J., 5:2445.
Michel, H., and Deisenhofer, J., 1988, "Relevance of the
    photosynthetic reaction center from purple bacteria to
    the structure of photosystem II", Biochemistry., 27:1.
Michel-Beyerle, M.E., Plato, M., Deisenhofer, J., Michel, H.,
    Bixon, M., and Jortner, J., 1988, "Unidirectionality of
    charge separation in reaction centers of photosynthetic
    bacteria", Biochim. Biophys. Acta. 932:52.
Robert, B., and Lutz, M., 1988, "Proteic events following
    charge separation in the bacterial reaction center :
    resonance raman spectroscopy", Biochemistry. 27:5108.
Rutherford, A.W., 1989, "Photosystem II, the water-splitting
    enzyme", TIBS. 14:227.
Trebst, A., 1986, "The three-dimensional structure of the
    herbicide binding niche on the reaction center
    polypeptides of photosystem II", Z.Naturforsch., 42c:742.

THE TOPOLOGY OF THE PHOTOSYSTEM II REACTION CENTRE AND ITS HERBICIDE BINDING SITE

Achim Trebst and Joachim Heinze

Dept. of Biology, Ruhr-University, D-Bochum 1, FRG

INTRODUCTION

Photosystem II of the photosynthetic electron flow system in the thylakoid membrane of chloroplasts splits water to oxygen and reduces plastoquinone. A purified photosystem II complex of about 300.000 molecular weight consists of six major integral and three peripheral polypeptide subunits. The complex binds the reaction centre P680 - likely a chlorophyll a dimer -, two pheophytins, one Fe, four Mn, Ca, two plastoquinones (Q_A and Q_B), two cytochromes b-559 and about 30 antenna chlorophylls (Babcock, 1987; Barber, 1987). The six integral polypeptides are encoded in the chloroplast genome by the genes psbA to F, the peripheral ones in the nucleus (Dyer, 1985). Recently, further polypeptides of photosystem II have been identified, encoded for by the psb genes H to N (Hallick, 1989). Their significance for structural and functional stability of the photosystem II complex is not yet clear. The integral polypeptides of 47 and 43 kDa molecular weight (the psbB and C products) carry the antenna chlorophyll, the 9 and 4 kDa subunits (the psbE and F products) together with the heme group of cytochrome b-559. It is now accepted that the reaction centre of photosystem II is bound to the D-1 and D-2 proteins. These two 32 kDa large polypeptides are encoded in the chloroplast by the psbA and psbD gene. They carry P680, two monomeric chlorophylls, two pheophytins, the plastoquinones and the Fe. Their function was predicted (Deisenhofer et al., 1985b; Trebst, 1986) from the homology between the amino acid sequences of the L and M subunits of the reaction centre of purple bacteria to that of the D-1 polypeptide (Williams et al., 1984; Youvan et al., 1984) and later also to the D-2 polypeptide when this had been sequenced as well (Rochaix et al., 1984).

It was known already that functionally photosystem II and the reaction centre of purple bacteria are very similar (Okamura et a., 1982; Rutherford, 1986; Babcock, 1987). The X-ray structure of the crystallized reaction centre of Rps.

viridis (Deisenhofer et al., 1984; Deisenhofer et al., 1985a; Deisenhofer et al., 1985b) provided for the first time detailed structural information on a photosystem and on the folding of a membrane protein. Specific amino acids involved in pigment and redox centre binding, particularly on the histidines in reaction centre and Fe binding could be directly observed in the X-ray structure (Deisenhofer et al., 1984). These functional amino acids in the bacterial reaction centre polypeptides were conserved at homologous positions in the D-1/D-2 protein. From this the prediction (Deisenhofer et al., 1985b; Trebst, 1986) on the nature of the reaction centre of photosystem II gained support and a structure of photosystem II could be proposed on that basis.

Important for the folding prediction (Trebst, 1986; Trebst and Draber, 1986; Trebst, 1987) of the D-1 polypeptide were data on amino acid changes in herbicide tolerant higher plants and algae (quoted in Trebst, 1986; Trebst and Draber, 1986; Rochaix and Erickson, 1988). It was known that the mode of action of such herbicides is in the displacement of Q_B from its binding site on the D-1 polypeptide and that this site is located on the matrix site of the thylakoid membrane (Velthuys, 1981; Kyle, 1985). These amino acid changes can therefore be used to describe the dimensions in the Q_B (=herbicide) binding niche and which side of the folding model is oriented towards which side of the membrane. The new folding model (Rao and Argos, 1986; Trebst, 1986; Trebst and Draber, 1986; Trebst, 1987) for the D-1 and D-2 polypeptide with five hydrophobic transmembrane helices interconnected by hydrophilic sequences extending on either side of the membrane, replaced an older version that had assumed seven such transmembrane helices (Rao et al., 1983). In the new folding and according to the herbicide binding site the N-terminus will be on the matrix, the C-terminus of the amino acid sequence of the D-1 protein on the lumen site of the membrane. In a successful preparation of a new reaction centre complex of photosystem II from higher plant thylakoids the core antenna chlorophyll binding proteins of 47 and 43 kDa had been removed (Barber, 1987; Nanba and Satoh, 1987). These polypeptides had been earlier assumed to form the reaction centre (quoted in Trebst, 1986). The new reaction centre of photosystem II consisted of only four polypeptide subunits: the D-1 and D-2 polypeptides with 4-5 chlorophylls and two pheophytins and the two subunits of cytochrome b-559 with one heme group (Nanba and Satoh, 1987; Barber, 1987). (In the meantime a fifth polypeptide - the product of the psbI gene - has been spotted as well in the new reaction centre preparations (Ikeuchi and Inoue, 1988; Webber et al., 1989)). This new photosystem II reaction centre preparation neither contains Q_A, nor Q_B and Fe. Nevertheless, the primary energy charge separation is identical to that of intact photosystem II preparations and therefore there are no longer doubts that the reaction centre of photosystem II has now finally been purified and identified.

As already stated, the histidines 198 for reaction centre chlorophyll binding on both the D-1 and D-2 subunits are conserved in photosystem II in homologous positions (Trebst, 1986; Michel and Deisenhofer, 1988) to those in the purple bacteria (Deisenhofer et al., 1985a). Advances for establishing the protein folding of photosystem II have been made in the modelling of the quinone binding sites (Trebst, 1987: Trebst et al., 1989). This is, as already mentioned, because the identification of binding sites of inhibitors competitive to the quinones is a very powerful experimental tool. Several new mutations in this site, including site directed in cyanobacteria, have been described (Ajlani et al., 1989; Horovitz et al., 1989; Ohad and Hirschberg, 1989), in addition to the earlier ones of 1985 (Rochaix and Erickson, 1988). All eight mutations identified so far are indicated in Fig. 1. They are relatively close to each other and all fall in a binding niche on the D-1 protein that is formed by transmembrane helices IV and V and a parallel helix connecting these two (see Fig. 1). This feature of the Q_B binding site follows that of the purple reaction centre (Deisenhofer et al., 1985a) and therefore new results on the purple bacterial

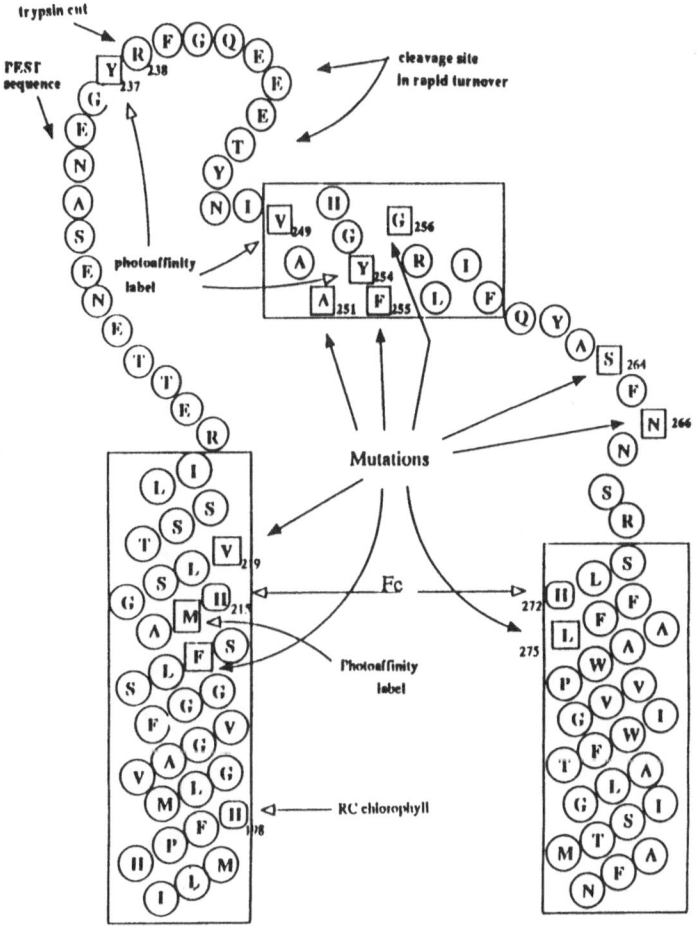

Fig. 1 Folding of the amino acid sequence of the D-1 protein as relevant for reaction centre chlorophyll and Q_B and herbicide binding. Indicated are amino acids tagged by mutations or photoaffinity labeling in helices IV and V and the interconnecting parallel helix on the matrix site of the membrane.

systems now also of <u>Rb. sphaeroides</u> (Allen et al., 1987) that
directly show Q_B in its site are again of great relevance for
discussing structure and function of photosystem II. Although
there are additional amino acids, two after the parallel helix
and about 14 before it, the principle folding seems to be very
similar. This may be concluded from the amino acid changes in
inhibitor tolerant purple bacteria mutants (Paddock et al.,
1987: Sinning et al., 1989b). The amino acid changes there
are almost identical to those found in higher plants.
Furthermore, the X-ray structure of the reaction centre of
<u>Rps. viridis</u> soaked in terbutryn or in phenanthroline (Michel
et al., 1986) can be transposed to that of the D-1 reaction
centre of photosystem II. The recent X-ray structure of a
terbutryn resistant mutant of <u>Rps. viridis</u> supports this in
further detail (Sinning et al., 1989b), showing also that
there are no major changes in the structure of the protein in
the herbicide tolerant mutant, as compared to the wildtype
(Sinning et al., 1989b).

Three-dimensional views are possible by computer drawing
(Crofts et al., 1987). These models give a better view from
different angles than the two dimensional schemes of Figs. 1
and 2, but not better results. The computer drawing of
photosystem II actually only copies the reaction centre of the
purple bacteria. The computing of the folding of the amino
acid sequences of the D-1/D-2 polypeptide still has some way
to go.

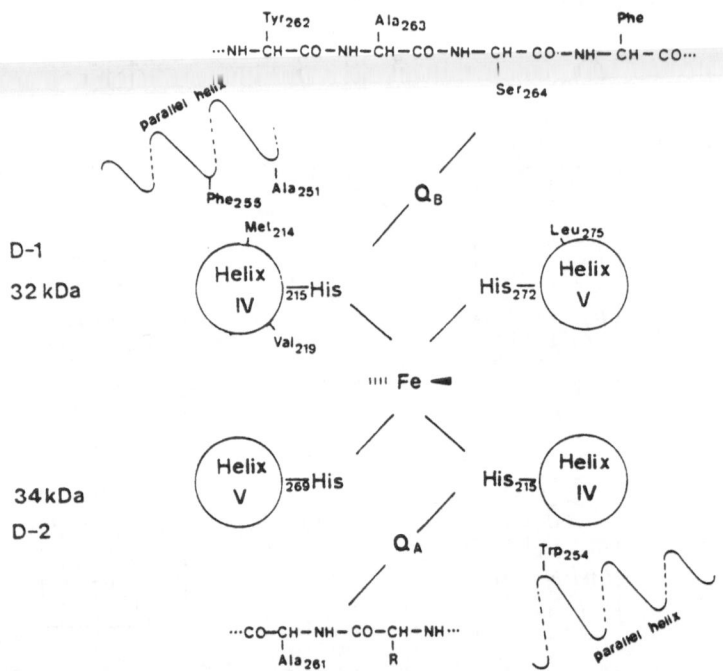

Fig. 2   Schematic representation of the quinone and
Fe binding site on the D-1 and D-2 polypeptides. View is on the top of the transmembrane helices with the parallel helices
that actually are perpendicular to the page
with ala 251 on the top.

The amino acid changes in the inhibitor tolerant mutants can be used to describe the Q_B site, as this is the site where the inhibitors bind. Nevertheless, as the mutants grow phototrophically, these amino acids are of lesser importance for Q_B than they are for inhibitor binding. This is quite likely different with another experimental approach. Photoaffinity labelling of the inhibitor binding site in the D-1 protein with radioactive azido-derivatives of inhibitors indicates amino acids in the Q_B binding niche, not yet accessed by mutations. Methionine 214 had been shown (Wolber et al., 1986) to be attacked by an azido-triazine; tyr 237 and tyr 254 by azido-monuron (Dostatni et al., 1988) and val 247 by an azido-ioxynil (Oettmeier et al., 1989). Except for tyr 237 these amino acids fit into the general conception on the folding of the Q_B site as discussed so far (see Fig. 1). Tyrosine 237, however, seemed off the binding niche. This tyrosine is in the sequence of the D-1 polypeptide that has no homology in the L-subunit of the purple bacteria, a major difference between the two polypeptides. This sequence then brings about a major change in the conformation of the Q_B site in photosystem II, as compared to the purple bacteria. It is proposed that this sequence that includes tyrosine 237 is part of the Q_B binding niche in photosystem II (Trebst et al., 1988). In proposing that this sequence is part of the Q_B site, a major difference in D-1 protein vs the L-subunit is assumed (Trebst et al., 1989). The folding of that amino acid sequence atop the Q_B binding site would not allow another subunit (D-2, equivalent to M) to do so in photosytem II. But this is the case in the bacterial reaction centre where the M-subunit provides a glutamic acid for Fe binding (Deisenhofer et al., 1985a).

The additional supporting evidence for the folding of those amino acids in the D-1 protein comes from data on the influence of the functional state of the Q_B site on a specific trypsin cut (Trebst et al., 1988) and on the sequence that is probably involved in the cleavage of the D-1 polypeptide in the "rapid turnover" (Greenberg et al., 1987).

Trypsin cuts the D-1 polypeptide in the membrane without prior denaturation quickly and specifically at arginine 238. The trypsin also does this on the D-2 polypeptide at arg 234 (see Trebst et al., 1988). These are equivalent positions in the folding of either polypeptide. The easy trypsin cut indicates that these two arginines are easily accessible on the matrix surface of the membrane and are already folded in the proper conformation to be recognized by the protease. That is, this sequence might fold similar to that of the trypsin inhibitor as it docks into trypsin. The X-ray structure of the inhibitor in the active site of trypsin (Ruhlmann et al., 1973) might be used for the modelling (Trebst et al., 1989). This trypsin cut at arg 238 or arg 234 respectively is controlled by the occupancy state of the Q_B site (Trebst et al., 1988). Either a quinone in the site, plastoquinone itself or a synthetic benzoquinone acting as artificial acceptor of photosystem II, or a specific inhibitor for the Q_B site of the DCMU/triazine family protect the D-1 polypeptide as well as the D-2 polypeptide from the trypsin cut at their equivalent arginines. The influence of the occupancy state of the Q_B site on also the conformation of the D-2 polypeptide (the Q_A binding polypeptide) indicates that there is a contact site between the two polypeptides close to the arginines involved in the trypsin cut. Inhibitors of the phenol-type do not protect from the trypsin (Trebst et al., 1988). This indicates that the phenol binding site is somewhat different from the DCMU binding site, as already indicated by photoaffinity labelling experiments (Oettmeier et al., 1989), by the new mutant (Ajlani et al., 1989), the cross-sensitivity in triazine/urea mutations as well as by the long-known functional differences between the two groups of inhibitors (Oettmeier and Trebst, 1983; Trebst and Draber, 1986).

The rapid turnover of the D-1 polypeptide has long been known (Marder et al., 1984), but its physiological significance is still not well understood. It is probably related to photoinhibition of photosynthesis (that is specifically damaging photosystem II) as well as to the mechanisms in the assembly of photosystem II. In rapid turnover a PEST sequence in the D-1 polypeptide appears to guide a protease to a cleavage site close to glu 242 as the recent experimentations of Edelman et al. (Greenberg et al., 1987 and 1989) strongly suggest. The rapid turnover of the D-1 polypeptide is inhibited by DCMU (Marder et al., 1984) again indicating that the herbicide binding niche controls the accessibility·of the PEST and cleavage site to the protease, involved in the degradation of the D-1 polypeptide.

It seems difficult to visualize the mechanism how the core of photosystem II, usually written schematically as a D-1/D-2 centre surrounded by the antenna polypeptides (Barber, 1987), can be taken out and replaced in photoinhibition. Indeed, the term core for the reaction centre polypeptide is an operational term and should not be mistaken for a structural core. It would be better to visualize photosystem II like a handle: the core antenna of the 47 and 43 kDa polypeptide being one half, the D-1/D-2 reaction centre the other half. Such a handle might separate easily in photoinhibition. Some support for such a model comes from a separation procedure in which we have separated photosystem II particles by a mild detergent into four complexes: two different light harvesting complexes and two green protein complexes. Of the latter the upper one consists of the 47 and 43 kDa integral core antenna polypeptides, but with the 33 kDa peripheral protein still attached. The other consists of the D-1 and D-2 polypeptide. Obviously the two parts of photosytem II can easily be separated. The attachment of the 33 kDa polypeptide shows affinity to the 47/43 kDa polypeptide that might come as a surprise, as presently this peripheral polypeptide is thought to help attach manganese to the D-1 and D-2 proteins. However, a cross link between the 33 kDa and the 47 kDa polypeptide had already showed this nearest neighbour relationship (Enami et al., 1989).

The topology of the reaction centre of photosystem II and of the D-1/D-2 polypeptide in particular has made progress in describing the quinone binding sites, as discussed above. The folding model has also been useful in the identification of the primary donor for P680. Iodine (Takahashi and Styring, 1987, 1989) and deuterium (Barry and Babcock, 1987) labeling had suggested that a tyrosine is likely oxidized by P680. Site specific mutations (Debus et al., 1988; Vermaas et al.,

1988) of tyrosine 160 in the D-2 polypeptide identified it as the donor for P680 called D. On the basis of symmetry and homology it is assumed that tyrosine 161 of the D-1 polypeptide is the functional donor Y for P680.

It is assumed now that the manganese binding site, that is stabilized by the attachment of the peripheral 33 kDa polypeptide onto the photosystem II complex, is also on the D-1 and D-2 polypeptide (see Barber, 1987). But the amino acids involved have not been clearly identified. There is the possibility that the Mn cluster is either on the hydrophilic amino acid sequences of the D-1 polypeptide protruding into the lumen space of the thylakoid, i.e. those connecting helices I and II or III with IV, or on the carboxy terminus of the D-1 polypeptide. Another possibility is that the Mn cluster is below helices IV and V on both the D-1 and D-2 polypeptide, similar to the attachment of the cytochrome c_2 subunit to the reaction centre of the purple bacteria (Deisenhofer et al., 1985a). This unresolved situation indicates that the understanding of the folding of the hyrophilic sequences on the lumen side of photosystem II is not at all advanced.

ACKNOWLEDGEMENTS

Work at Bochum is supported by the Deutsche Forschungsgemeinschaft and by the European Community.

REFERENCES

Allen, J.P., Feher, G., Yeates, T.O., Komiya, H. and Rees, D.C., 1987, Structure of the reaction centre from Rhodobacter sphaeroides R-26: The cofactors, Proc. Natl. Acad. Sci. USA, 84:5730-5734.

Ajlani, G., Meyer, I., Vernotte, C., Astier, C., 1989, Mutation in phenol-type herbicide resistance maps within the psbA gene in Synechocystis 6714, FEBS Lett., 246:207,210.

Babcock, G.T., 1987, The photosynthetic oxygen-evolving process, in: "New Comprehensive Biochemistry", Vol. 15. "Photosynthesis", J Amesz, ed., Elsevier, Amsterdam.

Barber, J., 1987, Composition, organization, and dynamics of the thylakoid membrane in relation to its function, in: "The Biochemistry of Plants". Vol. 10. "Photosynthesis", M.D. Hatch and N.K. Boardman, eds., Academic Press, New York.

Barber, J., Chapman, D.J. and Telfer, A., 1987, Characterization of a PS II reaction centre isolated from the chloroplasts of Pisum sativum, FEBS Lett., 220:67-73.

Barry, B.A. and Babcock, G.T., 1987, Tryrosine radicals are involved in the photosynthetic oxygen-evolving system, Proc. Natl. Acad. Sci. USA, 84:7099.

Crofts, A., Robinson, H., Andrews, K., van Doren, S. and Berry, E., 1987, Catalytic sites for reduction and oxidation of quinones, in: "Cytochrome Systems. Molecular Biology and Bioenergetics", S. Papa, B. Chance and L. Ernster, eds., Plenum Press, New York.

Debus, R.J., Barry, B.A., Babcock, G.T. and McIntosh, L., 1988, Site directed mutagenesis identifies a tyrosine radical involved in the photosynthetic oxygen-evolving system, <u>Proc. Natl. Acad. Sci. USA</u>, 85:427-430.

Deisenhofer, J., Epp, O., Miki, K., Huber, R. and Michel, H., 1984, X-ray structure analysis of a membrane protein complex. Electron density map at 3 Å resolution and a model of the chromatophores of the photosynthetic reaction centre from <u>Rhodopseudomonas viridis</u>, <u>J. Mol. Biol.</u>, 180:385-398.

Deisenhofer, J., Epp, O., Miki, K., Huber, R. and Michel, H., 1985, Structure of the protein subunits in the photosynthetic reaction centre of <u>Rhodopseudomonas viridis</u> at 3 Å resolution, <u>Nature</u>, 318:618-624.

Deisenhofer, J., Michel, H. and Huber, R., 1985, The structural basis of photosynthetic light reactions in bacteria, <u>Trends Biochem. Sci.</u>, 10:243-248.

Dostatni, R., Meyer, H.E. and Oettmeier, W., 1988, Mapping of two tyrosine residues involved in the quinone-(Q_B) binding site of the D-1 reaction centre polypeptide of photosystem II, <u>FEBS Lett.</u>, 239:207-210.

Dyer, T.A., 1985, The Chloroplast genome and its products, <u>Oxford Surv. Plant Mol. Cell Biol.</u>, 2:147:177.

Enami, I., Miyaoka, T., Mochizuki, Y., Shen, J.-R., Satoh, K. and Katoh, S., 1989, Nearest neighbour relationships among constituent proteins of oxygen-evolving photosystem II membranes: binding and function of the extrinsic 33 kDa protein, <u>Biochim. Biophys. Acta.</u>, 973:35-40.

Greenberg, B.M., Gaba, V., Mattoo, A.K. and Edelman, M., 1987, Identification of a primary <u>in vivo</u> degradation product of the rapidly turning-over 32 kd protein of photosystem II, <u>EMBO J.</u>, 6:2865-2869.

Greenberg, B.M., Gaba, V., Mattoo, A.K. and Edelman, M., 1989, Degradation of the 32 kDa photosystem II reaction centre protein in UV visible and far red light occurs through a common 23.5 kDa intermediate, <u>Z. Naturforsch.</u>, 44c:450-452.

Hallick, R., et al., 1989, Nomenclature for the genes for thylakoid polypeptides, proposed at the <u>Photosyn. Cong. at Stockholm</u>, Aug. 1989.

Horovitz, A., Ohad, N. and Hirschberg, J., 1989, Predicted affects on herbicide binding of amino acid substitutions in the D1 protein of photosystem II, <u>FEBS Lett.</u>, 243:161-164.

Ikeuchi, M. and Inoue, Y., 1988, A new photosystem II reaction centre component (4.8 kDa protein) encoded by chloroplast genome, <u>FEBS Lett.</u>, 241:99-104.

Kyle, D.J., 1985, The 32,000 Dalton Q_B protein of photosystem II, <u>Photochem. Photobiol.</u>, 41:107-116.

Marder, J.B., Goloubinoff, P. and Edelman, M., 1984, Molecular archtecture of the rapidly metabolized 32-kilodalton protein of photosystem II. Indications for COOH-terminal processing of a chloroplast membrane polypeptide, <u>J. Biol. Chem.</u>, 259:3900-3908.

Michel, H., Epp, O. and Deisenhofer, J., 1986, Pigment-protein interactions in the photosynthetic reaction centre from Rhodopseudomonas viridis, <u>EMBO J.</u> 5:2445-2451.

Michel, H. and Desidenhofer, J., 1988, Relevance of the photosynthetic reaction centre from purple bacteria to the structure of photosystem II, <u>Biochemistry</u> 27:1-7.

Nanba, O. and Satoh, K., 1987, Isolation of a photosystem II reaction centre consisting of D-1 and D-2 polypeptides and cytochrome b-559, <u>Proc. Natl. Acad. Sci. USA</u>, 84:109-112.

Oettmeier, W. and Trebst, A., 1983, Inhibitor and plastoquinone binding to photosystem II, <u>in</u>: "The Oxygen Evolving System of Photosynthesis", Y. Inoue, A.R. Crofts, Govindjee, N. Murata, G. Renger and K. Satoh, eds., Academic Press, Japan, Tokyo.

Oettmeier, W., Masson, K., Höhfeld, J., Meyer, H.E., Pfister, K. and Fischer, H.-P., 1989, [^{125}1]Azido-ioxynil labels Val$_{249}$ of the photosystem II D-1 reaction centre protein, <u>Z. Naturforsch</u>, 44c:444-449.

Ohad, N. and Hirschberg, J., 1989, A similar structure of the herbicide binding site in photosystem II of plants and cyanobacteria is demonstrated by site specific mutagenesis of the <u>psbA</u> gene, <u>Photosyn. Res.</u>, in press.

Okamura, M.Y., Feher, G. and Nelson, N., 1982, Reaction centres, <u>in</u>: "Photosynthesis". Vol. 1 "Energy Conversion by Plants and Bacteria", Govindjee, ed., Academic Press, New York.

Paddock, M.L., Williams, J.C., Rongey, S.H. Abresch, E.C., Feher, G. and Okamura, M.Y., 1987, Characterization of three herbicide resistant mutants of <u>Rhodopseudomonas sphaeroides</u> 2.4.1: Structure-Function relationship, <u>in</u>: "Progress in Photosynthesis Research", Vol. III, J. Biggins, ed., pp. 775-778, Martinus Nijhoff Publishers, Dordrecht.

Rao, J.K.M., Hargrave, P.A. and Argos, P., 1983, Will the seven-helix bundle be a common structure for integral membrane proteins? <u>FEBS Lett.</u>, 156:165-169.

Rao, J.K.M. and Argos, P., 1986, A conformational preference parameter to predict helices in integral membrane proteins, <u>Biochim. Biophys. Acta.</u>, 869:197-214.

Rochaix, J.D., Dron, M., Rahire, M. and Malnoe, P., 1984, Sequence homology between the 32k dalton and the D2 chloroplast membrane polypeptide, <u>Plant Molec. Biol.</u>, 3:363-370.

Rochaix, J.-D., Erickson, J., 1988, Function and assembly of photosystem II: genetic and molecular analysis, <u>Trends Biochem. Sci.</u>, 13:56-59.

Rühlmann, A., Kukla, D., Schwager, P., Bartels, K. and Huber, R., 1973, Structure of the complex formed by bovine trypsin and bovine pancreatic trypsin inhibitor. Crystal structure determination and stereochemistry of the contact region, <u>J. Mol. Biol.</u>, 77:417-436.

Rutherford, A.W., 1986, How close is the analogy between the reaction centre of photosystem II and that of purple bacteria? <u>Biochemical Society Transactions</u>, 14:15-18.

Sinning, I., Michel, H., Mathis, P. and Rutherford, A.W., 1989, Terbutryn resistance in a purple bacterium can induce sensitivity toward the plant herbicide DCMU, <u>FEBS Lett</u>. 256:192-196.

Sinning, I., Michel, H., Mathis, P. and Rutherford, A.W., 1989, Characterization of four herbicide-resistant mutants of Rhodopseudomonas viridis by genetic analysis, electron paramagnetic resonance, and optical spectroscopy, <u>Biochemistry</u>, 28:5544-5553.

Takahashi, Y. and Styring, S., 1987, A comparative study of the reduction of EPR signal II$_{slow}$ by iodide and the iodo-labeling of the D2-protein in photosystem II, <u>FEBS Lett.</u> 223:371-375.

Takahashi, Y., Satoh, K., 1989, Identification of the photochemically iodinated amino-acid residue on D1-protein in the Photosystem II core complex by peptide mapping analysis, <u>biochim. Biophys. Acta.</u> 973:138-146.

Trebst, A., 1986, The topology of the plastoquinone and herbicide binding peptides of photosystem II in the thylakoid membrane, <u>Z. Naturforsch.</u>, 41c:240-245.

Trebst, A. and Draber, W., 1986, Inhibitors of photosystem II and the topology of the herbicide and Q_B binding polypeptide in the thylakoid membrane, <u>Photosyn. Res.</u>, 10:381-392.

Trebst, A., 1987, The three-dimensional structure of the herbicide binding niche on the reaction centre polypeptides of photosystem II. <u>Z. Naturforsch</u>, 42c:742-750.

Trebst, A., Depka, B., Kraft, B. and Johanningmeier, U., 1988, The Q_B site modulates the conformation of the photosytem II reaction centre polypeptides, <u>Photosyn. Res.</u>, 18:163-177.

Trebst, A., Depka, B. and Kipper, M., 1989, The topology of the reaction centre polypeptides of photosystem II, <u>Proc. Photosyn. Congr. at Stockholm.</u>, Aug. 1989.

Velthuys, B.R., 1981, Electron-dependent competition between pastoquinone and inhibitors for binding to photosystem II, <u>FEBS Lett.</u>, 126:277-281.

Vermaas, W.F.J., Rutherford, A.W. andHansson, O., 1988, Site-directed mutagenesis in photosystem II of the cyanobacterium <u>Synechocystis sp</u>. PCC 6803: Donor D is a tyrosine residue in the D2 protein, <u>Proc. Natl. Acad. Sci. USA</u>, 85:8477-8481.

Webber, A.N., Packman, L., Chapman, D.J., Barber, J. and Gray, J.C., 1989, A fifth chloroplast-encoded polypeptide is present in the photosystem II reaction centre complex, <u>FEBS Lett</u>. 242:259-262.

Williams, J.C., Steiner, L.A., Feher, G. and Simon, M.I., 1984, Primary structure of the L subunit of the reaction centre from <u>Rhodopseudomonas sphaeroides</u>, <u>Proc. Natl. Acad. Sci. USA</u>, 81:7303-7307.

Wolber, P.K., Eilmann, M. and Steinback, K.E., 1986, Mapping of the triazine binding site to a highly conserved region of the Q_B-protein, <u>Arch. Biochim. Biophys.</u>, 248:224-233.

Youvan, D.C., Buylina, E.J., Alberti, M., Begusch, H. and hearst, J.E., 1984, Nucleotide and deduced polypeptide sequences of the photosynthetic reaction centre, B870 antenna, and flanking polypeptides from <u>R. capsulata</u>, <u>Cell</u>, 37:949-957.

PHOTOINACTIVATION OF THE ISOLATED PHOTOSYSTEM TWO REACTION
CENTRE AND ITS PREVENTION

J. Barber

AFRC Photosynthesis Research Group, Department of
Biochemistry, Imperial College of Science,
Technology and Medicine, London SW7 2AY, UK

INTRODUCTION

A remarkable feature of the photosystem two (PS2)
reaction centre is the fact that one of its components, the 32
kD D1 polypeptide, is rapidly turning over (Edelman et al.,
1984). This protein is encoded by the psbA gene of the
chloroplast genome and was first sequenced by Zurawski et al.
(1982). It is known to bind the secondary quinone acceptor Q_B
and be the site of attack of a wide range of herbicides (Kyle,
1985; Trebst, 1986). The reason for its rapid turnover seems
to be related to the vulnerability of photosystem two to
damage by bright light. Indeed, photoinhibition occurs when
the rate of synthesis of the D1 polypeptide is not matched by
the rate of its removal from the PS2 complex (Kyle et al.
1984). The removal process involves the attack by a
proteolytic enzyme to yield a fragment with an apparent
molecular weight of 23.5 kD (Greenberg et al. 1987).

Why the D1 polypeptide has been singled out as the only
component of the PS2 complex to be rapidly turned over has
been a mystery. Also it has not been clear why PS2 is
vulnerable to photodamage. Answers to these questions are
starting to emerge from studies with isolated reaction centres
of PS2.

This reaction centre was first isolated from higher
plants about three years ago (Nanba and Satoh, 1987; Barber et
al. 1987). Since then it has been shown to consist of five
polypeptides (Marder et al. 1987; Webber et al. 1989), one of
which is the D1 polypeptide. The others were the D2
polypeptide (product of the psbD gene), α and β-subunits of
cytochrome b559 (Products of the psbE and psbF genes) and the
product of psbI gene. By analogy with the L and M subunits of
the purple bacterial reaction centres, it seems likely that D1
and D2 form a heterodimer which binds the prosthetic groups
involved in primary and secondary electron flow (Barber 1987;
Michel and Deisenhofer, 1988). These groups consist of

chlorophyll $\underline{a}$, pheophytin $\underline{a}$ and plastoquinone, as determined by absorption and resonance spectroscopy (Rutherford 1986). Indeed, chemical analysis indicated that the isolated D1/D2 complex binds chlorophyll $\underline{a}$ and pheophytin $\underline{a}$ in a ratio of 2 to 1 (Barber et al. 1987). By analogy with the purple bacterial reaction centre it would seem likely that each PS2 reaction centre contains 2 pheophytins and 4 chlorophylls, with two of the latter existing as a special pair which acts as primary electron donor, P680. Taking this analogy further it has been speculated that the two chlorphylls of P680 are held in the reaction centre by forming ligands with histidine residues 198 in D1 and D2 (Barber, 1987; Michel and Deisenhofer, 1988). The isolated PS2 reaction centre does not, however, retain the secondary quinone acceptors Q_A and Q_B despite the fact that their binding sites should be available (but see Giardi et al. 1988). This means that on the absorption of light the oxidation of P680 is brought about by the reduction of pheophytin. This primary electron transfer occurs very rapidly (within 3 ps, see Wasielewski et al. 1989) and creates the state $P680^+ Pheo^-$ which has a lifetime of about 35 ns (Danielius et al. 1987; Takahashi et al. 1987). In vivo the reduction of $P680^+$ by water involves an intermediate component Z, recognised by a characteristic EPR signal of g= 2.0046 (Babcock, 1987). Using molecular biological techniques, it has recently been shown that Z is probably tyrosine 160 on the D1 polypeptide (Debus et al. 1988).

These new findings have emerged in the last two years and emphasise the key role played by the D1 polypeptide. Now there is little doubt that this polypeptide forms the heart of the PS2 reaction centre and is not only involved in secondary electron flow on the reducing side but also on the oxidising side (Barber, 1988). Like the L-subunit of the purple bacterial reaction centre it forms the active branch for primary charge separation but, unlike the bacterial polypep-tide, it is not stable. This paper describes a wide range of work that has been carried out by my colleagues and I which indicates that the photodamage sustained by the PS2 reaction centre, leading to the turnover of the D1 polypeptide, may be due to the formation of singlet oxygen via triplet quenching.

MATERIALS AND METHODS

The PS2 reaction centre has been prepared from pea chloroplasts as described in Chapman et al. (1988) with modifications given in Crystall et al. (1989). Steady state light induced changes were measured (as in Chapman et al. 1988; Telfer and Barber, 1989). Flash induced absorption measurements were made using a pulsed N_2-laser as described in Durrant et al. (1990), and fluorescence lifetimes determined with a single photon counting apparatus outlined in Crystall et al. (1989).

RESULTS

Aerobic conditions

In the absence of artificial electron donors and acceptors the effect of light on the isolated reaction centre

is to generate primary charge separation giving rise to the
P680$^+$Pheo$^-$ state (Wasielewski et al. 1989; Danielius et al.
1987; Takahashi et al. 1987). Normally the radical pair back-
reacts rapidly (Danielius et al. 1987; Takahashi et al. 1987)
but when the electrons are supplied to P680$^+$ by adding sodium
dithionite it is possible to photo-accumulate the state
P680Pheo$^-$ (Nanba and Satoh, 1987; Barber et al. 1987). On the
other hand, when silicomolybdate (SiMo) is present, it is
possible to accept electrons from Pheo$^-$ and observe an
accumulation of P680$^+$ (Barber et al. 1987; Gounaris et al.
1988). Spectra for these two states are shown in Fig. 1. EPR
studies indicate that reduction of SiMo can occur at liquid
helium temperatures suggesting that this acceptor can bind
very close to the reduced pheophytin (Nugent et al. 1989).

When, in addition to SiMo, artificial electron donors are
present it is possible to observe a light induced steady state
rate of reduction of the acceptor (Barber et al. 1987).
Suitable donors for this reaction are diphenylcarbazide (DPC)
and MnCl$_2$ (Chapman et al. 1988). Also observed with the
isolated reaction centre is quinone mediated light induced
reduction of cytochrome b559 (Chapman et al. 1988; Gounaris et
al. 1988).

All the above electron transfer processes can be observed
in reaction centres which have been isolated using Triton X-
100 and kept cold and dark. Such reaction centres are
characterised by an absorption spectrum which has a room
temperature red maximum at about 676 nm and a room temperature
emission spectrum at 683 nm. Degradation of the preparation
however was found to occur if it was allowed to warm up above
the normal working temperature of 4°C (Chapman et al. 1988).

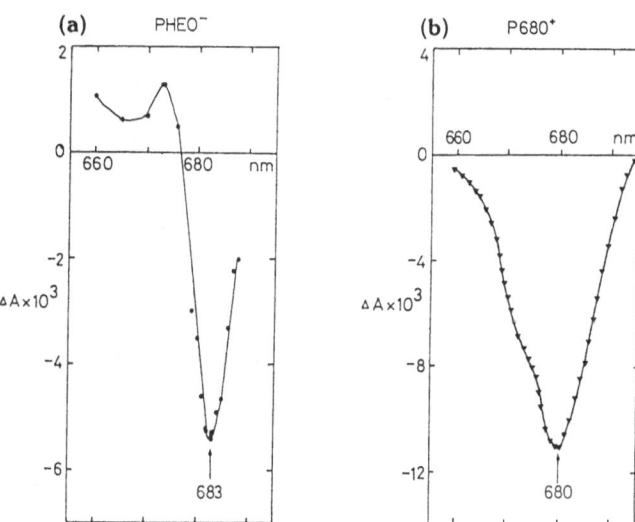

Fig. 1   Light minus dark difference spectra of
         reversible absorption changes observed in PS2
         reaction centres at 4°C in the presence of
         (a) sodium dithionite and (b) 0.5 mM SiMo.
         Chlorophyll 2μg ml^{-1}, in Triton X-100
         (see Gounaris et al. 1988).

This degradation occurred in the dark and resulted in blue shifts of the absorption and emission spectra and a loss of photochemical activity (see Fig. 2). Like others (Seibert et al. 1988) we have found that this sensitivity to temperature could be overcome by exchanging the isolated complex into other detergents and, as Fig. 3 shows, we have been using β-lauryl maltoside (Chapman et al. 1989).

This stabilisation however, does not protect the isolated complex against photodamage. The effect of preillumination is to cause a blue shift in the red band of the absorption spectrum due to the selective bleaching of chlorphylls which absorb at 680 nm, probably P680 itself (Fig. 4). Concomitant with this shift is a loss of photochemical activity (Fig. 5). Additional evidence that the photodestruction involves P680 comes from circular dichroism (CD) studies (He et al. 1990). Fig. 6 shows that the reaction centre gives positive and negative CD signals indicative of excitonic coupling between the chlorophylls, and as would be expected if P680 is a special pair. As can be seen the photoinhibition of sample causes an irreversible quenching of the optical activity. Associated with these changes is a blue shift in the emission spectra and an increase in the intensity of a 6.5 ns fluorescence decay component relative to a 36.5 ns component attributed to direct recombination between P680$^+$ and Pheo$^-$ (see Table 1 and Crystall et al. 1989).

The deactivation of P680$^+$Pheo$^-$ state can also occur via the formation of a P680 triplet state which can be detected in the isolated reaction centre at liquid helium temperatures using EPR (Okamura et al. 1987; Telfer et al. 1988) and at a lower yield at room temperature (Takahashi et al. 1987; Durrant et al. 1990). The yield of this triplet signal is reduced as the red absorption spectrum shifts to the blue as a consequence of pre-illumination treatments (Fig. 7).

All the above data indicate that the photo-susceptibility of the isolated reaction centre involves impairment of primary charge separation probably by a destruction or perturbation of the special pair of chlorophyll molecules which constitute P680.

Anaerobic Conditions

When oxygen is removed from the preparation using a glucose/glucose oxidase trap it becomes remarkably stable to preillumination with bright light. This stability is manifested in no blue shift of the red maximum of the absorption and emission spectra and in the persistence of its photochemical activity (e.g. Fig. 8).

Flash absorption studies (Durrant et al. 1990) show that the effect of removing oxygen is to lengthen the lifetime of the P680 triplet signal from 33 ± 3 μs to 1.0 ± 0.1 ms (see Fig. 9). The use of anaerobic conditions overcame serious problems which we experienced in those studies which required illumination of samples for reasonable periods of time. One such measurement was single photon counting to obtain fluorescent lifetimes which was difficult under aerobic conditions due to the instability of the sample (Crystall et al. 1989). With anaerobic samples a high yield (44%) of the

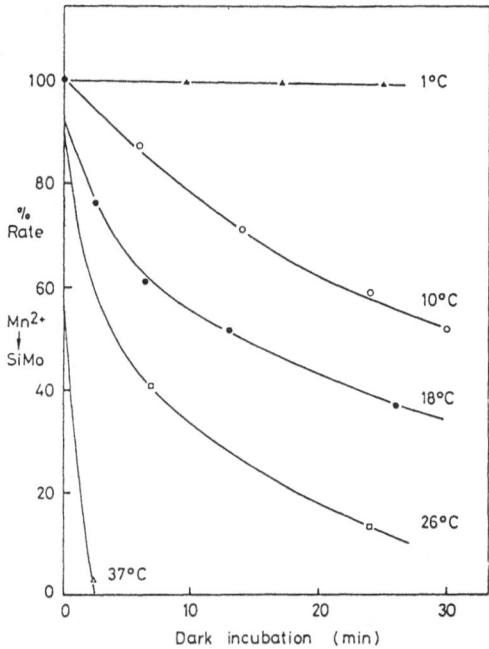

Fig. 2   Effects of temperature pretreatments on the
Mn^{2+} to SiMo rate measured at 4°C, in Triton
X-100, under aerobic conditions (see Chapman
et al. 1988).

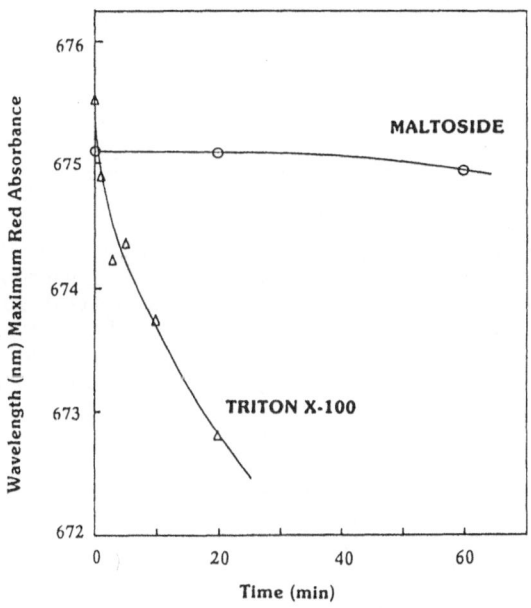

Fig. 3   Stabilisation of isolated reaction centre to
temperature sensitive destruction in the dark
by exchanging Triton X-100 with 2 mM β-lauryl
maltoside.   Incubation at 20°C and under
aerobic conditons.

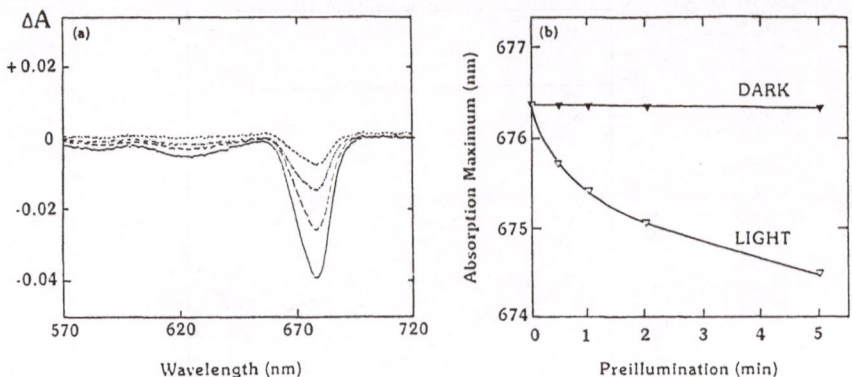

Fig. 4 Effect of various times of preillumination with white light (2000 μEinsteins m^{-2}s^{-1}) on the absorption spectrum of the PS2 reaction centre in β-lauryl maltoside (a) difference spectrum (b) blue shift. Under aerobic conditions.

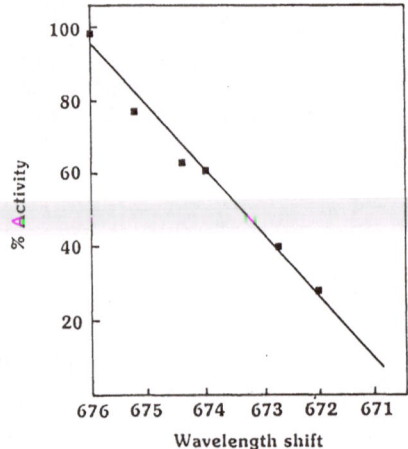

Fig. 5 Relationship between the blue shift of the red absorption peak and the photochemical activity (Mn^{2+} → SiMo) measured with PS2 reaction centres subjected to various periods of preillumination. In β-lauryl maltoside and under aerobic conditions.

Table 1 Changes in relative yields of the two major components of the fluorescence decay as a function of blue shift of the red absorption peak due to preillumination (from Booth et al. 1990).

λabs (nm)	λflu (nm)	Υ_1 (ns)	Υ_2 (ns)	%F$_1$	%F$_2$
675.9	682.5	36.5	6.5	44	40
672.9	678.0	*	6.2	*	90
670.0	673.0	*	6.3	*	98

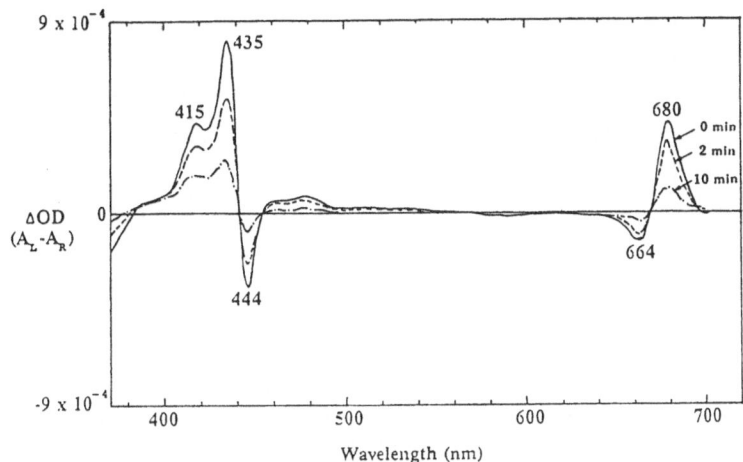

Fig. 6 Circular dichroism spectra of the PS2
reaction centre subjected to various periods
of preillumination (zero, 2 min and 10 min).
In β-lauryl maltoside and under aerobic
conditions.

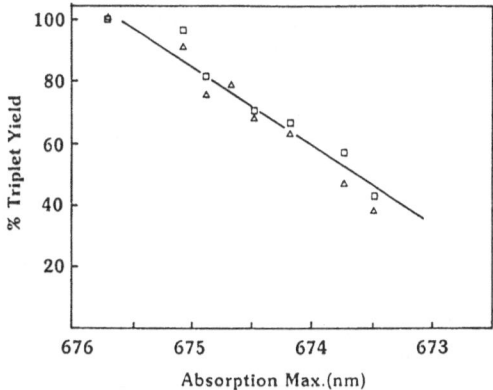

Fig. 7 Plot of relative triplet yield of P680 at
4 °C against blue shift of red absorption
maximum induced by varying the exposure of the
reaction centre to laser illumination (see
Durrant et al. 1990). Triangles measured at
680 nm, open squares measured at 740 nm.
In β-lauryl maltoside and under aerobic
conditions.

recombination fluorescent (36.5 ns) could be obtained at 40°C
(Fig. 10 and Table 2). Assuming the 6.5 ns component is from
free chlorophyll with an intrinsic fluorescence quantum yield
of 0.32 and that the quantum yield for emission from the
reaction centre is 0.02, then we estimate that 94% of the
chlorophyll in our samples is associated with functional
reaction centres (Booth et al. 1990). Removal of oxygen has
also allowed us to demonstrate that the CD signals thought to

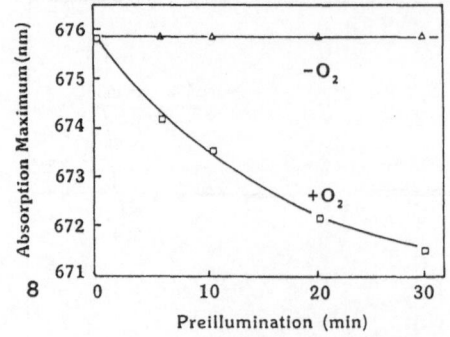

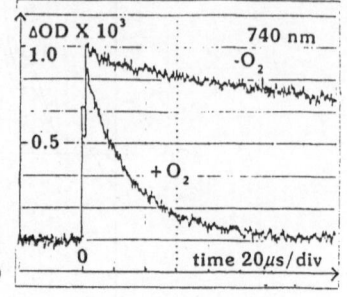

Fig. 8 Effect of removing oxygen on the blue shift
of the red absorption maximum induced by
various preillumination periods.

Fig. 9 Flash induced absorption changes observed in
PS2 reaction centres at 740 nm, (a) aerobic
and (b) anaerobic conditions measured at
4°C (see Durrant et al. 1990).

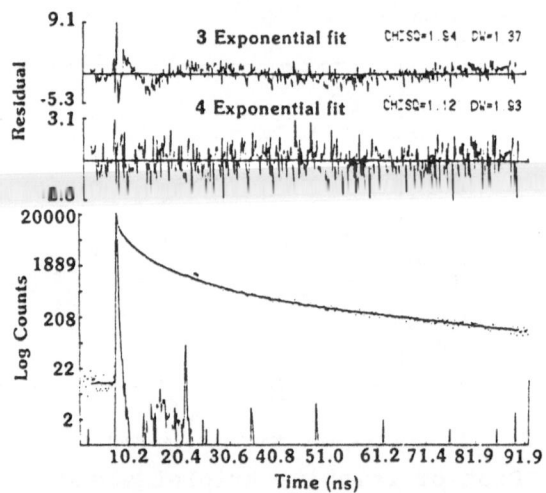

Fig. 10 A typical fluorescence decay of the PS2
reaction centre measured at 682 nm, in
β-lauryl maltoside and under anaerobic
conditions (see Crystall et al. 1989).

Table 2 Fluorescence lifetime components of PS2
reaction centres under anaerobic conditions
at 4°C (from Crystall et al. 1989).

T_x	Lifetime (ns)	Total Fluorescence emitted (%)
T_1	36.5 ± 2.5	44 ± 4
T_2	6.5 ± 1.0	40 ± 5
T_3	1.5 ± 0.5	11 ± 2
T_4	0.1 ± 0.1	5 ± 2

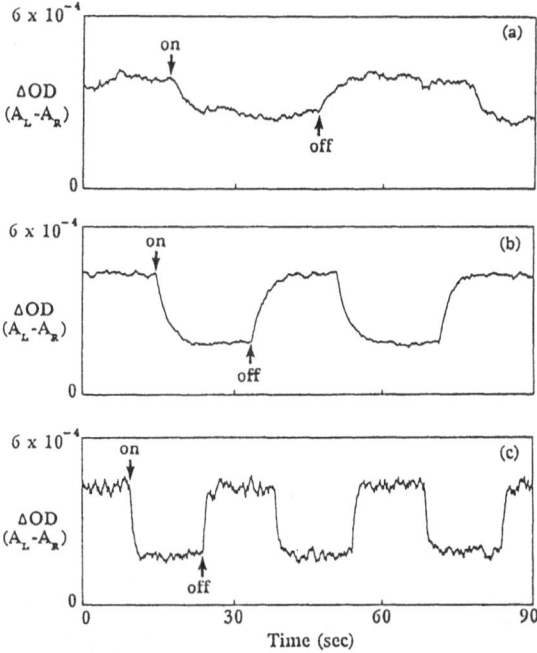

Fig. 11 Light induced reversible quenching of CD spectra of the PS2 reaction centre in β-lauryl maltoside, under anaerobic conditions (a) at 677.5 nm, (b) and (c) at 434 nm, measured in the presence of 1.0 mM SiMo (time constants 1 s for (a) and (b) and 0.25 s for (c)) (see He et al. 1990).

be due to P680 can be reversibly quenched by actinic light when SiMo is present as an electron acceptor (Fig. 11 and see He et al. 1990).

CONCLUSION

Our results clearly show that the removal of a dioxygen from a suspension of the isolated PS2 reaction centre protects it from photodamage. This effect of anaerobosis has also been noted by McTavish et al. (1989). In the ground state dioxygen exists in its triplet state ($^3O_2$) with two valence electrons from the two oxygen atoms each occupying the two antibonding molecular orbitals $2p\pi_x$ and $2p\pi_y$. They are in their highest energy level and their spins are parallel according to Hund's rule. Thus dioxygen is a biradical. However, most components in biological systems are in their singlet states in which the valence electrons are paired within their molecular orbitals and therefore have anti-parallel spins. As a consequence of this, reactions between triplet dioxygen and singlet state cell-components are forbidden so as to maintain sping conservation. This restriction is lifted if the cell component is also in a triplet state. From our experiments we conclude that when oxygen is present it quenches the P680 triplet and as a consequence is converted to singlet oxygen ($^1O_2$).

$$^3P680 + {}^3O_2 \rightarrow P680 + {}^1O_2$$

It is possible that the same mechanism may underlie the damage that occurs _in vivo_ during photinhibition, and which gives rise to the necessity for the D1 polypeptide to turn over so rapidly.

ACKNOWLEDGEMENTS

The work described here involves the work of several students and colleagues, the names of which appear on the references cited. All of us wish to acknowledge financial support from the AFRC, SERC, The Royal Society, EEC, BP plc and the Rowland Foundation.

REFERENCES

Barber, J., 1987, Photosynthetic reaction centres: A common link, _Trends in Biochem. Sci._, 12: 321-326.

Barber, J., 1988, Similarities and differences between the photosystem two and purple bacterial reaction centres, _in_: Light-Energy Transduction in Photosynthesis: Higher Plant and Bacterial Models (S. E. Stevens and D. A. Bryant, eds.), The American Soc. Plant Physiologists.

Barber, J., Chapman, D.J. and Telfer, A., 1987, Characterisation of a PSII reaction centre isolated from the chloroplasts of _Pisum sativum._, _FEBS Lett._, 220: 67-73.

Barber, J., Gounaris, K. and Chapman, D.J., 1987, Isolation of the photosystem two reaction centre and the location and function of cytochrome b559, _in_: Cytochrome Systems; Molecular Biology and Bioenergetics (Papa, S., Chance, B. and Ernster, L., eds.), Plenum Press, New York.

Babcock, G.T., 1987, The Photosynthetic oxygen-evolving process, _in_: Photosynthesis (J. Amesz, ed.), Pub. Elsevier, Amsterdam.

Booth, P.J., Crystall, B., Barber, J., Klug, D.R. and Porter, G., 1990, Thermodynamic properties of D1/D2/cyt b-559 reaction centres investigated by time-resolved fluorescence measurements. _Biochim. Biophys. Acta_, 1016: 141-152.

Chapman, D.J., Gounaris, K. and Barber, J., 1989, The D1/D2 cyt b559 photosystem two reaction centre from _Pisum sativum_ L: isolation, characterisation and damage by light, _Photosynthetica_, 23: 411-426.

Chapman, D.J., Gounaris, K. and Barber, J., 1988, Electron transport properties of the isolated D1-D2-cytochrome b559 photosystem II reaction centre, _Biochim. Biophys. Acta._ 933: 423-431.

Crystall, B., Booth, P.J., Klug, D.R., Barber, J. and Porter, G., 1989, Resolution of a long lived fluorescence component from D1/D2/cyt b559 reaction centres, _FEBS Lett._, 249: 75-78.

Danielius, R.V., Satoh, K., van Kan, P.J.M., Plijter, J.J., Nuijs, A.M. and van Gorkom, H.J., 1987, The primary reaction of photosystem II in the D1/D2 cytochrome b-559 _FEBS Lett._, 213: 241-244.

Debus, R.J., Barry, B.A., Sithole, I., Babcock, G.T. and McIntosh, L., 1988, Directed mutagenesis indicates that the donor to P680$^+$ in photosystem II is tyrosine-161 of th D1 polypeptide, <u>Biochemistry</u>, 27: 9071-9074.

Durrant, J.R., Giorgi, L.B., Barber, J., Klug, D.R. and Porter, G., 1990, Characterisation of triplet states in isolated photosystem two reaction centres: Oxygen quenching as a mechanism for photodamage. <u>Biochim. Biophys. Acta</u>, 1017: 167-175.

Edelman, M., Mattoo, A.K. and Marder, J.B., 1984, Three hats of rapidly metabolized 32 kD protein of thylakoids, <u>in</u>: Chloroplast Biogenesis (R.J. Ellis, ed), Cambridge University Press, Cambridge.

Giardi, M.T., Marder, J.B. and Barber, J., 1988, Herbicide binding to the isolated photosystem two reaction centre, <u>Biochim. Biophys. Acta</u>, 934: 64-71.

Gounaris, K., Chapman, D.J. and Barber, J., 1988, Reconstitution of plastoquinone in the D1/D2/cytochrome b559 photosystem 2 reaction centre complex, <u>FEBS Lett.</u>, 240: 143-147.

Greenberg, B.M., Gaba, V., Mattoo, A.K. and Edelman, M., 1987, Identification of a primary <u>in vivo</u> degradation product of the rapidly-turning-over 32kD protein of photosystem II, <u>EMBO J.</u>, 6: 2865-2869.

He, W.Z., Telfer, A., Drake, A., Hoadley, J. and Barber, J., 1990, Protection of the isolated PSII reaction centre against photodamage by removing oxygen or adding silicomolybdate. Proc. VIIIth Int. Congr. Photosynthesis, Stockholm.

Kyle, D.J., 1985, The 32,000 dalton Q_B protein of photosystem II, <u>Photochem. Photobiol.</u>, 41: 107-116.

Kyle, D.J., Ohad, I. and Arntzen, C.J., 1984, Membrane protein damage and repair, <u>Proc. Natl. Acad. Sci.</u>, (USA) 81: 4070-4074.

Marder, J.B., Chapman, D.J., Telfer, A., Nixon, P. and Barber, J., 1987, Identification of <u>psbA</u> and <u>psbD</u> gene products D1 and D2, as reaction centre proteins of photosystem 2. <u>Plant Mol. Biol.</u>, 9: 325-333.

McTavish, H., Picorel, R. and Seibert, M., 1989, Stabilisation of isolated photosystem II reaction centre complex in the dark and in the light using polyethylene glycol and an oxygen-scrubbing system, <u>Plant Physiol.</u>, 89: 452-456.

Michel, H. and Deisenhofer, J., 1988, Relevance of the photosynthetic reaction centre from purple bacteria to the structure of photosystem II, <u>Biochemistry</u>, 27: 1-7.

Nanba, O. and Satoh, K., 1987, Isolation of a photosystem II reaction centre containing D1 and D2 polypeptides and cytochrome b559, <u>Proc. Natl. Acad. Sci.</u>, (USA), 84: 109-112.

Nugent, J.H.A., Telfer, A., Demetriou, C. and Barber, J., 1989, Electron transfer in the isolated photosystem II reaction centre complex, <u>FEBS Lett</u>, 255: 53-58.

Okamura, M.Y., Satoh, K., Isaacson, R.A. and Feher, G., 1987, Evidence of the primary charge separation in the D1/D2 complex of PSII from spinach: EPR of the triplet state, <u>in</u>: Progress in Photosynthesis Res. (Vol. 1) (J. Biggins, Ed.), Pub. Martinus Nijhoff Publ., The Hague.

Rutherford, A.W., 1986, How close is the analogy between the reaction centre of photosystem II and that of purple bacteria? <u>Biochem. Soc. Trans.</u>, 14: 15-17.

Seibert, M., Picorel, R., Rubin, A.B. and Connolly, J.S., 1988, Spectral photophysical and stability properties of the isolated photosystem II reaction centre, <u>Plant Physiol</u>., 87: 303-306.

Takahashi, Y., Hansson, O., Mathis, P. and Satoh, K., 1987, Primary radical pair in the photosystem II reaction centre, <u>Biochim. Biophys. Acta</u>, 893: 49-59.

Telfer, A. and Barber, J., 1989, Evidence for the photo-induced oxidation of the primary electron donor P680 in the isolated photosystem II reaction centre, <u>FEBS Lett</u>., 246: 223-228.

Telfer, A., Barber, J. and Evans, M.C.W., 1988, Oxidation-reduction potential dependence of reaction centre triplet formation in the isolated D1/D2/cyt b559 photosystem II complex, <u>FEBS Lett</u>., 232: 209-213.

Trebst, A., 1986, The topology of the plastoquinone and herbicide binding peptides of photosystem 2 in the thylakoid membrane, <u>Z. Naturforsch</u>, 41c: 240-245.

Wasielewski, M.R., Johnson, D.G., Seibert, M. and Govindjee, 1989, Determination of the primary charge separation rate in isolated photosystem II reaction centers with 500-fs time resolution, <u>Proc. Natl. Acad. Sci</u>., (USA), 86: 524-528.

Webber, A.N., Packman, L., Chapman, D.J., Barber, J. and Gray, J.C., 1989, A fifth chloroplast-encoded polypeptide is present in the photosystem II reaction centre complex, <u>FEBS Lett</u>, 242: 259-262.

Zurawski, G., Bohnert, H.J., Whitfield, P.R. and Bottomley, W., 1982, Nucleotide sequence of the gene from the M_r 32,000 thylakoid protein from <u>Spinacia oleracea</u> and <u>Nicotiana debneyl</u> predicts a totally conserved primary translation of M_r 38,950, <u>Proc. Natl. Acad. Sci</u>., (USA), 79: 7699-7703.

MIGRATION AND TRAPPING OF THE ENERGY IN THE PHOTOSYNTHETIC
MEMBRANE OF Rhodobacter sphaeroides R26.1 PROBED BY
PHOTOVOLTAGE MEASUREMENTS

A. Dobek[1,2], J. Deprez[2], G. Paillotin[3],
H.-W. Trissl[4] and J. Breton[2]

[1]Institute of Physics, A. Mickiewicz University
Grunwaldzka 6, 60-780 Poznañ, Poland
[2]Service de Biopysique, Département de Biologie
Centre d'Études Nucléaires de Saclay
91191 Gif-sur-Yvette Cedex, France
[3]Commissariat a l'Énergie Atomique, Paris, France
[4]University of Osnabrück, F.B. Biologie/Chemie
Biophysik, Barbarastr. 11, D-4500 Osnabrück, FRG

INTRODUCTION

The analysis of subnanosecond fluorescence lifetimes is
the classical experimental approach to study the migration,
trapping and annihilation of excitations in the membrane of
photosynthetic bacteria. In the case of Rhodobacter
sphaeroides (Rb.sphaeroides), Rhodospirillum rubrum (R.rubrum)
or Thiocapsa roseopersicina, with open reaction centres (RC) a
fluorescence decay time of 60±10 ps has been reported (Sebban
and Moya, 1983; Borisov et al., 1985; Godik et al., 1985). It
increased to about 200 ps when the RCs are in the closed state
characterized by an oxidized primary donar P (Borisov et al.,
1985; Godik et al., 1985). In all cases, the fluorescence
decay kinetics, measured in the low energy limit, could be
well fitted by single exponentials. When the RCs are in the
open state before excitation, the lifetime of singlet excited
bacteriochrlorophyll antenna of R.rubrum chromaphores has been
estimated to 60±15 ps (Borisov et al., 1982; Razjivin et al.,
1982) and 80±10 ps (Nuijs et al., 1985) using picosecond
absorption spectroscopy. It increased to 200-400 ps when the
primary donar P was chemically oxydized (Nuijs et al., 1985).

Information about charge separation following the
excitation of isolated RCs of Rb.sphaeroides has been provided
by picosecond and subpicosecond absorption measurements
(Kirmaier and Holten, 1987). After formation of P*, the
excited state of the primary donar, charge separation between
P and the bacteriopheophytin electron acceptor H (P^+H^-),
occurs in about 3 ps (Woodbury et al., 1985; Martin et al.,
1986). The subsequent electron transfer from H to the first
stable quinon acceptor, Q_A takes place in about 200 ps
(Kirmaier and Holten, 1987).

The different location of the primary electron carriers with respect to the membrane normal gives rise upon excitation to a transmembrane potential which can be monitored as a photovoltage in light-gradient experiments (Trissl, et al., 1982; Trissl and Kunze, 1985). The photovoltage measurements can provide information about the kinetics of trapping and of charge stabilization as well as about the yield and the electrogenicity of the primary events (Leibl and Trissl, 1989: Trissl, 1983). This technique, developed in the nanosecond and in the picosecond time scales (Trissl and Kunze, 1985; Deprez et al., 1986), has been proposed as an alternative approach to fluorescence lifetime or absorption change measurements.

In this study, single and double (actinic and probe) ps flash experiments of photovoltage measurements are performed with Rb.sphaeroides R26.1 whole cells to probe the trapping time and the trapping efficiency for different redox states of RCs. The data are analyzed in terms of the competition between trapping and exciton-exciton annihilation in the lake model (Deprez et al., 1990) which has shown to apply to the photosynthetic membrane of purple bacteria (Monger and Parson, 1977; Campillo et al., 1977; Bakker et al., 1983).

MATERIALS AND METHODS

Rb.sphaeroides R26.1 cells were grown as described by Cohen-Bazire et al. (1957). For measurements, they were resuspended in Tris-HCl buffer (20 mM, pH 8) containing the electron mediator phenazine methosulfate (10 μM), yielding an optical density of 20 cm^{-1} at 520 nm.

The electric measurements of the light gradient type were carried out essentially as described (Trissl et al., 1987; Dobek et al., 1990). The excitation source was a mode locked Nd-YAG laser (YG-402, Quantel, France) delivering single 30-ps flashes. The 532 nm beam was divided into two flash beams P_1 and P_2. The first flash P_1 was sent directly to the sample while the second one, P_2, entered a constant delay path of $\Delta t=20$ ns. In single ps flash measurements, the P_1 beam was blocked and the photovoltage signal elicited by P_2 was recorded. In double flash experiments, P_1 served as an actinic preflash.

The experimental traces, recorded as single sweeps on a 4 GHz oscilloscope (TSN 660.2, Thomson, France), were compared with the results of a convolution calculation, assuming biexponential charge displacement current and taking into account the response time of the experimental set-up, the duration of the laser flash and the time constant of the discharge of the measuring cell capacitance into the 50-Ω input impedance of the preamplifier (Deprez et al., 1986; Deprez, 1986). The parameters of such fits are the time constants τ_1 and τ_2 of the two electrogenic phases of the photovoltage and the ratio A_2/A_1 of the electrogenicity parameters that represent the change of the dipole strength associated with the two steps of charge transfer.

RESULTS

Single ps Flash Excitation of Dark-Adapted Bacteria

The time course of the photovoltage elicited from whole cells of <u>Rb.sphaeroides</u> R26.1 by a single 30 ps flash is biphasic (Fig. 1, trace (a)). The electrogenic reactions associated with these two phases are attributed to the primary charge separation between the primary donor P and the intermediary acceptor H and to the further electron transfer from H to the quinone Q_A. In Fig. 1 ten single shot traces resulting from successive laser flashes of energy equal to $(1.9\pm0.2)*10^{14}$ photons.cm^{-2} are superimposed (b). The kinetic trace (c) obtained by averaging these traces is shown in the lower part of Fig. 1, where the dashed line is calculated by assuming biexponential displacement current. The best fit of curve (c) was obtained with the free running parameters τ_1, τ_2 and A_2/A_1 yielding the values of 40 ps, 190 ps and 1.66, respectively.

The amplitude of the photovoltage increased with the energy of the flash, Fig. 2a. Simultaneously the photovoltage rose faster. The result of the fit indicates that this acceleration is due to the shortening of the time constant τ_1 of the fast phase from 55 ± 10 ps to 30 ± 10 ps when energy increased from 10^{13} to 10^{15} photons.cm^{-2} (Fig. 2b). This

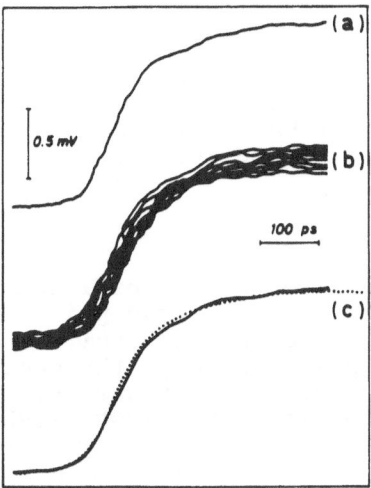

Fig. 1   Trace a: single sweep record of the photovoltage elicited by a 30 ps, 532 nm laser
flash at an energy $1.9*10^{14}$ photons.cm^{-2}
from <u>R.sphaeroides</u> R26.1 whole cells.
Traces b: ten single shot traces resulting
from ten successive laser flashes at an
energy of $(1.9\pm0.2)*10^{14}$ photons.cm^{-2}, re-
centred and superimposed. Trace c: Average
of the ten traces corresponding to the
individual sweeps depicted by trace b. The
dashed line is calculated by assuming bi-
exponential displacement current in the
measuring cell.

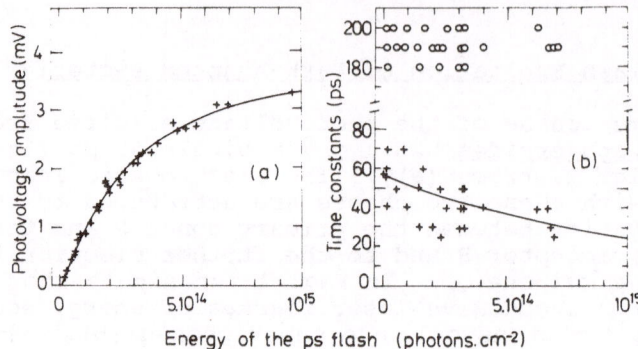

Energy of the ps flash (photons.cm⁻²)

Fig. 2 (a): Amplitude of the photovoltage from <u>Rb.</u> <u>sphaeroides</u> R26.1 whole cells elicited by a 30-ps, 532-nm laser flash as a function of the excitation energy. (b): Time constants τ_1 (+) and τ_2 (0) of the fast and slow phases of the photovoltage as a function of the energy of the 30-ps, 532-nm excitation laser flash. Continuous lines in (a) and (b) result from calculations described in the text.

indicates a faster trapping when the concentration of excitons is increased.

An increase of excitation energy has little effect on the relative electrogenicity of the two phases (A_1/A_2=1.6±0.2). The time constant τ_2 of the slower phase is equal to 190±10 ps in the whole range of excitation (Fig. 2b).

Double ps Flash Excitation

In order to characterize the efficiency of closure of the RCs by a ps flash, we have utilized pairs of ps flashes. The first flash P_1 was used to close a fraction of the RCs, while the photovoltage signal induced by the second flash P_2 incident at a time Δt=20 ns later was used to probe the RCs which were still open after the first flash. The energy of the probe flash P_2 was held constant at $1.9*10^{14}$ photons.cm⁻², while the energy E_1 of the actinic flash P_1 was varied.

The amplitude of the photovoltage created by P_2, normalized to the value obtained without P_1, is plotted as a function of E_1 of the actinic flash in Fig. 3a. The decrease of the photovoltage amplitude with increasing E_1 is due to the progressive increase of the fraction of RCs closed by the actinic flash.

The time constant τ_1 of the fast electrogenic component of the photovoltage induced by the probe pulse P_2 increased from 50±10 ps to 90±10 ps when the energy E_1 increased from 0 to about $1.6*10^{15}$ photons.cm⁻² (Fib. 3b). This increase of τ_1 reflects an increase of the mean trapping time of the exciton created by the probe flash when the RCs are progressively closed.

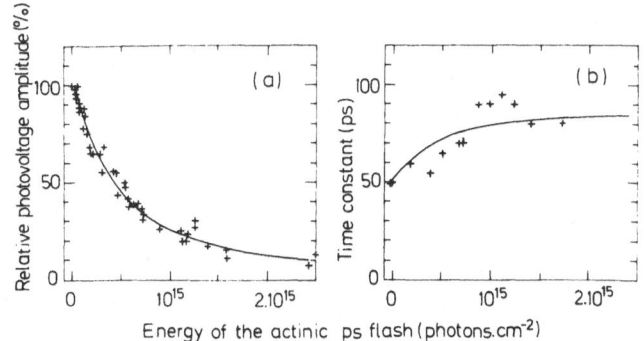

Energy of the actinic ps flash (photons.cm⁻²)

Fig. 3   (a): Amplitude of the photovoltage from <u>Rb.</u>
<u>sphaeroides</u> R26.1 whole cells elicited by a
30-ps, 532nm laser probe flash P_2 of constant
energy ($1.9*10^{14}$ photons.cm^{-1}) as a function
of the energy of the actinic flash P_1. The
amplitude is normalized to the value
measured with dark-adapted cells in absence
of P_1. The delay between P_2 and P_1 was 20ns.
(b): Time constant of the fast phase of the
photovoltage induced by P_2. Calculated fits
described in the text are drawn as
continuous lines in (a) and (b).

DATA ANALYSIS

        A macroscopic description of the exciton transfer and
trapping in the photosynthetic membrane of purple bacteria is
presented in Deprez et al. (1990). It considers two states of
the RCs during the lifetime of the excitons created by a ps
flash. The "open" state PHQ_A, in which the RC is able to
perform photochemistry, and the "closed" state in which the
primary donor is oxidized (P^+) are distinguished by their
different quenching rate constants, k_o and k_c, respectively.
These rate constants include radiative and non radiative
losses in addition to quenching by the RCs. We denote by q_o
an q_c the fraction of RCs in the open and closed states,
respectively. σ is the bimolecular exciton-exciton
annihilation rate constant, N is the average number of
bacteriochlorophyll molecules per RC and σ is the mean
absorption cross section per molecule at the excitation
wavelength. When the RCs are in the open state and when
exciton-exciton annihilation is neglected (low energy limit),
Γ is the probability for an exciton to be irreversibly trapped
by the RC, giving rise to the successive states $P^+H^-Q_A$ and
$P^+HQ_A^-$. For delta function excitation the time evolution of
the mean number of excitons per RC, n(t), is described by
(Paillotin et al., 1979; Paillotin et al., 1983; Deprez et
al., 1990):

(1)

$$n(t) = - \frac{1}{\Gamma k_o} \left(-\frac{q_o(t)}{Q_o} \right)^{-1} \frac{d}{dt} \left(\frac{q_o(t)}{Q_o} \right)$$

where:

$$\frac{d}{dt}\left(-\frac{q_0(t)}{Q_0}\right) = -\left(-\frac{q_0(t)}{Q_0}\right)^{\alpha+1}\left[\Gamma k_0 z + Q_0\frac{k_0-k_c}{\alpha-1} + \frac{k_c}{\alpha}\right]$$

$$+\left(-\frac{q_0(t)}{Q_0}\right)^2 Q_0\frac{k_0-k_c}{\alpha-1} + \left(-\frac{q_0(t)}{Q_0}\right)\frac{k_c}{\alpha} \tag{2}$$

Here z is defined as the initial number of excitons created per RC ($z=n(0)=\sigma NE$), $\alpha=\bar{\gamma}/2\Gamma k_0$ is the dimensionless parameter which measures the competition between singlet-singlet annihilation and exciton capture by open RCs and Q_0 is a fraction of RCs open before the flash. After the complete decay of the excitons in the antenna, $d(q_0(t)/q_0)/dt=0$. The fraction of RCs still open reaches its final value, q_{of}, which can be determined from the eq. (3):

$$\left(-\frac{q_{of}}{Q_0}\right)^\alpha\left[\Gamma k_0 z + Q_0\frac{k_0-k_c}{\alpha-1} + \frac{k_c}{\alpha}\right] = \left(-\frac{q_{of}}{Q_0}\right)Q_0\frac{k_0-k_c}{\alpha-1} + \frac{k_c}{\alpha} \tag{3}$$

Photovoltage amplitude

The ampitude of the photovoltage, $V(z)$, induced by the flash, is related to the fraction $\Delta q_c=Q_0-q_{of}$ of RCs closed by this flash (Leibl and Trissl, 1990):

$$V(z) = f V_0\left[\Delta q_{cu}(z) - \Delta q_{cl}(T_{eff} z)\right] \tag{4}$$

In eq. (4), V_0 denotes the effectively measured photovoltage when $\Delta q_c=\Delta q_{c,u}(Z)-\Delta q_{c,l}(T_{eff}\cdot z)=1$. The subscripts u and 1 indicate upper and lower membrane respectively and T_{eff}, the effective transmission of the "effective" vesicle. f denotes the proportionality factor. Using eqs. (3) and (4) with $\Gamma N\sigma$, k_0/k_c, α, fV_0 and T_{eff} as parameters, q_{of}, calculated as a function of Q_0 and E can be compared with experimental data and the measured amplitudes and kinetics of the photovoltage can be analyzed.

In single flash experiments with dark-adapted samples ($Q_0=1$), $V(z)$ is calculated as a function of E_2 (Fig. 2a). If actinic ps flashes (energy E_1) are used to partially close the RCs, the remaining fraction of open RCs, Q_0, can be determined using eq. (3). The amplitude $V(z_2)$ of the photovoltage elicited by the probe flash can then be calculated as a function of E_1 (Fig. 3a). All the continuous lines in Figs. 2a and 3a result from the fits using $\Gamma N\sigma=2.85*10^{15}$ cm^2, $k_0/k_c=3.0\pm0.2$, $\alpha=1.0\pm0.1$, $fV_0=4$ mV and $T_{eff}=0.003$.

Fast Phase Kinetics

The kinetics of the fast phase of the photovoltage, described by eq. (2), reflect the disappearance of the fraction of RCs in the open state, q_0. The calculated 1/e decay times are compared to the exponential rise time τ_1 of the fast phase. The lines drawn in Figs. 2b and 3b are obtained using the set of parmeters given above and a monomolecular rate constant of $k_0=(55$ ps$)^{-1}$.

DISCUSSION

Competition Between Trapping and Exciton Annihilation

The amplitude of the photovoltage elicted by the ps flash depends on the initial fraction of RCs in the open state (Fig. 3a). It can be correlated with Q_o which allows one to obtain, from the amplitude induced by the probe flash P_2, the initial fraction of RCs in the open state before excitation by P_2, and thus to deduce the fraction of RCs closed by the ps actinic flash P_1. This fraction increases with the energy P_1 and reaches 90% at $E_1=1.2*10^{15}$ photons.cm^{-2} corresponding to z=3.6 (Fig. 4).

It was found that even with the strongest flash used (z>5) only 65% of the RCs in R.rubrum and 80% of the RCs in Rb.capsulatus were oxidized (Bakker et al., 1983). This relatively low efficiency of trapping, corresponding to a value of $\alpha=4$ in R.rubrum (Deprez et al., 1990) reflects a severe competition between exciton-exciton annihilation and capture by the RCs, in contrast to the case of Rb.sphaeroides R26.1 ($\alpha=1$). This might be related to a different organization of peripheral and core antenna and/or to the presence of carotenoid (Kingma et al., 1985).

Slow Component of the Photovoltage Signal

The time course of the photovoltage signal measured in Rb.sphaeroides R26.1 whole cells demonstrates two distinct electrogenic events occurring during the first 500 ps. The time constant $\tau_2=190\pm15$ ps of the slower component of a bi-exponential displacement current (see appendix in Deprez et al. (1986)) attributed to the transfer of the electron from H to Q_A, is close to the time constant measured for the open state in isolated RCs of Rb.sphaeroides for the reoxidation of H$^-$ (Kirmaier and Holten, 1987). X-ray crystallography models of RC of Rb.sphaeroides R26 (Chang et al., 1986; Allen et al., 1987) give the location of H in the middle between P and Q_A. The ratio of the electrogenicities of the two reaction steps $A_2/A_1=1.6\pm0.2$ shows a higher dielectric constant of the protein environment between P and H than between H and Q_A.

Trapping Kinetics of RCs in the Open State

The time constant τ_1 of the fast component decreases upon increasing the flash energy (Fig. 2b). In principle, three phenomena may explain this decrease. If the kinetics of the trapping of the antenna excitation is a diffusion limited process (Pearstein, 1982), the decrease of the mean diffusion length of the excitons in the antenna at higher energy gives rise to a faster excitation of the RCs. The shortening of the mean trapping time may be also observed in the trap-limited model in which case the probability for an exciton to reside on the primary donor increases with the exciton density in the antenna. In addition, a further decrease of τ_1 is predicted for both models under conditions of exciton annihilation (Breton and Geacintov, 1980). In the present analysis, we have applied a model developed for the trap-limited case (Deprez et al., 1990). The weak dependency of the curve shown in Fig. 2b on the parameter α indicates that annihilation is not the dominant process which shortens the trapping time.

Trapping Kinetics of RCs in the Closed State

The rise time of the fast phase of the photovoltage measured after pre-excitation with an actinic ps flash (Fig. 3b) can be calculated as a function of the initial fraction $(1-Q_0)$ of RCs in the closed state. The trapping time constant increases from 55 ± 10 to 90 ± 15 ps when $(1-Q_0)$ increases from 0 up to 1. The time constants obtained from fluorescence measurements (Sebban and Moya, 1983; Borisov et al., 1985; Godik et al., 1985) are close to the value $\tau_1 = 55\pm10$ ps attributed to the trapping time in the present photovoltage studies. This value is also close to the lifetime of singlet excited bacteriochlorophyll in the antenna obtained from ps absorption spectroscopy (Borisov et al., 1982; Razjivin et al., 1982; Nuijs et al., 1985). Closing the RCs, the mean trapping time of the remaining RCs obtained from our measurements increases up to 90 ± 15 ps, whereas the exciton lifetime probed by spectroscopic studies increases to more than 200 ps. This difference can be explained by the intensity of the probe flash used to record the risetime of the photovoltage corresponding to the closure of about 40% of RCs when $Q_0 = 1$ (Fig. 4) which is not in the low energy limit. Nevertheless the theory developed by Deprez et al. (1990) together with the numerical value of the parameters deduced from these studies allows the trapping time in the low energy limit to be calculated. The value obtained (190 ps) is in good agreement with the spectroscopic data.

The analysis of our photovoltage data allows a discrimination between the lake and the puddle model. In both models, if $k_c < k_o$, the exciton lifetime is expected to increase upon closure of the RCs. In the puddle model, the photovoltage amplitude should be proportional to Q_0 and its kinetics should be independent of Q_0 and of k_o/k_c. In contrast, in the lake model and in the low energy limit the photovoltage amplitude is no longer proportional to Q_0 and both trapping and fluorescence kinetics are decribed by the same rate constant, whatever is the fraction of open RCs. The close analogy between the lengthening of the exciton lifetime probed by fluorescence decay measurements and of the trapping time determined in our photovoltage studies is thus compelling evidence in favour of the lake model description of the antenna organisation in Rb.sphaeroides R26.1.

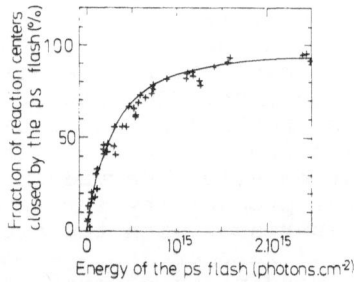

Fig. 4   Fraction of reaction centres of dark-adapted Rb.sphaeroides R26.1 whole cells closed by a 30-ps, 532-nm laser flash as a function of the exciton energy. The continuous line results from a fit described in the text.

ACKNOWLEDGEMENTS

The work at A.Mickiewicz University in Poznań was in part supported by Research Project Nr.RP.II.13.I.10. to A.D.; W.L. and H.-W.T. acknowledge the financial support of the Deutsche Forschungsgemeinschaft (SFB 171).

REFERENCES

Allen, J.P., Feher, G., Yeates,T.O., Komiya, H. and Rees, D.C., 1987, Structure of the reaction center from Rhodobacter sphaerodoides R-26: The cofactors, Proc. Natl. Acad. Sci. USA., 84:5730.

Bakker, J.C.G., van Grondelle, R. and den Hollander, W.T.F., 1983, Trapping, loss and annihilation of excitations in a photosynthetic system. II. Experiments with the purple bacteria Rhodospirillum rubrum and Rhodopseudomonas capsulata, Biochim. Biophys. Acta., 725:508.

Borisov, A.Yu., Gadonas, R.A., Danielius, R.V., Piskarskas, A.S. and Razjivin, A.P., 1982, Minor component B-905 of light-harvesting antenna in Rhodospirillum rubrum chromatophores and the mechanism of singlet-singlet annihilation as studied by difference selective picosecond spectroscopy, FEBS Lett., 138:25.

Borisov, A.Yu., Freiberg, A.M., Godik, V.I., Rebane, K.K. and Timpmann, K.E., 1985, Kinetics of picosecond bacteriochlorophyll luminescence in vivo as a function of the reaction centre state, Biochim. Biophys. Acta., 807:221.

Breton, J. and Geacintov, N.E., 1980, Picosecond fluorescence kinetics and fast energy transfer processes in photosynthetic membranes, Biochim. Biophys. Acta., 594:1.

Campillo, A.J., Hyer, R.C., Monger, T.G., Parson, W.W. and Shapiro, S.L., 1977, Light collection and harvesting processes in bacterial photosynthesis investigated on a picosecond time scale, Proc. Natl. Acad. Sci. USA., 74:1997.

Chang, C.H., Tiede, D., Tang, J., Smith, U., Norris, J. and Schiffer, M., 1986, Structure of Rhodopseudomonas sphaeroides R-26 reaction centre, FEBBS Lett., 205:82.

Cohen-Bazire, G., Sustrom, W.R. and Steiner, R.Y., 1957, Kinetics studies of pigment synthesis by non-sulfur purple bacteria, Cell. Comp. Physiol., 49:25.

Deprez, J., Trissl, H.-W. and Breton, J., 1986, Excitation trapping and primary charge stabilization in Rhodopseudomonas viridis cells, measured electrically with picosecond resolution, Proc. Natl. Acad. Sci. USA., 83:1699.

Deprez, J., 1986, Etude, dans l'échelle de temps subnanoseconds, de la migration et de la capture de l'excitation dans la membrane photosynthétique, Thesis, Université de Paris Sud, Orsay.

Deprez, J., Paillotin, G., Dobek, A., Leibl, W., Trissl, H.-W. and Breton, J., 1990, Competition between energy trapping and exciton annihilation in the lake model of the photosynthetic membrane of purple bacteria, Biochim. Biophys. Acta., 1015:295.

Dobek, A., Deprez, J., Paillotin, G., Leibl, W., Trissl, H.-W., and Breton, J., 1990, Excitation trapping efficiency and kinetics in Rb.sphaeroides R26.1 whole cells probed by photovoltage measurements in the picosecond time scale, <u>Biochim. Biophys. Acta.</u>, 1015:313.

Godik, V.I., Timpmann, K.E., Freiberg, A.M., Borisov, A.Yu., and Rebane, K.K., 1985, Picosecond kinetics of fluorescence decay during different states of reaction centres in <u>Thiocapsa roseopersicina</u>, <u>Dokl. Akad. Nauk. SSSR,</u> 284:491.

Kingma, H., van Grondelle, R. and Duysens, L.N.M., 1985, Magnetic-field effects in photosynthetic bacteria. II. Formation of triplet states in the reaction centre and the antenna of <u>Rhodospirillum rubrum</u> and <u>Rhodopseudomonas sphaeroides</u>. Magnetic field effects, <u>Biochim. Biophys. Acta.</u>, 808:383.

Kirmaier, C. and Holten, D., 1987, Primary photochemistry of reaction centres from the photosynthetic purple bacteria, <u>Photosynth. Res.</u>, 13:225.

Leibl, W. and Trissl, H.-W., 1990, Relationship between the fraction of closed photosynthetic reaction centres and the amplitude of the photovoltage from light-gradient experiments, <u>Biochim. Biophys. Acta.</u>, 1015:304.

Martin, J.L., Breton, J., Hoff, A.J., Migus, A. and Antonetti, A., 1986, Femtosecond spectroscopy of electron transfer in the reaction centre of the photosynthetic bacterium <u>Rhodopseudomonas sphaeroides</u> R26: Direct electron transfer from the dimeric bacteriochlorophyll primary donor to the bacteriopheophytin acceptor with a time constant of 2.8±0.2 psec, <u>Proc. Natl. Acad. Sci. USA.</u>, 83:957.

Monger, T.G. and Parson, W.W., 1977, Singlet-triplet fusion in Rhodopseudomonas sphaeroides chromatophores. A probe of the organization of the photosynthetic apparatus, <u>Biochim. Biophys. Acta.</u>, 460:393.

Nuijs, A.M., van Grondelle, R., Joppe, H.L.P., van Bochove, A.C. and Duysens, L.N.M., 1985, Singlet and triplet excited carotenoid and antenna bacteriochlorophyll of the photosynthetic purple bacterium <u>Rhodospirillum rubrum</u> as studied by picosecond absorbance difference spectroscopy, <u>Biochim. Biophys. Acta.</u>, 810:94.

Paillotin, G., Swenberg, C.E., Breton, J. and Geacintov, N.E., 1979, Analysis of picosecond laser-induced fluorescence phenomena in photosynthetic membranes utilizing a master equation approach, <u>Biophys. J.</u>, 25:513.

Paillotin, G., Geacintov, N.E. and Breton, J., 1983, A master equation theory of fluorescence induction, photochemical yield, and singlet-triplet exciton quenching in photosynthetic systems, <u>Biophys. J.</u>, 44:65.

Pearlstein, R.M., 1982, Exciton migration and trapping in photosynthesis, <u>Photochem. Photobiol.</u>, 35:835.

Razjivin, A.P., Danielius, R.V., Gadonas, R.A., Borisov, A.Yu. and Piskarskas, A.S., 1982, The study of excitation transfer between light harvesting antenna and reaction centres in chromtophores <u>R.rubrum</u> by selective picosecond spectroscopy, <u>FEBS Lett.</u>, 143:40.

Sebban, P. and Moya, I., 1983, Fluorescence lifetime spectra of in vivo bacteriochlorophyll at room temperature, <u>Biochim. Biophys. Acta.</u>, 722:436.

Trissl, H.-W., Kunze, U. and Junge, W., 1982, Extremely fast photoelectric signals from suspensions of broken chloroplasts and of isolated chromatophores, <u>Biochim. Biophys. Acta.</u>, 682:364.

Trissl. H.-W., 1983, Spatial correlation between primary redox components in reaction centres of <u>Rhodopseudomonas</u> <u>sphaeroides</u> measured by two electrical methods in the nanosecond range, <u>Proc. Natl. Acad. Sci. USA.</u>, 80:7173.

Trissl, H.-W., and Kunze, U., 1985, II. Primary electrogenic reactions in chloroplasts probed by picosecond flash-induced dielectric polarization, <u>Biochim. Biophys. Acta.</u>, 806:136.

Trissl, H.-W., Leibl, W., Deprez, J., Dobek, A. and Breton, J., 1987, Trapping and annihilation in the antenna system of Photosystem I, <u>Biochim. Biophys. Acta.</u>, 893:320.

Woodbury, N.W.T., Becker, M., Middendorf, D. and Parson, W.W., 1985, Picosecond kinetics of the initial photochemical electron-transfer reaction in bacterial photosynthetic reaction centre, <u>Biochemistry</u>, 24:7516.

THE NECESSITY OF CAROTENOIDS FOR THE ASSEMBLY OF ACTIVE PHOTOSYSTEM 2 REACTION CENTRES

Navassard V. Karapetyan, Yulia V. Bolychevtseva
and Marina G. Rakhimberdieva

A.N. Bakh Institute of Biochemistry, USSR Academy
of Sciences, Moscow 117071, USSR

INTRODUCTION

Carotenoids are the components of both the pigment apparatus and reaction centres of photosynthetic organisms. They function as accessory pigments and play a defence role protecting the reaction centres from photodestruction under high irradiance (Okayama and Butler, 1972; Searle and Wessels, 1978). However, it was found that the reaction centres of Photosystem I (PS1) and of purple bacteria retain the functional activity in the absence of carotenes that were extracted from lyophilized samples. At the same time, the β-carotene removal with hexane from Photosystem 2 (PS2), that is accompanied by extraction of quinones, leads to inactivation of PS2. Addition of plastoquinone together with β-carotene reactivates PS2, i.e. $P680^+$ and C550 signals appear again (Searl and Wessels, 1978). Therefore, β-carotene is closely connected with the PS2 reaction centre and seems to be essential for appearance of the reduced pheophytin signal.

An approach to elucidate the role of carotenoids in organization of the pigment-protein complex of PS2 (CPa) involves research into PS2 of plants with suppressed carotenoid biosynthesis (for instance under the effect of norflurazon). Unfortunately, such carotenoid-less plants have no PS2 activity (Öquist et al., 1980; Lehoczki et al., 1982). The present work was aimed at elucidation of the reasons for the absence of PS2 in carotenoid-deficient plants. For this we studied the pigment and protein-synthesizing apparatus, activity of photosystems, ultrastructure, fatty acid composition of lipids and polypeptide composition of chloroplasts from barley seedlings grown with high concentration of norflurazon in dim light that significantly diminishes or even eliminates chloroplast photodestruction. Such studies can help to determine whether the PS2 absence is caused by destruction of the photosynthetic apparatus or if it results from blocking the biogenesis.

Light in Biology and Medicine. Volume 2 Edited by R.H. Douglas *et al.*
Plenum Press, New York, 1991

MATERIAL AND METHODS

Barley seedlings were germinated for 5-7 days with 100 μM norflurazon (SAN 9789) at light intensities 10, 30 or 100 lux (continuous light) or under flash illumination. In the latter case, 7-days etiolated seedlings were illuminated during the day with 2,5 ms flashes every 12 min.; the total dose approximately corresponds to the dose received by the plant for 5 days at light intensity 10 lux. Chloroplasts were isolated from the seedlings as described elswhere (Bolychevtseva et a., 1987); since chloroplasts of the plants grown under flash irradience were less developed they were precipitated by centrifugation at 3500g for 8 min.

The activity of PS2 of leaves and chloroplasts and the presence of PS2 reaction centres were recorded by means of variable fluorescence at 20° and -196^{c}C, respectively, by using a difference fluorimeter designed in our laboratory (Rakhimberdieva et al., 1982). Photoinduced absorption changes of isolated chloroplasts at 695-700 nm were measured by means of a one-beam difference spectrophotometer (Karapetyan et al., 1987) to reveal the presence of PS1 reaction centres. Absorption spectra were registered by using a Hitachi 557 spectrophotometer, the low temperature fluorescence spectra were recorded by means of a Hitachi MPF-4 spectrofluorimeter and equipment described previously (Lebedev et al., 1988). The other techniques are described elsewhere: determining the pigments and the lipid fatty acid composition (Bolychevtseva et al., 1987), chloroplast ultrastructure relative contents of 70S and 80S ribosomes and rRNA of the ribosomes (Bolychevtseva et al., 1988). Electrophoresis in PAG was performed according to the modified Laemmli technique (Bolychevtseva et al., 1988).

RESULTS

To elucidate the role of carotenoids in photosynthetic apparatus biogenesis we should first characterize in detail the chloroplasts from plants grown with norflurazon that suppresses carotenoid biosynthesis. Norflurazon is attributed to bleaching herbicides that reduce the chlorophyll content (Bartels, 1970). At low concentrations (10-30 μM) norflurazon is able to inhibit the electron transport in the PS2 acceptor part but to a less degree than DCMU (Karapetyan et al., 1981). During long-term treatment of greening seedlings norflurazon brings about drastic changes in the photosynthetic apparatus (Rakhimberdieva et al., 1982). The same picture is observed when barley seeds and then seedlings are grown with 100 μM norflurazon. Since norflurazon almost completely suppresses carotenoid biosynthesis in the latter case, the plants should be grown at a very low irradiance (below 30 lux).

It was found earlier that long-term treatment of plants with norflurazon at high light intensities caused the bleaching of chlorophylls and carotenoids (Lichtenthaler and Kleudgen, 1972), photodestruction of lipids and a sharp decrease in the content of 70S ribosomes (Feierabend and Winkelhusener, 1982; Feierabend et al., 1982). We decided to determine whether photodestruction in carotenoid-less seedlings takes place at very low light intensities (10 or 30 lux) and whether it is only photodestruction that causes changes of the photosynthetic apparatus or whether the changes result from the blocked biogenesis of certain structures.

Pigments

Studying the absorption spectra of extracts from etiolated, barley seedlings grown with norflurazon showed that herbicide had a little effect on the protochlorophyllide: 7.4 and 7.0 g/g w.w. in control plants and in those grown with norflurazon, respectively. However, the content of long-chain carotenoids absorbing at 400-500 nm significantly decreases in etiolated seedlings, and products with UV absorbance appear that are considered to be phytoene and phytofluene. The results prove the norflurazon blocking of the carotenoid biosynthesis at the stage of phytoene desaturation. We shall show below that the barley seedlings grown with norflurazon under flash irradiation contain more chlorophyll than control plants, i.e. norflurazon fails to block the chlorophyll biosynthesis. Therefore, the decrease in the chlorophyll contents of seedlings grown with norflurazon at light intensities 10-30 lux must result from the chlorophyll bleaching caused by the absence of carotenoids.

The content of chlorophylls a, b and carotenoids in seedlings grown with norflurazon depends on the light intensity (Table 1): the contents of chlorphyll a and chlorophyll b decreased 3 and 7 times at 10 lux and 9 and 30 times at 30 lux, respectively, as compared to control plants. At 100 lux the pigment content diminished more than 100 times. The carotenoid content decreased down to 1% of control at 100 lux with light intensity increasing, the β-carotene content decreased to a greater degree.

The Fatty Acid Composition of Lipids

Besides chlorophyll bleaching, norflurazon also caused photodestruction of polyunsaturated fatty acids. The amount of linolenate decreased, while the amount of linoleate increased in etiolated seedlings. In seedlings grown at 10 lux the linolenate content increased as a result of light inactivation of the synthesis and the linoleate content decreased. The difference between control and herbicide-treated seedlings was minimal (Table 1). In norflurazon-treated seedlings grown at higher light intensities (30 and 100 lux) we observed selective destruction of linolenate and trans-hexadecenoic acid, which indicates the existence of oxidative destruction (Bolychevtseva et al., 1987). The conclusion is proved by accumulation of malondialdehyde (MDA) - the product of lipid peroxidation. MDA contents in etiolated control and herbicide-treated seedlings are almost equal at 10 lux, but at 30 lux the MDA content is 2.5 times higher in norflurazon-treated seedlings as compared to control plants (Table 1). The lower MDA content in seedlings grown with norflurazon at 100 lux is associated with the lower content of chlorophyll, that serves as a sensitizer of photodestruction, and with a high rate of MDA utilization in metabolic processes. The results obtained indicate that photodestruction of pigments and lipids occurs in carotenoid-less seedlings even at low light intensities.

47

Table 1 Characteristics of Chloroplasts from Barley Seedlings Grown with Norflurazon in Continuous Light

	10 lux		30 lux		100 lux	
	control	+norfl.	control	+norfl.	control	+norfl.
Chl a	295.0	126.5	343.4	33.3	677.0	4.1
Chl b	35.7	5.2	51.5	1.5	131.1	0.8
Carotenoids	97.5	5.2	118.3	3.3	194.1	1.6
$C_{18:2}$	16.0	20.9	13.9	28.5	13.8	36.5
$C_{18:3}$	64.0	63.0	70.1	55.3	68.0	44.7
$C_{18:2}/C_{18:3}$	0.2	0.3	0.2	0.5	0.2	0.8
MDA	11.2	14.7	12.5	30.0	12.2	16.4
70S/70S+80S	42.4	43.5	40.9	42.6	38.8	-
Chl/P700	210.0	100.0	230.0	50.0		-
$\Delta F/F_o$	2.7	-	3.2	-		-

Chl a, Chl b and carotenoids - $\mu g/g$ w.w.; $\Delta F/F_o$ - at $20°$ C; $C_{18:2}$ and $C_{18:3}$ - mol.% ; MDA - nmol/g w.w. ; 70S/70S+80S - %;

70S Ribosomes

70S ribosomes were found to disappear in chloroplasts of plants grown with norflurazon at high light intensities (Feierabend et al., 1982), but their content did not change at low light intensities. We found a limiting range of irradiance 30-100 lux, within the range we observed a sharp decrease of the 70S ribosome content in herbicide-treated seedlings. At 10 and 30 lux, the relative content of 70S ribosomes and rRNA from these ribosomes is almost the same for control and norflurazon-treated plants (Table 1). However, at 100 lux the homogenate of the treated seedlings gave no 70S sedimentation peak. Electrophoresis of the total RNA showed a sharp decrease of the chloroplast rRNA content (Karapetyan et al., 1987).

It is noteworthy that the disappearance of 70S ribosomes in treated seedlings grown at 100 lux is accompanied by the absence of the photosynthetic membrane and chloroplast envelope, which together with the low pigment content gives evidence for complete degradation of chloroplasts in carotenoid-deficient seedlings at 100 lux caused by photodestruction.

Photosystems

The low temperature fluorescence spectrum of norflurazon-treated barley seedlings gave no band at 695 nm and the long wavelength band at 740 nm shifted to 720 nm (Fig. 1), which indicates the disappearance of the nearest antenna of PS2 and the peripheric antenna of PS1. The band at 720 nm appears as a result of the PS1 core emission. The excitation spectra of 685 and 720 nm emission bands of norflurazon-treated plants showed a complete absence of carotenoid maxima at 491, 497 and 510 nm and of the pheophytin peak at 540 nm, which are well seen in the excitation spectrum of the control seedlings (Lebedev et al., 1988). The simultaneous disappearance of the emission band at 695 nm and the excitation band at 540 nm may indicate the absence of PS2 reaction centres. Chloroplasts of norflurazon-treated seedlings are significantly enriched with PS1: the chlorophyll/P700 ratio is 100-50 in contrast to 230-210 for chloroplasts of control seedlings (Table 1). The light-induced difference spectrum of chloroplasts isolated from treated seedlings showed a negative band at 696.5 nm typical of oxidized P700.

No variable fluorescence was observed at 20°C in barley seedlings grown with norflurazon at 10, 30 and 100 lux (Table 1), which indicates the absence of active PS2. The inability of the seedlings to fluorescence induction at -196°C gives evidence for the destruction of CPa, apparently caused by oxidative photodestruction or by blocking the PS2 biogenesis.

Chloroplast Polypeptides

Electrophoresis of membrane pigment-protein complexes of norflurazon-treated barley seedlings enabled us to determine the reasons for CPa destruction in these seedlings. The absence of CPa and the light-harvesting complex (LHC) in norflurazon-treated plants was earlier revealed by gel scanning at 677 nm (Öquist et al., 1980), which may be associated with the absence of chlorophylls but not polypeptides of the complexes. We performed the electrophoretic analysis of polypeptides from chloroplast thylakoids of treated plants that showed a significant decrease in the content of PS2 polypeptides (52, 45, 33-40 kD) and in that of the LHC (26-24 kD) (Fig. 2). At the same time polypeptides of CP1 and the coupling factor were present in membranes of both control and herbicide-treated plants. It is noteworthy that the soluable fraction of chloroplasts from treated plants contained a significant amount of Rubisco.

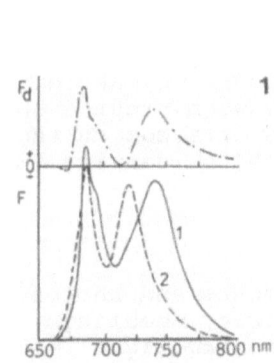

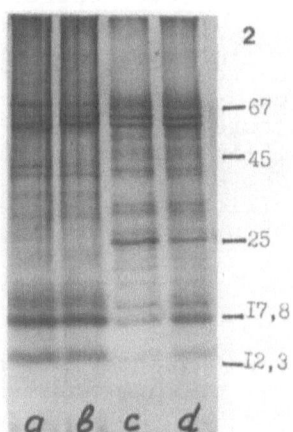

Fig. 1 Fluorescence spectra at -196° C of seedlings grown at 10 lux without (1) and with norflurazon (2); F_d-difference spectrum.

Fig. 2 SDS-electrophoresis of thylakoid polypeptides from seedlings grown at 30 (a,c) or 10 lux (b,d) without (c,d) and with (a,b) norflurazon.

It is well known that PS1 and coupling factor are located in stroma lamellae, while PS2 and LHC are in the thylakoids of granae. The results of the electrophoretic analysis correlate well with the data on chloroplast ultrastructure, i.e. the membrane system of chloroplasts from barley seedlings grown with norflurozon at 10 and 30 lux consisted of the stroma lamellae only, and granae were absent in contrast to control plants, which substantiates the results obtained previously (Axelsson et al., 1982). Thus, our results indicate that the PS2 content is significantly lower in herbicide-treated plants grown even at 10 lux, although they have enough pigments and active protein-synthesizing apparatus.

Flash-irradiated Seedlings

A sharp decrease in the polypeptide content of CPa and LHC of norflurazon-treated seedlings may indicate either their photodestruction or the mediated blocking of their biogenesis. Therefore, it is necessary to prevent photodestruction when growing norlfurazon-treated seedlings.

Seedlings grown with norflurazon under flash irradiance contained more chlorophyll than control plants; the chlorophyll a/b ratio was 11-13. Thus, the reason for the lower content of chlorophyll at high norflurazon concentrations is not the inhibition of chlorophyll synthesis, as was proposed (Axelsson et al., 1982). Control seedlings showed variable fluorescence at -196°C, while it was not observed in herbicide-treated plants (Table 2), i.e. the treated seedlings had no active PS2. The disappearance of the 695 nm band in the low temperature fluorescence spectrum also indicates the absence of CPa in flash-irradiated seedlings (Fig. 3). At the same time, treated seedlings accumulating not less chlorophyll than control plants are significantly enriched with PS1: the chlorophyll/P700 ratio was 60 in contrast to 150 for control seedlings (Table 2).

A significant difference was observed in the chloroplast membrane system which consisted mainly of primary granae composed of 2-3 lamellae in control plants; the chloroplast ultrastructure of herbicide-treated plants represented individual lamellae. Electrophoresis of thylakoid polypeptides showed no difference between treated and control seedlings (Fig. 4). The coincidence of bands on the electrophoregrams may be insufficient to identify PS2

Table 2 Characteristics of Pigments and Photosystems
of Chloroplasts from Barley Seedlings Grown
with Norflurazon in Flashing Light

	Chl a	Chl b	Carotenoids	Chl/P700	$\Delta F/F_o$
		(μg/g w.w)			
Control	51.8	4.1	103.3	150.0	0.5
+Norflurazon	59.3	5.3	3.0	60.0	-

$\Delta F/F_o$ - at -196°C

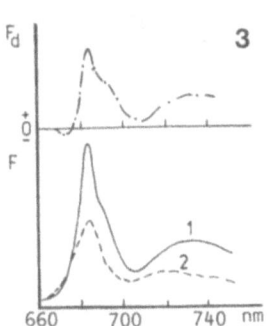

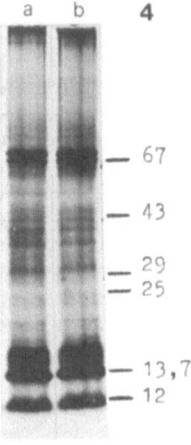

Fig. 3   Fluorescence spectra at -196° C of seedlings grown at flash light without (1) and with norflurazon (2); F_d - difference spectrum.

Fig. 4   SDS-electrophoresis of thylakoid polypeptides from seedlings grown at flash light without (a) and with (b) norflurazon.

polypeptides, nevertheless the complete similarity of polypeptide profiles allowed us to conclude that norflurazon had no direct effect on expression of the genes coding the thylakoid polypeptides, i.e. the biosynthesis of PS2 proteins was not suppressed.

Thus PS2 polypeptides from seedlings grown with norflurazon under flash illumination are incorporated into the membrane but fail to form an active CPa. According to our results, the chloroplast membranes of seedlings grown with norflurazon with flashing light contain no LHC proteins and the chlorophyll/protein ratio is very low (1/300 - 1/400) i.e. on average one chlorophyll molecule per 10 protein molecules. In spite of the presence of PS2 polypeptides in chloroplasts of flash-irradiated seedlings, no activity of PS2 reaction centres was revealed by low temperature fluorescence induction (Table 2), which may also result from violation of the PS2 ability to emit variable fluorescence at -196° C in case of carotenoid-deficient plants. However, as has been shown above, chloroplasts from flash-illuminated seedlings are characterized by an increased content of P700 - the primary donor of PS1 reaction centre. With an equal chlorophyll content and in the LHC absence in control and treated seedlings the pigments are redistributed in favour of PS1; the Chl binding to CPa is suppressed.

Apparently, the absence of carotenoids decreases the affinity of PS2 polypeptides for chlorophyll, which under conditions of photosystems competition for pigment molecules (Tzinas et al., 1987) leads to preferential chlorophyll binding to CP1.

DISCUSSION

The results obtained show that in barley seedlings that lost more than 97% of carotenoids, photodestruction of pigments and lipids occurs even at low light intensities. The protective effect of carotenoids results from their ability to inactivate chlorophyll triplet states, to quench singlet oxygen, and to catch free radicals (Mathews-Roth et al., 1974). The concentration of long-living chlorophyll triplets increases to a great extent in norflurazon-treated seedlings (Krasnovsky jr. et al., 1981). Phytoene accumulated in these seedlings has a much lower constant of singlet oxygen quenching and is unable to prevent photodestruction.

As has been mentioned above, within the limiting range of light intensities (30-100 lux) herbicide-treated seedlings are characterized by a decreased content of chloroplast ribosomes and by destruction of photosynthetic membranes, complete destruction being observed at 100 lux. Such light-dependent disturbances seem to result from pronounced photodestruction and are associated with formation of superoxide-radicals with a longer life-time than singlet oxygen.

It is noteworthy that even at low light intensities (10 lux) the pigment bleaching sensitizes not only peroxidation of lipids but also leads to destruction of PS2 polypeptides despite the absence of noticeable changes of the protein-synthesizing system. In this case the absence of LHC polypeptides may also be caused by photodestruction, although we failed to prove this as flash-illuminated plants accumulate little chlorophyll and have no LHC (Tzinas et al., 1987).

No Chl photobleaching was observed in barley seedlings grown with norflurazon under flash illumination. Surprisingly, despite the fact that the polypeptide composition of thylakoids from control and carotenoid-less seedlings was identical, no PS2 activity was detected in the latter. Carotenoid deficiency caused redistribution of Chl in favour of PS1 and decreased the affinity of PS2 polypeptides for Chl. At the same time it is obvious that PS2 protein destruction under continuous light must be sensitized by chlorophyll bleaching. Therefore, carotenoid deficiency affects the ability of PS2 polypeptides to bind chlorophyll, but does not block the process completely. As a result chlorophyll-poor CPa are formed that may be incapable of variable fluorescence. Probably, under continuous light at high rates of chlorophyll synthesis, the chlorophyll binding to CPa is less dependent on the carotenoid content, nevertheless, the complex is destroyed in the absence of carotenoid defence.

Thus besides light harvesting and having a protecting role, carotenoids are important as structural components for assembly of active CPa and its stabilization.

REFERENCES

Axelsson, L., Dahlin, C. and Ryberg, H., 1982, The function of carotenoids during chloroplast development. Y. Correlation between carotenoid content, ultrastructure and chlorophyll b to chlorophyll a ratio, <u>Physiol. Plant.</u>, 55:111.

Bartels, P.G. and Hyde, A., 1970, Chloroplast development in Sandoz 6706 treated wheat seedlings, <u>Plant. Physiol.</u>, 45:807.

Bolychevtseva, Yu.V., Chivkunova, O.B., Merzlyak, M.N. and Karapetyan, N.V., 1987 Effect of norflurazon on the contents of chlorophyll, fatty acid and lipid peroxidation products in barley seedlings grown under different illumination conditions, <u>Biochemistry</u> (USSR), 52:160.

Bolychevtseva, Yu.V., Turishcheva, M.S., Bezsmertnaya, I.N. and Karapetyan, N.V., 1988, Polypeptide composition and structure of chloroplasts from barley seedlings grown in the presence of norflurazon in dim light, <u>Biochemistry</u> (USSR), 53:677.

Feierabend, J. and Winkelhusener, T., 1982, Nature of photooxidative events in leaves treated with chlorosis-inducing herbicides, <u>Plant Physiol.</u>, 70:1277.

Feierabend, J., Winkelhusener, T., Kemmerich, P. and Schulz, U., 1982, Mechanism of bleaching in leaves treated with chlorosis-inducing herbicides, <u>Z. Naturforsch.</u>, 37C:898.

Karapetyan, N.V., Rakhimberdieva, M.G., Lehoczki, E. and Krasnovsky, A.A., 1981, Effect of pyridazinone herbicides on the photosynthetic electron transport chain of chloroplasts and Chlorella, <u>Biochemistry</u> (USSR), 46:2082.

Karapetyan, N.V., Bolychevtseva, Yu.V., Turishcheva, M.S. and Bezsmertnaya, I.N., 1987, Changes in photosystem activity and structure of chloroplasts in barley seedlings grown with norflurazon in dim light, <u>Proc. Indian natn. Sci. Acad.</u>, B54:369.

Krasnovsky, A.A., jr., Kovalev, Yu.V. and Lehoczki, E., 1981, Luminescence analysis of chlorophyll triplets in barley leaves with inhibited carotenoid biosynthesis, <u>Proc. Acad. Sci.</u>, (USSR), 256:726.

Lebedev, N.N., Pakshina, E.V., Bolychevtseva, Yu.V. and Karapetyan, N.V., 1988, Fluorescence characterization of chlorphyll-proteins in barley seedlings grown with norflurazon under low irradiance, <u>Photosynthetica</u>, 22:371.

Lehoczki, E., Rakhimberdieva, M.G. and Karapetyan, N.V., 1982, Effect of blocking of carotenoid synthesis by pyridazinones on photosystem 2 formation in barley leaves, <u>Plant Physiol.</u>, (USSR), 29:682.

Lichtenthaler, H.K. and Kleudgen, H.K., 1977, Effect of San 6706 on biosynthesis of photosynthetic pigments and prenylquinones in Raphanus and in Hordeum seedlings, <u>Z. Naturforsch.</u>, 32C:236.

Mathews-Roth, M.M., Wilson, T., Fujimori, E., and Krinsky, N.I., 1974, Carotenoid chromophore length and protection against photosensitization, <u>Photochem. Photobiol.</u>, 19:217.

Okayama, S. and Butler, W.L., 1972, Extraction and reconstitution of photosystem 2, <u>Plant Physiol.</u>, 49:769.

Öquist, G., Samuelsson, G. and Bishop, N.I., 1980, On the role of β-carotene in the reaction centre chlorophyll a antennae of photosystem 1, <u>Physiol. Plant.</u>, 50:63.

Rakhimberdieva, M.G., Lehoczki, E., Karapetyan, N.V. and Krasnovsky, A.A., 1982, Effects of pyridazinones and cerulenin on biosynthesis and functional state of photosystem 2 in barley leaves, <u>Biochemistry</u> (USSR), 47:637.

Searle, G.F.W. and Wessels, J.S.C., 1978, Role of β-carotene in reaction centres of photosystem 1 and 2 of spinach chloroplasts prepared in non-polar solvents, <u>Biochim. Biophys. Acta</u>, 504:84.

Tzinas, G., Argyroudi-Akoyunoglou, J.H. and Akoyunoglou, G., 1987, The effect of the dark interval in intermittent light on thylakoid development: photosynthetic unit formation and light harvesting protein accumulation, <u>Photosynthesis Res.</u>, 14:241.

ENERGY TRANSFER IN PHOTOSYSTEM II ANTENNA

Robert C. Jennings, Giuseppe Zucchelli and
Flavio M. Garlaschi

Centro CNR Biol. Cell. e Mol. Delle Piante, Dip. di
Biologia, Universita' di Milano, Via Celoria 26
20133 Milano, Italy

INTRODUCTION

The absorption of light by the photosystems (PS) of green
plants is achieved by a large array of pigment molecules,
bound to chl-protein complexes (Anderson, 1986; Thornber,
1986), which transfer the excitation energy to reaction
centres (RCs) where primary charge separation occurs. The
number of chlorophyll (chl) molecules which transfer energy to
PSII RCs is greater than for PSI due to the massive presence
of the light harvesting chl a/b protein complex (LHCII) which
is preferentially associated with PSII (Anderson, 1980).

The antenna matrix system is formed by two distinct chl
species, chla and chlb, with chlb constituting about 35% of
the PSII antenna system. Chla is present as at least five
different spectral forms (French et al., 1972; van Ginkel and
Kleinen Hammans, 1980), which can be arranged, with respect to
wavelength, in a series in which the absorption maxima differ
by 7nm-11nm. It is generally thought (e.g. Knox, 1975) that
energy is transferred within the antenna system by the Forster
R^{-6} dipole-dipole resonance interaction mechanism (Forster,
1959), though direct evidence for this is lacking. A number
of authors have suggested that energy may be transferred from
peripheral antenna towards RCs along a "downhill" energy
gradient in which the long wavelength spectral forms are
preferentially located in the inner antenna, close to RCs
(Seely, 1973; Shipman and Housman, 1979; Fetisova et al.,
1985). Experimental evidence in favour of this concept has
been presented for photosynthetic bacteria (Freiberg et al.,
1989) but not for higher plant photosystems.

In the present paper we address both these problems for
PSII antenna: (a) by analysing energy transfer efficiency to
RCs both within and between PSII units as a function of the
chl spectral forms, (b) by studying energy transfer to
artificial traps as a function of chl concentration.

Light in Biology and Medicine. Volume 2 Edited by R.H. Douglas *et al.*
Plenum Press, New York, 1991

55

MATERIALS AND METHODS

BBY-grana, containing LHCII and PSII core chl/protein complexes and free of PSI, were prepared as described by van Dorssen et al. (1987a). Final resuspension was in Tricine 30mM (pH 8), NaCl 10mM, $MgCl_2$ 5mM, sucrose 0.2M and the chl concentration was $4\mu g/ml$. LHCPII was prepared according to Ryrie et al. (1980). Final resuspension was in sucrose 0.05M, Tricine 5mM (pH 8). The chla/chlb ratio was 1.1. The F_o and F_m fluorescence levels in BBY grana were measured as described by Jennings et al. (1981), with DCMU and hydroxylamine addition after F_o determination. F_v was determined as the difference between F_m and F_o. Excitation was in the wavelength range 645nm-701nm (Oriel sharp cut off and Balzers interference filters), and fluorescence was measured at 745nm. The stray light to signal ratio was considerably less than 0.05. The absorbed light was approximately equal at all wavelengths giving an intensity dependent first order rate constant of $1-2s^{-1}$. We believe that the population of fluorescing molecules measured at 745nm is similar to that measured near the emission peak at 682nm as a) no significant difference in the F_v/F_o ratio for the 745nm emission with respect to the 682nm emission could be detected when fluorescence was excited at 430nm (unpublished data) and b) the 745nm emission quencher titration characteristics are similar to those measured at 682nm (Jennings et al., 1989).

Energy transfer between PSII units was determined by analysing the normalised variable fluorescence (emission wavelength 745nm) as a function of the area growth above the fluorescence induction curve (Melis and Duysens, 1979), measured in the presence of DCMU and hydroxylamine for excitation wavelengths between 645nm - 701nm.

Deconvolution of the room temperature absorption spectra into asymmetric gaussian component was performed as described by Jennings et al. (1989). The contribution of each gaussian components to total absorptance (Ag) is

$$Ag = \int d\lambda \ (1-T(\lambda))_n \ T(\lambda)_f$$

where $(1-T)_n$ is the absorptance of the nth gaussian band and T_f is the transmittance of the excitation filter combination.

Photobleaching was performed and room temperature absorption spectra determined as already described (Zucchelli et al., 1988).

Tritration of chl fluorescence of LHCII and BBY-grana was performed with dibrothymoquinone (DBMIB) as described by Jennings et al. (1983). The quencher additions were $0.9\mu M$ for LHCII and $0.7\mu M$ for BBY-grana. In the last case DCMU was present. Fluorescence was excited at 650nm and measured at 745nm. Flourescence values were normalised to the absorptance of the sample at 650nm.

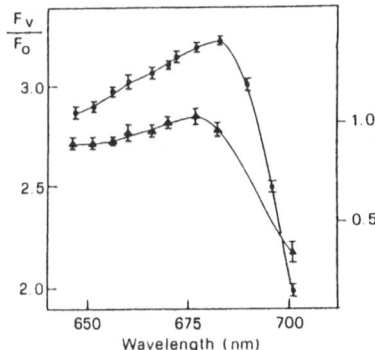

Fig. 1 F_v/F_o-trapping efficiency of PSII RCs as a function of absorption wavelength in BBY-grana (●) (left scale) and in thylakoids of the chlorina barley mutant (▲) (right scale). Each data point is the mean of 20 determinations with different preparations. Due to variability in the ratio with different preparations, the single values were normalised to the overall mean value. The bars indicate the interval estimate of the mean at 99% confidence level.

RESULTS

Energy transfer to PSII RCs from the chl spectral forms

Relative energy transfer efficiency to RCs was determined by measuring the fluorescence quenching by open RCs as the F_v/F_o ratio (F_v/F_o-trapping efficiency) for different absorption wavelengths (Fig. 1). Even though fluorescence lifetime studies have indicated that the PSII emission decay law is not a single exponential (Nairn et al., 1982; Holzwarth et al., 1985; France et al., 1988), abundant evidence exist that this ratio may be represented to a good first approximation by $k_T/\Sigma k$ (Melis and Duysens, 1979; Kitajima and Butler, 1975; Sonneveld et al., 1980). k_T is the averaged trapping rate by RCs, Σk is the sum of the other excited state decays within the antenna. The data in Fig. 1 indicate that transfer to PSII RCs is a function of the absorption wavelength, with maximal transfer occuring from antenna absorbing around 683nm. While a lower transfer efficiency to RCs on the long wavelength side is expected for energetic reasons, this is not the case for wavelengths below 683nm. We have analysed energy transfer to RCs in terms of the spectroscopic chl species for wavelengths below 682nm (Fig. 2). The F_v/F_o values have been related with the fractional absorptance of the chl spectroscopic species (Fig. 2). Within each of the three areas, indicated by the broken lines, more than 95% of the total absorptance is by three spectral forms. By selecting F_v/F_o intervals in which one of these spectral forms has an unchanging absorptance contribution it is possible to establish the relative order of energy transfer efficiency to RCs of the other two chl species. The following qualitative conclusions are established:

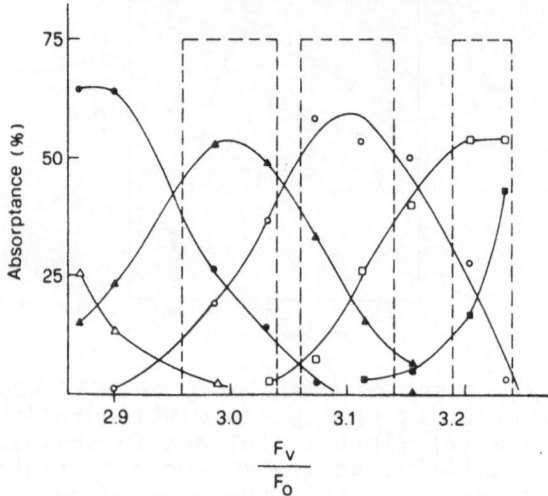

Fig. 2　F_v/F_o-trapping efficiency of PSII RCs as a
function of the fractional absorptance
(see Materials and Methods) by the major
spectroscopic chl forms from 650nm to 684nm.
Δ-638nm, ●-650nm, ▲-661nm, o-670nm,
□-678nm, ■-684nm.

The F_v/F_o data have been taken from Fig. 1.
The areas delimited by the broken lines
evidence parameter spacings within which
the transfer efficiency sequences of the chl
spectral forms (see text) were established.

$$684nm > 670nm > 650nm \qquad\qquad (1)$$
$$677nm > 661nm \qquad\qquad (2)$$

Though it is not possible to compare neighbouring bands
directly due to spectral overlap between them, energy transfer
to RCs seems to be a direct function of the peak wavelength of
the different chl spectroscopic species up to 684nm.

The sense of the 650nm band in this scheme is not
straightforward as chlb is not thought to emit flourescence
(van Metter, 1977; Ide et al., 1987; Gillbro et al., 1985;
Marchiarullo and Ross, 1985). While is is not known to which
chla spectral forms energy is preferentially transferred from
chlb within LHCII complexes, it would seem from the sequences
above that significant energy acceptor contributions can be
made only by either/or both the 661nm and the 677nm species.
It is interesting to note that, within the F_v/F_o intervals
indicated by the broken lines in Fig. 2, the changes in
absorption contribution of the spectral forms correlating with
a given change in F_v/F_o are rather similar. This leads us to
think that the differences in transfer efficiency between the
various spectral components in the above transfer sequences
are similar. It would therefore seem likely that sequence 2
lies somewhere to the right of the 684nm species in sequence
1, indicating the 684nm species as the most efficient at
transferring energy to PSII RCs.

The PSII units are thought to be organised in a kind of limited matrix within the grana membranes, in which energy transfer between the different units is possible (Joliot and Joliot, 1964; Butler, 1980). To establish whether the relative transfer efficiencies of the spectral forms represent transfer to RCs within or between photosystems, we have measured excitation transfer between PSII units as a function of the absorption wavelength (see Materials and Methods).

The normalised variable fluorescence at a particular concentration of open RCs is a function not only of energy transfer between PSII units but also of the F_v/F_o-transfer efficiency to RCs (Butler, 1980). We have calculated the PSII connection parameter Ψ_{22} (eq. 10, Butler, 1980) for the various excitation wavelengths (Table 1). The data show that PSII-PSII energy transfer from the shorter wavelength spectral forms is greater than from the longer wavelength absorbing forms.

As the bulk of PSII antenna is made up of LHCPII (about 80% of the total chl) which contains all the spectral bands (Thornber, 1986), it is reasonable to expect that these relative transfer efficiencies involve excitation transfer from LHCPII to RCs. We have asked the question as to whether the different chl species which compose the core antenna of PSII might also transfer energy to RCs with different efficiencies. We have measured F_v/F_o as a function of wavelength in thylakoids of the chlorina barley mutant, which lacks LHCII, containing only the core antenna of PSII. The data (Fig. 1) are qualitatively rather similar to those for BBY-grana with energy transfer to RCs decreasing significantly at absorption wavelengths lower than about 677nm. The maximal F_v/F_o was detected at lower wavelengths than for BBY-grana. This difference could be due to an increasing contribution of PSI fluorescence to the F_o value at excitation wavelengths above 680nm in the barley mutant thylakoids.

Energy transfer to artificial traps as a function of chl concentration

The fluorescence yield (F) of an excitation energy-transferring chl antenna matrix may be represented by $F = k_f/(\Sigma k + Ck_q)$, where k_f is the fluorescence rate constant, Σk is the sum of all intramolecular decay processes, k_q is the

Table 1   Energy transfer between PSII units as a function of absorption wavelength

Absorption (nm)	683	678	673	666	662	651
$F_v(t)$	.357	.352	.347	.338	.337	.341
	(.005)	(.007)	(.007)	(.007)	(.006)	(.010)
Ψ_{22}	.32	.33	.36	.39	.40	.40

$F_v(t)$: normalised variable fluorescence at 50% area growth. In parentheses are the interval estimates of the mean at the 95% confidence interval. Ψ_{22} : Butler-PSII connection parameter, calculated as in Materials and Methods.

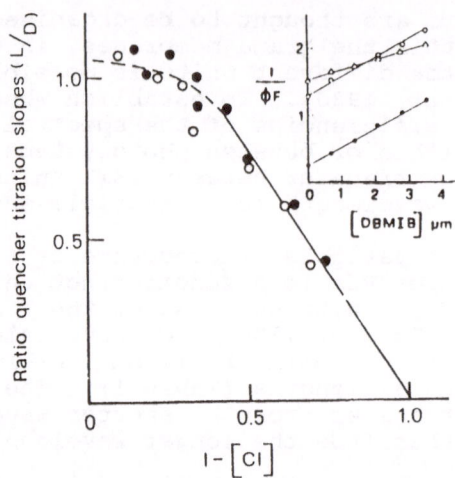

Fig. 3   The influence of different levels of chl
         photobleaching on the titration slope of
         the reciprocal of the chl fluorescence
         with the artificial energy trap
         dibromothymoquinone.   The abscissa is the
         level of the chl photobleaching.   Data are
         presented as the ratio of the titration
         slopes of photobleached samples (L) with
         respect to the dark controls (D).   (●)
         LHCPII: (o), BBY-grana.   Insert: the
         effect of different extents of chl
         photobleaching of BBY-grana on the
         titration of the reciprocal of the chl
         fluorescence with dibromothymoquinone.   (●),
         dark control; (o), sample photobleached by
         32%; (Δ), sample photobleached by 71%.

trapping constant of the artificial traps averaged over the
matrix and is a function of the energy-transfer process, C is
the concentration of the artificial traps in the pigment
matrix (Kitajima and Butler, 1975; Sonneveld et al., 1980;
Jennings, 1984; Hodges and Barber, 1984; Hodges et al., 1987;
Karukstis et al., 1988).   When the fluorescence is titrated
with an artificial trap, the slope of 1/F versus the
concentration of the quencher is a function of k_q/k_f.   We have
used this approach to examine the influence of decreasing the
chl concentration by photobleaching on energy transfer in both
spinach LHCPII and the BBY grana preparations.   Examples of
such titrations are in Fig. 3 (insert) for BBY grana.   After
30% photobleaching of chl, the slope of the plot is similar to
that of non photobleached samples, while a considerable slope
decrease was observed after 70% photobleaching.   The different
1/F intercepts are due to a substantial fluorescence quenching
induced by the high light intensity treatment.

     The decrease of the quencher titration slope after
pronounced photobleaching could be explained by an increased
fluorescence rate constant (k_f) or a decreased trapping
constant (k_q).   The gaussian decomposition of the room
temperature absorption spectrum of the BBY-grana after about

50% photobleaching of the total chl did not significantly change either the peak position or the half-band width of the major bands (Jennings et al., 1989). It would therefore seem unlikely that either the overall fluorescence rate constant or the quinone-chl interaction was significantly altered. Small changes in the relative contribution of the major gaussian components after photobleaching are observable in the BBY-grana preparation while such changes are almost absent with LHCII (Zucchelli et al., 1988). On the basis of these observations we interpret the decrease in the quencher titration slope after pronounced photobleaching as being due to a decreased energy transfer rate to the artificial traps within these antenna systems. The relative decrease in the transfer rate is given by the slope of the photobleached sample normalised to the slope of the non-photobleached sample.

In Fig. 3 data are presented for a large number of such quencher titration experiments performed with both BBY-grana and LHCII. Energy transfer to the artificial traps shows a biphasic decline with decreasing chl concentration. The second phase is linear, extrapolating to a slope value of zero at 100% photobleaching. The first phase shows a much lesser dependence on chlorophyll concentration.

DISCUSSION

The data on the F_v/F_o-trapping efficiency suggest that energy transfer to RCs is a function of the wavelength position of the spectral band with the longer wavelength bands being more efficient than the shorter wavelength bands. The 684nm band seems to be the most efficient. The opposite situation, in which the shorter wavelength bands are the most efficient, applies to energy transfer between PSII units.

The transfer rate from antenna to RCs is expected to be determined, in general terms, by the strength of energy coupling between antenna chls and by the number of transfer steps (N) in the case of a localised transfer mechanism. Two antenna models may be considered to explain the present observations.

1)   The long wavelength spectral forms, and in particular the 684nm form, have a closer topological relationship with RCs than the short wavelength bands. This is the kind of antenna model analysed theroetically by several authors (Seely, 1973; Shipman and Housman, 1979; Fetisova et al., 1985) and forms the basis of the so called "funnel" concept of antenna organisation. Experimental evidence in favour of such a model exists for photosynthetic bacteria (Freiberg et al., 1989). In this case, for the long wavelength forms, N will be smallest for transfer to RCs within the PSII unit and greatest for transfer to other PSII units.

2)   There is no particular macroscopic topological distribution of the spectral forms with respect to RCs. In this case transfer to RCs would be determined essentially by the transfer microparameters of the donor-acceptor antenna chlorophylls. The short wavelength

spectral forms are expected to be able to transfer energy to a greater variety of antenna chlorophylls than the longer wavelength forms. In this context the shorter wavelength forms will limit the number of transfer possibilities available to the longer wavelength forms, thus bringing about a situation in which N is smaller for the longer wavelength forms. This suggestion is similar to the "antitrap" model suggested by Knox (1977). The greater efficiency of energy transfer between PSII units for the shorter wavelength forms can be explained by their greater transfer probability to other antenna chl forms than the long wavelength forms.

On the basis of the above energy transfer data alone it is not possible to distinguish between these two models. The data about the relative amounts of the different spectral forms in both LHCII and BBY-grana (Fig. 4) show that the 684nm form, associated with the highest F_v/F_o transfer efficiency, is present at high levels in LHCII. This observation suggests that the 684nm component is not enriched in the core antenna complexes and seems to be in agreement with van Dorssen et al (1987b) who do not report the presence of the 684nm component in the CP47 core complex. Thus it is difficult to explain the high F_v/F_o efficiency of this spectral form in terms of the "funnel" model. We suggest that "antitrap" considerations best account for the high F_v/F_o transfer efficiency of the 684nm spectral form. As this component is present at relatively high levels in LHCII (15% - 20% of Q_v absorption) we feel that "antitrapping" may play an important role in energy transfer within the LHCII outer antenna matrix of PSII.

From Fig.4 it can also be seen that the shorter wavelength bands (650nm and 661nm) are enriched in LHCII with respect to BBY-grana and thus presumably also with respect to the core antenna complexes. On the other hand both the 670nm and 677nm forms are present at high levels in BBY-grana, indicating their enrichment in core complexes with respect to

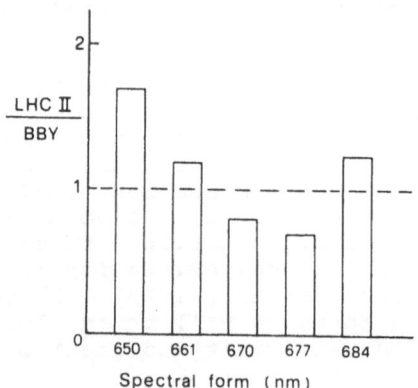

Fig. 4   The extent of the ratio between the fractional contribution of each spectral species to the room temperature absorption spectra of spinach-LHCPII and BBY-grana preparations.

LHCII, in agreement with van Dorssen et al (1987b) who found high levels of these two spectral forms in the CP47 complex. We therefore feel that while the "antitrap" concept best explains the 684nm data, energy transfer from LHCII to the core complexes may well have a "funnel" component.

The Q_y transition dipole strength of chlb in organic solvents is weaker than that of chla (Goedheer, 1966). Deconvolution of the room temperature absorption spectra into the asymmetric gaussian bands for both BBY-grana and purified LHCII (Fig. 5) show that this is also the case within the PSII antenna. In the BBY-grana chlb constitutes about 35% of the total chl and in LHCII almost 50% whereas its contribution to

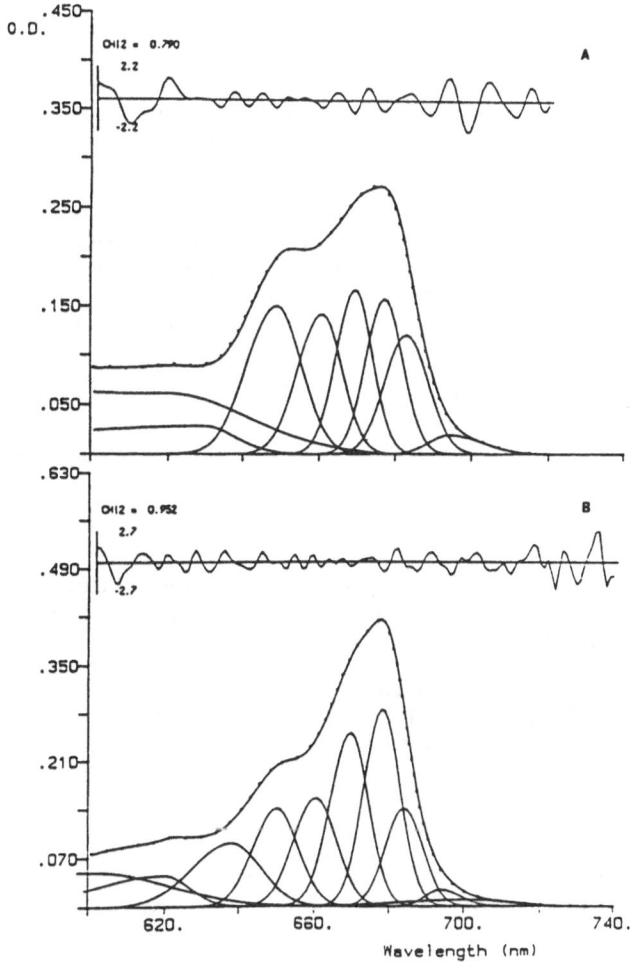

Fig. 5 Room temperature absorption spectra of spinach-LHCPII (A) and spinach-BBY (B) suspensions. The experimental data range over the 600-740nm wavelength interval and are the dotted curves while and full lines are the sum of the gaussian components. Plots of the residuals are also shown with the λ^2 values.

absorption (areas under the gaussian OD bands) is much less (15% and 25% respectively). The low Q_y absorption "efficiency" of such an important antenna component as chlb is difficult to understand in terms of the light harvesting economy. On the other hand it can be understood in terms of the "antitrap" properties of chlb as a weaker transition dipole will further lower the energetically "uphill" transfer rate from the chla spectral forms by decreasing the matrix interaction element. This would have the effect of enhancing the "antitrap" properties of chlb.

In this context it is interesting to note that linear dichroism measurements of both PSII and LHCII preparations show that the Q_y transitions of chlb and the lower wavelength absorbing chla spectral forms have an average orientation which is rather different from that of the longer wavelength chla spectral forms (Breton, 1986). This is also expected to decrease the matrix interaction element and hence the transfer rate from the longer wavelength chlorophyll species thus enhancing the "antitrap" role of the shorter wavelength chlorophyll species.

The data presented on energy transfer in PSII antenna to artificial traps show that this parameter decreases in a biphasic manner with decreasing chl concentration. In the first phase, photobleaching of up to 30% - 40% of the total chl in BBY grana and LHCII has little effect on energy transfer. In the second phase this declines in a linear fashion with decreasing chl concentration. These data may be interpreted in terms of the analysis performed by Burshtein (1984) for non-coherent energy transfer to fluorescence quenching centres. The rate of trapping (k) may be represented by

$$k = M_t^{3/n} \cdot M_d^{1-(3/n)} \cdot C^{(n/3)-1}$$

where M_t and M_d are the donor-trap and donor-donor transfer microparameters, respectively; C is the donor concentration, n is the exponent of the distance (R) dependence of the transfer process. For the Forster R^{-6} mechanism the trapping rate is linear with respect to donor concentration. This explains the second, linear, phase in Fig. 3. On the other hand the first phase is best described by an n value(s) considerably less than 6. Thus we suggest that in the non-photobleached antenna energy transfer may have a significant non-Forster component. This suggestion is not at all theoretically unreasonable (Kenkre and Knox, 1984). On the basis of this analysis, using an average band spread of 30nm, a chl-chl distance of 1nm, and an average orientation factor of 1, one may calculate a transfer process with an $R^{-3.5}$ to R^{-4} distance dependence.

REFERENCES

Anderson, J.M., 1986, Photoregulation of the composition, function and structure of thylakoid membranes, <u>Ann. Rev. Plant Physiol.</u>, 37: 93.
Anderson, J.M., 1980, Chlorophyll-protein complexes of higher plant thylakoids: distribution, stoichimometry and organization in the photosynthetic unit, <u>FEBS Lett.</u>, 117: 327.

Breton, J., 1986, Molecular orientation of the pigments and the problem of energy trapping in photosynthesis, in: "Encyclopedia of Plant Physiology N.S." Vol. 19, L. A. Staehelin and C.J. Arntzen eds., Springer-Verlag, Berlin-Heidelberg.

Burshtein, A.I., 1984, Concentration quenching of noncoherent excitation in solutions, Sov. Phys. Usp., 27: 579.

Butler, W.L., 1980, Energy transfer between photosystem II units in a connected package model of photochemical apparatus of photosynthesis, Proc. Natl. Acad. Sci. USA, 77: 4679.

Fetisova, Z.G., Borisov, A.Yu. and Fok, M.V., 1985, Analysis of structure-function correlations in light-harvesting photosynthetic antenna: structure optimization parameters, J. Theor. Biol., 112:41.

Forster, T., 1959, Transfer mechanisms of electronic excitation, Discussions Faraday Soc., 27:7.

France, L., Geacintov, N.E., Lin, S., Wittmershaus, B.P., Knox, R.S. and Breton, J., 1988, Fluorescence decay kinetics and characteristics of bimolecular exciton annhilation in chloroplasts, Photochem. Photobiol., 48:333.

Freiberg, A., Godick, V.I., Pullerits, T. and Timpman, K., 1989, Picosecond dymanics of directed excitation transfer in spectrally heterogeneous light-harvesting antenna of purple bacteria, Biochim. Biophys. Acta, 973:93.

French, C.S., Brown, J.S., and Lawrence, M.C., 1972, Four universal forms of chlorphyll a, Plant Physiol., 49:421.

Gillbro, T., Sundstrom, V., Sandstrom, A., Sprangfort, M. and Andersson, B., 1985, Energy transfer within the isolated light-harvesting chlorophyll a/b protein of photosystem II (LHC-II), FEBS Lett., 193:267.

Goedheer, J.C., 1966, Visible absorption and fluorescence of chlorphyll and its aggregates in solution, in: "The Chlorophylls", Vernon, L.P. and Seely, G.R. eds., Academic Press, New York and London.

Hodges, M. and Barber, J., 1984, Analysis of chlorophyll fluorescence quenching by DBMIB as a means of investigating the consequences of thylakoid membrane phosphorylation, Biochim. Biophys. Acta., 767:102.

Hodges, M., Briantais, J-M and Moya, I., 1987, The effect of thylakoid membrane reorganisation on chlorphyll fluorescence lifetime components: a comparison between state transitions, protein phosphorylation and the absence of Mg^{2+}, Biochim. Biophys. Acta, 893:480.

Holzwarth, A.R., Wendler, J. and Haehnel, W., 1985, Time-resolved picosecond fluorescence spectra of the antenna chlorophyll in Chlorella vulgaris. Resolution of photosystem I fluorescence, Biochim. Biophys. Acta, 807:155.

Ide, J., Klug, D.R., Kuhlbrandt, W., Girogi, L.B. and Porter, G., 1987, The state of detergent solubilized light-harvesting chlorophyll a/b protein complex as monitored by picosecond time-resolved fluorescence and circular dichroism, Biochim. Biophys. Acta, 893:349.

Jennings, R.C., Garlaschi, F.M., Gerola, P.D., Etzion-Katz, R. and Forti, G., 1981, Proton induced grana formation in chloroplasts. Distribution of chlorophyll-protein complexes in photosystem II photochemistry, Biochim. Biophys. Acta, 638:100.

Jennings, R.C., Garlaschi, F.M. and Gerola, P.D., 1983, A
    study on the lateral distribution of the plastoquinone
    pool with respect to photosystem II in stacked and
    unstacked spinach chloroplasts, Biochim. Biophys. Acta,
    722:144.
Jennings, R.C., 1984, Independent effects of magnesium ions on
    energy-transfer processes in chloroplasts, Biochim.
    Biophys. Acta, 766:102.
Jennings, R.C., Zucchelli, G. and Garlaschi, F.M., 1989, The
    influence of reducing the chlorophyll concentration by
    photobleaching on energy transfer to artificial traps
    within photosystem II antenna systems, Biochim. Biophys.
    Acta., 975:29.
Joliot, A. and Joliot, P., 1984, Etude cinetque de la rection
    photochimique liberant l'oxygene au cours de la
    photosynthese, C. R. Acad. Sci. Paris, 258:4622.
Karukstis, K.K., Gruber, S.M., Fruetel, J.A. and Boegeman,
    S.C., 1988, Quenching of chlorophyll fluorescence by
    substituted anthraquinones, Biochim. Biophys. Acta,
    932:84.
Kenkre, V.M. and Knox, R.S., 1974, Theory of fast and slow
    excitation transfer rates, Phys. Rev. Letters, 33:803.
Kitajima, M. and Butler, W.L., 1985, Quenching of chlorophyll
    fluorescence and primary photochemistry in chloroplasts
    by dibromothymoquinone, Biochim. Biophys. Acta, 376:105.
Knox, R. S., 1975, Excitation energy transfer and migration:
    theoretical considerationsj, in: "Bioenergetics of
    Photosynthesis", Govindjee ed., Academic Press, New York.
Knox, R.S., 1977, Photosynthetic efficiency and exciton
    transfer and trapping, in: "Topics in Photosynthesis"
    (vol 2), Barber J. ed., Elsevier, Amsterdam.
Marchiarullo, M.A. and Ross, R.T., 1985, Resolution of
    component spectra for spinach chloroplasts and green
    algae by means of factor analysis, Biochim. Biophys.
    Acta, 807:52.
Melis, A. and Duysens, L.N.M., 1979, Biphasic energy
    conversion kinetics and absorbance difference spectra of
    photosystem II of chlorplasts. Evidence for two
    different photosystem II reaction centers, Photochem.
    Photobiol., 29:373.
Nairn, J.A., Haehnel, W., Reisburg, P. and Sauer, K., 1982,
    Picosecond fluorescence kinetics in spinach chloroplasts
    at room temperature, Biochim. Biophys. Acta, 682:420.
Ryrie, I.J., Anderson, J.M. and Goodchild, D.J., 1980, The
    role of the light harvesting chlorophyll a/b protein
    complex in chloroplast membrane stacking, Eur. J.
    Biochem., 107:345.
Seely, G.R., 1973, Effects of spectral variety and molecular
    orientation on energy trapping in photosynthetic unit: a
    model calculation, J. Theor. Biol., 40:173.
Shipman, L.L. and Housman, D.L., 1979, Forster transfer rate
    for chlorophyll a, Photochem. Photobiol., 29:1163.
Sonneveld, A., Rademaker, H. and Duysens, L.N.M., 1980,
    Transfer and trapping of excitation energy in photosystem
    II as studied by chlorophyll a_2 fluorescence quenching by
    dinitrobenzene and carotenoid triplet. The matrix model,
    Biochim. Biophys. Acta., 593:272.
Thornber, J.B., 1986, Biochemical characterization and
    structure of pigment-proteins of photosynthetic organism,
    in: "Encyclopedia of Plant Physiology N.S." Vol. 19, L.

A. Staehelin and C.J. Arntzen eds., Springer-Verlag, Berlin-Heidelberg.

van Dorssen, R.J., Plijter, J.J., Dekker, J.P., den Ouden, A., Amesz, J. and van Gorkom, H.J., 1987a, Spectroscopic properties of chloroplast grana membranes of the core of photosystem II, Biochim. Biophys. Acta, 890:134.

van Dorssen, R.J., Breton, J., Plijter, J.J., Satoh, K., van Gorkom, H.J. and Amesz, J., 1987b, Spectroscopic properties of the reaction center and of the 47kDa chlorophyll-protein of photosystem II, Biochim. Biophys. Acta, 893:267.

van Ginkel, G. and Kelinen Hammans, J.W., 1980, Action spectra of photophosphorylation -II. ATP formation catalized by phenazinemethosulphate suggesting the involvement of long wavelength pigment forms in the light-harvesting process for PS II and PS I, Photochem. Photobiol., 31:385.

van Metter, R.L., 1977, Excitation energy transfer in light-harvesting chlorophyll a/b protein, Biochim. Biophys. Acta, 462:642.

Zucchelli, G., Garlaschi, F.M. and Jennings, R.C., 1988, Influence of electrostatic screening by cations on energy coupling between photosystem II reaction centres and the light-harvesting chlorophyll a/b protein complex II, Biochim. Biophys. Acta, 934:144.

NON-LINEAR PROCESSES IN PHOTOSYNTHETIC LIGHT ABSORPTION

Paul Hoffmann[1] and Dieter Leupold[2]

[1]Humboldt University, Dept. of Biology
Reinhardtstr. 4, GDR-1040 Berlin
[2]Central Institute of Optics and Spectroscopy
Academy of Sciences, Rudower Chaussee 5; GDR-1199
Berlin

INTRODUCTION

In the centre of the fundamental process of
photosynthesis, the entropic basis of life on the earth, are
the chlorophylls (chl). Their role in photosynthesis rests on
the specific combination between their complex reactivity
(review by Höxtermann et al., 1986) and structural
organisation in membranes leading to an irreversible charge
separation (Hoffmann, 1987; Renger, 1987; Witt, 1987). We
report on comparative laser spectroscopic characterization of
chl _in vivo_ in order to contribute to a better understanding
of light conversion in the primary photosynthetic process.

On the basis of our knowledge concerning the functional
heterogeneity of chl and the molecular structure of the
photosynthetic membrane (Hoffman 1987) special attention is
paid to the "absorption unit", its integration in the
thylakoid and a proposed term scheme of chl _in vivo_.

NON-LINEAR ABSORPTION MEASUREMENTS

Characterization of an absorption unit

If the transmission of a sample (e.g. chl a in solution)
is measured at a fixed wavelength λ within an absorption band
with (laser-) light pulses of low intensity, a constant ("low
signal") value T_q is obtained irrespective of variations of
the pulse intensity. This linear absorption takes place as
long as the change of intensity during the pulse interaction
with the sample is proportional to this intensity I_λ itself.

For example when measuring the transmission of a solution
of chl a in dioxane within the Soret or red absorption band
with nanosecond (ns) pulses, this region of linear absorption

(constant transmission) is extended up to intensities of about 10^5 W cm^{-2} (corresponding to photon flux density of some 10^{24} cm^{-2} s^{-1}, or photon densities per pulse of some 10^{13} cm^{-2}). Further increase in intensity of the measuring pulse effects a non vanishing population of excited state (s), this means the absorption becomes non-linear, i.e. the transmission becomes a function of pulse intensity. In the mentioned chl a example non-linear absorption is manifested in a (transient) bleaching. The course of transmission vs intensity in the non-linear region is very informative because it reflects implicitily all excited state absorption, relaxation and interaction processes which are involved in the light matter interaction process under investigation (Leupold, 1986; Leupold et al., 1988, 1989a). The measurements on chl _in vivo_ (primary leaves of _Triticum aestivum_ seedlings) were carried out at 694 nm with photon densities per pulse between 5 x 10^8 and 5 x 10^{18} cm^{-2} (photon flux densities ranging from 2.5 x 10^{16} to 2.5 x 10^{26} cm^{-2} s^{-1}, or intensities ranging from 8 x 10^{-3} to 8 x 10^7 W cm^{-2}). In this case the beginning of nonlinearity lies at photon densities per pulse as low as about 4 x 10^9 cm^{-2} (photon flux density of 2 x 10^{17} cm^{-2} s^{-1}).

At the lowest intensities the transmission is constant, corresponding to the low signal level T_o. For intensities above 8 x 10^{-2} W cm^{-2} a decrease of transmission is observed. This course of the curve shows that a second absorption after the ground state absorption becomes effective, the cross section of which, σ_2, is greater than that of the first transition (σ_1).

Remarkably, this stepwise two photon absorption occurs even in an intensity range comparable with that of sunlight (maximum of 10^{17} quanta cm^{-2} s^{-1}). It should be noted in addition that in the higher intensity range between 8 x 10^1 and 8 x 10^7 W cm^{-2} several local extrema of transmission occur (Leupold et al., 1976); however, T (I) < T_o is satisfied up to the highest measured intensities. In comparison with chl _in vitro_ the nonlinearity of the absorption of chl _in vivo_ begins at intensities several orders of magnitude lower. This led to the assumption that the absorbing unit _in vivo_ consists of a great number of molecules, whereas in dilute solution it consists of single molecules, dimers, or relatively small oligomers.

This number N of molecules per absorption unit - as well as the absorption cross section, σ_2 - can be determined from the course of the non-linear function. In the framework of the coupled system of differential equations for the population densities of the energy levels and the photon transport through the sample an absorption cross section of the order of 10^{-10} cm^2 was calculated (Hoffmann et al., 1978: Leupold, 1986). The corresponding value of the chl monomer is $\sigma_m \simeq 10^{-16}$ cm^2. Supposing a linear relationship between σ and the number of molecules collectively interacting in a transition, it follows that in the primary absorption process more than 10^5 chl molecules cooperate. This collective is defined as an absorption unit (Hoffmann and Leupold, 1980; Leupold, 1986). In the absorption process, this unit acts like a supermolecule. That is the cross section is a genuine cross section for a single absorbing object. The size of the absorbing unit depends on the excitation intensity, but is relatively independent of whether excitation occurs in the red or Soret region (Fig. 1).

Biological relevance of the absorption unit

In order to get more information about the physiological relevance of this functional unit, experiments were undertaken to quantify the absorption unit during its biogenesis. In etiolated primary Triticum leaves greening was induced by 4 light flashes (T = 1.25 ms; E = 160 Ws; L IV) and by continuous illumination of different duration. In etioplasts illuminated with four light flashes the nonlinearity at 435 nm begins at 7×10^4 W cm^{-2} (Fig. 2). For comparison in monomeric chl solutions this point lies at about 10^5 W cm^{-2}.

The absorption unit comprises in this case about 5 chl molecules. This number corresponds to the number of protochlorophyllid molecules bound per one NADPH+H$^+$ - Pchlid - Oxidoreductase (Canaani and Sauer 1977; Apel et al., 1980). So only one isolated excitation state limited to one protein can be assumed (Seifert 1984).

After 5 min illumination nonlinearity starts at nearly 4 $\times 10^4$ W cm^{-2}. After 48 h illumination the onset of non-linear transmission lies at intensities lower by more than one order of magnitude (2×10^3 cm^{-2}). The occurence of larger absorption units correlates with the red shift and the integration of the synthesized chl molecules in their carrier membrane (Seifert 1984), underlining the biological basis of

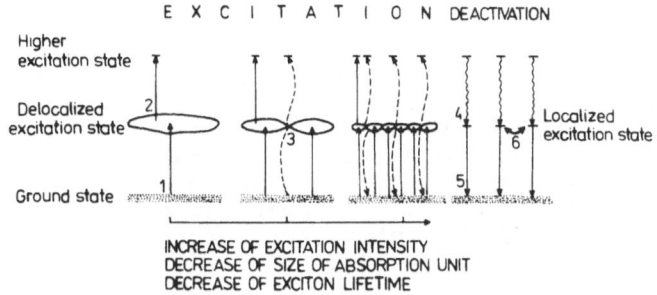

E X C I T A T I O N   DEACTIVATION

Fig. 1   Scheme of excitation and deactivation processes of chl in vivo under special consideration of absorption unit.

1. Absorption of a photon by an absorption unit; excitation delocalization; coherent exciton.
2. Interaction of a second photon with the delocalized excited state of the absorption unit ($T(I) < T(I_o)$)
3. Annihilation of coherent excitons
4. Deactivation of the exciton annihilation product
5. Fluorescence from localized S_1
6. Annihilation of incoherent excitons

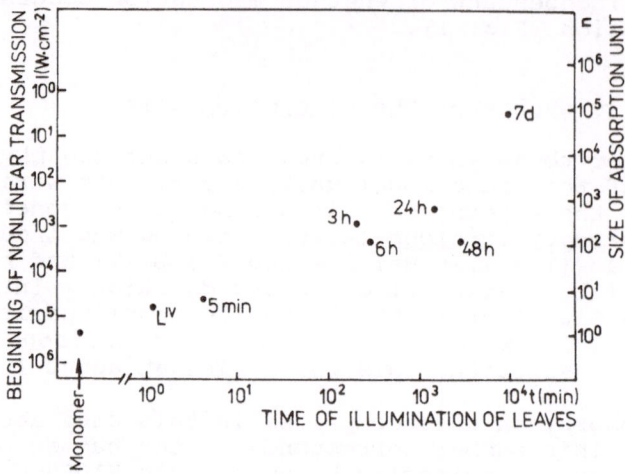

Fig. 2   Biogenesis of absorption unit in etiolated
          primary leaves of <u>Triticum aestivum</u> during
          greening.

the absorption unit. Similar conclusions were drawn from
comparable measurements of intensity dependent transmission of
DCMU - treated leaves (Hieke et al., 1979) and the
fluorescence enhancement effect of chl <u>in vivo</u> (Leupold and
Hoffmann 1986). The absorption unit is also underlined by
dichroism measurements showing that the pigment protein
complexes can be assembled into large helically organized
macrodomains (Garab 1989). There is no evident correlation
between the size of the absorption unit and the PSU
(Photosynthetic unit). The absorption unit describes the
interaction of pigment molecules during the absorption of a
photon. PSU, however, comprises molecules acting collectively
in the ns to μs range, both in energy conversion and oxygen
evolution. Absorption unit and PSU are different levels -
with different time constants - in the hierarchical structure
of the photosynthetic apparatus (Hoffmann and Leupold 1980;
Hoffmann 1987). From a structural point of view $10^4 - 10^6$ chl
molecules correspond to the chl content of one thylakoid.

ON THE PROBLEMS OF CHLOROPHYLL FORMS

     The hypothesis of "chlorophyll forms" was developed on
the basis of a substructure analysis on the red absorption
band (Gauss-/Lorentz shape analysis, derivative spectroscopy
in connection with low temperatures) (French 1971, Litvin et
al., 1971; Meister 1986). Though this is an indirect
procedure, the French - Brown result of 4 universal forms and
several minor forms obtained by this way is widely accepted.
To characterize these chl forms, Resonance Raman Scattering
and Coherent Anti-Stokes Raman Scattering methods were applied
(Höxtermann et al., 1982). At least six discrete chl a
vibrational subspecies could be distinguished. But whether
these subspecies correspond to the forms mentioned above was
still an open question. Non-linear absorption laser
spectroscopy with two synchronous but spectrally different

pulses allows a direct check. The principle of the method is the following. With a laser pulse at λ_1 in the red absorption band a non-linearity in absorption is induced (that means a transmission different from the small signal value). A second synchronous laser pulse detects as test beam eventually existing nonlinearities of the absorption at another wavelength λ_2 inside the red absorption band. In case of a homogeneous broadening an induced nonlinearity at one wavelength λ_1 is measurable in the whole band profile (i.e. the wavelength variable test beam λ_2 measures everywhere deviations from the small signal transmission). In the case of an inhomogeneous absorption band with substructure (i.e. in case of existence of different chl forms) the induced nonlinearity remains limited to the special subband (form) around λ_1, and outside of this band the test beam λ_2 measures an unchanged small signal transmission. At first measurements were made in the intensity range of initial nonlinearity at about 10 W cm^{-2} with λ_1 = 690 nm and λ_2 = 670 nm (Voigt et al., 1979). The probe pulse λ_2 was located at the short - wave side of the λ_1 pulse in order to avoid an absorption change at λ_2 by simple rapid energy migration after λ_1 excitation. Due to the early beginning of nonlinearity the absorption at λ_2 = 670 nm is already changed by the probe pulse intensity, but - more important - an additonal change takes place as a result of the simultaneous action of the λ_1 pulse (690 nm). This testifies to the homogeneous broadening of the red chl a band between 670 nm and 690 nm. To compare this result with a recent conventional chl form analysis at primary leaves of <u>Triticum aestivum</u>, such a shape fitting was made in cooperation with Litvin and his coworkers at his laboratory. In the region between 620 nm and 720 nm 8 subbands were obtained as the most probably hypothetical variant. The above mentioned result of homogeneous broadening between 670 and 690 nm makes this subband analysis questionable, but does not yet enable a final rejection, because of relatively broad subbands and subband overlap at 690 nm. Such a final decision demands a wider spacing of λ_1 and λ_2 in the laser experiment, i.e. a greater part of the whole band should be checked. This has been done in two ways: Firstly, also under low intensity conditions, the pulses were tuned to 665 nm and 694 nm and again a homogeneous behaviour could be stated (Hoffmann et al., 1978). Secondly, in an experiment at high intensities (up to 10^6 W cm^{-2}) nonlinearity was induced with λ_1 = 700 nm and the probe pulse tuned to 622 nm. The absorption at 622 nm is changed under the action of the 700 nm pulse. That means there is at least one component with a homogeneously broadened absorption between 622 and 700 nm. Obviously this contradicts the conventional analysis result of subbands with small absorption regions. Newer experimental results with spectrally broad probe pulses confirm this statement once again (Leupold et al., 1989b; see also Fig. 3). They show that the overall S_o - S_1 absorption band of antenna chlorophyll <u>in vivo</u> behaves like a mixture of homogenous bands of the known different antenna types (Core, LHC), at least for time intervals $\Delta t \geq 0.5$ ns (probe pulse duration).

In summary, the laserspectroscopic results give no hint to any such substructure of the red absorption band which could be the base for declaration of "forms" with individual lifetimes $\geq$ ns. Therefore, the several subspecies identified

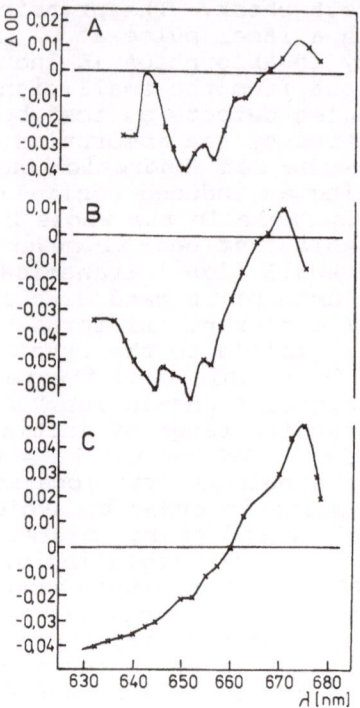

Fig. 3   Non-linear light absorption (differential
optical density spectra; exc. at 683 nm)
of Hordeum systems.
A: Light harvesting complex
B: Thylakoid preparations from Hordeum
   vulgare seedlings.
C: Thylakoid preparations from Hordeum
   vulgare, chl b less mutant chlorina
   f2; 2800
Measurements together with E.Brecht,
Gatersleben.

by the Raman methods cited above have either picosecond
lifetims or they are not reflected in the shape of the red
absorption band.   The absence of any forms is consistent with
our result of collective action of all light harvesting chl in
the primary absorption process and the definition of an
absorption unit consisting of $\geq 10^5$ chl a molecules (Leupold
1986; Hoffmann et al., 1990).

REFERENCES

Apel, K., Santel, H.-J., Redlinger, and Falk, H., 1980, The
    protochlorophyllide holochrome of barley (Hordeum vulgare
    L.).   Isolation and characterisation of the NADPH-
    protochlorophyllide oxidoreductase, Eu. J. Biochem.,
    111:251.
Canaani, O.D. and Sauer, K., 1977, Analysis of subunit
    structure of protochlorophyllide holochrome by sodium
    dodecyl sulfate-poly-amide gel electrophoresis, Plant
    Physiol., 60:422.

French, C.S., 1971, The distribution and action in photosynthesis of several forms of chlorophyll, Proc. Nat. Acad. Sci. USA 68:2893.

Garab, G., 1989, Helically organized macrodomains of the pigment-protein complexes in chloroplast thylakoid membranes: conclusions from macroscopic and microscopic dichroism measurements, Photobiol. and Biotechnol. Proceed. Int. Symp., Poznan, Poland, June 27th - 30th:42.

Hieke, B., Hoffmann, P., Leupold, D., Mory, S. and Schotte, J., 1979, Das Absorptionsverhalten von Chlorophyll in vivo und in vitro bei Anregung mit Laserimpulsen von 694, 3 nm, Photosynthetica, 13:37.

Höxtermann, E., Werncke, W., Stadnitchuk, I.N., Lau, A. and Hoffmann, P., 1982, Resonance coherent antistokes Raman scattering (CARS) of chlorophyll. I-III, Studia biophys., 92:147.

Höxtermann, E., Wiesner, B., Hoffmann, P. and Leupold, D., 1986, Zur Chemie und Photophysik des Chlorophylls, Biol. Rundsch., 24:27; 119.

Hoffmann, P., 1987, "Photosynthese" Akademieverlag, Berlin.

Hoffmann, P., Leupold, D., Hieke, B., and Voigt, B., 1978, Laserspectroscopic characterization of absorption behaviour of chlorophyll in vitro and in vivo, Biochem. Physiol. Pflanzen, 173:364.

Hoffmann, P. and Leupold, D., 1980, Primärprozesse der photosynthetischen Energiekonvertierung bei hoheren Pflanzen, Colloqu. Pflanzenphysiologie, Humboldt-Universitat zu Berlin, 3:21.

Hoffmann, P., Pfarrherr, A., Teuchner, K. and Leupold, D., Chlorophyll b - photophysical properties and its role in photosynthetic energy conversion, Proced. VIth Int. Conference on Energy and Electron Transfer. Prague, 14th - 18th August 1989, (in press).

Leupold, D., 1986, Der Informationsgehalt der nichtlinearen Absorption und - Emission organischer Molekule. Demonstration am Chlorophyll, Diss. B., Humboldt-Universitat zu Berlin.

Leupold, D., Mory, S. and Hoffmann, P., 1976, Nonlinear absorption and stimulated emission of chlorphyll a, Conf. on Luminescence, Szeged, Acta Phys. Chem., 23:33.

Leupold, D., Voigt, B., Mory, S., Hoffmann, P. and Hieke, B., 1978, Low-intensity two-step absorption of chlorophyll a in vivo, Biophys. J., 21:177.

Leupold, D. and Hoffmann, P., 1986, Large absorption unit reflected in the fluorescence enhancement effect of chlorophyll in vivo, Photobiochem. Photobiophys., 12:33.

Leupold, D., Stiel, H. and Hoffmann, P., 1988, Non-linear absorption spectroscopy of antenna chlorophyll a in higher plants, in: "Photosynthetic light-harvesting Systems," H. Scheer and S. Schneider, ed., Walter de Gruyter, Berlin, New York.

Leupold, D., Ehlert, J., Oberlander, S., Klose, E., Mory, S. and Winkelmann, G., 1989a, Nonlinear laser chemistry of maleic acid, Laser Chem., 10:73.

Leupold, D., Stiel, H., Klose, E. and Hoffmann, P., 1989b, Search of "chlorophyll forms" in vivo by non-linear laser sepctroscopy, Ber. Bunsenges. Phys. Chem., 93:371.

Litvin, F., Gulyaev, B.A. and Sineschchenkov, V.H., 1971, Spectralnye characteristici otnositelnaya konzentrazija i koeffizienty migrazii energii 10 nativnych form chlorophilla a, Dokl. Acad. Nauk SSSR, 199:1428.

Meister, A., 1986, Formen von Chlorophyll-Nachweis und Interpreation, Diss. B, Adademie d. Wissensch, DDR, Berlin.

Renger, G., 1987, Biologische Sonnenenergienutzung durch photosynthetische Wasserspaltung <u>Angew. Chemie</u>, 99:660.

Seifert, B., 1984, Spectroskopische Charakterisierung der Entwicklung des Photosyntheseapparates in Keimpflanzen ausgewählter Weizen-Evolutionsformen, Diss. A, Humboldt-Universitat zu Berlin.

Voigt, B., Leupold, D., Hoffmann, P. and Hieke, B., 1979, Laserspectroscopic investigations on the S_0 - S_1 subbands of chlorophyll a <u>in vivo</u>, Proc. III. Conf. of Luminescence, Szeged, II, 343.

Witt, H.T., 1987, Examples for the cooperation of photons, excitons, electron electric fields and protons in the photosynthetic membrane, <u>New Journ. Chem.</u>, 11:91.

DIFFERENTIAL POLARIZATION IMAGING OF CHLOROPLASTS: MICROSCOPIC AND MACROSCOPIC LINEAR AND CIRCULAR DICHROISM

Gyozo Garab[1,2] Laura Finzi[1] and Carlos Bustamante[1]

[1]Department of Chemistry, University of New Mexico Albuquerque, N.M. 87131, USA and
[2]Biological Research Centre, Hungarian Academy of Sciences, Szeged, H-6701, Hungary

INTRODUCTION

Differential polarization imaging (DPI) is a newly emerging non-invasive structure analysis technique. It provides two-dimensional mapping of the polarization-dependent optical properties of the object (Tinoco et al., 1987: Kim et al., 1987a). Digitally recorded differential polarization images, taken at different focal depths in the object ("optical slicing"), permit a reconstruction of a three dimensional image (Kim et al., 1987b; Juang et al., 1988). Hence, domains that have different optical properties and different structures can be distinguished and spatially resolved. Further information can be gained by recording microscopic polarization spectra in different locations in the object.

Most biological organelles, membranes, macrocomplexes and macromolecules contain anisotropic domains. The transition dipoles in these domains are not at random, but are oriented with respect to a reference axis. Thus, the sample may show linear dichroism (LD) and/or polarized fluorescence emission. Biological samples also often exhibit long-range chirality in their ultrastructure which can induce an intense circular dichroism (CD) signal (Keller and Bustamante, 1986). Digital scanning confocal LD and CD microscopy and microspectrometry can thus be applied in various fields of biology and medicine.

In this paper, we investigate the correlations between the macrosocopic and microscopic LD and CD parameters in chloroplasts. We also show that "optical slicing" combined with DPI can successfully be applied to those organelles exhibiting a complex membrane ultrastructure. Furthermore, it is shown that CD images and microspectrometry reveal important information about the molecular organization of the pigment-protein complexes in the thylakoid membranes.

The efficiency of photosynthetic conversion of light

Light in Biology and Medicine. Volume 2 Edited by R.H. Douglas *et al.*
Plenum Press, New York, 1991

77

energy into chemical energy is largely determined by the macromolecular organization of the pigment molecules. The antenna pigments in chloroplasts, chlorophyll (Chl) $\underline{a}$ and $\underline{b}$ and carotenoids, which are non-covalently bound to the pigment-protein complexes, absorb the light and "funnel" its energy into the reaction centres where primary charge separation takes place. As revealed by spectroscopy of plane polarized light, the complexes and the pigment molecules are embedded in the thylakoid membranes in a highly ordered fashion (Breton and Vermeglio, 1982; Garab et al., 1987). Preferential in-plane orientation of the Q_y transition dipoles of Chls favours long-range diffusion of the excitation energy in a direction parallel to the membrane plane (Garab et al., 1987). Recent CD and CIDS (circular intensity differential scattering) measurements strongly suggested the existence of an additional organization of the antenna system: the pigment-protein complexes in chloroplast thylakoid membranes are assembled in macrodomains whose dimensions are a sizeable fraction of the wavelength of visible light (Garab et al., 1988a,c). The large chirally organized macrodomains could be directly observed by DPI and microspectropolarimetry (Finzi et al., 1989). These macrodomains can provide the structural basis for a long-range diffusion of the excitation energy in the "lake" of antenna molecules, and thus can play an important role in the efficient utilization of solar radiation in chloroplasts.

MATERIALS AND METHODS

Chloroplasts were isolated from spinach (Spinacea oleracea) leaves (Chylla et al., 1987) and trapped in aligned position in polyacrylamide gel between two coverslips in 1.5 T magnetic field. By placing the cover slips perpendicularly or parallel to the field vector, chloroplasts can be face-aligned or edge-aligned, respectively; in the microscope showing preferentially the plane and the edge of their membranes.

Using a confocal scanning differential polarization microscope (Juang et al., 1988), the sample was illuminated point by point through a pinhole by polarized monochromatic light. Polarization was alternating between orthogonal states (p1 and p2) at a frequency of 50.3 kHz (CD) or 100.6 kHz (LD). The transmitted light emerging from the sample traversed the objective and a top pinhole and was detected by a photomultiplier tube. The signal was sent to an integrator to measure $(I_{p1}+I_{p2})/2$, the average transmitted intensity, while a lock-in amplifier measured $I_{p1}-I_{p2}$. The average transmitted intensity and the dichroic ratio $dr=(I_{p1}-_{p2})/(I_{p1}+I_{p2})$, were recorded in a computer, constituting one pixel of the transmission and LD or CD images, respectively. In LD images, p1 and p2 are linearly polarized light at 90° and 0° with respect to the laboratory frame of reference. (In edge-aligned chloroplasts the membranes are preferentially aligned at 0°.) In some experiments p1 and p2 were adjusted to 45° and -45°, respectively. In CD images, p1 and p2 stand for right and left circularly polarized light, respectively.

In the images shown in Figs. 1, 3, 5, 6 and 7, which were reproduced from black and white pictures, white and light grey represent positive dr values, grey represents zero and dark grey and black represent negative dr values. In Figs. 2 and 4, which are produced from colour-coded images, white, grey and black represent positive, negative and zero dr values, respectively. (For the magnitude of dr see Figure legends.)

Local CD spectra were recorded using a Jasco 40-C spectropolarimeter attached to the microscope. The top pinhole of the microscope determined the pixel-area from which the transmitted light was collected by the photomultiplier tube. CD measurements in macroscopic samples were performed in a Jasco 600 spectropolarimeter.

RESULTS AND DISCUSSION

Linear dichroic images

Chloroplasts exhibit intense LD signals in almost all spectral regions (reviewed by Breton and Vermeglio, 1982; Garab et al., 1987). This is because (i) the orientation angle of the pigment dipoles is well defined with respect to the main protein-axis; (ii) the protein complexes are embedded in the membranes in an ordered fashion and with little precessional freedom; (iii) the thylakoid membranes run nearly parallel to each other and the equatorial plane of the chloroplast. This explains why the LD spectrum in a suspension of edge-aligned chloroplasts is maximum while, LD

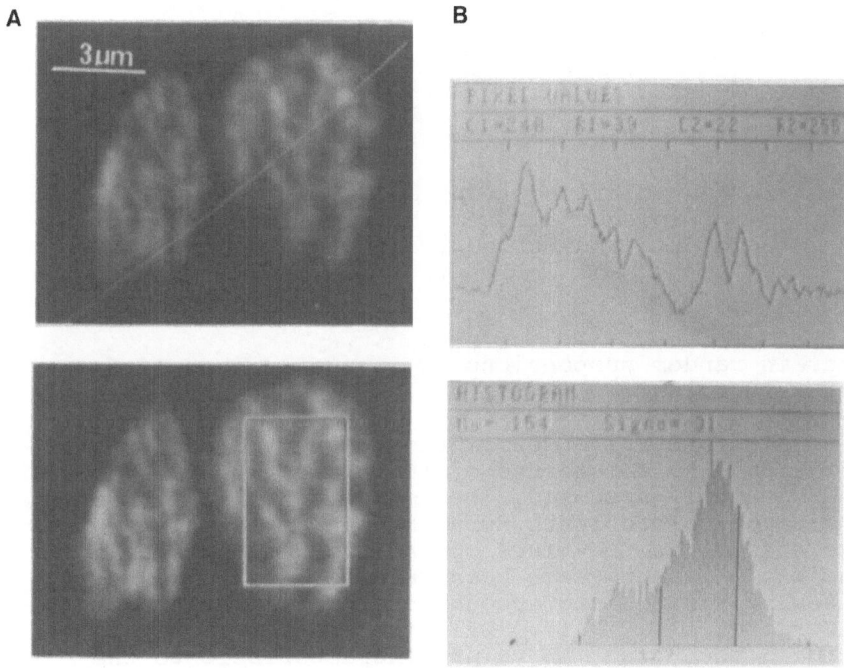

A   B

Fig. 1 LD image at 435 nm of two edge-aligned chloroplasts showing the variation of dr values along a line (A) and the histogram of dr values (B) in a frame in the image; dr is found between -0.001 and 0.011.

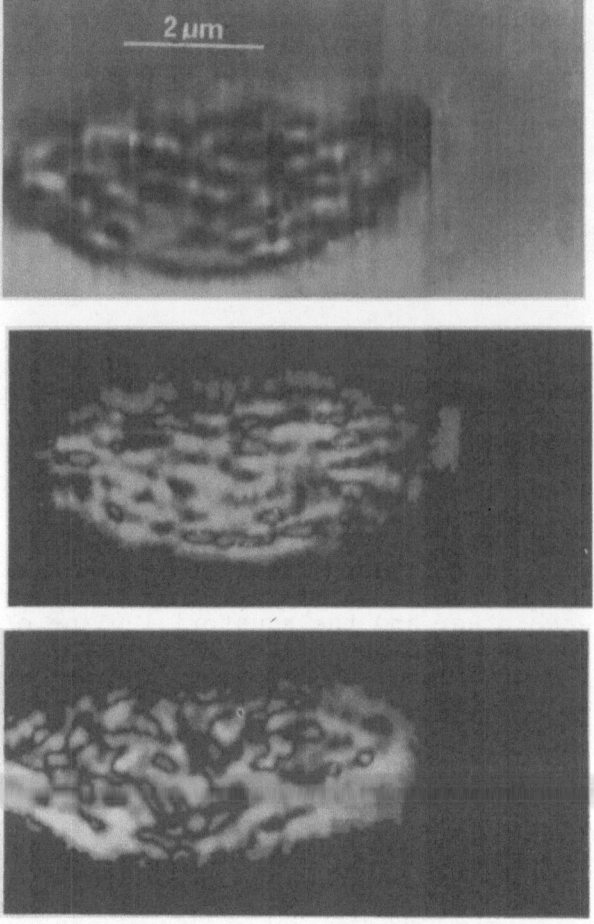

Fig. 2   Non-polarized (top) and LD images at 435 nm
of an edge-aligned chloroplast; in the
middle and bottom image p1 and p2 were 90°
and 0° and - 45° and 45°, respectively.

vanishes in random suspensions or in suspensions of face-
aligned chloroplasts.

For a quantitative description of the molecular
organization, the dependence of LD on the factors listed above
must be carefully assessed. LD measurements performed on a
suspension of chloroplasts furnish only an LD value which is
the average of the LD values coming from all particles in the
suspension. Attempts have been made to calculate the _in situ_
LD values by taking into account the membrane ultrastructure
(e.g. Garab et al., 1981). DPI offers a direct method for
determining the "true" LD values characteristic to the
orientation angle of the absorbance dipoles with respect to
the membrane plane.

Fig. 1A shows how the LD varies strongly inside the
chloroplasts although, in this particular orientation, it
takes almost exclusive positive values. LD is most intense

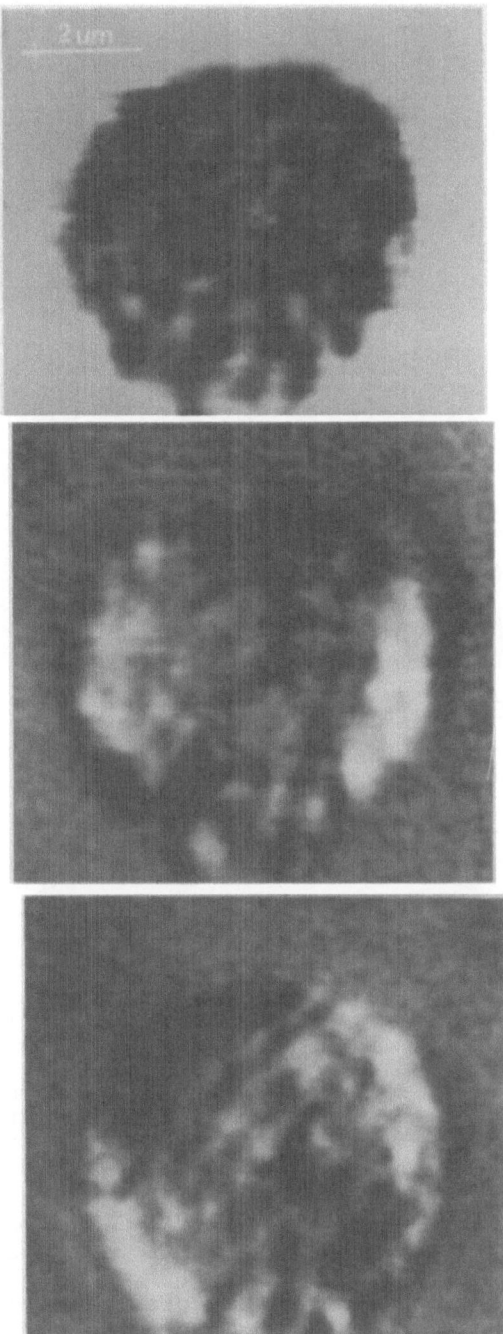

Fig. 3 Non-polarized and LD images of a face-
aligned chloroplast; p1 and p2 as in Fig. 2.

for membranes running parallel to the equatorial plane and
diminished for membrane sections tilting out of this plane.
(Part of the variation is correlated to local differences in
the strength of absorbance; this can also be calculated and
corrected for from the nonpolarized image.) As a consequence
of the complex membrane ultrastructure, the dr values are

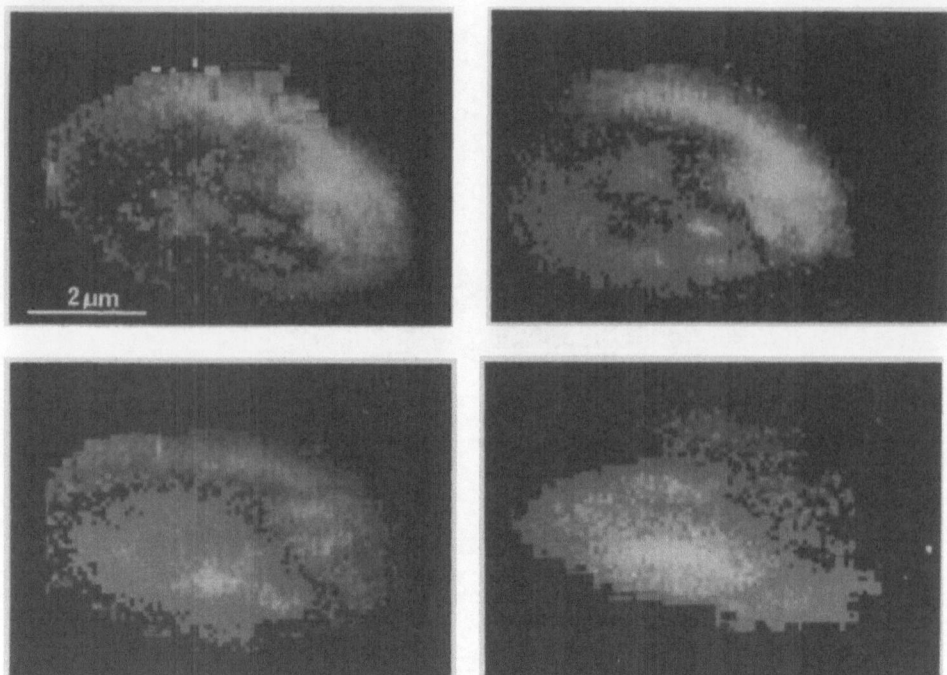

Fig. 4   Optical sections of a chloroplast imaged
with linearly polarized light. The distance
between the first and second and between the
consecutive layers as 0.5 μm and 1 μm,
respectively, dr ranged between -0.019 and
0.019.

distributed over a relatively broad range (Fig. 1B). The
ability to determine the local LD values can be especially
important upon changes in the lipid membrane affecting both
the membrane ultrastructure and the precessional freedom of
the complexes.

Correlation of the LD images and the chloroplast
ultrastructure can clearly be seen in Fig. 2. LD images taken
with polarization planes perpendicular and parallel to the
long axis of the chloroplast ($p1=90°$ and $p2=0°$, see Methods)
emphasise the membranes running parallel to the long axis
whereas the image taken at 45° and -45° emphasises those
membranes which tilt out from the equatorial plane.

Fig. 3 shows a face-aligned chloroplast. The LD is
nearly zero in the centre of the chloroplast while it assumes
large non-zero values (dr 0.01) toward the edges of the
chloroplast, showing four sectors of intense LD. This
symmetric arrangement is maintained upon a 45° rotation of the
polarizer/modulator unit. The membranes toward the edges of
the chloroplast tilt out more from the idealized face-aligned
position and thus show strong LD. However, because of the
cylindrical symmetry of the ultrastructure the positive and
negative LD values tend to cancel each other, and the LD when
averaged for the entire image is very close to zero. This is
in harmony with the macroscopic observation that in the

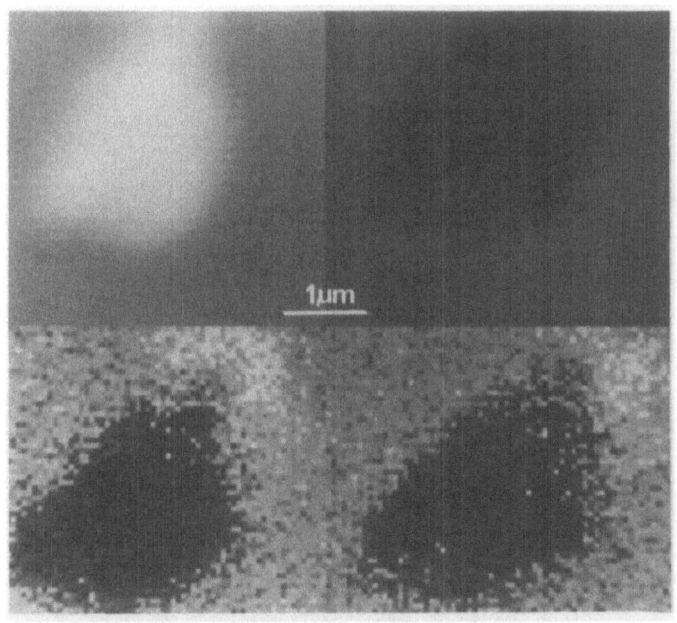

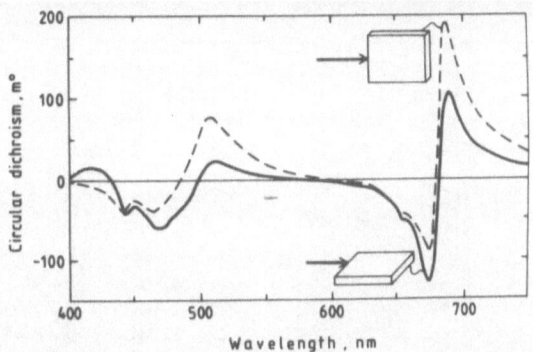

Fig. 5 Low resolution LD (upper pair) and CD images
(lower pair) at 670 nm and CD spectra of
edge-aligned chloroplasts. The images were
recorded at two orthogonal positions of the
polarizer/modulator unit; LD and CD dr
values ranged between -0.02 and 0.02 and
-0.009 and 0.007, respectively. The alignment of an idealized membrane plane in the
CD measurements is shown in the figure; the
Chl content of the geltrapped edge-aligned
suspension of chloroplasts was adjusted to
20 μg/ml.

suspension of face aligned chloroplasts LD=0. We must
emphasise, however, that microscopically even a perfectly
face-aligned chloroplast can show strong local LD values. As
a corollary, data of fluorescence depolarization due to energy
transfer must be interpreted very carefully. As pointed out
by Breton et al. (1973) residual anisotropy of the pigment
dipoles contributes substantially to the measured degree of
fluorescence polarization.

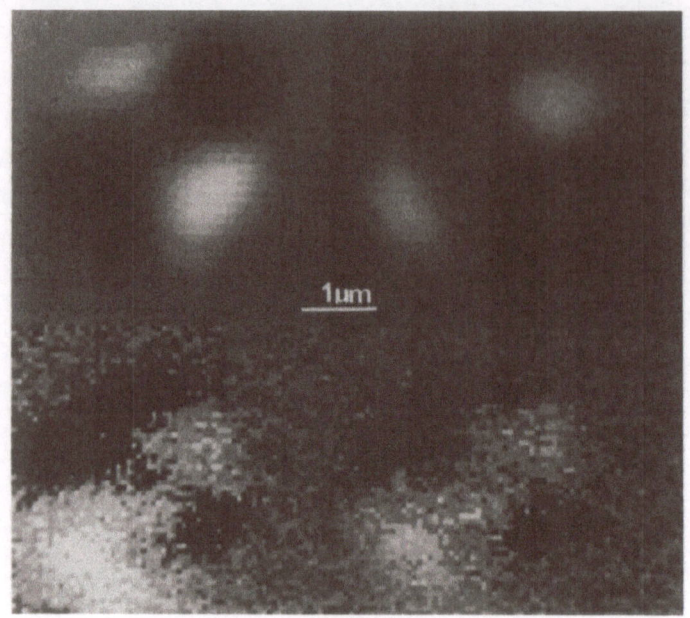

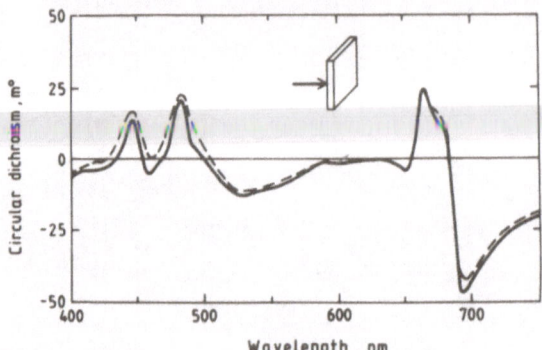

Fig. 6   Low resolution LD (upper pair) and CD
         images (lower pair) at 670 nm and CD spec-
         tra of face-aligned chloroplasts.   In the
         images dr values of LD and CD ranged between
         -0.017 and 0.017 and -0.007 and 0.008,
         respectively.   The CD spectra were recorded
         in vertical and horizontal positions of the
         1cm x 1cm optical cell containing the gel-
         trapped suspension of face-aligned chloro-
         plasts.   For other details see Fig. 5.

     In Fig. 4 we illustrate that LD images can be recorded at
different focal planes in the chloroplast.   These images are
suitable for reconstructing the three dimensional structure of
chloroplasts (L. F., C. B. and G. G., in preparation).

     The three-dimensional structure of chloroplasts has
recently been reconstructed from fluorescence images (van

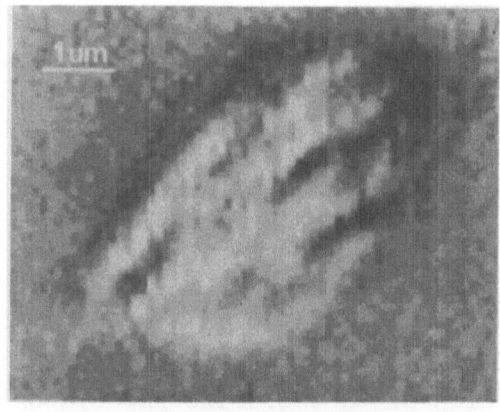

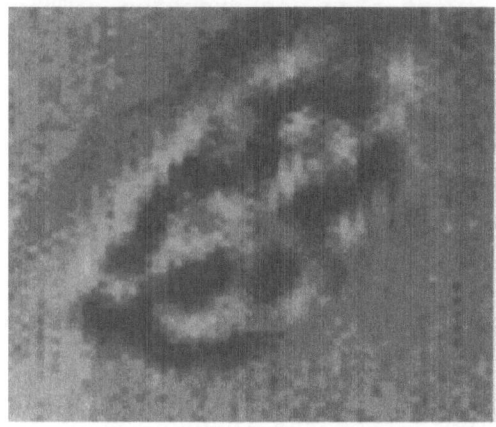

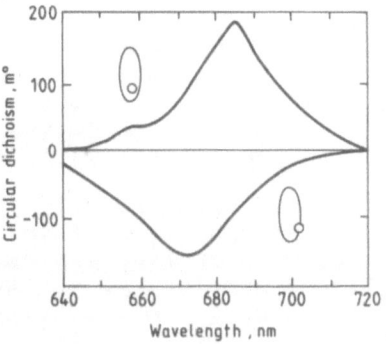

Fig. 7   High resolution LD and CD images at 670 nm
of an edge-aligned chloroplast (upper and
lower image, respectively) and local CD
spectra recorded at two pixels of an edge-
aligned chloroplast (not identical with
that shown in the image). CD dr values
in the image ranged between -0.009 and 0.01.
The maximum LD dr value was 0.024.

Spronsen et al., 1988). A three-dimensional "LD structure"
may reveal further details of the membrane ultrastructure and
can be used for determining the alignment of chloroplasts
inside the cell.

Circular dichroic images and spectroscopy

Recent circular dichroism studies in chloroplasts
indicated that circular differential scattering and psi-type
CD (Keller and Bustamante, 1986) play an important role in
defining the static and dynamic CD characteristics of
thylakoid membranes (Garab et al., 1988a,b). CIDS
measurements indicated that the pigment-protein complexes are
assembled into helically organized macrodomains with a
diameter of 200-400 nm (Garab et al., 1988c). Recently these
discrete macrodomains which give rise to the anomalous CD

signals have been detected by DPI and microspectropolarimetry (Finzi et al., 1989). Since the LD is at least an order of magnitude more intense than CD (see Garab et al., 1987) and the measuring beam can contain some ellipticity, a careful approach is required when measuring the CD of chloroplasts.

Tests for LD contributions to CD measurements can be performed based on the fact that LD inverts sign upon 90° rotation of the polarizer/modulator unit with respect to the sample, while CD is unaffected (Kim et al., 1987b). Figs. 5 and 6 show examples of such a test performed on an edge-aligned and a face-aligned chloroplast, respectively. For clarity the image was recorded at low resolution and the wavelength was chosen so that both CD and LD signals be relatively intense. It is noticed that in the CD images no signal inverts sign upon rotation of the polarizer/modulator unit, while the LD images do invert sign, as expected. This proves the absence of LD artefacts from CD images.

Corroborating data were obtained in macroscopic samples (see the spectra in Figs. 5 and 6). Face aligned samples contain no appreciable amount of LD contribution. In edge-aligned samples, which are aligned to give maximal LD, some distortion of CD occurs. However, the basic characteristics of the CD spectra remain unchanged.

CD images, as discussed in more detail by Finzi et al. (1989), reveal anomalous features of the microscopic CD. Local dichroic values are much higher than anticipated on the basis of macroscopic measurements. Both positive and negative CD can be observed at 670 nm. They are found to originate from different regions of the chloroplast and tend to cancel each other. However, as concluded from the inspection of a large number of images, in face-aligned and edge-aligned chloroplasts respectively the positive and the negative CD signal prevail after averaging over the entire membrane system. This observation is consistent with the macroscopic data. Edge-aligned chloroplasts exhibit a negative band peaking at 672 nm whereas face-aligned chloroplasts display a positive signal at 670 nm (Figs. 5 and 6; see also Garab et al., 1988a).

A CD image taken at high resolution (Fig. 7) shows that CD is more discontinuous and inhomogenous than revealed by low resolution images. CD clearly appears to originate from domains of very strong ellipticities. They exhibit anomalous local CD spectra which do not resemble the split 'excitonic-like' CD spectrum of chloroplasts. However, as shown by Finzi et al. (1989), the microscopic CD spectra can be combined to yield the macroscopic spectrum. These observations provide strong evidence for the existence of large chirally organized macrodomains in the thylakoid membranes.

The domains revealed by microscopic and macroscopic CD might be correlated with the large absorbance units observed in fully developed membranes (Hoffmann and Leupold, 1991). Other studies also suggest that the light harvesting antenna pigments are assembled into "lakes" where long-range diffusion of the excitation energy is facilitated (Fetisova et al., 1988; Hunter et al., 1989; Kolubayev et al., 1985).

The existence of macrodomains in thylakoid membranes permit the delocalization of the excitation energy throughout the entire aggregate, and thus their presence can be advantageous in the efficient photosynthetic utilization of light energy. In addition, it can be hypothesized that these macrodomains, which protrude into the lumen, determine the dielectric and protolytic properties of the thylakoid membranes, for they represent the bulk membrane proteins of granal chloroplasts.

ACKNOWLEDGEMENTS

This work was supported by grants from the National Institute of Health (Grant GM-32543), the National Science Foundation (Grant DMB - 8609654), the Center for High Technology and Materials (UNM), and the Minority Biomedical Research Support (Grant 5-S06-RR08139-1f5). G.G. was in part supported by a grant from the National Foundation of Technical Development ((T/t)222/1988 (Hungary)).

REFERENCES

Breton, J., Becker, J.F. and Geacintov, N.E., 1973, Fluorescence depolarization study of randomly oriented and magneto-oriented spinach chloroplasts in suspension, Biochem. Biophys. Res. Commun. 54:1403.

Breton, J., and Vermeglio, A., 1982, Orientation of photosynthetic pigments in vivo, in: "Energy Conversion by Plants and Bacteria" Govindjee, ed., Academic Press, New York.

Chylla, R.A., Garab, G. and Whitmarsh, J., 1987, Evidence for the slow turnover in a fraction of Photosystem II complexes in thylakoid membranes, Biochim. Biophys. Acta., 894:562.

Festisova, Z.G., Freiberg, A.M. and Timpmann, K.E., 1988, Long-range molecular order as an efficienct strategy for light harvesting in photosynthesis, Nature, 334:633.

Finzi, L., Bustamante, C., Garab, G. and Juang, Ch-B., 1989, Direct observation of large chiral domains in chloroplast thylakoid membranes by differential polarization microscopy, Proc. Natl., Acad. Sci. USA, (In press).

Garab, G., Kiss, J.G., Mustardy, L.A. and Michel-Villaz, M., 1981, Orientation of emitting dipoles of chlorophyll a in thylakoids. Considerations on the orientation factor in vivo, Biophys. J., 34:423.

Garab, G., Szito, T. and Faludi-Daniel, A., 1987, Organization of pigments and pigment-protein complexes of thylakoids revealed by polarized light spectroscopy, in: "The Light Reactions," J. Barber, ed., Elsevier, Amsterdam.

Garab, G., Faludi-Daniel, A., Sutherland, J.C. and Hind, G., 1988, Macroorganization of chlorophyll a/b light-harvesting complex in thylakoids and aggregates: Information from circular differential scattering, Biochem., 27:2425.

Garab, G., Leegood, R.C., Walker, D.A., Sutherland, J.C. and Hind, G., 1988, Reversible changes in macrooriganization of the light-harvesting chlorophyll a/b pigment-protein complex detected by circular dichroism, Biochem., 27:2430.

Garab, G., Wells, S., Finzi, L. and Bustamante, C., 1988, Helically organized macroaggregates of pigment-protein complexes in chloroplasts: Evidence from circular intensity differential scattering, <u>Biochem.</u>, 27:5839.

Hoffman, P. and Leupold, D., 1991, Non-linear processes in photosynthetic light absorption, <u>in</u>: "Light in Biology and Medicine, Vol. II," R. H. Douglas, J. Moan, and Gy. Ronto, ed., Plenum Press, London.

Hunter, C.N., van Grondelle, R. and Olsen, J.D., 1989, Photosynthetic antenna proteins: 100 ps before photochemistry starts, <u>Trends in Biochem. Sci.</u>, 14:72.

Juang, Ch-B., Finzi, L. and Bustamante, C.J., 1988, Design and application of a computer-controlled confocal scanning differential polarization microscope, <u>Rev. Sci. Instrum.</u>, 59:2399.

Keller, D. and Bustamante, C., 1986, Theory of the interaction of light with large inhomogeneous molecular aggregates. II. Psi-type cirucular dichroism, <u>J. Chem. Phys.</u>, 84:2972.

Kim, M., Keller, D. and Bustamante, C., 1987a, Differential polarization imaging. I. Theory, <u>Biophys. J.</u>, 52:911.

Kim, M., Ulibarri, L. and Bustamante, C., 1987b, Differential polarization imaging. II. Symmetry properties and calculations, <u>Biophys. J.</u>, 52:929.

Kolubayev, T., Geacintov, N.E., Paillotin, G. and Breton, J., 1985, Domain sizes in chloroplasts and chlorophyll-protein complexes probed by fluorescence yield quenching induced by singlet-triplet exciton annihilation, <u>Biochim. Biophys. Acta.</u>, 808:66.

Tinoco, I. Jr., Mickols, W., Maestre, M.F. and Bustamante, C., 1987, Absorption, scattering, and imaging of biomolecular structures with prolarized light, <u>Ann. Rev. Biophys. Biophys. Chem.</u>, 16:319.

van Spronsen, E.A., Sarafis, V., Brakenho, G.J., Vandervo, H.T. and Nanninga, N., 1989, 3-Dimensional structure of living chloroplasts as visualized by confocal scanning laser microscopy, <u>Protoplasma</u>, 148:8.

ORGANIZATION OF THE RHODOBACTER CAPSULATUS CAROTENOID BIOSYNTHESIS GENE CLUSTER

Gregory A. Armstrong[1,2], Marie Albetti[2], Francesca Leach[2] and John E. Hearst[1,2]

[1]Department of Chemistry, University of California and [2]Division of Chemical Biodynamics, Lawrence Berkeley Laboratory, Berkeley, CA 94720, USA

INTRODUCTION

Carotenoids are a major class of pigment molecules synthesized in all photosynthetic organisms, and some non-photosynthetic bacteria, fungi and yeasts (reviewed in Goodwin, 1980). In photosynthetic organisms carotenoids are not only essential physical quenchers of excited state triplet chlorophyll and bacteriochlorophyll (Bchl) and singlet oxygen generated by these species, but also serve as accessory light-harvesting pigments (reviewed in Cogdell and Frank, 1987). The isolation of the R-prime plasmid pRPS404, containing a 46 kb region from the Rhodobacter capsulatus chromosome which complemented all known point mutation defects in photosynthesis, suggested that the genes encoding structural photosynthetic polypeptides and the enzymes of carotenoid and bacteriochlorophyll biosynthesis were clustered (Marrs, 1981). The genes encoding the reaction centre and light-harvesting I polypeptides, flanking the pigment biosynthesis genes, were subsequently located and sequenced (Youvan et al., 1984a), as were the unlinked genes encoding the light -harvesting II antenna polypeptides (Youvan and Ismail, 1985). No DNA sequences were previously available for the genes encoding carotenoid biosynthetic enzymes from any carotenogenic organism. Thus, the determination of the nucleotide sequence and the organization of the crt genes from R. capsulatus was essential both to further studies of the gene products and of gene regulation. We have focused our attention on the characterization of the subcluster of crt genes within the photosynthesis gene cluster (for a description of the carotenoid biosynthesis pathway see Armstrong et al., 1989). Seven of the eight previously identified R. capsulatus crt genes were known to be clustered on the BamHI-H, -G, -M, and -J fragments of pRPS404 in the order crtA, I, B, C, D, E, F from left to right on the genetic-physical map (Fig 1) (Taylor et al. 1983; Zsebo and Hearst, 1984; Giuliano et al., 1988). These studies established that mutations causing Bchl⁻ phenotypes map within

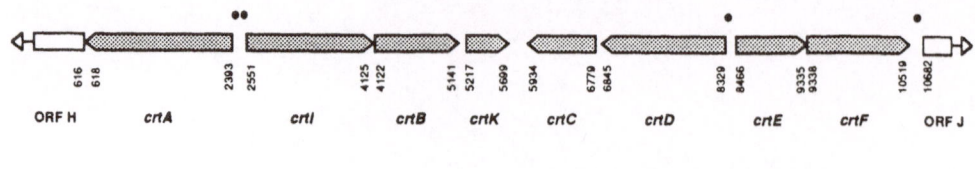

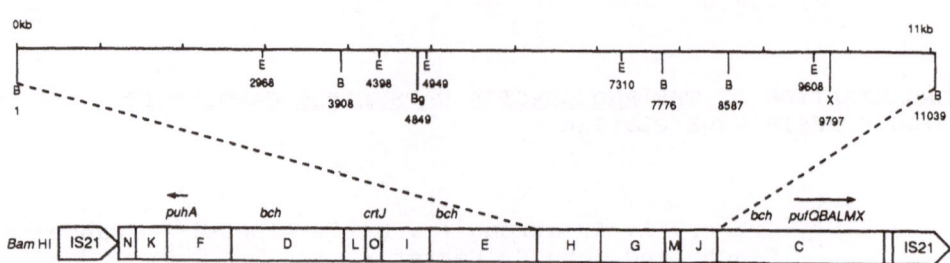

Fig. 1   Organization of the carotenoid biosynthetic
          gene cluster. Polarities of the crt genes
          (shaded) and ORFs (unshaded) are shown and
          putative transcriptional regulatory sites
          (Fig. 5) are indicated by (·). Numbers
          below the genes show the putative nucleo-
          tide positions of translational starts and
          stops. ORF H and ORF J extend beyond the
          region shown here, indicated by the detached
          arrowheads. A new start site has been
          assumed for crtI (Fig. 2A), replacing our
          previous proposal (Armstrong et al., 1989).
          Restriction sites referred to here or in
          previous genetic-physical mapping studies
          are indicated below the genes. B,BamHI;E,
          EcoRI;Bg,BglII;X,XhoI. Boxes containing a
          letter indicate specific restriction frag-
          ments from the photosynthesis gene cluster
          of pRPS404 (Zsebo and Hearst, 1984), while
          the IS21 elements derived from the vector
          are indicated to the left and right. The
          locations of photosynthesis genes outside
          the crt gene cluster are shown above the
          boxes. The pufB, A, L and M genes encode
          the LH-I β, α and the RC-L, RC-M polypep-
          tides, respectively, while the puhA gene
          encodes the RC-H polypeptide (Youvan et al.,
          (1984). Regions containing Bchl bio-
          synthetic genes are indicated by bch. The
          structure of crtJ, identified by a Tn5.7
          insertion (Zsebo and Hearst, 1984) and
          separated from the other crt genes by about
          12 kb (Fig. 1), is currently under study.

these four BamHI fragments, flanking both ends of the crt gene
cluster. We have determined the nucleotide sequence of an
11039 bp region encompassing the BamHI-H, -G, -M, and -J
fragments of pRPS404 (Armstrong et al., 1989). The nucleotide
sequence reveals the presence of a new gene, crtK, not
described in previous studies. We present here a
comprehensive analysis of the DNA sequence and the gene
organization, and discuss nucleotide sequences potentially

involved in the initiation, regulation and termination of transcription within this region.

RESULTS

Alignment of the Nucleotide Sequence with Genetic-Physical Maps Identifies a New Gene, crtK

Sequencing across the BamHI sites (Fig. 1) demonstrated that the BamHI-J, -M, -G and -H fragments are indeed contiguous. Fig. 1 shows the genes located within the 11039 base pair (bp) sequenced region (Armstrong et al., 1989). Because pRPS404 carries the crtD223 point mutation (Marrs, 1981), the nucleotide and deduced polypeptide sequences determined reflect this deviation from the R. capsulatus wildtype sequences. The sequenced region contains three additional ORFs, designated crtK, ORF H and ORF J, distinct from any of the previously described crt genes. Interposon mutations introduced at SalI (bp 5583) and NruI (bp 6723) sites (Fig. 1) have both been proposed to lie within crtC because they result in the accumulation of neurosporene, a CrtC⁻ phenotype (Giuliano et al., 1988). Based on the DNA sequence, however, the interposons interrupt two distinct genes, which cannot be cotranscribed because of their convergent transcriptional orientations. Genetic-physical maps (Taylor et al., 1983; Zsebo and Hearst, 1984) have shown crtC to be bounded by crtB and crtD, with a gap left between crtB and crtC. On the basis of these studies, we designate the previously undetected gene found in this gap as crtK (Fig. 1).

Ribosome Binding Sites and Start Codons

The proposed amino acid sequences of the crt gene products and the ORFs (data not shown) correspond to the longest possible translations of ORFs possessing ribosome binding sites and typical R. capsulatus codon usage, located in the appropriate regions of the crt gene cluster. Translation of the nucleotide sequence in any of the alternative forward or reverse reading frames, with respect to a given gene, results in the frequent appearance of stop codons and the few alternative ORFs which do have ribosome binding sites show atypical codon usage (data now shown). ATA, CTA and TTA are never found amoung the 3038 predicted crt codons, while GTA (Val) appears only once, within crtK (Armstrong et al., 1989). All of the start codons proposed for the crt genes are preceded by purine-rich stretches containing possible ribosome binding sites (Fig. 2A) showing complementarity to the 3' end of the R. capsulatus 16 S rRNA (Fig. 2B) (Youvan et al., 1984b). An ATG start preceded by a ribosome binding site was not observed for ORFs in the region genetically mapped to crtF, although a possible GTG start was found (Fig. 2A). We therefore propose that the coding region of crtF begins with a GTG start codon. GTG start codons are used in about 8% of Escherichia coli genes (Stormo, 1986), and both the fbcF gene from R. capsulatus (Gabellini and Sebald, 1986), and the pucB gene of Rhodobacter sphaeroides have GTG starts. We originally proposed that the 5' end of the crtI coding region was a GTG codon found at bp 2650 (Armstrong et al., 1989), but more recent evidence suggests that the

A

Gene		5'		3'	Residues	MW
crtA		tcac<u>aggGGAGG</u>actgag	ATG		591	64761
crtB		ccgggcc<u>AAGGcGG</u>cgca	ATG		339	37299
crtC		ggc<u>gaAAAGG</u>ccttctcg	ATG		281	31855
crtD		tgcgtgc<u>gGGAG</u>cgagcg	ATG		494	52309
crtE		gcagc<u>GGAGG</u>gctctgtc	ATG		289	30004
crtF		cgcc<u>gaGAGGg</u>Gctgact	GTG		393	43004
crtI	(I)	gaaactacc<u>gaAGAAA</u>cc	ATG		524	57974
crtI	(II)	tcc<u>aAGAA</u>cac<u>AGAA</u>ggt	ATG		517	57226
crtK		ccacaacc<u>GGAGG</u>ccatg	ATG		160	17607

B   5' AGAAAGGAGGTGAT..3'

3' $_{HO}$UCUUUCCUCCACUA..5'

Fig. 2A,B Ribosome binding sites, start codons and
the predicted gene products. (A) Sequences
to the left of the ATG/GTG start codons
contain purine-rich stretches (underlined),
including nucleotides matching the predicted
ribosome binding site (uppercase). The
length of the gene product in amino acids
and its calculated molecular weight are
given to the right. The ribosome binding
sites preceding two possible crtI start
codons are shown, based on a revision of
our orginial proposal for the 5' end of the
crtI gene (Armstrong, et al. 1990a).
(B) shows the predicted ribosome
binding site (above) as the DNA complement
of the 3' end of the <u>R. capsulatus</u> 16 S rRNA
(below) (Youvan et al., 1984b).

translation start corresponds to one of two upstream ATG
codons located at bp 2551 and bp 2572, respectively (Armstrong
et al., 1990a). We have assumed a <u>crtI</u> start at bp 2551
throughout the text and figures. Absolute confirmation of the
deduced amino acid sequences will require the isolation of the
gene products.

Organisation of the Carotenoid Gene Cluster

The <u>crt</u> genes must form at least four distinct operons
because of the inversions of transcriptional orientation which
occur between <u>crtA</u>-<u>crtI</u>, <u>crtK</u>-<u>crtC</u>, and <u>crtD</u>-<u>crtE</u> (Fig. 1).
<u>crtA</u> cannot be cotranscribed with the other <u>crt</u> genes because
of its divergent orientation at one end of the gene cluster.
An interposon insertion at an <u>Eco</u>RV site (bp 1303) and a
transposon insertion (between bp 999-1244) cause Bchl[-]
phenotypes, although both of these mutations lie within the 3'
end of the <u>crtA</u> gene (Giuliano et al., 1988; Armstrong et al.,
1990b). Mutations at the 5' end of <u>crtA</u> cause a CrtA[-]
phenotype but do not affect Bchl synthesis. The most likely
explanation for these effects is the polar inactivation of ORF

H or a downstream gene in the same operon (Fig. 1) in the 3'
insertion mutants. This suggests that crtA is not
cotranscribed with ORF H, although the promoter(s) for ORF H
may overlap crtA. ORF H may thus be part of an operon
required for Bchl biosynthesis. Groups of genes which could
also form operons are crtIBK, crtDC and crtEF. A mutant
bearing an interposon insertion at an ApaI site (bp 10713)
within ORF J exhibits a Bchl⁻ phenotype, suggesting that this
ORF may also belong to an operon required for Bchl
biosynthesis (Giuliano et al., 1988). ORF J, located
downstream from crtF (Fig. 1), does not appear to be
transcribed as part of an operon including crt genes
(Armstrong et al., 1990b). The coding regions of the crt
genes are closely spaced. In the most extreme case, the TGA
stop codon of crtI overlaps the putative ATG start of crtB,
reminiscent of the overlap between the coding regions of the
R. capsulatus pufL and pufM genes (Youvan et al., 1984a).

5' Non-coding Regions are A + T-rich

Fig. 3 illustrates the extreme asymmetry in % A + T
content within the region encoding the crt gene cluster.
Although the entire genome of R. capsulatus has an average A +
T content of 34%, the 5' flanking regions of crtA-crtI, crtD-
crtE and ORF J are unusually A + T-rich, ranging up to 53% in
A + T content averaged over a 151 bp window. These A + T-rich
regions contain DNA sequences which may bind transcription
factors or serve as E. coli-like σ^{70} promoters. The presence
of A + T-rich islands in the 5' control regions of genes from
an organism with a low average A + T content suggests a
compelling selective pressure for the preservation of the
nucleotide bias. Non-coding regions surrounding prokaryotic
transcription initiation and termination points are A + T-rich
compared to the coding regions, as determined using a data
base composed predominantly of E. coli genes (Nussinov et al.,
1987). Specific structural features in A + T-rich regions of
the chromosome may alert DNA-binding proteins to the presence
of potential sites of action (Nussinov et al., 1987).

E.coli-like σ^{70} Promoter Sequences

We have located three sequences closely resembling the
σ^{70} consensus promoter, TTGACA N_{15-19} TATAAT (N = any
nucleotide) used by the major RNA polymerase of E. coli
(McClure, 1985). The R. capsulatus sequences, found 5' to
crtI, crtD and ORF J, are compared to the canonical E. coli
promoter in Fig. 4. An optimal spacing of 17 bp is observed
between the -35 and -10 regions in E. coli (McClure, 1985).
The putative crtD and crtI promoters show spacings of 16 bp,
while the putative ORF J promoter has a spacing of 17 bp. No
other sequences with a total of nine or more nucleotide
matches to the E. coli σ^{70} consensus promoter, including five
of the six most conserved nucleotides (Fig. 4), were found
within the crt gene cluster (Fig. 1). We allowed a variable
spacing of N_{15-19} between the -35 and -10 regions for these
homology searches.

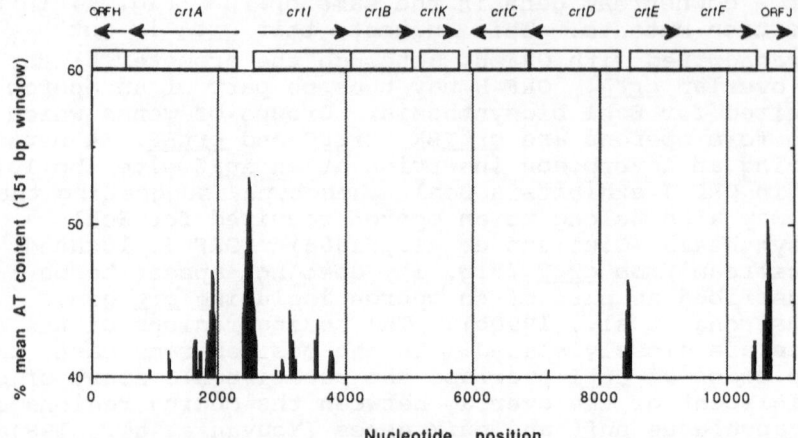

Fig. 3 Average A + T nucleotide content within the carotenoid gene cluster. Analysis of percentage A + T content throughout the DNA sequence was also performed using programs described by Pustell and Kafatos (1982). Locations and polarities of the <u>crt</u> genes are indicated at the top. ORF H and ORF J extend beyond the boundaries of the region shown. Nucleotide positions are as in Fig. 1. Percentage A + T content was calculated by averaging over a 151 bp window at 10 bp intervals, and values exceeding 40% were plotted (average genomic percentage A + T content is 34%). Note the unusually high mean A + T content in 5' flanking regions of genes.

	-35		-10	
2489	**TTGtaA**	atcggaattgac-gacc	**TATcAT**..	34bp..*crtI*
8434	**TTGgcA**	ttcgcacctacctgtg-	**TAaAcT**...77bp..*crtD*
10599	**TTGACA**	gtcgggcgtgtaagttc	**aATgAT**...54bp..ORF J
	*** * ***	*	* *	
	<u>TTGACA</u>	N15-19	<u>TATAAT</u>	

Fig. 4 Comparison of sequences found 5' to <u>crtI</u>, <u>crtD</u> and ORF J with the <u>E</u>. <u>coli</u> σ^{70} consensus promoter. The numbers on the left indicate the position of the 5' nucleotide of each sequence (Fig. 1). The distance in bp from each putative promoter to the start codon of the 3' gene is shown at the right. The consensus <u>E</u>. <u>coli</u> promoter is shown below with the six most highly conserved nucleotides underlined (McClure, 1985). Putative -35 and -10 regions (above) are indicated in boldface, and uppercase letters show matches to the <u>E</u>. <u>coli</u> consensus. Gaps (-) were placed between these regions to maximise the nucleotide alignment. Nucleotides absolutely conserved in all three <u>R</u>. <u>capsulatus</u> sequences are indicated by (*).

94

A Conserved Palindromic Motif is Related to a Recognition Site for DNA-binding Regulatory Proteins

We have identified a highly conserved palindromic nucleotide sequence, found four times in 5' flanking regions within the crt gene cluster. This motif occurs twice in the crtA-crtI 5' flanking region, and once each in the crtD-crtE and ORF J 5' flanking regions (Fig. 1). A search among other published R. capsulatus nucleotide sequences also revealed the presence of this palindrome 5' to the coding region of the puc

A

| | 2551 | 2434 | | 2410 | 2393 |
| <--crtI | aga**TGTAAAT**atcocg**TTACAC**atc | crtA--> |

```
        2551    2434                   2410    2393
<--crtI  agaTGTAAATatcocgTTACACatc  crtA-->

        2393    2487                   2511    2551
<--crtA  agtTGTAAATcggAatTgACgacct  crtI-->

        8329    8394                   8418    8466
<--crtD  gggTGTAAGTttcAgtTTACACagg  crtE-->

       10519   10610                  10634   10682
-->crtF  gcgTGTAAGTtcaAtgaTACACaca  ORF J-->

                24                      58     198
         cacTGTAAGoccgActTTACACttg  pucB-->
                               ••••••
```

```
         +++++               ++
B        TGTAART  N₃ A N₂  TTACAC
         |||               ||||
C        TGTGT        N₆₋₁₀    ACACA
```

Fig. 5A-C Comparison of a palindromic motif found 5' to photosynthesis genes with a consensus regulatory protein binding site. (A) The genes flanking each palindrome are indicated to the left and right, respectively. Arrows show the directions of transcription. Numbers above each sequence show the nucleotide positions (as in Fig. 1, except for pucB (see Youvan and Ismail, 1985)) of the 5' or 3' ends of the flanking genes with respect to the location of the palindrome. Possible puc operon transcription initiation signals (Zucconi and Beatty, 1988) are indicated by (·). Complementary nucleotides in the two halves of the palindromes are underlined. (B) Nucleotides which match the R. capsulatus consensus are given in uppercase, while those that occur in positions defined by the consensus are shown in boldface. (+) indicates an absolutely conserved nucleotide in the palindrome. (C) The R. capsulatus consensus sequence is compared to a consensus derived from the recognition sites of the transcription factors NifA, AraC, CAP, LacI, GalR, LexA, TnpR, LysR and λcII. Nucleotides conserved between the two consensuses are indicated by (|) between the sequences.

operon (Youvan and Ismail, 1985). Based on these five examples, the consensus sequence is TGTAARTN$_3$AN$_2$TTACAC (R = purine) (Fig. 5B). The palindromes are centered anywhere from 162 bp (pucB) to 29 bp (crtA) from the start codon of the nearest gene (Fig 5A). Each of the three putative E. coli-like σ^{70} promoters located within the crt gene cluster overlaps one of the palindromes (compare Figs. 4, 5). No additional palindromes were found when we required matches to each absolutely conserved nucleotide in the consensus (Fig. 5B) in a search of the coding and flanking regions of other published R. capsulatus nucleotide sequences encoding proteins (Armstrong et al., 1989), as well as from the 5' end of ORF J to the 5' end of the pufQ gene (Fig. 1) (M. Alberti, unpublished data). The R. capsulatus consensus palindrome shows strong similarity to a consensus sequence, TGTGTN$_{6-10}$ ACACA, derived from the recognition sites of a collection of prokaryotic transcription factors containing examples of both positive and negative regulators (Fig. 5C) (Gicquel-Sanzey and Cossart, 1982; Buck at el., 1986). The R. capsulatus consensus palindrome is, in fact, very similar to the E. coli TyrR protein consensus recognition sequence TGTAAAN$_6$TTTACA (Yang and Pittard, 1987). TyrR is known to be a transcriptional regulator of genes required for aromatic amino acid metabolism. Based on these sequence similarities to the sites of action of known DNA-binding proteins we propose that the R. capsulatus palindromic motif represents the binding site for a transcription factor.

Rho-independent Transcription Termination Signals

The region shown in Fig. 1 was searched for regions of dyad symmetry with the potential to form stem-loop structures in RNA. Possible secondary structures found between crtK and crtC include two GC-rich stem-loops, one 3' to crtK and the other 3' to crtC, each followed by a run of three thymidines (Armstrong et al., 1989). Single regions of dyad symmetry followed by thymidine-rich stretches were found 3' to the crtI, crtB and crtF genes, respectively. The combination of a GC-rich dyad symmetrical region, followed by several thymidine residues is characteristic of rho-independent transcriptional terminators in bacteria (Platt, 1986). Possible rho-indepdendent termination signals were previously noted close to the 3' ends of the R. capsulatus puf and puc operon mRNAs as mapped by nuclease protection experiments (Zucconi and Beatty, 1988; Chen et al., 1988).

DISCUSSION

The minimum four operons in the crt gene cluster are crtA, crtIBK, crtEF and crtDC based on the polarities of the genes, although the latter operon seems unlikely because of the phenotypes of polar Ω interposon insertions within crtD (Giuliano et al., 1988). Possible rho-independent transcriptional terminators have been found 3' to the crtI, B, K, C and F genes (Armstrong et al., 1989), suggesting that the former three genes could form separate operons. The R. capsulatus crt gene cluster (Fig. 1) is bounded by genetic loci required for Bchl biosynthesis (Taylor et al., 1983; Zsebo and Hearst, 1984; Giuliano et al., 1988). The correspondence between these loci and specific ORFs has not yet been established, although mutations within ORF J and 5' to ORF H, in the 3' end of crtA, cause Bchl⁻ phenotypes (Giuliano et al., 1988; Armstrong et al., 1990b). We therefore propose that ORF J and ORF H are part of two operons which include genes required for Bchl but not carotenoid biosynthesis, and which are transcribed outwards away from the crt gene cluster. crtK was not identified previously by interposon mutagenesis, presumably because of the similarity of CrtC⁻ and CrtK⁻ phenotypes and the fact that crtC and crtK are adjacent (Giuliano et al., 1988).

E. coli-like σ^{70} promoters have never been observed in Rhodobacter (Kiley and Kaplan, 1988), nor have detailed data on any R. capsulatus promoters been available until recently. We have, however, found possible E. coli-like σ^{70} promoters (McClure, 1985) 5' to crtI, crtD and ORF J (Fig. 4). Within the constraints of our homology search (see Results), no other E. coli-like σ^{70} promoter sequences were found within the crt gene cluster, although the gene organization suggests that there must be promoters 5' to both crtA and crtE (Fig. 1). Whether these promoters have a weaker match to the σ^{70} consensus or perhaps have an entirely different structure remains to be determined. Neither Bchl nor carotenoids accumulate in E. coli strains harboring the R. capsulatus photosynthesis gene cluster carried on pRPS404 (Marrs, 1981). Our observation that the R. capsulatus crtD and crtI genes may have E. coli-like σ^{70} promoters, thus, was not anticipated. E. coli may fail to recognise at least one R. capsulatus crt promoter or lack the proper transcription factors required for crt gene expression. In addition, post-transcriptional regulation could also differ between the two species.

We have found five examples of a conserved nucleotide motif (Fig. 5A) in the 5' flanking regions of R. capsulatus photosynthesis genes. One example of the R. capsulatus palindromic motif occurs 5' to the puc operon (Fig. 5A), which encodes the LH-II antenna polypeptides. Zucconi and Beatty (1988) mapped the 5' triphosphate-containing ends of puc operon mRNAs and have suggested that a direct repeat of ACACTTG, located 5' to each of the two mapped mRNA start sites, may be involved in transcription initiation. The palindrome 5' to the puc operon overlaps the upstream ACACTTG sequence (Fig. 5A), and is located ~35 and ~50 nucleotides, respectively, upstream from the two 5' ends of the puc mRNAs.

Three other examples of the palindrome overlap the putative E. coli-like σ^{70} promoter sequences found 5' to crtI, crtD and ORF J (compare Figs. 4, 5). We propose a role for the palindromes in transcriptional regulation because of the extraordinary conservation of the motif and its sequence similarity to binding sites of known transcription factors, and because of its overlap with three putative E. coli-like σ^{70} promoters in the crt gene cluster. Overlap of the R. capsulatus palindromes with promoter sequences could be consistent with either positive or negative gene regulation. The regulatory sites may also be widely separated from the promoters with which they interact. Further experiments are in progress to define the interaction between the putative regulatory palindromes and sequences involved in transcription initiation.

The puc operon is highly regulated at the transcriptional level in response to oxygen tension (Klug et al., 1985). We have recently shown that expression of several crt genes is strongly induced during a shift from aerobic to photosynthetic growth (Armstrong, G. A. and Hearst, J. E., unpublished data), while Giuliano et al (1988) have found an increase in the steady-state levels of 5' ends from crtA, C and E mRNAs in anaerobic versus aerobic cultures. The common feature of anaerobic gene expression would seem a reasonable explanation for the unexpected presence of identical transcriptional regulatory signals 5' to both the puc and crt operons. On the other hand, these regulatory sequences are not found close to the puf and puh operons (Fig. 1), whose expression is also induced by reduction of the oxygen tension (Clark et al., 1984; Klug et al., 1985). The palindromes (Fig. 5A) may thus bind a transcription factor involved in the regulation of a subset of the R.capsulatus photosynthesis genes. Whether a linkage exists between the expression of the puc operon and the regulated crt genes remains to be tested. We have determined the first nucleotide and deduced amino acid sequences of genes and gene products involved in carotenoid biosynthesis, and have also identified possible promoter, terminator and transcriptional regulatory signals which govern crt gene expression. Previous studies of crt gene regulation in R. capsulatus have been hampered by the lack of gene-specific probes (Clark et al., 1984; Klug et al., 1985; Zhu and Hearst, 1986; Zhu et al., 1986). The work presented here will facilitate an examination of the regulation of individual crt genes.

REFERENCES

Armstrong, G.A., Alberti, M., Leach, F., and Hearst, J.E., 1989, Nucleotide sequence, organization and nature of the protein products of the carotenoid biosynthesis gene cluster of Rhodobacter capsulatus, Mol. Gen. Genet., 216: 254-268.

Armstrong, G.A., Schmidt, A., Sandman, G. and Hearst, J.E., 1990a, Genetic and biochemical characterization of carotenoid biosynthesis mutants of Rhodobacter capsulatus, J. Biol. Chem., 265: 8329-8338.

Armstrong, G.A., Alberti, M. and Hearst, J.E., 1990b, Conserved enzymes mediate the early reactions of carotenoid biosynthesis in nonphotosynthetic and photosynthetic prokaryotes, Proc. Nat. Acad. Sci., USA, in press.

Buck, M., Miller, S., Drummond, M., and Dixon R., 1986, Upstream activator sequences are present in the promoters of nitrogen fixation genes, Nature, 320: 374-378.

Chen, C.Y., Beatty, J.T., Cohen, S.N. and Belasco, J. G., 1988, An intercistronic stem-loop structure functions as an mRNA decay terminator necessary but insufficient for puf mRNA stability, Cell, 52: 609-619.

Clark, W.G., Davidson, E., and Marrs, B.L., 1984, Variation of levels of mRNA coding for antenna and reaction center polypeptides in Rhodopseudomonas capsulata in response to changes in oxygen concentration, J. Bacteriol., 157: 945-948.

Cogdell, R.J. and Frank, H.A., 1987, How carotenoids function in photosynthetic bacteria, <u>Biochim. Biophys. Acta</u>, 895: 63-79.

Gabellini, N. and Sebald, W. 1986, Nucleotide sequence and transcription of the <u>fbc</u> operon from <u>Rhodopseudomonas sphaeroides</u>, <u>Eur. J. Biochem.</u>, 154: 569-579.

Gicquel-Sanzey, B. and Cossart, P., 1982, Homologies between different procaryotic DNA-binding regulatory proteins and between their sites of action, <u>EMBO J.</u>, 1: 591-595.

Giuliano, G., Pollock, D., Stapp, H., and Scolnik, P.A., 1988, A genetic-physical map of the <u>Rhodobacter capsulatus</u> carotenoid biosynthesis gene cluster, <u>Mol. Gen. Genet.</u>, 213: 78-83.

Goodwin, T.W., 1980, The Biochemistry of the Carotenoids: Plants, Chapman and Hall, Ltd., New York, New York.

Kiley, P.J. and Kaplan, S., 1988, Molecular genetics of photosynthetic membrane biosynthesis in <u>Rhodobacter sphaeroides</u>. <u>Microbiol. Rev.</u>, 52: 50-69.

Klug, G., Kaufmann, N. and Drews, G., 1985, Gene expression of pigment-binding proteins of the bacterial photosynthetic apparatus: transcription and assembly in th membrane of <u>Rhodopseudomonas capsulata</u>, <u>Proc. Nat. Acad. Sci., USA</u>, 82: 6485-6489.

Marrs, B., 1981, Mobilization of the genes for photosynthesis from <u>Rhodopseudomonas capsulata</u> by a promiscuous plasmid, <u>J. Bacteriol.</u>, 146: 1003-1012.

McClure, W.R., 1985, Mechanism and control of transcription initiation in prokaryotes, <u>Annu. Rev. Biochem.</u>, 54: 171-204.

Nussinov, R., Barber, A., and Maizel, J.V., 1987, The distributions of nucleotides near bacterial transcription initiation and termination sites show distinct signals that may affect DNA geometry, <u>J. Mol. Evol.</u>, 26: 187-197.

Platt, T., 1986, Transcription termination and the regulation of gene expression, <u>Annu. Rev. Biochem.</u>, 55: 339-372.

Pustell, J., and Kafatos, F., 1982, A convenient and adaptable package of DNA sequence analysis programs for microcomputers, <u>Nucl. Acids. Res.</u>, 10: 51-59.

Stormo, G.D., 1986, Translation initiation, <u>in</u>: "Maximizing Gene Expression," W. Reznikoff and L. Gold, eds., Butterworths, Stoneham, Massachusetts, pp. 195-224.

Taylor, D.P., Cohen, S.N., Clark, W.G., and Marrs, B.L., 1983, Alignment of the genetic and restriction maps of the photosynthesis region of the <u>Rhodopseudomonas capsulata</u> chromosome by a conjugation-mediated marker rescue technique, <u>J. Bacteriol.</u>, 154: 580-590.

Yang, J. and Pittard, J., 1987, Molecular analysis of the regulatory region of the <u>Escherichia coli</u> K-12 tyrB gene, <u>J. Bacteriol.</u>, 169: 4710-4715.

Youvan, D.C., Bylina, E.J., Alberti, M., Begusch, H. and Hearst, J.E., 1984a, Nucleotide and deduced polypeptide sequences of the photosynthetic reaction-center, B870 antenna, and flanking polypeptides from <u>R. capsulata</u>, <u>Cell</u>, 37: 949-957.

Youvan, D.C., Alberti, M., Begusch, H., Bylina, E.J. and Hearst, J.E., 1984b, Reaction center and light-harvesting genes from <u>Rhodopseudomonas capsulata</u>, <u>Proc. Nat. Acad. Sci. USA</u>, 81: 189-192.

Youvan, D.C., and Ismail, S., 1985, Light-harvesting II (B800-B850 complex) structural genes from <u>Rhodopseudomonas capsulata</u>, <u>Proc. Nat. Acad. Sci., USA</u>, 82:58-62.

Zhu, Y.S., and Hearst, J.E., 1986, Regulation of the expression of the genes for light-harvesting antenna proteins LH-I and LH-II: reaction center polypeptides RC-L, RC-M, and RC-H; and enzymes of bacteriochlorophyll and carotenoid biosynthesis in <u>Rhodobacter capsulatus</u> by light and oxygen, <u>Proc. Nat. Acad. Sci., USA</u>, 83: 7613-7617.

Zhu, Y.S., Cook, D.N., Leach, F., Armstrong, G.A., Alberti, M. ad Hearst, J.E., 1986, Oxygen-regulated mRNAs for light-harvesting and reaction centre complexes and for bacteriochlorophyll and carotenoid biosynthesis in <u>Rhodobacter capsulatus</u> during the shift from anaerobic to aerobic growth, <u>J. Bacteriol.</u>, 168: 1180-1188.

Zsebo, K.M., and Hearst, J.E., 1984, Genetic-physical mapping of a photosynthetic gene cluster from <u>R. capsulata</u>, <u>Cell</u>, 37: 937-947.

Zucconi, A.P. and Beatty, J.T., 1988, Posttranscriptional regulation by light of the steady-state levels of mature B800-850 light-harvesting complexes in <u>Rhodobacter capsulatus</u>, <u>J. Bacteriol.</u>, 170: 877-882.

PHOTOSYNTHETIC MEMBRANE ENERGIZATION AND ENERGY STORAGE:
STUDIES WITH ELECTROCHROMIC, FLUORESCENCE AND PHOTOACOUSTIC
SPECTROSCOPY

Wim J. Vredenberg, Jan F.H. Snel and
Jaap J.J. Ooms

Dept. of Plant Physiological Research, Wageningen
Agricultural University, Gen Foulkesweg 72, 6703
BW Wageningen, The Netherlands

INTRODUCTION

Photosynthetic energy conversion requires a dynamic
membrane organization which in plant mesophyll and algal cells
is provided by the chloroplast thylakoid membrane. The
function of this membrane and its physical and physico-
chemical characteristics can be studied on a ms timescale in
intact leaves with non invasive techniques such as absorbance,
fluorescence and photoacoustic spectroscopy. This
communication deals with the application of P515 flash
spectroscopy and of combined fluorescence (pulse-modulated)
and photoacoustic (sinewave-modulated) spectroscopy on
isolated intact chloroplasts and spinach leaves.

The photosynthetic machinery starts with light absorption
and excitation energy transfer in light-harvesting pigment
complexes and results in primary photochemical reactions in
the photosynthetic reaction centers of photosystem 1 (PS1) and
2 (PS2). The primary reactions comprise a fast (of the order
of ps) trans-membrane charge separation from chlorophyll-donor
molecules (P700 and P680) of PS1 and PS2, respectively to
primary and secondary acceptors. This is followed by electron
transport and transmembrane proton transfer, ultimately
leading to photolysis of water (O_2 evolution) and the
production of reducing power (NADPH) and ATP which are
required for CO_2 assimilation in the so called reductive
pentosephosphate cycle (Calvin cycle) in C-3 plants.

The primary charge separation leads to the genesis of a
transmembrane electric field which results in a delocalised
transmembrane electric potential (difference). This potential
can be measured with glass micro-capillary electrodes when
inserted in granum stacks of giant chloroplasts of Peperomia
metallica (Vredenberg, 1976). When single turnover light
flashes are used to initiate single charge separations in the
reaction centers, the electrochromic bandshift in the 480-540

Light in Biology and Medicine, Volume 2 Edited by R.H. Douglas et al.
Plenum Press, New York, 1991

101

nm wavelength region, better known as the P515 response, is an elegant tool to follow the generation and subsequent dark decay of transmembrane electric fields (Vredenberg, 1981), caused by charge transfer either through the reaction centers (Reaction RC/1), or through the Q-b-f protein complex (Reaction 1/Q) (Ooms et al., 1989). Furthermore it enables amongst others to discriminate between electrogenic and non electrogenic electrochromic events in the membrane. The non-electrogenic event, denoted as Reaction 2, so far is the less understood component. However it is in general neglected, or erroneously taken as a measure of secondary electrogenic charge transfer across the membrane, i.e. assumed to be similar to or even identical with Reaction 1/Q. We will discuss the characteristics of Reaction 2 in terms of charge stabilization in localized membrane domains. These will be shown to be different from the domains identified by Dilley's group (Dilley et al., 1987), but are (also) apparently easily accessible by protons generated by light or by ATP-hydrolysis.

The momentary efficiency of the energy conversion is quantitatively related to the yield of dissipative processes such as radiative (fluorescence) and radiationless (thermal) energy dissipations in the pigment systems and (passive) charge dissipation via non-specific leaky channels in the membrane. Thus 680-720 nm chlorophyll fluorescence emission, at room temperature mainly originating from chlorophyll associated with PS2, is used as a tool to investigate the efficiency and responsiveness of the photosynthetic system under a large variety of environmental conditions in intact organisms and tissues. These parameters and time-dependent changes therein, if treated properly (Krause et al., 1988), bear quantitative relations to changes in fluorescence emission (yield) caused by so called photochemical (q_P) and non-photochemical (q_N) fluorescence quenching. For instance it has been derived (Weiss and Berry, 1987) that the net rate of oxygen evolution (J_O) is related as follows to q_P, q_N and to the intrinsic rates J_{PO} and J_{PE} of 'open' PS2 reaction centers in the absence and presence, respectively of energy-dependent, non-photochemical, quenching:

$$J_O = q_P \quad ((1 - q_N) \, J_{PO} + q_N \, J_{PE}) \tag{1}$$

Here we will show the effectiveness of photoacoustic spectroscopy (PAS) in measuring O_2 evolution in intact leaf discs simultaneously with fluorescence measurements using the modulation principle for both (Schreiber, 1986; Bicanic et al., 1989). The relation above can easily be tested and the method enables a relatively easy means of studying photo-synthetic performance of intact leaves under a variety of environmental (stress) conditions.

MATERIAL AND METHODS

All measurements were done with spinach leaf discs (approx 1 cm in diameter, harvested from a green house grown variety (Amsterdams Breedblad)), or with fresh chloroplasts isolated thereof as described elsewhere (Ooms et al., 1989). Chloroplast samples were taken from stock suspensions after dark adaptation of at least 2 hours. Unless stated otherwise the reaction medium contained (in mmol/liter) 330 sorbitol,

10 NaCl, 5 MgCl$_2$, 2 EDTA, 5 ascorbic acid, 75 Hepes/KOH pH 7.5 and chloroplasts equivalent to 25 μg chlorophyll per ml.

P515 absorbance changes were measured using a modified Aminco Chance spectrophotometer (Schapendonk et al., 1979) with signal processing as described by Snel (1985). Single turnover saturating light flashes were from a Xe flash lamp (half width duration approx 6 μs). The fluorescence and PAS measuring set-up was essentially as described in detail by Bicanic et al (1989). In brief, the leaf sample, contained in a sealed chamber is illuminated with modulated light. Absorption of light in the leaf causes (modulated) fluorescence emission, heat production and oxygen evolution. The latter two give rise to (modulated) pressure changes which are detected by a circular array of nine microphones in the side wall of the cylindrical cell chamber. The acoustic signal thereof can be, after appropriate frequency handling and electronic manipulation, separated in the photothermal and photobaric component. By application of short, usually 500 ms, saturating light pulses on top of the modulated measuring beam, the photochemical rate in the measuring beam has become zero during the pulse. Consequently the instantaneous photochemical quenching and the photobaric components are removed from the respective fluorescence and the photoacoustic signals. Thus changes in fluorescence quenching parameters (q_N and q_P) can be measured simultaneously with oxygen evolution with a time resolution of less than 100 ms. This time resolution for the measurement of oxygen evolution is much faster than can be obtained with a conventional Clark-type electrode.

RESULTS AND INTERPRETATION

Fluorescence and photoacoustics

Figure 1 shows the kinetics of the increase in fluorescence (yield) and the (transient) photoacoustically detected oxygen evolution during the first 4 s of an illumination period (approximately 30 W/m^2, 660 nm) of a spinach leaf which has been dark-adapted for 5 min. The fluorescence induction shows the well known O-I-P pattern (Papageorgiou, 1975). At the intensity used the fluorescence yield has increased from level O (q_P = 0) to the maximum level P (q_P = 1) with a half time of approximately 125 s. The maximum of oxygen production is far below 100 ms and is reached with a half time of less than 30 ms at the intensity used. This is, to our knowledge, the first experiment in which oxygen evolution and fluorescence in an intact leaf have been measured simultaneously at such high time resolution. Most interestingly the experiment shows that the rate of oxygen evolution declines simultaneously with the rise in q_N. This has not been observed before in this part of the induction curve. It should be mentioned that the relaxation time of the photoacoustically detected oxygen (evolution) is determined by the relaxation in the oxygen evolving complex (OEC) and by the diffusion path in the leaf. Thus our experiments do not allow conclusions as yet with respect to the relaxation time of the OEC in vivo, which is under debate (Plijter et al., 1988).

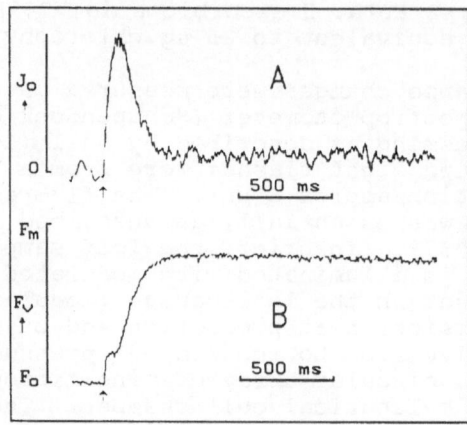

Fig. 1 Kinetics of photoacoustic oxygen (A) and
fluorescence (B) signal transients from a
dark-adapted (5 min) spinach leaf. PAS was
detected at a modulation frequency of 177
Hz; light intensity (A,B) was approx 30 W/m^2
(660 nm). Time constants: PAS, 10 ms;
fluorescence 7.5 ms.

Figure 2 shows a plot of the intrinsic yield of oxygen
evolution in open PS2 centers (J_0/q_P) versus q_N. Data were
obtained from simultaneous time recordings of
photoacoustically detected oxygen evolution and fluorescence
induction using the pulse modulated technique (Bicanic et al.,
1989). Experiments were performed with spinach leaves at
different (initial) CO_2 concentrations in the range between
160 and 1260 ppm. Dark adaptations of the leaf were 20
minutes. The data show a reasonable fit for a linear
relationship between q_N and J_0/q_P. This result is in line
with the model proposed by Weiss and Berry (1987) and
represented by eq. 1. Extrapolation of the line to q_N = 0 and
q_N = 1 gives values of J_{PO} and J_{PE} of 24.6 and 1.88,
respectively. It appears that the intrinsic yield (J_{PE}) of
open reaction centers under conditions of non-photochemical
quenching is only a fraction of the yield (J_{PO}) in the
absence of this quenching. The low value of J_{PE}, combined
with the fact that under normal steady state light conditions
(i.e. in the absence of stress factors) q_N is about 0.3 and q_P
is approximately 0.8 causes the second term in eq. 1 to be
negligible when compared to the first term. This results in
a first approximation with unstressed leaves

$$J_0 = q_P \ (1 - q_N) \ J_{PO} \qquad\qquad (2)$$

If eq. 2 would have been used for the data of figure one would
have calculated J_0 with an underestimation of about 3%.

P515 measurements

Figure 3 shows the flash-induced P515 signal in dark-
adapted intact chloroplasts. The interpretation of the
complex kinetic pattern is still a matter of much debate (c.f
Rich, 1988). Here we adopt the multi-component deconvolution,
which was originally documented by Schapendonk et al., (1979)

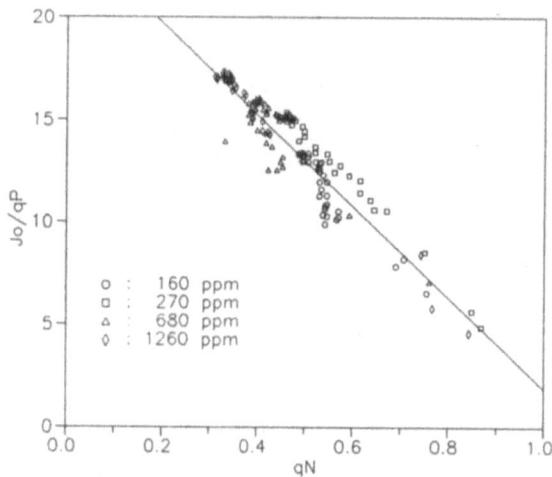

Fig. 2   Dependence of the intrinsic yield of oxygen
          evolution in open PS2 centers J_0/q_p) on non-
          photochemical chlorophyll fluorescence
          quenching (q_N) at the indicated CO_2
          concentrations. Dark adaptation of the leaf
          was for 20 minutes.

in which 4 different components, R1/RC, R1/Q, R2 and R3, can
be distinguished. We will discuss the first three components
in more detail. R3, formerly denoted as component phase d
(Schapendonk et al., 1979), has been identified as non-
electrochromic and gramicidin-insensitive. Recent experiments
have shown that R3 is related to electron transport in PS1. It
has been proposed that the reduction of primary and secondary
acceptors of PS1 cause slowly relaxing conformational changes
in proteins at the stromal side of the thylakoid membrane
which may cause scattering changes giving rise to R3
(Vredenberg et al., 1989).

R1/RC reflects the primary charge separation in PS1 and
PS2 induced by a single turnover flash, and the subsequent
decay of the generated trans-membrane electric field in the
dark (Schapendonk et al., 1979). Its rise time is less than
0.5 ms (instrument-limited), the decay is single exponential

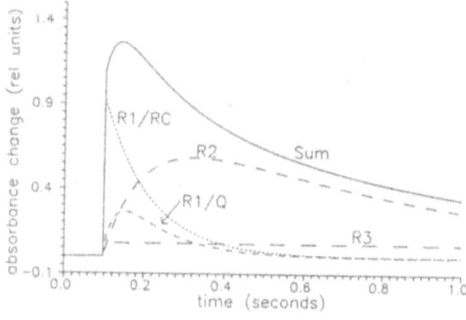

Fig. 3   Illustration of different components in the
          flash-induced P515 signal of intact
          chloroplasts.

with a half time of 50 - 100 ms. R1/RC is normally completely
hidden in the overall signal (see figure 3) which shows, after
a fast initial rise (which is the rise of R1/RC), a secondary
slow rise and a kinetically complex decay with a half time of
hundreds of ms. R1/RC is measured as the predominant
component under, amongst others, the following conditions:

i)   upon a second flash fired 100 ms after a first one
      (Schapendonk et al., 1979; van Kooten et al., 1983; Ooms
      et al., 1989;

ii)   under conditions at which ATP-hydrolysis occurs
      (Schreiber et al., 1982; Peters et al., 1984; Kramer et
      al., 1989), and/or

iii) in the presence of non-uncoupling (nmol) concentrations
      of CCCP (or valinomycin) plus DBMIB (to inhibit R1/Q, see
      below) (Schapendonk, 1980; Peters, 1985; Hope et al.,
      1987).

      R1/Q reflects an electrogenic secondary charge separation
due to a Q-cycle. Its rise occurs in the order of 10 ms and
the decay occurs with a similar, if not identical, half time
to that of R1/RC. R1/Q also is usually hidden in the overall
response of dark-adapted chloroplasts or intact leaves. R1/Q
can clearly be distinguished from the other slow component R2
(see below) on basis of its (much) faster dark decay. One
would predict indeed, as outlined elsewhere in more detail
(Vredenberg, 1981), that the decay of the field generated by
secondary electron flow is identical to that of R1/RC. R1/Q
can only be measured simultaneously with R1/RC, provided the
other components have been saturated or inhibited. Isolation
of R1/Q has been obtained under one of the following
conditions:

i)   the DQH2 stimulated signal either in a 0.1-0.5 Hz flash
      train, or upon a single flash under ATP-hydrolyzing
      conditions (Ooms et al., 1989);

ii)   the DBMIB-sensitive signal in a single flash in the
      presence of non-uncoupling (nM) concentrations of
      valinomycin (or CCCP) (Hope et al., 1987; Ooms et al.,
      (1989);

iii) the stigmatellin-sensitive signal in the presence of DQH2
      and NQNO, or of PMS (Hope et al., 1989).

      R2 is a component, largely determining the slow decay of
the P515 response in dark-adapted chloroplasts (Schapendonk et
al., 1979; van Kooten et al., 1983; Ooms et al., 1989). It
has been evidenced to be non-electrogenic (Schapendonk et al.,
1979; Vredenberg, 1981), and proposed to be associated with
the liberation and stabilization of protons in localized inner
membrane domains (e.g. Vredenberg, 1981; van Kooten, 1988).
R2 is largely, if not completely, absent from the overall P515
response under any of the following conditions:

i)   after pre-illumination under conditions which lead to
      activation of the ATPase and subsequent ATP hydrolysis
      (Peters et al., 1983, 1984);

ii) in the presence of low non-uncoupling (nM) concentrations of valinomycin (Schapendonk, 1980; Hope et al., 1987), or CCCP (Peters et al., 1984); and,

iii) in second and following flashes of a 6 - 10 Hz flash train (e.g. Schapendonk et al., 1979), and see also figure 4).

Figure 4 gives results of experiments to show that the protonation state of local membrane domains of the kind identified by Dilley et al., (1987), is not directly linked with the appearance of R2 in light flashes. Using broken low salt chloroplasts, it has been shown that addition of CCCP followed by BSA sets proton domains in an unprotonated state, whereas addition of BSA followed by CCCP sets the domains in a protonated state (Dilley et al., 1987). The unprotonated state of the CCCP/BSA-, relative to BSA/CCCP-treated chloroplasts has been illustrated by the delayed flash induced activation of the ATP-synthase (Dilley et al., 1987; Ooms et al., 1989). P515 signals have been measured under the same experimental conditions in a 1Hz flash-train. These signals are shown respectively for CCCP/BSA-treated chloroplasts in the presence of ADP (Fig. 4A) and for BSA/CCCP- treated chloroplasts in the presence and absence of ADP (Fig. 4B and 4C). It should be stressed that with CCCP alone the P515 response upon the first flash did not contain an R2 component, as mentioned before, and showed R1/RC dark kinetics; addition of BSA restored the R2 component (van Vliet, unpublished observations). In the second flash the P515 decay is

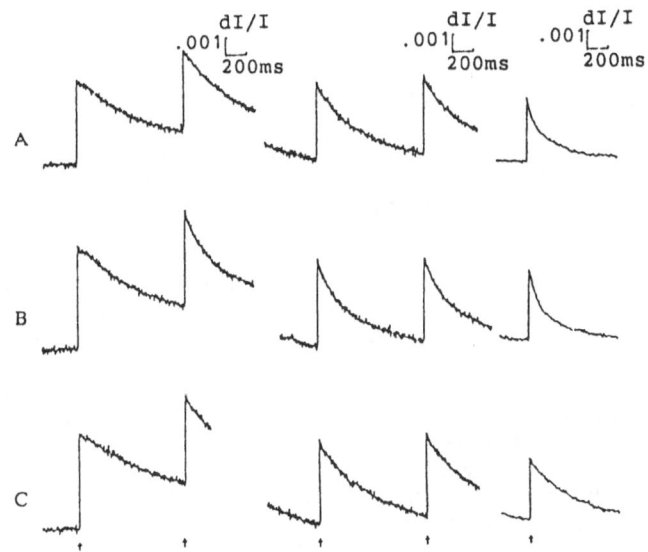

Fig. 4 Flash-induced P515 signals in broken pea chloroplasts monitored during a flash train. Shown are P515 signals of flashes 1, 2, 3, 4 and 60. A: CCCP/BSA-treated low salt chloroplasts, with ADP in the assay medium. B and C, BSA/CCCP- treated low salt chloroplasts with and without ADP in the assay medium respectively.

definitely accelerated under the conditions used here. This is caused, as discussed before, by the suppression of reaction 2 due to its induction by the preceding flash. This suppression of reaction 2 is independent of the ATP-ase activity, as it occurs in the presence as well as in the absence of ADP (Fig. 4B and 4C).

Moreover the apparent acceleration of the decay is seen after the first flash whereas ATP-synthase activity was found to start after the second and twelfth flash for respectively BSA/CCCP- and CCCP/BSA-treated chloroplasts (data not shown, but see Ooms et al., 1989). The P515 signals in the 60th flash are likely to originate mainly, if not exclusively, from R1/RC. The decay of R1/RC shows a clear acceleration in the presence of ADP (Fig. 4A, B) as compared to the decay seen in the first flashes, or in the 60th flash in the absence of ADP (Fig. 4C). A correlation was found between the ATP-ase activity and the half decay times of reaction 1/RC (Ooms et al., 1989).

Comparing the signals in the first flashes in Fig. 4A and 4B, it is clear that the kinetics of P515 and consequently of reaction 2 are hardly, if at all influenced by the protonation state of the domains. It has been checked whether the localized or delocalized coupling mode of the chloroplasts (Dilley et al., 1987), influences the appearance of reaction 2 by recording the P515 signal in the first flash in respectively low salt (localized coupling) and high salt chloroplasts (delocalized coupling). The difference between these signals was found to be minimal which suggests no major influence of the coupling mode on the appearance and kinetics of reaction 2 in dark adapted chloroplasts.

CONCLUSIONS AND PERSPECTIVES

The high time resolution of simultaneous measurements of oxygen evolution and fluorescence (induction) in intact tissue (c.f. Fig. 1) provides a useful means to study the dynamics of the thylakoid membrane.

It is still not possible to reach a definite conclusion with respect to the process that is responsible for the appearance of R2 in the flash-induced P515 response in dark-adapted leaves and chloroplasts. Unfortunately its contribution to the overall signal is often neglected which sometimes leads to erroneous conclusions. For instance the transition from a slow decay of the overall P515 signal to a response with a relatively fast decay is often taken as an indicator of the transition of the ATP-ase into its activated state (Ort et al., 1989; Kramer et al., 1989). This is only correct for the situation at which the activated ATP-ase works as a hydrolase. ATP hydrolysis then causes a (dark-) saturation of R2, and consequently its absence in flashes. It certainly cannot be concluded that activation of the ATP-ase is accompanied by an increase in membrane conductance. This is only true when the threshold potential for ATP-synthesis has been passed (Fig. 4A, B) and the ATP-ase functions as a proton-conducting channel. This indeed shows up as a faster decay of the transmembrane potential as reflected by the decay of R1/RC. Although different from the domains characterized by Dilly et al (1987), as shown here, we keep thinking that R2 is a reflection of loading and stabilization of charges in localized membrane domains as proposed earlier. It has to be verified whether the absence of R2 after the addition of low concentrations of ionophores is also due to the loading of the domains by charges, which have become less tightly bound in the presence of the chaotropics.

ACKNOWLEDGEMENTS

We thank Rienk Bouma, Michael Hegeman, Jan van Kreel and Gerrit van de Zande for their technical assistence during the construction of the photoacoustic cell and Wim van Ieperen, Wilma Versluis and Pieter van Vliet for their experimental contribution. This research was in part supported by the Stichting Scheikundig Onderzoek Nederland (SON), financed by the Nederlandse organisatie voor Wetenschappelijk Onderzoek (NWO).

REFERENCES

Bicanic, D. Harren, F., Reuss, J., Woltering, E. Snel, J., 1989, Comparison of chlorophyll fluorescence and photoacoustic transients in spinach leaves, in: 'Photoacoustic, Photothermal and Photochemical Processes in Gases', P.Hess ed., Springer Verlag Berlin, p.213

Dilley, R.A., Theg S.M., and Beard, W., 1987, Membrane-proton interactions in chloroplast bioenergetics, Annu. Rev. Plant Physiol., 38:347

Hodges, M., Cornic, G. and Briantais, J.-M., 1989, Chlorophyll fluorescence from spinach leaves: resolution of non-photochemical quenching, Biochim. Biophys. Acta., 974:289

Hope, A.B. and Rich P.R., 1989, Proton utake by the chloroplast cytochrome bf complex, Biochim. Biophys. Acta., 975:96

Hope, A.B. and Matthews D.B. 1987, The slow phase of the electrochromic shift in relation to the Q-cycle in thylakoids, Austr. J. Plant Physiol. 14: 29

Kramer D.N. and Crofts, A.R. 1989, Diurnal pattern of chloroplast coupling oxidation kinetics in leaves of intact sunflower. Physiol. Plant., 76:A157

Krause, G.H. and Weiss, E., 1989, in: "Applications of Chlorophyll Fluorescence in Photosynthesis Research, Stress Physiology, Hydrobiology and Remote Sensing", H.K. Lichtenthaler ed., Kluwer Acad Publ, Dordrecht

van Kooten, O. 1988, Photosynthetic free energy transduction. modelling electrochemical events, PhD Thesis, Wageningen Agricultural University

van Kooten, O., Gloudemans, A.G.M. and Vredenberg, W.J., 1983, On the slow component of P515 and the flash-induced reduction of cyt b-563 in chloroplast membranes, Photobiochem. Photobiophys., 6:19

Ooms, J.J.J., Vredenberg, W.J. and Buurmeijer, W.F., 1989, Evidence for an electrogenic and a non-electrogenic component in the slow phase of the P515 response in chloroplasts, Photosynth Res. 20:119 1

Ooms, J.J.J., van Vliet P.H., and Vredenberg, W.J., 1989, The slow P515 signal in relation to the status of inner

membrane proton domains Proc VIIIth Int. Congr. on
Photosynthesis, Stockholm, M. Baltscheffsky, ed. Kluwer
Acad. Publ., Dordrecht, in press

Ort, D.R., Grandoni, P. and Ortiz-Lopez A. 1989, The effect of
reduction on the energetics of CF1 factor activation and
on the efficiency of ATP formation, Physiol. Plant. 76:
A156

Papageorgiou, G., 1975, Chlorophyll fluorescence: an intrinsic
parameter of photosynthesis, in: "Bioenergetics of
Photosynthesis", Govindjee, ed.

Peters, R.L.A., Bossen, M.M., van Kooten, O and Vredenberg
W.J., 1983, On the correlation between the activity of
the ATP-hydrolase and the kinetics of the flash-induced
P515 response in spinach chloroplasts J.Bioenerg.
Biomembr., 15: 337

Peters, R.L.A., van Kooten, O. and Vredenberg W.J. 1984, The
effect of uncouplers (F)CCCP and NH_4Cl on the kinetics of
the flash-induced P515 response in spinach chloroplasts
FEBS Lett., 177:11

Plijter, J.J., Aalbers, S.E., Barends, J.P.F., Vos, M.H. and
van Gorkom H. 1988, Oxygen release may limit the rate of
photosynthetic electron transport; the use of a weakly
polarized oxygen electrode, Biochim. Biophys. Acta.,
935:235

Rich, P.R. 1988, A critical examination of the supposed
variable proton stoichiometry of the chloroplast
cytochrome bf complex, Biochim. Biophys. Acta., 932:33

Schapendonk, A.H.C.M., Vredenberg, W.J. and Tonk, W.J.M. 1979,
Studies on the kinetics of the 515 nm absorbance change
in chloroplasts. Evidence for the induction of a slow
and a fast P515 response upon saturating light flashes,
FEBS Lett. 100:325

Schapendonk, A.H.C.M. 1980, Electrical events associated with
primary photosynthetic reactions in chloroplast
membranes, PhD Thesis, Wageningen Agricultural
University

Schreiber, U. 1986, Detection of rapid induction kinetics with
a new type of high-frequency modulated chlorophyll
fluorometer, Photosynth Res., 9:261

Snel, J. 1985, Regulation of photosynthetic electron flow in
isolated chloroplasts by nicarbonate, formate and
herbicides. PhD Thesis, Wageningen Agricultural
University

Vredenberg, W. J. 1976, Electrical interactions and gradients
between chloroplast compartments and cytoplasm, in: "The
Intact Chloroplast", J. Barber ed., Elsevier Publ.
Amsterdam

Vredenberg, W.J. 1981, P515. A monitor of photosynthetic
energization in chloroplast membranes, Physiol. Plant.,
53:598

Vredenberg, W.J. Versluis, W. and Ooms, J.J.J. 1989, Flash-
induced absorbance changes in thylakoid membranes in the
490 - 550 nm wavelength region. The gramicidin-
insensitive component. Proc. VIIIth Int. Congr. on
Photosynthesis, Stockholm, M. Baltscheffsky, ed., Kluwer
Acad. Publ., Dordrecht, in press

Weiss, E. and Berry, J.A. 1987, Quantum efficiency of
photosystem II in relation to 'energy'-dependent
quenching of chlorophyll fluorescence, Biochim. Biophys.
Acta., 894:198

REDOX PROPERTIES OF THE CYTOCHROME bf COMPLEX FROM A THERMOPHILIC CYANOBACTERIUM

F. Koppenaal and K. Krab

Biological Laboratory, Vrije Universiteit, De
Boelelaan 1087, 1081 HV Amsterdam, The Netherlands

INTRODUCTION

Both in photosynthetic and in respiratory electron
transfer a quinol oxidizing and cytochrome c (or plastocyanin)
reducing enzyme plays a central role. To this group of
enzymes belong ubiquinol:cytochrome c oxidoreductases from
mitochondria, photosynthetic bacteria and from non-
photosynthetic bacteria such as <u>Paracoccus denitrificans</u>, and
the plastoquinol:plastocyanin oxidoreductases from
chloroplasts and cyanobacteria. Research carried out during
the last few decades has brought to light that there is a very
strong functional and structural similarity between enzyme
complexes from different sources (Hauska et al., 1983;
Gabellini, 1988).

Although it has been a fruitfull approach to the study of
these enzymes to emphasize the similarities between the
enzymes in this group, there are also some clear differences
between the two types: the ubiquinol:cytochrome c
oxidoreductases or bc_1-type on the one hand, and the
pastoquinol:plastocyanin oxidoreductases or bf-type on the
other hand. These differences include the E_m's of the
cytochromes (Hauska et al., 1983) and the different
sensitivity to "i-site" inhibitors such as antimycin.
Recently it was demonstrated that there is a firm basis in the
primary structure of the enzyme for the antimycin
insensitivity of the bf-type (DiRago and Colson, 1988).

The reation catalyzed by the bf-complex in the thylakoid
membrane (of chloroplasts or cyanabacteria) is:

$$PQH_2 + 2PCy^{ox} + nH^+_s - > PQ + 2PCy^{red} + 2H^+_L + nH^+_L$$

(S: stroma, L: thylakoid lumen). In this equation the n
reflects the uncertainly of the proton translocating role of
the enzyme. n = 0 implies that there is no proton
translocation associated with electron transfer through the

Light in Biology and Medicine. Volume 2 Edited by R.H. Douglas *et al.*
Plenum Press, New York, 1991

111

enzyme itself, and n = 2 implies that there is. The first situation is thought to be prevalent during linear electron transfer in chloroplasts; the second situation would be similar to the function of the bc_1-complex in respiration.

A central question with respect to the function of the bf-complex is to what extent n is variable. It has been proposed that either the membrane potential $\Delta\Psi$ (Hope et al., 1985), the competition between the Rieske Fe/S and cytochrome b-563 (Moss and Bendall, 1984) or, more specifically, the reduction state of the Fe/S (Rich, 1984) determines whether all reducing equivalents coming from plastoquinol are passed on to the high-potential branch (Rieske Fe/S and cytochrome c-554) or half of the electrons reduce cytochrome b-563. Only in the latter case proton translocation by a Q-cycle type mechanism could occur. Rich recently discussed this and made a case for an invariable n of 2 (Rich, 1988). A key-role in this plays the oxidant-induced reduction of cytochrome b-563: its explanation requires that reducing equivalents are split between cytochrome b-563 and the high-potential branch.

Up to now, the bf complex has been purified from chloroplasts (Hurt and Hauska, 1981, Hurt and Hauska 1982a) and from the cyanobacterium Anabaena variabilis (Krinner et al., 1982). We have isolated the enzyme from a thermophilic cyanobacterium, Synechococcus 6716. It was hoped that by analogy to the ATP synthase purified from the same organism the enzyme would be very stable and well-suited for reconstitution into proteoliposomes together with the native lipids of Synechococcus 6716 (Van Walraven et al., 1983). This communication describes the purification procedure for the complex, together with a characterization of its protein composition, and some spectral and redox properties of its cytochromes b and f. A method is described for deconvolution of steady-state spectra in terms of components from redox titrations. The method is applied to detect oxidant-induced reduction of cytochrome b-563.

MATERIALS AND METHODS

Isolation of complex

Synechococcus 6716 was grown at 50°C as described by Lubberding et al. (1981). Cells grown for one week after inoculation (15-30 g wet weight) were harvested and resuspended in mannitol buffer. To prepare spheroplasts, lysozyme was added and the suspension was incubated for 2 hours at 50°C. The resulting spheroplast suspension was washed with mannitolbuffer and resuspended at a total volume of 100 ml. The spheroplasts were osmotically shocked in 5 liters of 10 mM tricine-NaOH (pH 8.0) and centrifuged (5 min 18,000 g).

The pellet of membrane vesicles was resuspended in 0.4 M sucrose, 2 M NaBr, 10 mM tricine-NaOH (pH 8.0) to a chlorophyll concentration of 1 mg.ml^{-1}. After 30 min. incubation, the suspension was diluted with an equal volume of water and centrifuged for 30 min at 18,000 g. This treatment with NaBr was repeated once. The resulting pellet was washed in 800 ml extraction buffer, containing 0.2 M $(NH_4)_2SO_4$, 0.1 M sucrose, 1.5 mM KCl, 1.5 mM MgCl$_2$ and 20 mM tricine-NaOH (pH 8.0) and centrifuged (30 min. 18,000 g). The washed membrane vesicles were stored overnight at 4°C.

The pellet of membrane vesicles was resuspended in the above mentioned extraction buffer supplemented with 30 mM octylglucoside and 12 mM sodium cholate, to a chlorophyll concentration of 1.5 mg.ml^{-1}. After 30 min. of incubation the detergent extract was centrifuged at 280,000 g. The supernatant of the detergent extract was then stepwise saturated with a saturated ammoniumsulphate solution (pH 7.0) to 45 and 55% respectively. After addition of the ammoniumsulphate, the solution was thoroughly stirred for 30 min, followed by 10 min centrifugation at 32,000 g. The material precipitating between 45 and 55% saturation (the P$_{45-55}$ fraction) was dissolved in 1-2 ml of 20 mM tricine-NaOH (pH 8.0) containing 12 mM sodium cholate. To remove the remaining traces of ammoniumsulphate the P$_{45-55}$ fraction was dialyzed for 2 hours. The dialysed material was then loaded on a linear sucrose density gradient (7-30 % w/v) containing 20 mM tricine-NaOH (pH 8.0), 12 mM sodium cholate and 10 mM octylglucoside. The gradients were centrifuged in an MSE 6 x 14 titanium swing-out rotor at 220,000 g for 18 hours at 4°C.

The greenish brown bands at approximately 13% sucrose concentration, enriched in the b$_6$f complex, were pooled from the centrifuge tubes with a syringe and concentrated to a volume of 0.5-1.0 ml. The concentrated fraction was then frozen and stored in liquid nitrogen. When stored under these conditions, the oxidoreductase activity could be retained for several months when repeated thawing and freezing was avoided. In some cases the fraction was purified further by gel filtration on a Superose-6 HR 10/30 column using an FPLC system (Pharmacia).

Potentiometric titrations

Redox titrations of the cytochromes of the purified cytochrome bf complex were performed in a home made 3 ml glass cuvet continuously flushed with ultrapure argon gas (O$_2$ < 0.1 ppm). The redox titrations were carried out in a medium containing 30 mM MOPS (pH 7.0), 12 mM sodiumcholate, 13% sucrose, 20 mM KCl, at 25°C. During the titration spectra were recorded. The redox mediators used were: 20 μM p-benzoquinone, 20 μM 1,2-anthraquinone, 20 μM 1,4-anthraquinone, 2 μM pyocyanine, 10 μM juglone, 10 μM anthraquinone, 10 μM 2,5-dihydroxy-1,4-benzoquinone, 20 μM diaminodurene, 40 μM phenazine methosulphate, 40 μM phenazine ethosulphate, 15 μM 2-hydroxy-1,4-naphthoquinone, 15 μM anthraquinone-2-sulphonate, 15 μM anthraquinone-2,6-disulphonate, 5 μM benzyl viologen and 40 μM duroquinone. A platinum electrode was used to measure the ambient redox potential against a calomel reference electrode. Both electrodes were connected to a Knick digital pH/mV-meter. Before and after each titration, the electrode combination was calibrated by measuring the potential of a saturated solution of quinhydrone in 50 mM hydrogen phtalate at 25°C.

Reductive titrations were performed starting with a ferricyanide-oxidized complex. First, cytochrome c-554 was stepwise reduced by the addition of a concentrated ascorbate

solution. The amounts of ascorbate added were such that the
changes in redox potential were between 5 and 10 mV. If
overshoot situations could not be avoided the preparation was
oxidized with ferricyanide to the proper redox potential.
When the ambient redox potential was stabilized a full
spectrum was recorded between 500 and 600 nm with a scan speed
of 2 $nm.s^{-1}$. As a baseline for the titration of cytochrome c-
554 the fully oxidized spectrum was taken. The redox state of
cytochrome c-554 was determined from the absorbance difference
between 554 and 540 nm. For the reductive titration of
cytochrome b-563 the ascorbate-reduced spectrum was taken as a
baseline. Solutions of concentrated dithionite (prepared
freshly frequently) were used to titrate cytochrome b-563 in
steps of 5-10 mV. Again, after stabilization of the ambient
redox potential a spectrum was recorded and the absorbance
difference between 563 and 575 nm was taken to estimate the
redox state of cytochrome b-563.

Oxidant-induced reduction

Oxidant-induced reduction of cytochrome b-563 and
oxidation of cytochrome c-554 was measured in a reaction
mixture containing 60 mM KCl, 5 mM $MgCl_2$ and 20 mM Tricine-
NaOH (pH 8.0), at 25°C (Krinner et al., 1982). The
concentrations of ubiquinol-2 and the b_6f complex were as
indicated in the figures.

Spectrophotometry and deconvolution of spectra

Spectra and kinetic absorbance changes were measured with
an Aminco DW-2A UV/Vis spectrophotometer, connected to a PDP
11/03 microcomputer. Spectral data were processed in an HP
1000F minicomputer.

Spectra were deconvoluted into component spectra
essentially as described for kinetic traces by De Wolf et al.
(1988). M component spectra (see results) were arranged in a
L x M matrix E, in which L is the number of wavelength points
in a spectrum. Then the contributions of the components to a
spectrum S (a column vector of dimension L) are calculated as:

$$C = (E^T \ X \ E)^{-1} \ X \ E^T \ X \ S$$

(in which C is a column vector of dimension M). The residual
spectrum R is calculated as:

$$R = S - E \ X \ C$$

Difference spectra were used with A(575 nm) set to zero. The
spectra consisted of 127 points between 530 and 580 nm.

RESULTS

Isolation of the complex

The procedure followed in isolating the complex yielded a
preparation (sucrose gradient fraction) with a spectrum shown
in Fig. 1, upper panel lower trace, and a protein composition
as shown in Fig. 2b. The cytochrome-carrying polypeptides
were identified with the help of heme staining (Fig. 2a). In

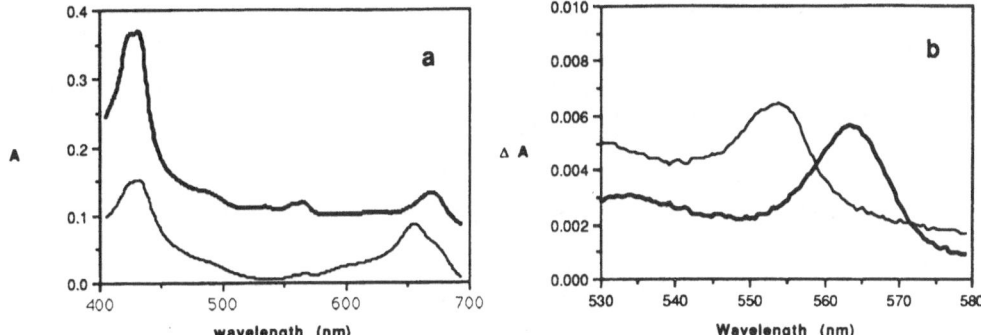

Fig. 1   Spectra of isolated cytochrome bf complex.
a. thin line, dithionite reduced sucrose
gradient fraction; fat line, dithionite
reduced gelfiltration fraction.   b. thin
line, difference ascorbate reduced minus
ferricyanide oxidized; fat line, dithionite
reduced minus ascorbate reduced.   Concentration of complex: 0.12 μM (based on cytochrome c-554).

addition to the cytochrome c polypeptide (30.2 kDa) and the
cytochrome b polypeptide (23.4 kDa) a third peptide (ca. 22
kDa) was present (probably the Fe/S protein), as well as some
phycocyanin (bands at 19 and 20 kDa).

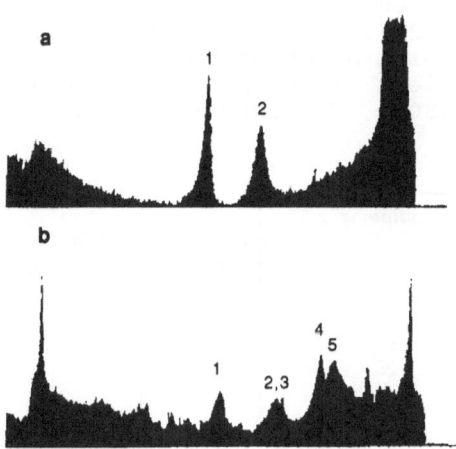

Fig. 2   Polypeptide composition of the sucrose gradient fraction.   Electrophoresis of 0.1 nmol
complex was carried out on a 15% SDS gel
according to Laemmli (1970) a.   Densitogram
after heme staining.   b.   Densitogram after
protein staining.   1: cytochrome c-554 polypeptide, 2: cytochrome b-563 polypeptide, 3:
22 kDa polypeptide (Fe/S), 4,5: phycocyanin
polypeptides.   Electrophoresis from left to
right.

Separate spectra of cytochrome c-554 (ascorbate reduced minus ferricyanide oxidized) and cytochrome b-563 (dithionite reduced minus ascorbate reduced) are shown in Fig. 1b. The absolute spectrum of the preparation showed in addition to the cytochrome features phycocyanin and some chlorophyll (Fig. 1a, lower trace). The phycocyanins could be removed by gel filtration (Fig. 1, upper panel upper trace).

Activity of the enzyme was measured as ubiquinol-2:cytochrome c oxidoreductase. The turnover number of the sucrose gradient fraction was 40 s^{-1}. On a protein basis this fraction is enriched 19-fold with respect to the supernatant of the detergent extract.

Summarizing, the enzyme preparation resembles very much the preparations obtained from chloroplasts (Hurt and Hauska, 1981; Hurt and Hauska, 1982a) and from the mesophilic cyanobacterium <u>Anabaena variabilis</u> (Krinner et al., 1982).

Selection of a set of basic spectra for deconvolution

Redox changes in the isolated complex that result in absorbance changes in the wavelength range from 530 to 580 nm may be caused by cytochrome c-554 (cytochrome f) and by cytochrome b-563 (cytochrome b_6). Spectra of these components were obtained from redox titrations of the complex, as illustrated in Fig 3.

For cytochrome c-554 the complex was titrated in the potential range from 450 to 100 mV. Spectra were recorded at each point: the "basic" spectrum of cytochrome c-554 for deconvolution was calculated from the difference between 419 mV and 189 mV (the black dots in Fig. 3). The spectrum was shifted and multiplied in such a way that A(575 nm) = 0.0000 and A(554 nm) = 0.0247. This corresponds to a reduced minus oxidized spectrum of 1 μM cytochrome c-554 (using an extinction coefficient of 20 mM^{-1}.cm^{-1} at 554 nm, as derived from Wasserman, 1980). The spectrum is shown in Fig. 4, top left panel.

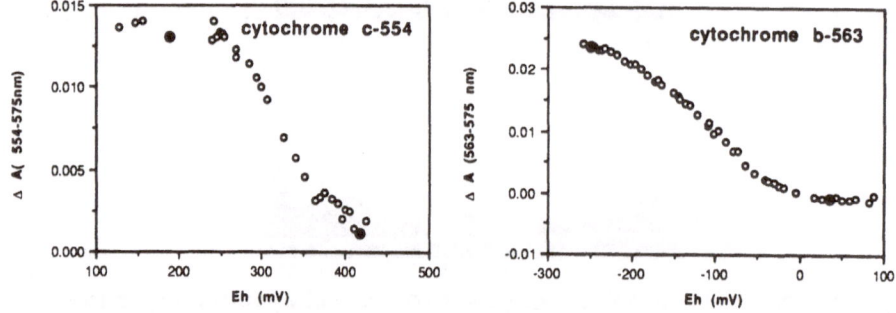

Fig. 3 Redox titration of cytochrome c-554 and cytochrome b-563. The wavelength pairs 563-575 nm and 554-540 nm were used to follow the redox changes of cytochrome b-563 and cytochrome c-554, respectively. Complete spectra were recorded at the indicated points in the dual-wavelength plots between 530 and 630 nm with 575 nm as the reference.

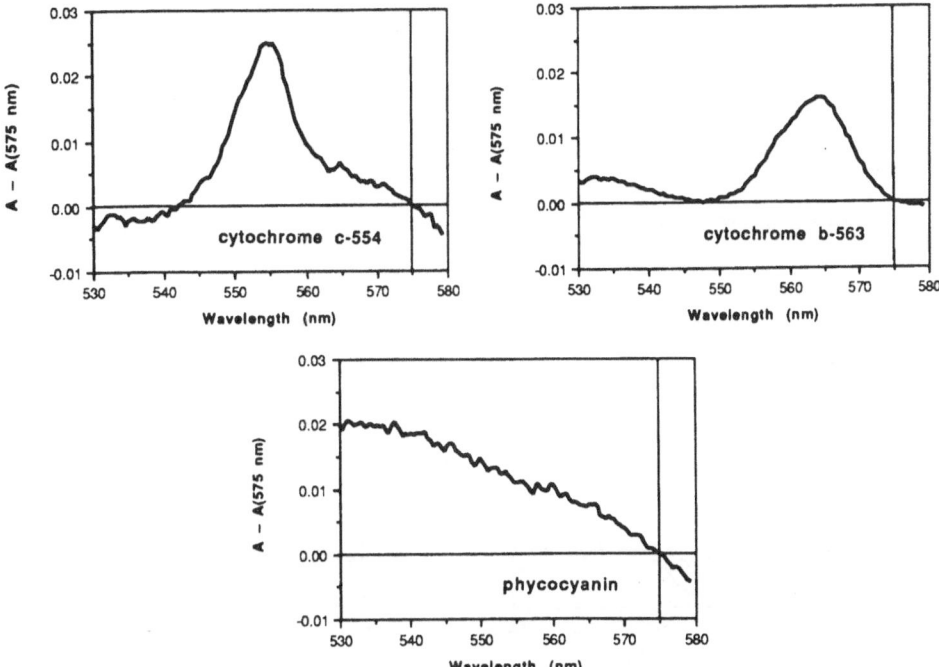

Fig. 4   Basic spectra used for deconvolution (see
         text).

     For cytochrome b-563 the complex was titrated from 100 to
-250 mV. The "basic" spectrum of cytochrome b-563 was
calculated from the difference between -250 mV and 34 mV.
Again, the spectrum was treated such that A(575 nm) = 0.0000
and A(563 nm) = 0.0159. With e(563 nm) = 14.5 mM^{-1}.cm^{-1},
derived from Cramer and Whitmarsh (1977), this spectrum is the
reduced minus oxidized spectrum of 1 μM cytochrome b-563. The
spectrum is shown in Fig. 4, top right panel.

     Deconvolution of spectra with this set of two basic
spectra indicated absorbance changes due to a third component.
This problem manifested itself as non-linear residual spectra
at higher wavelengths.

     Figs. 1 and 2 show that our preparations are contaminated
with phycocyanin. A basic phycocyanin spectrum was obtained
from one of the wash fractions (difference ferricyanide
oxidized minus untreated), and normalized such that A(575 nm)
= 0.0000 and A(530 nm) = 0.0200 (Fig. 4, lower panel).

     With this set of three basic spectra straight, near zero
residual spectra were obtained, except in cases where in
addition redox changes of horse-heart cytochrome c occurred
(Koppenaal and Hotting, unpublished).

Oxidant-induced reduction of cytochrome b

     Fig. 5 shows an experiment in which oxidant-induced
reduction of cytochrome b-563 is measured in the absence of
added inhibitor. A baseline is recorded, and after each

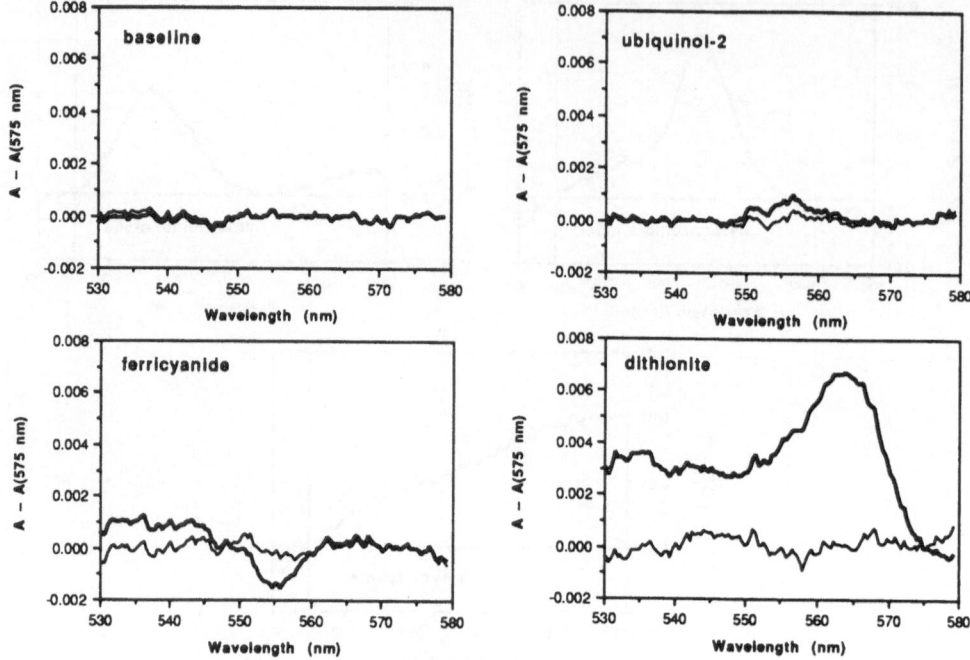

Fig. 5   Steady-state spectra of isolated sucrose
          gradient fraction.   Both spectra (fat lines)
          and residuals after deconvolution (thin
          lines) are shown.

addition a baseline-corrected spectrum.   Addition of
ubiquinol-2 induced little change; subsequent addition of
ferricyanide revealed that in our preparation cytochrome c-554
is reduced.   The spectrum of the fully reduced complex is
measured after a final addition of dithionite.

Addition of ferricyanide results in a transient increase
of absorbance at 563 - 575 nm (not shown), followed by a
steady state in which absorbance is still higher than before
ferricyanide addition.

In fig. 5 the spectra are shown together with the
residuals remaining after deconvolution.   The deconvoluted
results are shown in Fig. 6.   Addition of ferricyanide indeed
leads to reduction of cytochrome b-563.   The total amounts of
cytochrome c-554 (dithionite minus ferricyanide) and
cytochrome b-563 (dithionite minus baseline) are 0.12 and 0.34
μM, respectively.   9% of cytochrome b-563 is reduced in the
steady state obtained after addition of ferricyanide.

DISCUSSION

Purification of the cytochrome bf complex according to
the general principles of Hauska et al. (Hauska et al., 1983)
in this case yields a preparation contaminated by some chlorophyll (similarly to the complexes from chloroplasts (Hurt and
Hauska, 1982a) and <u>Anabaena variabilis</u> (Krinner et al., 1982))
and some phycocyanin.   Contamination by phycocyanin

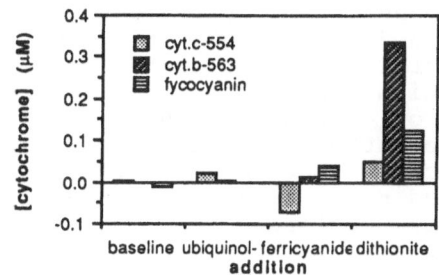

Fig. 6   Component contributions to steady state
         spectra. Cytochromes are given in μM,
         phycocyanin in arbitrary units.

could be removed by gelfitration on Superose-6. This
treatment decreases the yield of the procedure by only ca.
10%, but also results in dilution of the preparation.
However, as illustrated by the oxidant-induced reduction
experiments it is also possible to carry out spectroscopic
studies in the presence of phycocyanin.

Differences in the procedure with respect to the method
used by Krinner et al. to isolate cytochrome bf from Anabaena
variabilis (Krinner et al., 1982) include the method to
prepare the membrane vesicles (in our case osmotic shock) and
the higher temperature at which the procedure can be carried
out (incubation with lysozyme at 50°C, and further
manipulations at room temperature).

The advantage of the present method above the one
published before (Koppenaal et al., 1987) is that
contamination with chlorophyll is much less.
It is possible to store spheroplasts, membrane vesicles
and purified complex frozen for long periods of time without
affecting the activity.

As in cytochrome bf from other sources, oxidant-induced
reduction of cytochrome b-563 can easily be observed at room
temperature in the absence of inhibitors of "site i" (Krinner
et al., 1982; Hurt and Hauska, 1982b; Hurt and Hauska 1982c).
This is different from the situation with cytochrome bc_1
complex of mitochondria, where low temperature or fast kinetic
resolution is required (Erecinska and Wilson, 1972).

The phenomenon is a prerequisite for the occurrence of a
Q-cycle type of mechanism.

As electron transfer from the semiquinone at the "o-site"
to the low-potential cytochrome b-563 does apparently not
contribute to the generation of $\Delta\Upsilon$ (Jones and Whitmarsh,
1985), oxidant-induced reduction in principle also could occur
under conditions where a high $\Delta\Upsilon$ is thought to prevent
operation of a Q-cycle type of mechanism to translocate extra
H^+ (Hope et al., 1985). However, electron transfer to
cytochrome b would then be a dead-end, with apparently no
energetic consequences.

It is interesting to note that not only a transient

oxidant-induced reduction of cytochrome b-563 can be observed in the absence of a "site i" inhibitor, but that net reduction of cytochrome b persists in the steady state that follows ferricyanide addition under these conditions. This indicates that re-oxidation of cytochrome b-563, at least in isolated cytochrome bf preparations, is relatively slow. A possibility suggested by a Q-cycle (but not directly by the semiquinone cycle, Wikström and Krab, 1986) is that addition of plastoquinone would accelerate re-oxidation. However, no large accelerating effects of plastoquinone on the oxidation rate have been found (Hurt and Hauska, 1982c).

ACKNOWLEDGEMENTS

The authors would like to thank Dr. J. G. Fernandez Velasco for his advice with respect to redox titrations and Dr. R. Kraayenhof for critically reading the manuscript. This work was supported financially by the Foundation for Chemical Research (SON) under the auspices of the Netherlands Organization for Scientific Research (NWO).

REFERENCES

Cramer, W.A. and Whitmarsh, J., 1977, Photosynthetic cytochromes, Annu. Rev. Plant Physiol., 28, 133-172.

De Wolf, F.A., Krab, K., Visschers, R.W., De Waard, J.H. and Kraayenhof, R., 1988, Studies on well-coupled photosystem I-enriched subchloroplast vesicles - characteristics and reinterpretation of single-turnover cyclic electron transfer. Biochim. Biophys. Acta., 936, 487-503.

DiRago, J.P. and Colson, A.-M., 1988, Molecular basis for resistance to antimycin and diuron, Q-cycle inhibitors acting at the Qi site in the mitochondrial ubiquinol-cytochrome c reductase in Saccharomyces cerevisiae, J. Biol. Chem., 263, 12564-12570.

Erecinska, M. and Wilson D.F., 1972, Kinetic studies on cytochrome b-c_1 interaction in the isolated succinate-cytochrome c reductase, FEBS Lett, 24, 269-272.

Gabellini, N., 1988, Organization and structure of the genes for the cytochrome b/c_1 complex in purple photosynthetic bacteria. A phylogenetic study describing the homology of the b/c_1 subunits between prokaryotes, mitochondria and chloroplasts, J. Bioenerg. Biomembr., 20, 59-83.

Hauska, G., Hurt, E., Gabellini, N. and Lockau, W., 1983, Comparative aspects of quinol-cytochrome c/plastocyanin oxidoreductase, Biochim. Biophys. Acta, 726, 97-133.

Hope, A.B., Handley, L. and Matthews, D.B., 1985, Further studies of proton translocations in chloroplasts after single-turnover flashes. III. Conditions for the operation of an apparent Q-cycle in thylakoids, Aus. J. Plant Physiol., 12, 387-394.

Hurt, E. and Hauska, G., 1981, A cytochrome f/b_6 complex of five polypeptides with plastoquinol-plastocyanin-oxidoreductase activity from spinach chloroplasts, Eur. J. Biochem., 117, 591-599.

Hurt, E. and Hauska, G., 1982a, Identification of the polypeptides in the cytochrome b_6/f complex from spinach chloroplasts with redox-centre-carrying subunits, J. Bioenerg. Biomembr., 14, 405-424.

Hurt, E. and Hauska, G., 1982b, Oxidant-induced reduction of cytochrome b_6 in the isolated cytochrome b_6/f complex from chloroplasts, Photobiochem. Photobiophys., 4, 9-15.

Hurt, E. and Hauska, G., 1982c, Involvement of plastoquinone bound within the isolated cytochrome b_6-f complex from chlrooplasts in oxidant-induced reduction of cytochrome b_6, Biochim. Biophys. Acta., 682, 466-473

Jones, R.W. and Whitmarsh, J., 1985, Origin of the electrogenic reaction in the chloroplast cytochrome b_6/f complex, Photobiochem. Photobiophys., 9, 119-127.

Koppenaal, F., Krab, F. and Kraayenhof, R., 1987, Isolation and characterization of the Qbc-complex from the thermophilic cyanobacterium Synechococcus 6716. In: Progress in photosynthesis research II (Biggins, J., ed.) Martinus Nijhoff Publishers, Dordrecht.

Krinner, M., Hauska, G., Hurt, E. and Lockau, W., 1982, A cytochrome f-b_6 complex with plastoquinol-cytochrome c oxidoreductase activity from Anabaena variabilis, Biochim. Biophys. Acta., 681, 110-117.

Laemmli, U.K., 1970, Cleavage of structural proteins during the assembly of the head of bacteriophage T4, Nature, 227, 680-685.

Lubberding, H.J., Offerijns, F., Vel., W.A.C. and De Vries, P.J.R., 1981, Characterization of the ATPase of the thermophilic cyanobacterium Synechococcus lividus. In: Photosynthesis II. Photosynthetic electron transport and photophosphorylation (Akoyunoglou, G., ed.) Balaban Int. Sci. Services, Philadelphia.

Moss, D.A. and Bendall, D.S., 1984, Cyclic electron transfer in chloroplasts. The Q-cycle and the site of action of antimycin, Biochim. Biophys. Acta., 767, 389-395.

Rich, P.R., 1984, Electron and proton transfer through quinones and cytochrome bc complexes, Biochim. Biophys. Acta., 768, 53-79.

Rich, P.R., 1988, A critical examination of the supposed variable proton stoichiometry of the chloroplast cytochrome bf complex, Biochim. Biophys. Acta., 932, 33-42.

Van Walraven, H.S., Lubberding, H.J., Marvin, H.J.P. and Kraayenhof, R., 1983, Characterization of reconstituted ATPase complex proteoliposomes prepared from the thermophilic cyanobacterium Synechococcus 6716, Eur. J. Biochem., 137, 101-106.

Wasserman, A.R., 1980, Chloroplast cytochromes f, b-559 and b_6, Methods Enzymol., 69, 181-202.

Wikström, M. and Krab, K., 1986, The semiquinone cycle. A hypothesis of electron transfer and proton translocation in cytochrome bc-type complexes, J. Bioenerg. Biomembr., 18, 181-193.

PHOTOSYNTHETIC ELECTRON TRANSPORT IN THE ANOXYGENICALLY GROWN
CYANOBACTERIUM OSCILLATORIA LIMNETICA

Christiaan Sybesma

Biophysics Laboratory, Vrije Universiteit Brussel
Pleinlaan 2, B-1050 Brussels, Belgium

INRODUCTION

Until about a dozen years ago, photosynthesis in
cyanobacteria was considered to be similar to higher plant
photosynthesis: In a process driven by two photosystems PSII
and PS I, water, serving as an electron donor, becomes
oxidized, thereby evolving oxygen, and a low potential
reductant, NADP, becomes reduced. This electron transport
occurs in a number of transmembrane protein complexes (Murphy,
1986), such as the reaction centre complexes of PS I and PS II
and a complex involved in the electron transport between the
two photosystems, the cytochrome b_6/f complex (which is
analogous to the b/c_1 complex of mitochondria, cf. Rich,
1984). The reaction centre of PSII delivers its electrons to
a plastoquinone pool that reacts with the cytochrome b_6/f
complex. The electron transport connection between the
cytochrome b_6/f complex and reaction centre of PS I is made by
a peripheral, loosely bound copper compound, plastocyanin. In
some organisms, especially when grown in a medium that lacks
copper (like Oscillatoria limnetica grown in our laboratory)
plastocyanin is replaced by a c-type cytochrome, cytochrome c
553 (Namba and Katoh, 1983).

Although it was already known for a long time that
cyanobacteria occur in anaerobic habitats (cf. Carr and
Whitton, 1973), a facultative anoxygenic form of
photosynthesis occurring under anaerobic circumstances in the
presence of large concentrations of sulphide, was reported
only fairly recently, first in the cyanobacterium Oscillatoria
limnetica, an isolate from the Solar Lake near Elath in Israel
(Cohen et al., 1975), and somewhat later in a number of other
species of cyanobacteria as well (Garlick et al., 1977). In
this form of photosynthesis PS II, the photosystem that causes
water oxidation concomittant with oxygen evolution, is
switched off and the organisms grow in the light, using
sulphide as the electron donor in a reaction that involves
only PS I. Sulphide electrons are transported through a PS I-

driven reaction and terminate either in CO_2 assimilation or in hydrogen evolution, the latter, however, at a much lower rate and only in the absence of CO_2 (Belkin and Padan, 1978a). Both, sulphide-dependent CO_2 assimilation and H_2 evolution are inhibited by plastoquinone analogues and by the ferredoxin inhibitor DSPD (Belkin and Padan, 1983), thus indicating that, at least from plastoquinone to ferredoxin, the sulphide electrons follow the same pathway in both processes.

The switch from oxygenic to anoxygenic photosynthesis is a genetic one: When O. limnetica is transferred from an aerobic medium in which it performs "normal" photosynthesis with oxygen evolution, to an anaerobic medium that contains high concentrations of sulphide, photosynthesis stops. After an induction period of several hours in the light, photosynthesis resumes but it is then insensitive to the PS II inhibitor DCMU and no oxygen is evolved, thus indicating that only PS I is operative (Oren and Padan, 1978). The organisms, under those circumstances, grow normally with a doubling time of about 36 to about 48 hours (Oren and Padan, 1978; Belkin and Padan, 1978b; Slooten et al., 1989). The addition of chloramphenicol, an inhibitor of protein synthesis, prevents the transformation to anoxygenic photosynthesis (Oren and Padan, 1978). This would indicate that de novo synthesis of protein is required for the operation of the anoxygenic system. A remarkable fact is, however, that the induction period can be diminished, and even eliminated, by strongly reducing conditions, such as the presence of sodium dithionite (Belkin and Padan, 1983). When anoxygenically grown cells are tranferred to a sulphide-free medium, oxygenic photosynthesis resumes instantly, thus demonstrating that oxygenic photsynthesis is present constituitively (Oren and Padan, 1978).

Recently, the group of E. Padan of the Hebrew University in Jerusalem demonstrated the sulphide-induced synthesis of a couple of proteins in O. limnetica (Arieli et al., 1989). As the evidence shows, however, these proteins are not membrane-bound; they seem to occur in the periplasmic phase of the cell. Although Arieli et al. (1989) mention a tentative identification of a sulfide-induced protein fraction in the membrane moiety of their preparations as well, the occurrence seems to be minor and the evidence so far is not very strong. Yet, as all the evidence available to date shows, a membrane-bound sulphide oxidase operates in the induced organisms (Sybesma and Slooten, 1987; Shahak et al., 1987; Slooten et al., 1989).

In this paper I shall review the experimental results obtained in our laboratory, putting them in perspective together with results obtained in other laboratories. Part of our experimental results are published elsewhere (Sybesma et al., 1986; Sybesma and Slooten, 1987; Slooten, et al., 1989).

MATERIALS AND METHODS

Oscillatoria limnetica, a kind gift from Dr. E. Padan, was grown as described previously (Slooten et al., 1989). Anoxygenic cells were prepared by adding 10 μM DCMU to oxygenically grown cells, flushing them with nitrogen and adding 4 mM Na₂S. They were harvested after 2 days of anaerobic growth. Thylakoids were prepared either from spheroplasts as described by Sybesma et al, (1986) or directly by sonication as described by Slooten et al. (1989).

Hydrogen evolution was measured with an Intersmat 10C gas chromatograph (Sybesma et al., 1986). NADP reduction was followed by measuring light-induced absorbance changes at 340 nm, as described by Sybesma and Slooten (1987). The continuous light-induced P700 and cytochrome redox reactions were measured as described by Slooten et al. (1989). Flash-induced absorbance changes were measured with a single-beam spectrophotometer constructed in the laboratory. The measurements were controlled by an Olivetti M24SP. 3-ns flashes were delivered by a tuneable dye laser, pumped by an eximer laser (FL2000 and EMG 102, "Lambda Physik"). The dye laser was tuned at 680 nm. The flashes were saturating. The time courses were averages from 25 to 100 flashes separated by 12 s, unless otherwise indicated.

RESULTS AND DISCUSSION

Hydrogen Production

Light induced sulphide-dependent hydrogen production in O.limnetica occurs via hydrogenase. This follows from the fact that the process does not require ATP and is not inhibited by uncouplers (Belkin and Padan, 1978a). Hydrogenase activity in whole cells and in cell-free preparations was determined by measuring the hydrogen evolution in the dark mediated by MV reduced by Na-dithionite (Adams et al., 1981). In whole cells, grown oxygenically, a hydrogenase activity of about 26 μmol H₂/mg chl a. hr was measured; in cell free preparations this was about half of that amount. This is in agreement with the results of Belkin and Padan (1978). In cell-free preparations from anoxygenically grown cells the hydrogenase activity is increased by a factor of about 8, most of it occurring in the soluble fraction. However, only in anoxygenically grown cells could light-induced hydrogen production in the presence of sulphide be detected. None occurs in oxygenically grown cells. In anoxygenically grown cells sulphide dependent light-induced hydrogen production starts immediately after the induction period of about 2 hours at a low rate; but after about one doubling period the rate increased to about the amount of maximum hydrogenase activity.

Since, in cell-free preparations, most of the hydrogenase activity occurs in the soluble phase, it was of interest to look for sulphide-dependent light-induced hydrogen production in membrane fractions (thylakoids) from both cell types in the presence of each of the supernatants. Table 1 gives the results of such an experiment. These results show three things: First, membranes from oxygenic cells seem to be capable of producing hydrogen in the light in the presence of sulphide if presented with the soluble fraction from anoxygenic cells; second, membranes from anoxygenic cells produce more than double the amount produced by oxygenic membranes, even when presented with the soluble fraction from oxygenic cells; and finally, most of the light-induced

Table 1 Sulphide-dependent hydrogen production in the light in $\mu M/mg$ chl a.h in membrane fractions (columns) suspended in supernatants (rows) in the presence of 10 μM DCMU, 10 mM dithionite and 5 mM Na_2S.

	Membr. from oxygen. cells	Membr. from anoxyg. cells
Supern. from oxygen. cells	0	3.8
Supern. from anoxyg. cells	1.5	6.7

sulphide-dependent hydrogen production occurs in membranes from anoxygenic cells in the presence of their own soluble fraction. Since all these amounts still fall short of the maximum hydrogenase activity in cell-free preparations, it is clear that the hydrogenase activity cannot be the limiting factor in the light-induced sulphide-dependent process, and that, therefore, some other factor, only present in anoxygenic cells is required. It is not clear from these results, where that factor is located. If it is exclusively in the soluble fraction, the so much higher rate in anoxygenic membranes, even in the presence of the soluble fraction from oxygenic cells, is not quite understandable. If it is exclusively in the membrane fraction, the light-induced sulphide-dependent hydrogen production in membranes from oxygenic cells in the presence of the anoxygenic soluble fraction cannot be explained. Further experiments on the photoreduction of NADP (see below) have thrown more light on this observation, however.

Sulphide-dependent NADP Photoreduction

Fig. 1 shows the rate of NADP photoreduction as a function of the sulphide concentration in thylakoids from oxygenically grown cells. As the figure shows, pretreatment with BAL, which has been shown to inhibit electron transport in the cytochrome b_6/f complex (Belkin et al., 1984), had no effect on these kinetics. DBMIB (5 μM) did not inhibit NADP photoreduction in these preparations, neither at high (1.44 mM), nor at low (0.38 mM) sulphide concentrations. Thylakoids prepared from anoxygenically grown cells behaved completely differently, however (Fig. 2). While the oxygenic thylakoids show simple Michaelis-Menten kinetics (with a K_m around 1 mM), the saturation curve of the anoxygenic thylakoids was strongly bi-phasic. These data could be reconstructed as the sum of two independent Michaelis-Menten components, one with a K_m of about 23.6 μM and the other with one of about 5.5 mM. These somewhat surprising results indicate two independent electron pathways from sulphide to NADP in thylakoids from anoxygenic cells, one with a high affinity for sulphide and the other, apparently also occurring in thylakoids from oxygenic cells, with a much lower sulphide affinity. If the data of Table 1 are considered in the context of these observations, one could

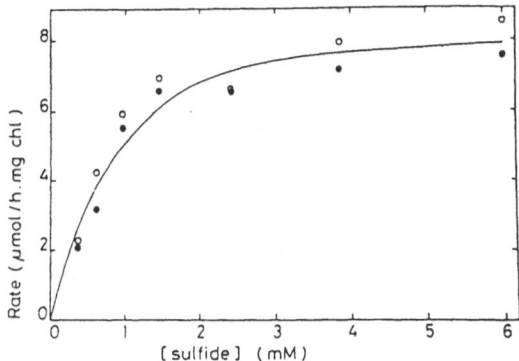

Fig. 1 Sulphide dependence of NADP photoreduction
        in thylakoids from oxygenically grown cells.
        Open circles: no additions. Close circles:
        BAL-pretreated. The complete system con-
        tained 20 mM glucose, 15 units/ml glucose
        oxidase, 0.1 mM NADP, 7 μg/ml chlorophyll
        and an equivalent amount of soluble fraction.

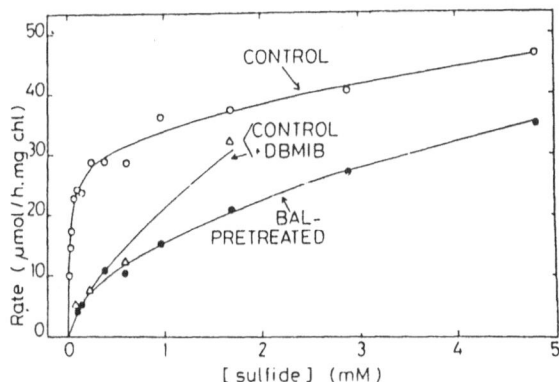

Fig. 2 Sulphide dependence of NADP photoreduction
        in thylakoids from anoxygenically grown
        cells. Conditions as in Fig. 1. 5 μM
        DBMIB was present where indicated.

conclude that light-induced sulphide-dependent hydrogen
production could also occur along these two pathways, the non-
specific one in thylakoids from both cell types (explaining
the production of hydrogen by thylakoids from oxygenic
membranes in the presence of the soluble fraction from
anoxygenic cells) and the specific one only in thylakoids from
anoxygenic cells. The lack of hydrogen production by oxygenic
membranes in the presence of the soluble fraction from
oxygenic cells is probably due to the small amount of
hydrogenase present in that soluble fraction.

    The effect of pretreatment with BAL, an effect somewhat
similar to that of addition of DBMIB (Belkin et al., 1984)
seems to be only inhibition of the specific electron pathway
in thylakoids from anoxygenically grown cells. This fact, and

also the behaviour of the inhibitor DBMIB reveal that the specific, high affinity pathway comprises the cytochrome b_6/f complex while the non-specific pathway bypasses this complex.

Sulphide-dependent light-induced electron transport

Light-induced redox reactions of the primary reaction centre component of PS I, P700, could be followed by looking at absorbance changes at 435 nm, a wavelength which is almost isosbestic for other electron transport components, such as cytochromes. Flash-induced reaction kinetics of P700, measured under the same conditions as those under which the data of Table 1 were measured, are shown in Fig. 3. Also these results could be explained assuming a non-specific electron pathway from sulphide, driven by P700, occurring in membranes from both types of cells and a specific pathway from sulphide occurring only in membranes from anoxygenically grown cells. While in the membrane fractions of oxygenically grown cells the reduction of P700 takes place in an apparent single phase with a half-time of approximately 400 ms, in membranes from anoxygenically grown cells this slower re-reduction is preceded by a more rapid phase of approximately 60 ms half-time, thus reflecting an additional electron pathway specific for these membranes.

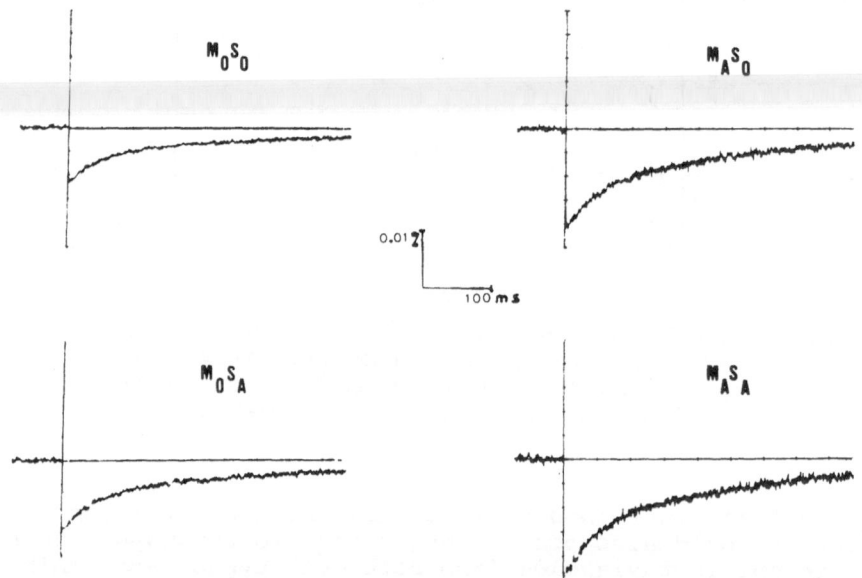

Fig. 3 Flash-induced absorbance changes at 435 nm in membrane fractions from anoxygenically and oxygenically grown cells in the presence of the soluble fraction from anoxygenically and oxygenically grown cells: M means membrane fraction, S means soluble fraction. The subscripts O and A mean oxygenic and anoxygenic respectively. 4 mM Na_2S was added in all four cases. The chlorophyll a concentration was 4 to 5 μg/ml. Averages of 25 signals with flashes fired every 12 s.

More evidence for the participation of electron transport
components in each of the two pathways is obtained from
experiments in which the kinetics of the absorbance changes
induced by switching on and off of continuous light are
measured. These measurements were carried out with thylakoids
in a suspension without the soluble fraction. Fig. 4 gives an
example of such kinetics measured at 435 nm at two different
concentrations of sulphide. At low sulphide concentrations
(0.02 mM), thylakoids from oxygenically grown cells show a
slow recovery after light-off. Thylakoids from anoxygenically
grown cells, however, show an initial very fast recovery
(limited by the time resolution of the instrument) followed by
a slower phase. At higher sulphide concentrations (0.2 mM)
this very rapid phase in anoxygenic thylakoids does not
change; rather, the slower phase begins to show a biphasic
character. The latter is also the case with thylakoids from
oxygenic cells. Similar kinetics were measured at 415 nm, a
wavelength at which the absorbance changes reflect
predominantly the redox changes of cytochrome c 553 (Namba and
Katoh, 1983). Both the time constant and the amplitude of
these two phases depended on the sulphide concentration. We
concluded that these phases reflect the nonspecific pathway
(Slooten et al., 1989).

The absorption difference spectra given in Fig. 5 give
evidence of the involvement of specific components in the
electron transport pathways. Fig. 5A gives spectra that show
an "immediate" light-induced reduction of cytochrome b_6 and
oxidation of cytochrome c 553. When the light was still on
both cytochromes became re-oxidized and re-reduced
respectively, while the oxidation of cytochrome f becomes
apparent (this is clearly seen in curve 4, which is the
difference between curves 2 and 1). When the light is turned
off, another b-type cytochrome, cytochrome b559, seemed to
become oxidized (curve 3). Oxidized cytochrome f is still
seen in this spectrum.

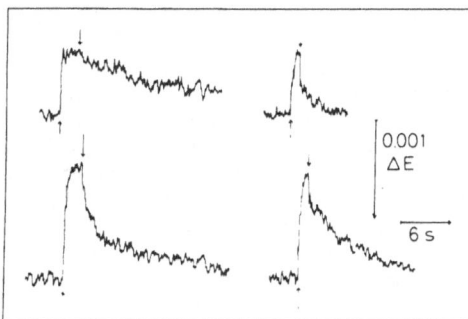

Fig. 4   Kinetics of light-induced absorbance changes
         at 435 nm in thylakoids for oxygenically
         (left) and anoxygenically (right) grown
         cells in the presence of 0,02 mM (top) or
         0,2 mM (bottom) Na$_2$S. Other additions: 0.1
         mM MV, 20 mM glucose, 20 u/ml glucose oxi-
         dase. Chlorophyll $\underline{a}$ concentration was about
         5 μg/ml. Up and downward arrows indicate on
         and off of the light respectively.

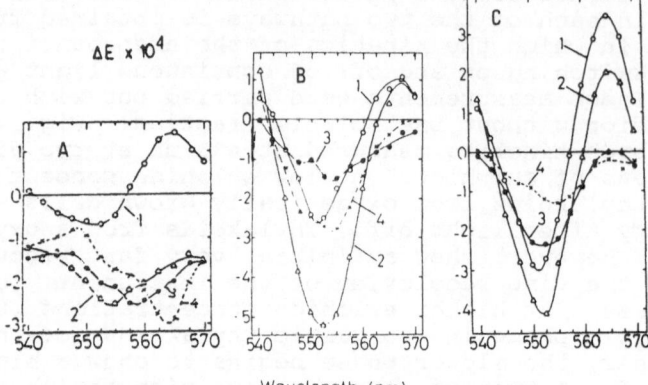

Fig. 5   Spectra of light-induced absorbance changes
         obtained with thylakoids from anoxygenically
         grown cells in the presence of 75 μM Na_2S,
         20 mM glucose, 0.1 mM MV, 10 μ/ml glucose
         oxidase; chlorophyll a concentration was
         18.3 μg/ml.   A: no further additions; B:
         with 2μM DBMIB: C: with 10 μM NQNO. Curve
         1: 0.15 s after light-on; curve 2: 1 s after
         light-on, curve 3: 0.6 s after light-off;
         curve 4: difference between curves 2 and 1.

The inhibitors DBMIB and NQNO are typical cytochrome b_6/f
complex inhibitors. The effect of these two inhibitors on the
light-induced sulphide-dependent electron transport in
thylakoids from anoxygenically grown cells is shown in Figs.
5B and C. The light-induced reduction of cytochrome b was
inhibited by DBMIB (Fig. 5B) but stimulated by NQNO (Fig. 5C).
This is in agreement with the present consensus about the
action of these inhibitors (O'Keefe, 1986; Rich et al., 1987).
This is also true for the larger amount of cytochrome f that
becomes oxidized in the presence of DBMIB. What seemed
somewhat anomalous, however, is the fact that the amount of
cytochrome f that stays oxidized in the dark for the time of
the measurement does not seem to have been changed much by the
addition of the inhibitors. The slow re-reduction of
cytochrome f is also apparent from the kinetics of the flash-
induced absorbance changes in the α-bands of the cytochrome
spectra (Fig. 6A). Cytochrome c 553 and cytochrome b_6 seem to
recover relatively rapidly while cytochrome f remains
oxidized, at least partly, within the 1 s measuring time.
This also shows up in the time-resolved spectra of the flash-
induced absorbance changes (Fig. 6B).

The light induced absorbance difference spectrum obtained
with thylakoids from oxygenically grown cells (not shown) has
an apparent single peak centred at about 553.5 nm. It seems
to show predominantly the oxidation of cytochrome c 553 but it
might also have a contribution from cytochrome f. DBMIB
seems to have little effect, although the results are not
conclusive because of the larger concentration of sulphide
used in these thylakoids.

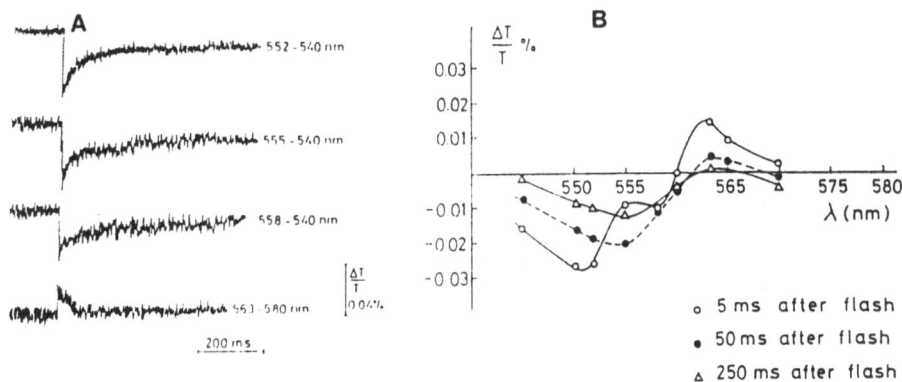

Fig. 6   A: Kinetics of flash-induced absorbance
difference changes at wavelengths as indicated and B: absorbance difference spectra of
those absorbance changes, measured, however,
with a different sample. Conditions as in
Fig. 5. Averages of 100 flashes fired
every 3 s.

CONCLUSIONS

     There are two main conclusions which can be taken from
these results:

     (1) In thylakoids from anoxygenically grown cells,
sulphide electrons can follow two pathways to the reduction of
NADP (or to the hydrogenase), a "nonspecific" one, with a low
affinity for sulphide, donating electrons directly to cyt c
553 and/or P700 and a "specific" one with a high affinity for
sulphide involving the cyt b_6/f complex; and

     (2) The sulphide oxidase activity ("specific" pathway)
seems to be located in the thylakoid membrane. The
"nonspecific" pathway seems also to occur in thylakoids from
oxygenically grown cells. Since no in vivo sulphide-dependent
electron transport, either to NADP or to hydrogenase that is
insensitive to cyt b_6/f complex inhibitors could be detected,
its physiological significance seems doubtful. The "specific"
pathway is the one which uses a sulphide oxidase system
induced by sulphide under anaerobic conditions.

     The results strongly suggest that this sulphide oxidase
system is located in the membrane. This is substantiated by
the finding by Shahak et al. (1987) of a sulphide-dependent
light-induced proton translocation only in thylakoids from
induced anoxygenically grown cells. However, the only
sulphide-induced de novo synthesis of protein that so far has
been detected unequivocally (Arieli et al., 1989) involved
only periplasmic proteins. It is doubtful that such proteins
would participate in electron transport. In the absence of
any further evidence of sulphide-induced membrane protein
synthesis, we would offer the suggestion that the sulphide-
induced oxidase system is formed by a functional modification
of the cyt b_6/f complex. The somewhat anomalous behaviour of
cyt f may point in that direction. Some of our evidence also

131

suggests an involvement of cyt b 559 (Fig. 5A). The role of cyt b 559 in higher plant and algal photosynthesis is still not very clear (cf. Namba and Satoh, 1987).

ABBREVIATIONS

BAL:2,3-dimercaptopropan-1-ol (British Antilewisite); DBMIB: 2,5-dibromothymoquinone; DCMU: 3-(3',4'-dichlorophenyl)-1,1-dimethylurea; DSPD: disalicylidenpropanediamine; NADP: nicotinamide adenine dinucleotide; MV: methylviologen; NQNO: 2-n-nonyl-4-hydroxyquinoline N-oxide.

ACKNOWLEDGEMENTS

The research described was supported by grant number 2.9010.84 from the Belgian Fund for Collective Fundamental Research (FKFO). The technical assistance of Mrs. S. Vandenbranden is highly appreciated.

REFERENCES

Adams, N.H.W., Mortenson, L.E. and Chen, J.S., 1981, Hydrogenase, Biochim. Biophys. Acta, 594:105.
Arieli, B., Binder, B., Shahak, Y. and Padan, E., 1989, Sulfide induction of synthesis of a periplasmic protein in the cyanobacterium O. limnetica, J. Bacteriol., 171:699.
Belkin, S. and Padan, E., 1978a, Hydrogen metabolism in the facultative anoxygenic cyanobacteria (blue-green algae) Oscillatoria limnetica and Aphanothece hyalophytica, Arch. Microbiol., 116:109.
Belkin, S. and Padan, E., 1978b, Sulfide-dependent hydrogen evolution in the cyanobacterium Oscillatoria limnetica, FEBS Lett., 94:291.
Belkin, S. and Padan, E., 1983, Na-dithionite promotes photosynthetic sulfide utilization by the cyanobacterium Oscillatoria limnetica, Plant Physiol., 72:825.
Belkin, S., Siderer, Y., Shahak, Y., Arieli, B. and Padan, E., 1984, 2,3-dimercaptopropaan-1-ol.(BAL), an aerobic electron-transport inhibitor, but an anaerobic photosynthetic electron donor, Biochim. Biophys. Acta, 766:563.
Carr, N.G. and Whitton, B.A. (eds.), 1973, "The biology of blue-green algae", Blackwell Scientific Pub., Oxford.
Cohen, Y., Padan E. and Shilo, M., 1975, Facultative anoxygenic photosynthesis in the cyanobacterium Oscillatoria limnetica, J. Bacteriol., 123:855.
Garlick, S., Oren, A. and Padan, E., 1977, Occurrence of facultative anoxygenic photosynthesis among filamentous and unicellular cyanobacteria, J. Bacteriol., 129:623.
Murphy, D.J., 1986, The molecular organisation of the photosynthetic membranes of higher plants, Biochim. Biophys. Acta, 864:33.
Nanba, M. and Katoh, S., 1983, Reaction kinetics of P700, cytochrome c553 and cytochrome f in the cyanobacterium, Synechococcus sp., Biochim. Biophys. Acta, 725:272.
Nanba, O. and Satoh, K., 1987, Isolation of a photosystem II

reaction center consisting of D-1 and D-2 polypeptides and cytochrome b 559, <u>Proc. Natl. Acad. Sci. U.S.</u>, 84:109.

O'Keefe, D.P., 1988, Structure and function of the chloroplast cytochrome bf complex, <u>Photosyn. Res.</u>, 17:189.

Oren, A. and Padan, E., 1978, Induction of anaerobic photoautotrophic growth in the cyanobacterium <u>Oscillatoria limnetica</u>, <u>J. Bacteriol.</u>, 133:558.

Rich, P.R., 1984, Electron and proton transfers through quinones and cytochrome bc complexes, <u>Biochim. Biophys. Acta</u>, 768:53.

Rich, P.R., Heathcote, P. and Moss, D.A., 1987, Kinetic studies of electron transfer in a hybrid system constructed from the cytochrome bf complex and photosystem I, <u>Biochim. Biophys. Acta</u>, 892:138.

Selak, M.A. and Whitmarsh, J., 1982, Kinetics of the electrogenic step and cytochrome b6 and f redox changes in chloroplasts: Evidence for a Q-cycle, <u>FEBS Lett.</u>, 150:586.

Shahak, Y., Arieli, B., Binder, B. and Padan, E., 1987, Sulfide-dependent photosynthetic electron flow coupled to proton translocation in thylakoids of the cyanobacterium <u>Oscillatoria limnetica</u>, <u>Arch. Biochem. Biophys.</u>, 259:605.

Slooten, L., De Smet, M. and Sybesma, C., 1989, Sulfide-dependent electron transport in thylakoids from the cyanobacterium <u>Oscillatoria limnetica</u>, <u>Biochim. Biophys. Acta</u>, 973:272.

Sybesma, C., Schowanek, D., Slooten, L. and Walravens, N., 1986, Anoxygenic photosynthetic hydrogen production and electron transport in the cyanobacterium <u>Oscillatoria limnetica</u>, <u>Photosyn. Res.</u>, 9:149.

Sybesma, C. and Slooten, L., 1987, Sulfide-dependent electron transport in thylakoids from the cyanobacterium <u>Oscillatoria limnetica</u>, <u>in</u>: "Progress in photosynthesis", J. Biggings, Ed., Martinus Nijhoff Publishers, Dordrecht, The Netherlands.

BIOSYNTHESIS OF CHLOROPHYLL IN A CHEMOHETEROTROPHIC CYANOBACTERIUM, SYNECHOCYSTIS PCC 6714

Barbara Hinterstoisser, Margit Cichna, Christian Obinger and Gunter A. Peschek

Biophysical Chemistry Group, Institute of Physical Chemistry, University of Vienna, Wahringerstrasse 42, A-1090 Wien, Austria

INTRODUCTION

One of the pathways following porphyrin biosynthesis, which starts from succinyl coenzyme A and glycin in animals and bacteria (Jacobs, 1974; Shemin and Russell, 1953), but glutamate in plants and cyanobacteria (Avissar, 1980; Kipe-Nolt and Stevens, 1980), leads to chlorophyll a in all types of oxygenic phototrophic organisms (Bogorad, 1966; Castelfranco and Beale, 1983). In general, the biosynthesis of chlorophyll is under strict control by light and oxygen. Exceptions, however, must be the facultatively anoxygenic cyanobacteria (growing anaerobically on sulfide in the light; Padan, 1979) and the facultatively chemoheterotrophic cyanobacteria (growing aerobically on e.g. glucose in the dark but still actively synthesizing chlorophyll).

With respect to the intracellular location of chlorophyll precursors and enzymes it was recently suggested that some of them, viz. Mg-chelatase (Fuesler et al., 1984), Mg-protoporphyrin methyl ester (Johanningmeier and Howell, 1984), and protochlorophyllide as well as chlorophyllide (Pineau et al., 1986) might be situated in the (chlorophyll-free) envelope membranes of the chloroplast, and likewise chlorophyll-free plasma membranes from cyanobacteria were also shown to contain chlorophyllide and protochlorophyllide (Hinterstoisser et al., 1988; Peschek et al., 1989). Conversion of protochlorophyllide into chlorophyllide by some (plasma or envelope) membrane-bound NADPH-protochlorophyllide reductase was found to be light independent in Anacystis nidulans (Peschek et al., 1989) and green algae but light dependent in spinach chloroplasts (Pineau et al., 1986).

The present paper reports on chlorophyll a synthesis in chemoheterotrophically growing cyanobacterium Synechocystis 6714. This species goes on synthesizing up to 50% of its phototrophic chlorophyll a level even after prolonged growth in strict darkness. It will be shown that isolated and

Light in Biology and Medicine. Volume 2 Edited by R.H. Douglas *et al.*
Plenum Press. New York. 1991

purified plasma membranes, which are completely devoid of chlorophyll per se, do contain a small but significant pool of protochlorophyllide and chlorophyllide. These two chlorophyll precursors could be reversibly and stoichiometrically transformed into each other by incubating the plasma membrane preparation with NADP(H) in strict darkness, apparently by virtue of some endogenous NADP(H): protochlorophyllide oxidoreductase which does not require light.

EXPERIMENTAL

Synechocystis 6714 (Pasteur Culture Collection of Cyanobacteria, Paris, France) was grown either photoautotrophically at 34°C in medium BG-11 supplemented with 1mM sodium carbonate and 10 mM sodium bicarbonate and gassed with 1.5% CO_2 in sterile air in axenic batch cultures (pH 7.9-8.7) illuminated with 10-15 w/m^2 warm white fluorescent light as measured with a YSI Radiometer, model 65, at the surface of the vessel (Wastyn et al., 1987) or chemoheterotrophically at 34°C in medium BG-11 supplemented with 40mM glucose and gassed with rapid stream of sterile air in vigorously agitated fermenter cultures in strict darkness. For certain experiments (cf. Table 1) "salt-adapted" Synechocystis 6714 (both photoautotrophic and chemoheterotrophic) was obtained by including 0.4-0.5 M NaCl in the respective growth media (cf. Wastyn et al., 1987). Cells were harvested at the desired stage of growth (cf. Fig. 1 and Table 1) by centrifugation at room temperature, washed twice with sterile medium BG-11 (carbonate and glucose omitted), and subjected to lysozyme and

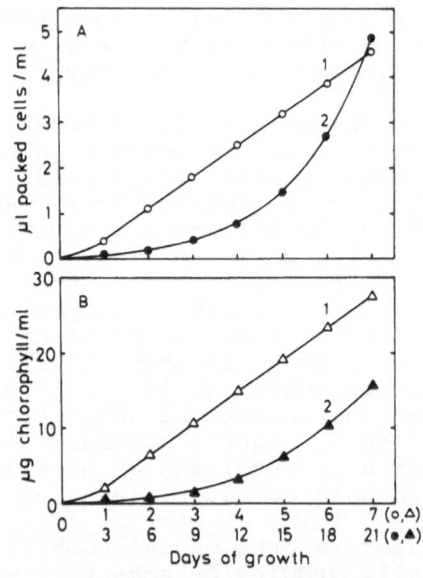

Fig. 1 Growth curves of photoautotrophically (1) and chemoheterotrophically (2) growing Synechocystis 6714 as monitored by the increase of packed cell mass (A) and chlorophyll content (B) per ml (cf. Table 1).

Table 1 Some properties of isolated and purified plasma (CM) and thylakoid (ICM) membranes from <u>Synechocystis</u> PCC 6714 grown photo-autotrophically or chemoheterotrophically (40 mM glucose) at 34°C in the presence (+) or absence (−) of 0.5M NaCl.

Growth conditions (Days of growth)	Cytochrome[a] oxidase		Chl/prot[b] (w/w)	Whole cell[c] chl content	Protochl.[d] chl.
	CM	ICM	ICM		CM
Photo (−)					
(4)	4.3	20.9	0.18	2.8	0.30
(7)	6.2	19.3	0.18	2.7	0.36
Photo (+)					
(7)	32	167.0	0.15	−	0.38
Hetero (−)					
(12)	15.5	114.4	0.11	1.3	−
(21)	16.8	113.3	0.08	0.9	0.45
Hetero (+)					
(21)	117.5	378.1	0.07	−	0.48

Protochl. = Protochlorophyllide
Chl. = Chlorophyllide

[a] oxidation of horse heart cytochrome c (nmol/min per mg protein) followed by dual wavelength spectrophotometry at room temperature; the reaction was fully inhibited by 2-5 μM KCN

[b] isolated and purified CM was virtually devoid of chlorphyll

[c] % methanol extractable chlorophyll <u>a</u> (w/w) per dry weight

[d] ratio of HPLC peak areas corresponding to protochlorophyllide (R_f = 13.06 min) and chlorophyllide (R_f = 13.64 min) in acetonic CM extracts.

French pressure cell treatment followed by isolation, separation and purification of plasma (CM) and thylakoid (ICM) membranes by discontinuous sucrose density gradient centrifugation as described (Murata and Omata, 1988, Peschek et al., 1988). Spectrally discernible, acetone or methanol extractable chlorophyll was practically absent from our purified CM preparations. Protein was measured according to Bradford (1976) and chlorophyll was measured according to Mackinney (1941).

CM and ICM from <u>Synechocystis</u> grown under different conditions and for different time spans were assayed for the oxidation of horse heart cytochrome <u>c</u> by dual wavelength spectrophotometry (Molitor and Peschek, 1986), and for pigment content by room and low temperature spectrophotometry and

spectrofluorimetry (both emission and excitation), combined
with differential solvent extraction, and by high performance
liquid chromatography (HPLC), as described (Peschek et al.,
1989). Isolated CM was incubated in the presence of 3.5 mM
NADP(H) in the dark as previously described (Peschek et al.,
1989), and the protochlorophyllide-chlorophyllide
transformation was followed by both spectrofluorimetry (Fig.
7) and HPLC (Fig. 8).

RESULTS

Fig. 1 shows the growth curves of Synechocystis 6714
under photoautotrophic (1) and chemoheterotrophic (2)
conditions as followed by the increase of cell mass (A) and
chlorophyll (B). Seemingly linear instead of logarithmic
photoautotrophic growth results from increasing light
limitation (self-shadowing of the cells) with progressive
growth at rather low light intensities; accordingly,
chemoheterotrophic growth did stay logarithmic for a much
longer time, viz, until exhaustion of the growth substrate
(glucose). Comparing cell mass (A) and chlorophyll (B) curves
it is evident that, despite a generally similar kinetic
appearance under phototrophic (1) and chemotrophic (2)
conditions, the latter (B/2) considerably lags behind the
former (A/2), in particular during the first ten days of
growth or so. This retarded chlorophyll synthesis, which
might be due, for example, to a shortage of reducing
equivalents needed for the protochlorophyllide-chlorophyllide
transformation, clearly results in a lower specific
chlorophyll content of chemotrophic cells compared to
phototrophic cells (Table 1). Consistent with this notion
would be the higher protochlorophyllide/chlorophyllide ratio
in the CM from such cells (Table 1) as judged from a
comparison of the corresponding relative HPLC peak areas (cf.
Figs. 5 and 9).

77K absorption spectra of CM and ICM are given in Fig. 2,
the CM spectrum being dominated by carotenoids and the ICM
spectrum by chlorophyll. The very weak absorption by CM
around 665-700 nm is likely to be due to chlorophyll
precursors but no chlorophyll per se (also cf. Fig. 3). 77K
fluorescence emission and excitation spectra of CM and ICM are
depicted in Fig. 3. Again it is seen that chlorophyll is
absent from CM while protochlorophyllide and chlorophyllide
(fluorescence emission at 636 and 682 nm, respectively; cf.
Fig. 4) appear to be absent from ICM. Results from solvent
partition experiments using CM differentially extracted with
(polar) acetone and (nonpolar) hexane are shown in Fig. 4.
Clearly, the 674-676 nm emission peak (corresponding to 682 nm
in aqueous membrane suspensions; Fig. 3) was distributed
between the polar and the nonpolar phase while the 628 nm
emission peak (corresponding to 636 nm in aqueous membrane
suspensions; Fig. 3) could not be extracted into hexane (Fig.
4C). Slightly different wavelengths of the peaks apparent in
Figs. 3 and 4 are attributed to the different physical
environment, viz. aqueous membrane suspensions (Fig. 3) and
solvent extracted pigments (Fig. 4). It is noted that spectra
nearly identical to those of Figs. 2-4 were previously
obtained with membranes isolated and purified from

(photoautotrophically grown) <u>Anacystis nidulans</u>
(Hinterstoisser et al., 1988; Peschek et al., 1989).

HPL chromatograms of acetonic or methanolic CM and ICM extracts together with those of authentic chlorophyll <u>a</u> and chlorophyllide <u>a</u> (Peschek et al., 1989) are given in Fig. 5. HPLC thus revealed chlorophyll <u>a</u> as the major ICM pigment, and chlorophyllide <u>a</u> as the major CM pigment (apart, of course, from carotenoids which, however, did not interfere in our more polar HPLC conditions). In particular, the markedly different, and opposite, ratios of peak areas 2 (chlorophyll <u>a</u>) and 3 (chlorophyllide <u>a</u>) in CM and ICM, respectively, indicate that the former does not contain any significant amount of chlorophyll <u>a</u> while the very small chlorophyllide <u>a</u> peak in the latter might originate from slight contamination with CM or as a minor degradation product of chlorophyll <u>a</u>. However, while pheophorbide (chlorophyll minus Mg and phytol; peak no. 5) and pheophytin (chlorophyll minus magnesium; peak

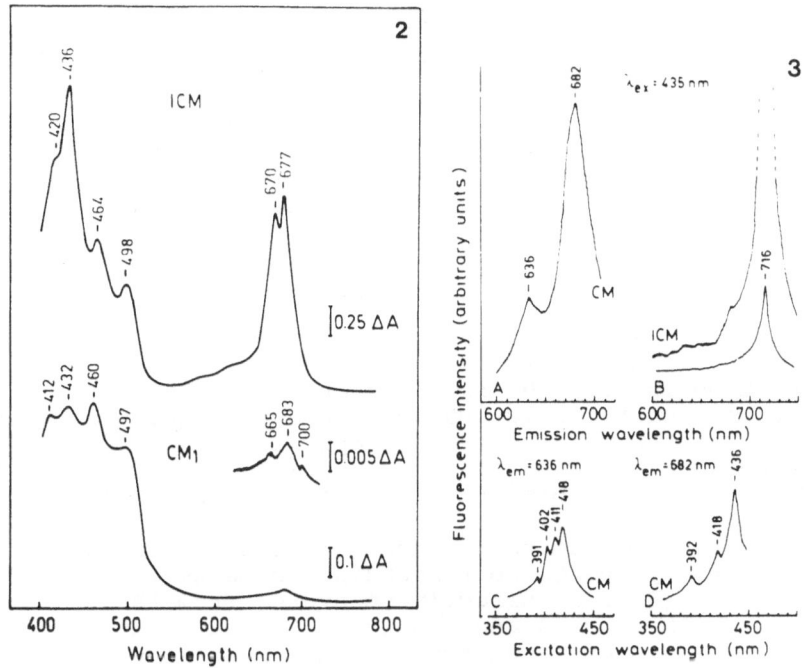

Fig. 2   Absorption spectra of aqueous suspensions of
         CM (44 μg of protein/ml) and ICM (116 μg of
         protein/ml) recorded at 77K. Automatic
         base-line correction was made with a
         Shimadzu-Sapcom computer.

Fig. 3   77K fluorescence emission (A,B) and
         excitation (C,D) spectra of aqueous
         suspensions of purified plasma (CM) and
         thylakoid (ICM) membranes. Protein
         concentration was 2.3 mg/ml (except for the
         inset of Fig. 3B, corresponding to 0.23
         mg/ml).

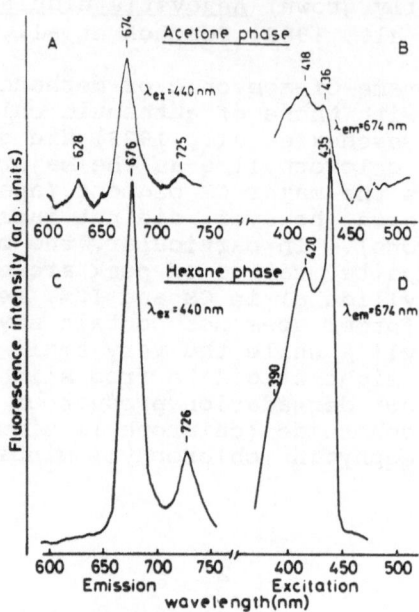

Fig. 4 77K fluorescence emission (A,C) and
excitation (B,D) spectra of the n-hexane-
extracted acetone phase (A,B) and the n-
hexane phase (C,D) of CM pigments
partitioned between acetone and hexane (for
details cf. Peschek et al., 1989; Pineau
et al., 1986)

no. 4) were found as marginal nonenzymatic degradation
products also in pure chlorphyll a and chlorphyllide a
preparations, characteristically no pheophorbide was detected
in CM nor was the polar protochlorophyllide found in ICM.
This argues against a major nonenzymatic contribution to the
strikingly different pigment ratios in CM and ICM (Fig. 5).

Fig. 6 shows room temperature fluorescene emission
spectra of the pigments corresponding to peaks 1 and 2 on HPL
chromatograms (Fig. 5) directly recovered from the column;
comparison with Figs. 3 and 4 appears to leave little doubt
that the two main CM fluorescence peaks must indeed be
assigned to protochlorophyllide and chlorphyllide (620-630
and 668-682 nm emission, respectively, depending on the
physical environment of the pigments).

Incubation of CM preparations of chemoheterotrophically
grown Synechocystis in the presence of NADPH produced an
increase of the 77K fluorescence emission peak at 682 nm and a
concomitant decrease of the peak at 636 nm (Fig. 7, curve 2)
$NADP^+$ leading to the opposite change (Fig. 7, curve 3); the
same transition could be followed by HPLC quantitating the
fractions eluting at 13.64 and 13.06 min retention time (Fig.
8; cf. Fig. 5). Based on HPLC and fluorescence
characteristics of these fractions (Fig. 9), and on their
relative peak areas, Fig. 10 gives a time course for the
transformation which most probably reflects the stoichiometric

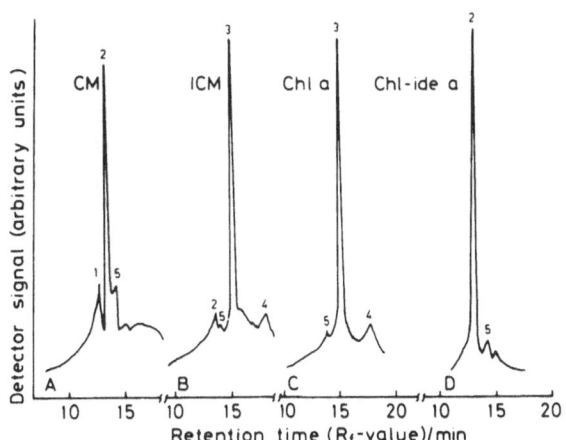

Fig. 5 Separation of CM (A) and ICM (B) pigments,
extracted with acetone or methanol, and
authentic chlorophyll a (C) and chlorophyllide a (D) standards dissolved in acetone
or methanol, by reversed phase HPLC (for
details cf. Peschek et al., 1989). Peak
numbers (retention time in min) correspond
to the following compounds (cf. Falkowski
and Sucher, 1981): 1=protochlorophyllide
(13.06), 2=chlorophyllide (13.64), 5=pheo-
phorbide (14.07), 3=chlorphyll (15.5) and
4=pheophytin (18.1), in the order of
decreasing polarity on elution with a
gradient of 70-98% (v/v) aqueous methanol.

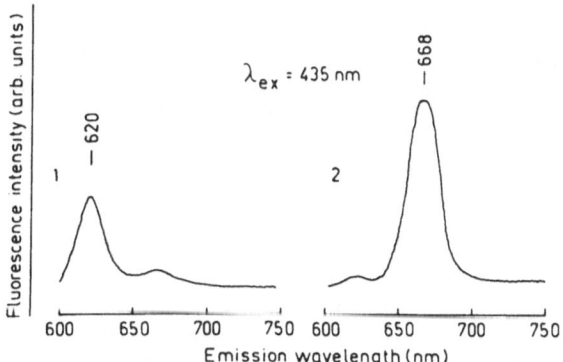

Fig. 6 Room temperature fluorescence emission
spectra of the pigments corresponding to
peaks 1 and 2 of Fig. 5 individually
recovered from the HPLC column.

conversion of protochlorophyllide into chlorphyllide (in the
presence of NADPH) and vice versa (in the presence of NADP$^+$)
in the plasma membrane of chemoheterotrophic Synechocystis,
very similar to what was recently shown with CM preparations
from the obligately photoautotrophic Anacystis nidulans
(Peschek et al., 1989).

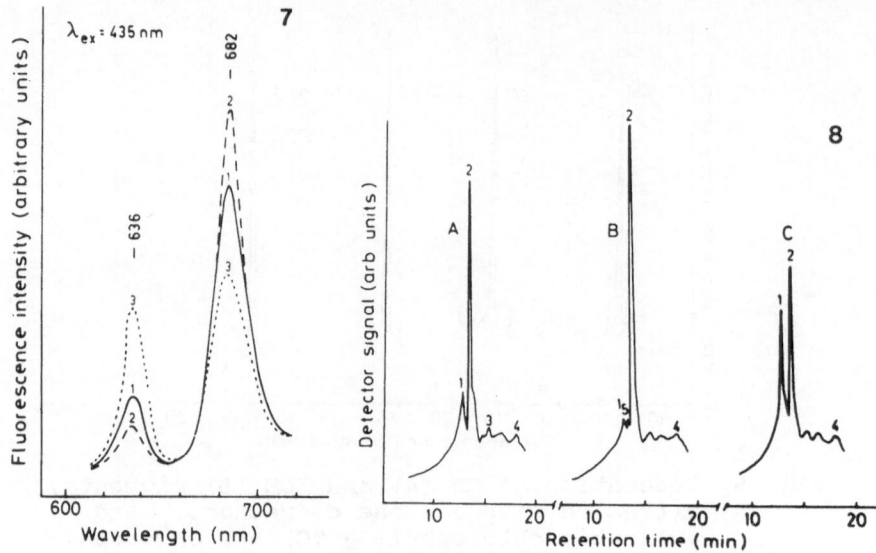

Fig. 7   77K fluorescence emission spectra of CM preparations (3.1 mg protein/ml) either freshly prepared (1) or after 10 min incubation with 3.5 mM NADPH (2) or $NADP^+$ (3) at 30°C in complete darkness. No changes in peak heights at 636 and 682 nm were seen when the membranes had been heated at 90°C for 5 min prior to the addition of pyridine nucleotides (controls not shown). Illumination with tungsten light (5-20 w/m²) was without influence on the reaction.

Fig. 8   Reversed phase high performance liquid chromatograms of acetonic extracts of CM freshly prepared (A), after incubation with NADPH (B), and after incubation with $NADP^+$ (C). Peak numbers are specified in Fig. 5. Controls were performed as in Fig. 7 (not shown).

DISCUSSION

Spectrofluorimetric investigation of purified envelope membranes from spinach chloroplasts recently gave evidence for the occurrence of NADPH-protochlorophyllide reductase in these membranes (Pineau et al., 1986). The conversion of protochlorophyllide to chlorophyllide in the chloroplast envelope membranes was light dependent. Similarly, we detected a small pool of protochlorophyllide and chlorophyllide in purified and likewise chlorophyll-free plasma membranes from the cyanobacteria <u>Anacystis nidulans</u> (Hinterstoisser et al., 1988; Peschek et al., 1989) and in (both photoautotrophically and) chemoheterotrophically grown <u>Synechocystis</u> 6714 as described in the present paper. In addition to spectrofluorimetry we used high performance liquid chromatography for identification of the two chlorophyll

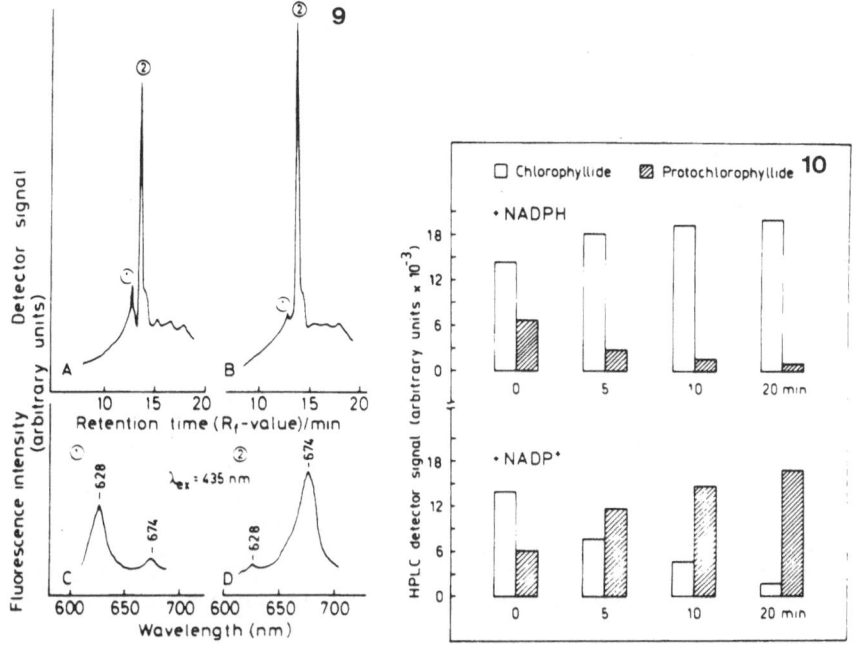

Fig. 9 HPL chromatograms of acetonic or methanolic
extracts of CM preparations before (A) and
after (B) 10 min incubation with 3.5 mM NADPH
at 30°C in strict darkness. (C) and (D)
represent 77K fluorescence emission spectra
of the pigments corresponding to peaks 1 and
2, respectively, as collected directly from
the HPLC column. Controls were performed as
in Figs. 7 and 8 (not shown). For the assignment of peaks cf. Figs. 5 and 6 (note, however,
the different assay conditions). Also cf.
Fig. 7.

Fig. 10 Time course of the protochlorophyllide-
chlorophyllide transformation in plasma
membranes from chemoheterotrophic <u>Synecho-
cystis</u> 6714. The isolated and purified
membranes were incubated with 3.5 mM NADPH
or NADP$^+$ at 30°C in strict darkness (cf.
Fig. 7). The ordinate gives relative peak
areas of peaks no. 1 and 2 (Fig. 9) as
displayed by the Chromotopac Integrator of
the HPLC system (Peschek et al., 1989).
Controls were the same as in Figs. 7-9.

precursors. Most interestingly, we found that in the
cyanobacterial plasma membranes, in contrast to the
chloroplast envelope membranes, the NADPH-dependent conversion
of protochlorophyllide to chlorophyllide did not require
light; moreover we found that NADP$^+$ was able to bring about
the reverse transition, as might be expected from a truely

enzymatic reaction. (Similar to the chloroplast envelope
membrane preparations, NAD(H) was much less effective than
NADP(H) also with cyanobacterial CM, giving only 20%
efficiency (not shown)). Since all chemoheterotrophically
growing cyanobacteria are known to synthesize appreciable
amounts of chlorophyll even in strict darkness our findings
might contribute to explaining this rather strange fact by
identifying a usually light-dependent step among the
complicated chain of biosynthetic reactions leading to
chlorophyll a in higher plants and algae, as a light-
independent step in cyanobacteria, similar to green algae.

Another problem, however, has still to be resolved: It
is not clear so far why the more or less immediate chlorophyll
precursors protochlorophyllide and chlorophyllide (Pineau et
al., 1986; Peschek et al., 1989) are located in the envelope
(chloroplasts) or plasma (cyanobacteria) membranes while the
final product, chlorophyll a, is entirely confined to the
thylakoid membranes. It is not known in which type of
membrane the very last step(s) of chlorophyll biosynthesis
occur and how the necessary intracellular communication
between the membranes is achieved. It would be easier to
explain the situation topographically if there were some

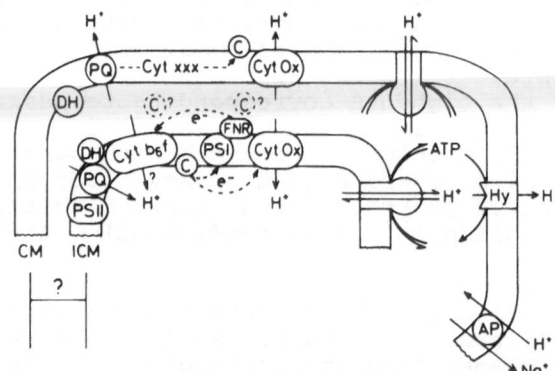

Fig. 11 Polarity and sidedness of proton transloc-
        ation, electron transfer, and ATP synthesis/
        hydrolysis in the energy-transducing
        membranes of a cyanbacterium. CM, plasma
        membrane; ICM, thylakoid membrane; DH,
        dehydrogenase(s); PSI, PSII, Photosystem I
        and II; PQ, plastoquinone; CytOx, cytochrome
        oxidase; c, 'soluble' c-type cytochromes of
        intrathylakoid, periplasmic, and/or
        cytosolic (?) location; Cyt xxx, hitherto
        uncharacterized cytochromes in the plasma
        membrane; FNR, ferredoxin:NADP-oxidoreduct-
        ase; Hy, unidirectional ATP hydrolase (?);
        AP, proton-sodium antiporter. Broken lines
        indicate the path of electrons (e⁻). The
        question mark reminds of a possible physical
        continuity of CM and ICM in view of the
        functionally opposite polarity of the two
        membranes in vivo.

continuity between the "outer" and the "inner" membranes of the chloroplast or cyanobacterium; the lipophilic precursors and/or products could then move laterally within the membrane from, e.g. the periphery to the interior of the cell (organelle). However, while in chloroplasts such continuity is assumed to exist at least at an early ('proplastidic') stage of development (Kirk and Tilney-Bassett, 1978) such continuity has never been (electronmicroscopically) observable in cyanobacteria at any stage of development (Nierzeicki-Bauer et al., 1983). On the other hand, a functional comparison of CM and ICM within a cyanobacterium (Fig. 11) would suggest that some continuity must exist: The polarities of proton translocation, ATP synthesis and electron transfer are opposite to each other in CM and ICM, respectively, which is most easily explained on the grounds of a (developmental) continuity between both membranes (Fig. 11; also cf. Peschek, 1984, 1987).

ACKNOWLEDGEMENTS

This work was supported by grants from the Austrian Science Foundation. Skilful technical assistance by Miss Irene Steininger and Mr. Otto Kuntner is gratefully acknowledged. We thank Dr. B. Pineau, CNRS Gif-sur-Yvette, France, for performing low temperature fluorescence spectra.

REFERENCES

Avissar, Y.J., 1980, Biosynthesis of 5-aminolevulinate from glutamate in Anabaena variabilis, Biochim. Biophys. Acta., 613: 220-228.

Bogorad, L., 1966, The biosynthesis of protochlorophyll, in: "The Chlorophylls", L. P. Vernon and G. R. Seely, eds., Academic Press, New York and London.

Bradford, M.M., 1976, A rapid and sensitive method for the quantitation of microgram quantities of protein utilizing the principle of protein-dye binding, Anal. Biochem., 72 248-254.

Castelfranco, P.A., and Beale, S.I., 1983, Chlorophyll biosynthesis: Recent advances and areas of current interest, Annu. Rev. Plant Physiol., 34: 241-275.

Falkowski, P.G., and Sucher, J., 1981, Rapid quantitative separation of chlorophylls and their degradation products by high performance liquid chromatography, J. Chromatogr., 213: 349-351.

Fuesler, T.P., Wong, Y.S. and Castelfranco, P.A., 1984, Localization of Mg chelatase and Mg-protoporphyrin IX monomethylester (oxidative) cyclase activity within isolated developing cucumber chloroplasts, Plant Physiol., 75: 662-664.

Hinterstoisser, B., Missbichler, A., Pineau, B. and Peschek, G.A., 1988, Detection of chlorophyllide in chlorophyll-free plasma membrane preparations from Anacystis nidulans, Biochem. Biophys. Res. Commun., 154: 839-846.

Jacobs, N.J., 1974, Biosynthesis of heme, in: Microbial Iron Metabolism: A Comprehensive Treatise, J.B. Neilands, ed., Academic Press, New York and London.

Johanningmeier, U. and Howell, S.H., 1984, Regulation of light-harvesting chlorophyllbinding protein mRMA

accumulation in <u>Chlamydomonas reinhardi</u>. Possible
involvement of chlorophyll synthesis precursors, <u>J.
Biol. Chem.</u>, 259: 13541-13549.

Kipe-Nolt, J.A. and Stevens, S.E., 1980, Biosynthesis of 5-
aminolevulinic acid from glutamate in <u>Agmenellum
quadruplicatum</u>, <u>Plant Physiol.</u>, 65: 126-128.

Kirk, J.T.O. and Tilney-Bassett, R.A.E., 1978, <u>The Plastids</u>,
2nd ed., Elsevier/North Holland Biomedical Press,
Amsterdam.

Mackinney, G., 1941, Absorption of light by chlorophyll
solutions, <u>J. Biol.Chem.</u>, 140: 315-322.

Molitor, V. and Peschek, G.A., 1986, Respiratory electron
transport in plasma and thylakoid membrane preparations
from the cyanobacterium <u>Anacystis nidulans</u>, <u>FEBS Lett.</u>,
195: 145-150.

Nierzwicki-Bauer, D.A., Balkwill, D.L., and Stevens, S.E.,
1983, Three-dimensional ultrastructure of a unicellular
cyanobacterium, <u>J. Cell Biol.</u>, 97: 713-722.

Padan, E., 1979, Facultative anoxygenic photosynthesis in
cyanobacteria, <u>Annu. Rev. Plant Physiol.</u>, 30: 27-40.

Peschek, G.A., 1984, Structure and function of respiratory
membranes in cyanobacteria (blue-green algae), <u>in</u>:
Subcell. Biochem., 10, D.B. Roodyn, ed., Plenum Press,
New York and London.

Peschek, G.A., 1987, Respiratory electron transport, <u>in</u>: "The
Cyanobacteria", P. Fay and C. Van Baalen, eds.,
Elsevier/North Holland Biomedical Press, Amsterdam.

Peschek, G.A., Molitor, V., Trnka, M., Wastyn, M., and Erber,
W., 1988, Characterization of cytochrome-c oxidase in
isolted and purifield plasma and thylakoid membranes from
cyanobacteria, <u>Methods Enzymol.</u> 167: 437-449.

Peschek, G.A., Hinterstoisser, B., Wastyn, M., Kuntner, O.,
Pineau, B., Missbichler, A. and Lang, J., 1989,
Chlorophyll precursors in the plasma membrane of a
cyanobacterium, <u>Anacystis nidulans</u>: Characterization of
protochlorophyllide and chlorophyllide by
spectrophotometry, spectrofluorimetry, solvent partition,
and high performance liquid chromatography, <u>J. Biol.
Chem.</u>, 264:11827-11832.

Pineau, B., Dubertret, G., Joyard, J. and Douce, R., 1986,
Fluorescence properties of envelope membranes from
spinach chloroplasts. Detection of protochlorophyllide,
<u>J. Biol. Chem.</u> 261: 9210-9215.

Shemin, D. and Russel, C.S., 1953, 5-Aminolevulinic acid, its
role in the biosynthesis of porphyrins and purines, <u>J.
Am. Chem. Soc.</u>, 75: 4873-4874.

Wastyn, M., Achatz, A., Trnka, M. and Peschek, G.A., 1987,
Immunological and spectral characterization of partly
purified cytochrome oxidase from the cyanobacterium
Synechocystis 6714, <u>Biochem. Biophys. Res. Commun.</u>, 149:
102-111.

NEW ASPECTS OF BIOSYNTHESIS OF CHLOROPHYLLS FROM PROTOCHLOROPHYLLIDES IN SCENEDESMUS

Horst Senger and Kiriakos Kotzabasis

Fachbereich Biologie/Botanik der Philipps Universität Marburg, Lahnberge, D-3550 Marburg F.R.G.

INTRODUCTION

Knowledge about the last steps of chlorophyll (Chl) biosynthesis derives mostly from work with greening angiosperm seedlings. Chlorphyll biosynthesis in these plants can be easily controlled since angiosperms need light for the photoreduction of protochlorophyllide (PChlide) to chlorophyllide (Chlide). In contrast, green algae are able to perform the complete Chl biosynthesis in darkness, if an organic carbon source is provided for heterotrophic growth. Only mutants, in which Chl biosynthesis becomes light-dependent, are suitable to study Chl biosynthesis in detail. For such studies we applied pigment mutant C-2A' of Scenedesmus obliquus, induced by X-ray (Bishop 1971). This mutant synthesizes Chl only upon transfer to light (Senger and Bishop, 1972).

CHLOROPHYLL PRECURSORS IN DARKNESS

When cells of Scenedesmus obliquus mutant C-2A' are transferred into fresh culture medium, they accumulate precursors of Chl (Oh-hama and Hase, 1980). Further analysis revealed the presence of 3 different PChlides (Kotzabasis and Senger, 1986) and a small amount of protochlorophyll (PChl) (Kotzabasis et al., 1989b). The pigments were characterized by their chromatographic behaviour, absorption spectra and fluorescence properties. The 3 PChlides were identified as monovinyl (MV), divinyl (DV) and an unknown PChlide of high polarity.

The kinetics for PChl and Chlide formation are different. The addition of 5-aminolevulinate (ALA) stimulates their biosynthesis but influences their kinetics in a different way (Kotzabasis and Senger, 1990). Biosynthesis of PChl is faster and saturates at lower ALA concentrations than the PChlides. When the dark-grown cells of mutant C-2A' reach the stationary growth phase, the precursers of Chl are degraded without being transformed into Chl (Senger and Brinkmann, 1986).

REDUCTION OF PROTOCHLOROPHYLLIDES AND PROTOCHLOROPHYLL

Upon transfer of dark-grown cells of Scenedesmus mutant C-2A' into light, the PChlides were immediately photoreduced to Chlide (Senger and Brinkmann, 1986; Kotzabasis et al., 1989b). Only one Chlide, namely Chlide a could be recovered after photoreduction. About the fate of the different PChlides no statement can be made at the moment. In addition to the light-dependent conversion of PChlides to Chlide a it was found that lowering of the temperature from 30° to 20°C causes a reduction of PChlides (Oh-hama et al., 1987; Kotzabasis et al., 1990b).

Not only PChlides but also PChl is photoreduced (Kotzabasis et al., 1989b). This is in contrast to the situation in angiosperms, in which PChl is not photoreducible according to the literature (Griffiths, 1980; Shioti and Sasa, 1983). The photoreduction of PChl to Chl in Scenedesmus obliquus mutant C-2A' follows the same fast kinetics as the photoreduction of PChlide to Chlide (Kotzabasis et al., 1989b).

Since the fast kinetics of photosynthetic reaction center formation in light rather parallels the fast kinetics of PChl to Chl reduction than the slower kinetics of phytylation of the Chlide, we assume that the Chl that is derived by photreduction of PChl is preferentially incorporated into the reaction centers of the photosynthetic apparatus (Kotzabasis et al., 1989b; Kotzabasis et al., 1990c). By lowering the growth temperature for mutant C-2A' of Scenedesmus from 30° to 20°C PChlide can be reduced independently from light (Oh-hama et al., 1987). At 20°C the reduction reaches about 13% of light-dependent chlorophyll biosynthesis. The Chls synthesized at the lower temperatures are assembled into the pigment-protein complexes, and concomitantly the capacities for photosystems I and II are developed. The light-independent PChlide reduction at lower temperatures is not limited by the enzyme photochlorophyllide oxidoreductase, but rather by the availability of ALA (Kotzabasis et al., 1990b).

POROTOCHLOROPHYLLIDE-NADPH-OXIDOREDUCTASE (PCR)

PCR occurs in dark-grown mutant C-2A' in high amounts of low activity and the amount decreases whereas the specific activity increases in the light (Kotzabasis, 1985). Calcium was shown to have an effect on the amount of PChlide formed and the activity of PCR (Kotzabasis, et al., 1990a). Depletion of Ca^{2+} caused an increase in PChlide, but a decrease in PCR activity, whereas addition of Ca^{2+} had the opposite effect. A stimulation of PCR by Ca^{2+} was also reported for Anacystis nidulans (Peschek et al., 1989).

Thioredoxin, which is known to control enzyme activities also influences the reduction of PChlides in pigment mutant C-2A' (Kotzabasis et al., 1989c). Reduced plastidal thioredoxin f, which is formed in this mutant during greening prior to Chl synthesis, stimulates PCR activity _in vitro_ in light and darkness. The cytoplasmic thioredoxins and _E.coli_ thioredoxin are less active. This system is considered to be a new case of thioredoxin-mediated control of a chloroplastic enzyme, and a new factor in the regulation of chlorphyll biosynthesis.

Intermediates of Chlides bound to PCR are described as "active" PChlides with specific absorption spectra (Oliver and Griffiths, 1982; Griffiths et al., 1984). Aggregates of monovinyl-PClide and divinyl-PChlide in toluene show absorption maxima and fluorescence characteristics similar to those "active" PChlides (Kotzabasis et al., 1990d). It is speculated that the reduction of PChlide to Chlide on the PCR includes aggregated forms of monovinyl and divinyl-PChlides.

BIOSYNTHESIS OF CHLOROPHYLL _b_

The pathway leading to Chl _b_ is still a matter of controversy. It is generally assumed that Chl _b_ derives directly from Chl _a_. We were able to extract for the first time Chlide _b_ from the wild type of _Scenedesmus obliquus_ and its pigment mutant C-2A'. Its identity was proven by absorption and fluorescence spectroscopy and by a positive hydroxylamine test. Chlide _b_ could be transformed into pheophorbide _b_ and methylpheophorbide _b_. The formation of Chlide _b_ by dephytylation with chlorophyllase could be ruled out (Kotzabasis and Senger, 1989a).

The branching point for Chlide _b_ synthesis must be at the level of Chlide, since no PChlide _b_ was detectable. In addition to the formation from Chlide _b_ Chl _b_ can be directly formed from Chl _a_. We have indications that the enzyme oxidizing Chl _a_ to Chl _b_ also accepts non-phytylated 17,18-dihydroporphyrins and is not restricted to chlorphylls. Preparations of Chlide _a_ and Chl _a_ could both be transferred with the same enzyme fraction to Chlide _b_ and Chl _b_ respectively. Preliminary experiments show this enzyme to be membrane-bound and light-independent (Kotzabasis and Senger, 1989b).

REGULATION OF CHLOROPHYLL BIOSYNTHESIS

The fact that PChlides are only synthesized in small amounts in darkness and that no overproduction of Chls in the light occurs demands separate regulating mechanisms. It was demonstrated that accumulated PChlides act directly with a negative feed-back on the ^{glu}t-RNA-ligase (one of the first steps in ALA biosynthesis, (Kannangara et al., 1988)) blocking further synthsis of ALA and subsequently PChlide (Dornemann et al., 1989). By increasing the amount of ligase at constant concentrations of PChlide and ^{glu}t-RNA it could clearly be demonstrated that PChlide directly inhibits the ligase activity and does not act on the t-RNA. The inhibitory effect of other tetrapyrroles like Chl _a_, pheophytin _a_ and protoporphyrin IX was much less effective even at oversaturating concentrations.

A different regulatory mechanism is possible in light: The accumulation of PChlide is almost completely blocked in darkness by the addition of 4,5-dioxovalerate, one of the early intermediates of ALA biosynthesis <u>in vivo</u>. Likewise, light-dependent chlorphyll biosynthesis is strongly inhibited by the addition of this compound during greening. The considerable increase of PChlide formation in darkness upon addition of ALA is also drastically reduced by external 4,5-dioxovalerate. It is shown by <u>in vitro</u> experiments that concentrations of dioxovalerate, above the physiological relevant level, inhibit ALA dehydratase. The K_i-value was determined to be 60 ± 5 μM. From these results it might be considered that besides the predominant control of ^{glu}t-RNA-ligase by PChlide a second regulatory mechanism could be involved in Chl biosynthesis (Kotzabasis et al., 1989b).

In addition to these feed-back mechanisms direct regulatory effects of Ca^{2+} and thioredoxins on the PCR have been described above.

CONCLUDING REMARKS

The results reported here are mainly obtained with a special pigment mutant. We do not claim that all these intermediates and reactions occur in the Chl biosynthesis of wild-type cells as well. But we would like to emphasize that all the reported reactions are possible in nature and that attention should be drawn to them in future research on Chl biosynthesis.

REFERENCES

Bishop, N.I., 1971, Preparation and properties of mutants : <u>Scenedesmus</u>, <u>in</u>: Methods in Enzymology, Vol. 23, Part A, A. San Pietro, Ed., Academic Press, London.

Dörnemann, D., Kotzabasis, K., Richter, P., Breu, V. and Senger, H., 1989, The regulation of chlorophyll biosynthesis by the action of protochlorophyllide on ^{glu}t-RNA-ligase, <u>Bot. Acta.</u>, 102: 112-115.

Griffiths, W.T., 1980, Substrate-specificity studies on protochlorophyllide reductase in barley (<u>Hordeum vulgare</u>) etioplast membranes, <u>Biochem. J.</u>, 186: 267-278.

Griffiths, W.T., Oliver, R.P., and Kay, S.A., 1984, A critical appraisal of the role and regulation of NADPH-protochlorophyllide oxidoreductase in greening plants, <u>in</u>: protochlorophyllide Reduction and Greening. C. Sironval and M. Brouers, Martinus Nijhoff/DR W. Junk Publishers, The Hague, Boston, Lancaster.

Kannangara, C.G., Gough, S.P., Bruyant, P., Hoober, J.K., Kahn, A. and von Wettstein, D., 1988, tRNA^{glu} as a cofactor in δ-aminolevulinate biosynthesis: steps that regulate chlorophyll synthesis, <u>Trends. Biochem. Sci.</u>, 13: 139-143.

Kotzabasis, K., 1985, Zur Isolierung, Wirkungsweise und Regulation der Protochlorophyllid-Oxidoreduktase, Diplomarbeit. Fachbereich Biologie der Philipps-Universität, Marburg, F.R.G.

Kotzabasis, K. and Senger, H., 1986, Isolation and characterization of 3 protochlorophyllides from pigment mutant c-2A' of <u>Scenedesmus obliquus</u>, <u>Z. Naturforsch.</u> 41c: 1001-1003.

Kotzabasis, K. and Senger, H., 1989a, Evidence for the presence of chlorophyllide b in the green alga Scenedesmus obliquus, Bot. Acta., 102: 173-178.

Kotzabasis, K. and Senger, H., 1989b, Biosynthesis of chlorphyll b in pigment mutant C-2A' of Scenedesmus obliquus, Physiol. Plant., 76: 474-478.

Kotzabasis, K. and Senger, H., 1990, The influence of 5-aminolevulinic acid on the protochlorphyllide and protochlorophyll accumulation in dark grown Scenedesmus, Z. Naturforsch., 45c: 71-73.

Kotzabasis, K., Breu, V. and Dörnemann, D., 1989a, The inhibitoy effect of 4,5-dioxovalerate on 5-aminolevulinate-dehydratase and its implication in the regulation of light-dependent chlorphyll formation and in pigment mutant C-2A' of Scenedesmus obliquus. Biochim. Biophys. Acta., 977: 309-314.

Kotzabasis, K., Schüring, M.-P. and Senger, H., 1989b. Occurrence of protochlorophyll and its phototransformation to chlorophyll in mutant C-2A' of Scenedesmus obliquus, Physiol. Plant., 75: 221-226.

Kotzabasis, K., Senger, H., Langlotz P.and Follmann, H., 1989c Stimulation of protochlorophyllide oxidoreductase by thioredoxin, J. Photochem. Photobiol. B., 3: 333-339.

Kotzabasis, K., Miyachi, S. and Senger, H., 1990a, Influence of calcium on protochlorophyllide reduction in the pigment mutant C-2A' of Scenedesmus obliquus, Plant Cell Physiol., 31:419-422.

Kotzabasis, K., Römer, S. and Senger, H. 1990b, The light-independent protochlorophyllide reduction and the assembly of the chlorphylls to active photosystems in pigment mutant c-2A' of Scenedesmus obliquus, Physiol, Plant., 92:622-629.

Kotzabasis, K., Humbeck, K. and Senger H., 1991, Incorporation of photoreduced protochlorphyll into reaction centers. J. Photochem. Photobiol. b., Vol. 8(3).

Kotzabasis, K., Senge, M., Seyfried, B. and Senger, H. 1990d, Aggregation of monovinyl- and divinyl-protochlorophyllide in organic solvents, Photochem. Photobiol., 52:95-101.

Oh-hama, T. and Hase, T., 1980, Formation of protochlorophyll(ide) in wild type and mutant C-2A' cells of Scenedesmus obliquus, Plant Cell Physiol., 21: 1263-1272.

Oh-hama, T., Kotzabasis, K. and Senger, H., 1987, Temperature inducible protochloraphyllide reduction in darkness in a pigment mutant of Scenedesmus obliquus, Physiol. Plant, 69: 29-34.

Oliver, R. and Griffiths, W.T., 1982, Pigment-protein complexes of illuminated etiolated leaves Plant Physiol., 70: 1019-1025.

Peschek, G.A., Hinterstoisser, B., Pineau, B. and Missbichler, A., 1989, Light-independent NADPH-protochlorophyllide oxidoreductase activity in purified plasma membrane from the cyanobacterium Anacystis nidulans, Biochim. Biophys. Res. Comm., 162: 71-78.

Senger, H. and Bishop, N.I., 1972, The development of structure and function in chloroplasts of greening mutants of Scenedesmus. I. Formation of chlorophyll, Plant Cell Physiol., 13: 633-649.

Senger, H. and Brinkmann, G., 1986, Protochlorophyll(ide) accumulation and degradation in the dark and

photoconversion to chlorphyll in the light in pigment
mutant C-2A' of <u>Scenedesmus obliquus</u>, <u>Physiol. Plant.</u>,
68: 119-124.
Shioi, Y. and Sasa, T., 1984, Chlorophyll formation in the YG-
6 mutant of <u>Chlorella regularis</u>, Spectral
characterisation of protochlorophyllide
phototransformation, <u>Plant Cell Physiol.</u>, 25: 139-149.

ULTRAWEAK LUMINESCENCE FROM PLANT TISSUE: SPECTRAL
CHARACTERISTICS AND EFFECTS OF ULTRAVIOLET RADIATION,
ANAEROBIOSIS AND AGEING

L.O. Björn, I. Panagopoulos and G.S. Björn

Department of Plant Physiology, University of Lund
Box 7007, S-220 07 Lund, Sweden

INTRODUCTION

Besides the well-known red luminescence from the
photosynthetic system ("afterglow" or "delayed fluorescence",
Strehler & Arnold 1951), plant tissue (even non-photosynthetic
tissue) exhibits other forms of luminescence (Colli and
Facchini 1954; Colli et al., 1955). This has been studied by
many groups (for reviews see Abeles 1986, 1987, Campbell 1988,
Slawinska and Slawinski, 1983), but we hope that we will be
able to add some new information. We have used a cooled
photomultiplier in the photon counting mode, and Schott cut
off filters for crude spectral characterization of the
luminescence. Counts were averaged over 2 minute periods.

RAPIDLY DECAYING LUMINESCENCE FOLLOWING ULTRAVIOLET
IRRADIATION

After irradiation of non-chlorophyllous plant tissue by
ultraviolet-C radiation, a fairly strong but rapidly decaying
glow can be observed. Most of the photons have wavelengths
between 400 and 600 nm. This may be simple phosphorescence
and was not investigated further.

OXYGEN DEPENDENT LUMINESCENCE FROM CARROT ROOT

In a series of experiments using either sliced or
shredded carrot root tissue, which as judged from the colour,
was free of chlorplasts, a luminescence was observed which
rapidly decreased to a lower steady level when oxygen was
removed by flushing the cuvette with nitrogen (Figures 1 and
2). When ordinary air was readmitted, the intensity rose very
rapidly and showed an overshoot before reaching the high
steady level typical for aerobic conditions. The steady level
was higher when the air was flowing rapidly through the
cuvette than in stationary or slowly moving air (Figure 1).

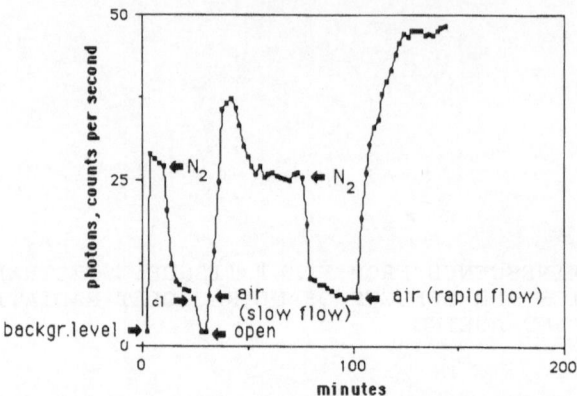

Fig. 1   Ultraweak luminescence from a slice of
carrot root. The experiment was started
under aerobic conditions. The gas phase
was changed as indicated. cl indicates
shutter closed for checking background
level.

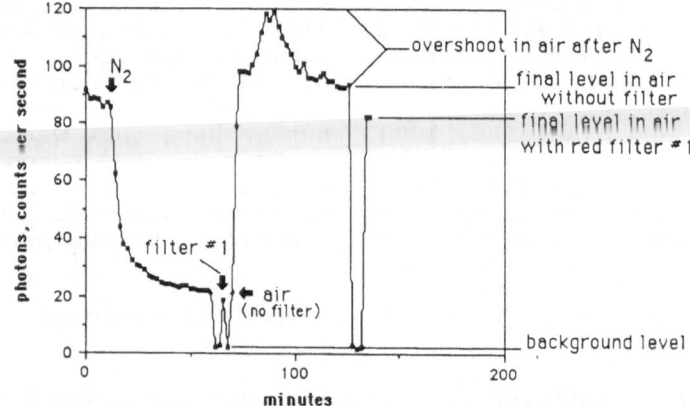

Fig. 2   Ultraweak luminescence from shredded carrot
root. A filter cutting off below 612 nm
was introduced and removed as indicated,
otherwise as Figure 1.

Using the computer program Stella^{TM} (High Performance Systems,
13 Dartmouth College Hwy, Lyme, N.H., 03768 USA) we could
construct a model (Figure 3) showing qualitively the same
behaviour (Figure 4). In this model the substrates for the
luminescence reaction are oxygen and an intermediate formed
from a stored material by an oxygen independent reaction.

Most of this luminescence consists of photons between 600
and 730 nm wavelength. With our crude resolution we do not
see much spectral difference between the aerobic and the
anaerobic luminescence (Figure 5).

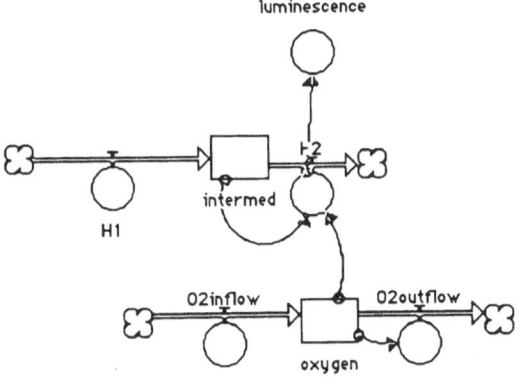

luminescence

intermed

H1

O2inflow   O2outflow

oxygen

☐ intermed = intermed + dt * (-H2 + H1)
   INIT(intermed) = 0
☐ oxygen = oxygen + dt * (O2inflow - O2outflow)
   INIT(oxygen) = 0
○ H1 = 30
○ H2 = 5*intermed*oxygen
○ luminescence = .70*(2*H2+20)
○ O2inflow=step(0.21,3)
○ O2outflow = oxygen

Fig. 3   Block diagram and equations for
         luminescence simulation generated by the
         computer program Stella^{TM} run on a
         Macintosh^{TM} computer. HI is the rate of
         the oxygen-independent reaction, and H2
         that of the luminescent, oxygen dependent
         reaction.

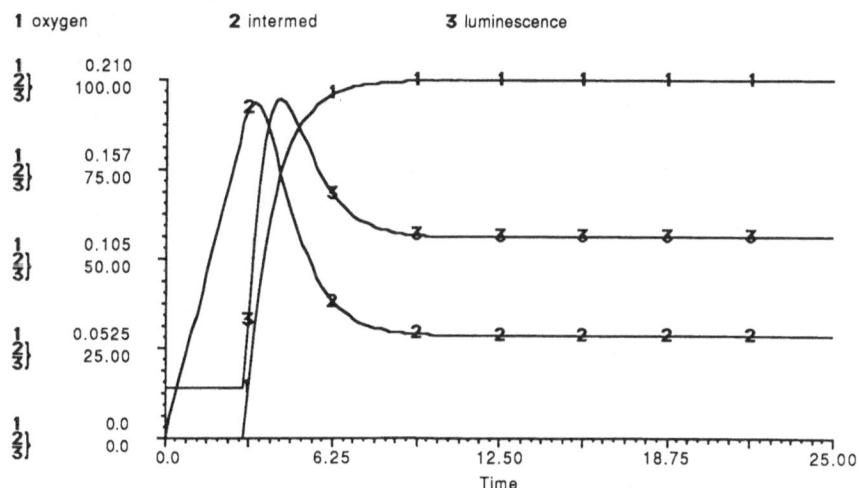

Fig. 4   Simulation of the time courses of oxygen
         concentration in the sample (curve 1,
         ordinate 0 to 0.21), intermediate
         concentration (curve 2, ordinate 0 to 100)
         and luminescence intensity (curve 3,
         ordinate 0 to 100) using the model in
         Figure 3.

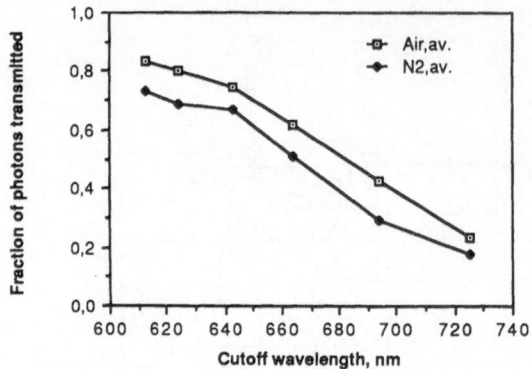

Fig. 5   The spectral composition of ultraweak
         luminescence from carrot root under aerobic
         and anaerobic conditions, recorded using
         cut off filters. The ordinate is the
         fraction of photons transmitted by filters
         cutting out light below the wavelengths
         shown on the abscissa.

LUMINESCENCE FROM LEAVES DURING NATURAL SENESCENCE

     For this series of experiments, beech leaves from a tree
growing outdoors in Lund were used. The leaves were picked at
various times in September and October preceding natural leaf-
fall. Before measurement of luminescence the leaves were kept
in darkness for 24 hours to allow the photosynthetic afterglow
to decay.

     At the start of the experiment about 90% of the photons
were passed by a filter cutting off below 612 nm. A notable
increase in luminescence intensity was noted when the time for
natural leaf-fall approached (Figure 6). We suspect that this
increase reflects increasing peroxidation of membrane lipids,
and will try to test this hypothesis in the future.

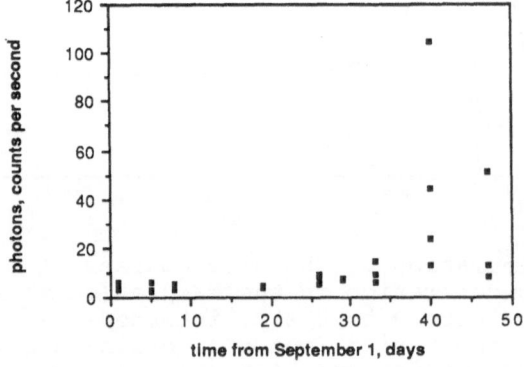

Fig. 6   Ultraweak luminescence from senescing beech
         leaves.

EFFECT OF A PULSE OF ULTRAVIOLET RADIATION ON ULTRAWEAK
LUMINESCENCE FROM GREEN LEAVES

Leaves of <u>Hibiscus</u> were used, because they are known to
be ultraviolet sensitive, and we found them to remain in good
condition for several days in an excised state.

In the experiments the leaves were exposed either to 10
minutes of ultraviolet-C radiation or 30 minutes of
ultraviolet-B radiation, or kept untreated. After irradiation
they were kept in darkness for 24 hours before the first
luminescence measurement was done, and the measurements were
then repeated during several days.

After both UV-B and UV-C treatments, but not in the
controls, a gradual increase in luminescence intensity was
noted (Panagopoulos et al., 1989), followed by a decline after
72 hours (Figure 7). The total peroxidase activity in leaf
extracts was also assayed at different times after the UV
pulse. This activity rises and declines in a similar way as
does the luminescence intensity (Figure 8). Although we know
that light is emitted during peroxidase catalyzed reactions,
we are not convinced of a causal connection between the
changes in luminescence and peroxidase activity. On the other
hand, it is known that ultraviolet irradiation might result in
formation of hydrogen peroxide in plant tissue. Therefore,
ultraviolet radiation may increase the rate of peroxidase-
catalyzed peroxidation even without an increase in enzymatic
activity itself. On the other hand, the enzyme lipoxygenase
might be of more significance than peroxidase in the present
connection.

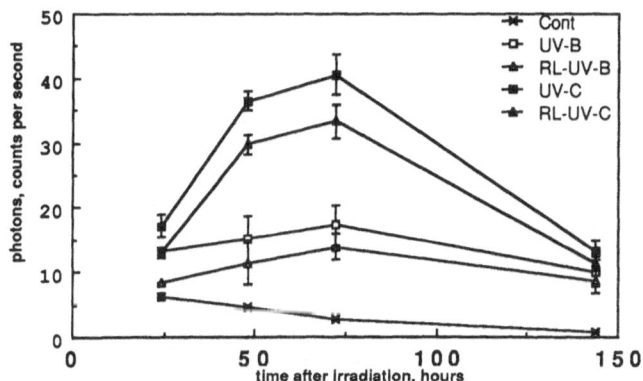

Fig. 7   Ultraweak luminescence from <u>Hibiscus</u> leaves
         irradiated with ultraviolet-C (254 nm, 3.6
         mmol photons per m^2 delivered in 10
         minutes, top two traces), or with ultra-
         violet-B (around 300 nm, 25 mmol per m^2
         delivered in 30 minutes, middle two traces),
         or nonirradiated control (bottom trace).
         The lower traces for UV-C and UV-B
         irradiated leaves (triangles) show the
         radiation passed by a 612 nm cut off filter
         (RL-UV-C and RL-UV-B, respectively) (from
         Panagopoulos et al., 1989).

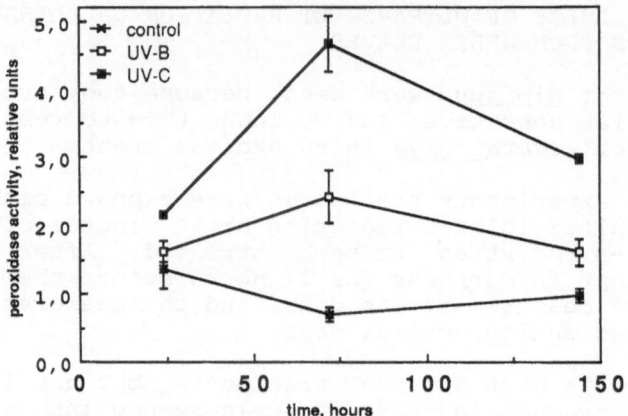

Fig. 8 Peroxidase activity in <u>Hibiscus</u> leaves
treated in the same way as the leaves in
Figure 7 (from Panagopoulos et al., 1989).

After UV irradiation, the fraction of photons with
wavelengths exceeding 612 nm gradually increases (Figure 9).
This means that the UV-induced luminescence has, at least in
part, another origin than the basal, irradiation-independent
luminescence. In both cases, however, a large part of the
luminescence has a wavelength exceeding 612 nm.

The effect of oxygen on ultraviolet induced luminescence
from green leaves is strikingly different from the effect of
oxygen on luminescence from carrot root tissue. Anaerobiosis
increases the intensity many times. Although ultraviolet-C
radiation induces more luminescence in air than ultraviolet-B
radiation does, the increase induced by anaerobiosis is much
more dramatic after ultraviolet-B irradiation.

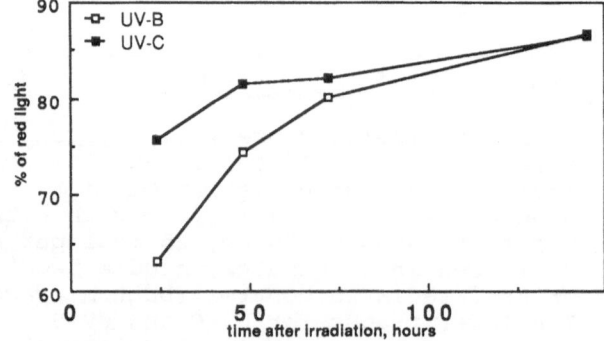

Fig. 9 Percent of photons from <u>Hibiscus</u> leaves
with wavelengths >612 nm (from
Papangopoulos et al., 1989).

DISCUSSION

The stimulation of luminescence of carrot roots by oxygen is in agreement with observations on other organisms by earlier investigators. The inhibition of luminescence from irradiated leaves by oxygen is more surprising, but could be due to non-radiative quenching of triplet excited molecules by triplet (ground-state) oxygen.

The most likely red-light emitters are singlet oxygen and chlorophyll, while ultraweak luminescence of shorter wavelength has often been attributed to excited carbonyl groups. Even leaves devoid of chlorophyll, however, emit light. We are now trying to measure singlet oxygen luminescence from the gas phase in contact with the plant material.

ACKNOWLEDGEMENTS

Different parts of the investigation were sponsored by Lars Hierta Foundation, the Swedish Environment Protection Board, the Swedish Natural Science Research Council, and Nestec Ltd.

REFERENCES

Abeles, F.B., 1986, Plant Chemiluminescence, Annu. Rev. Plant Physiol., 37: 49-72.

Abeles, F.B., 1987, Plant Chemiluminescence: An overview, Physiol., Plantarum, 71: 127-130.

Campbell, A.K., 1988, Chemiluminescence. Principles and applications in biology and medicine. Ellis Horwood, Chichester/VCH Verlagsgesellschaft, Weinheim.

Colli, L. and Facchini, U., 1954, Light emission from germinating plants, Nuovo Cimento, 12: 150-153.

Colli, L., Facchini, U., Guidott, G., Dugnani Lolati, R., Orsenigo, M. and Sommariva, M., 1955, Further measurements on the bioluminescence of the seedling, Experientia, 11: 479-481.

Panagopoulos, I., Bornman, J.F. and Björn, L.O., 1989, The effect of UV-B and UV-C radiation of Hibiscus leaves determined by ultraweak luminescence and fluorescence induction, Physiol. Plant., 76: 461-465.

Slawinska, D. and Slawinski, J., 1983, Biological chemiluminescence, Photochem. Photobiol., 37: 709-715.

Strehler, B.L. and Arnold, W.J., 1951, Light production by green plants, Gen. Physiol., 34: 809-820.

PHOTOSENSITIZED ACTION ON THE CELL MEMBRANE AND ITS CONSTITUENTS BY FUROCOUMARINS

Francesco Dall'Acqua and Daniela Vedaldi

Department of Pharmaceutical Sciences of the Padua University and Centro di Studio sulla Chimica del Farmaco del CRN, Via Marzolo 5, 35131 Padova Italy

INTRODUCTION

Furocoumarins (psoralens and angelicins) are naturally occurring compounds, present especially in the Umbelliferae, Rutaceae and Leguminosae (Musajo and Rodighiero, 1962). Other furocoumarins have also been prepared by chemical synthesis (Murray et al., 1982).

Many plant psoralens show marked photosensitizing activities, of which skin photosensitization followed by dark pigmentation is the best known (Musajo and Rodighiere, 1962). The ancient Hindus, Turks, Egyptians and other orientals have exploited this property in popular medicine, by using plants or herbs containing psoralens, for the treatment of vitiligo since ancient times (Scott et al., 1976). In 1948 El Mofty

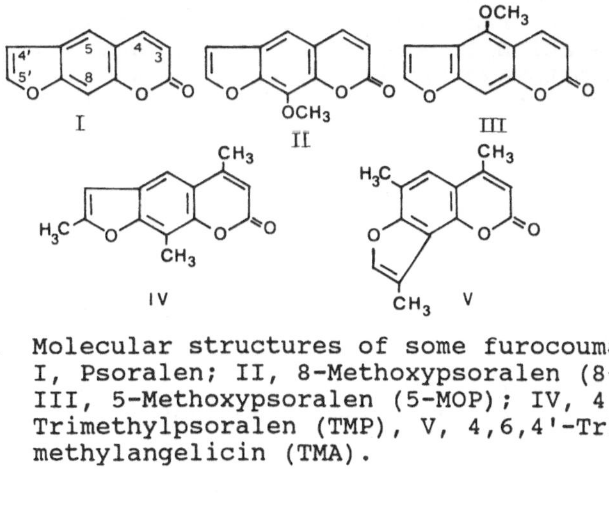

Fig. 1 Molecular structures of some furocoumarins: I, Psoralen; II, 8-Methoxypsoralen (8-MOP); III, 5-Methoxypsoralen (5-MOP); IV, 4,5',8-Trimethylpsoralen (TMP), V, 4,6,4'-Trimethylangelicin (TMA).

rationalised this type of therapy by using the active components of <u>Ammi majus</u> (mainly 8-MOP but also 5-MOP), isolated in a chemically pure state.

Scientific interest in psoralens has markedly grown over the last fifteen years, after the clinical introduction of 8-MOP followed by UV-4 irradiation in the treatment of psoriasis by Parrish et al., (1974). These authors introduced the term "photochemotherapy" to indicate this therapeutic approach. Psoralen photochemotherapy is now currently called PUVA. Various other skin diseases are treated with PUVA therapy and different psoralens are also used (Parrish et al., 1982). New studies directed toward better knowledge of their mechanism of action and a more precise evaluation of their toxicity have been developed.

New lines of research have also been developed to prepare and study new monofunctional furocoumarins, with the aim of obtaining new photochemotherapeutic agents as alternatives to psoralens (Rodighiero et al., 1988).

Quite interesting appear the studies on the furocoumarin excited states and free radicals (Song and Tapley, 1979; Anders et al., 1983; Craw et al., 1983) and their interactions with nucleic acid bases, aminoacids (Bensasson et al., 1978; Land and Truscott, 1979) and 3,4-dihydroxyphenylalanine (Craw et al., 1984). Also the reactivities of furocoumarin excited states with DNA in solution have been studied (Beaumont et al., 1980).

Psoralens have also been widely studied as photoactive probes of nucleic acid structure and function (Cimino et al., 1985).

MECHANISM OF PHOTOSENSITIZATION

Scheme 1 shows that three different pathways may be involved in furocoumarin photosensitization (Laustriat, 1986). Types I and II involve oxygen and are generally defined as photodynamic pathways. Type I involves substrate photooxidation by radicals (Cadet et al., 1984). Type II involves generation of singlet oxygen by energy transfer. Photodynamic pathways are biologically important since they may affect membranes (oxidation of unsaturated lipids),

Scheme I

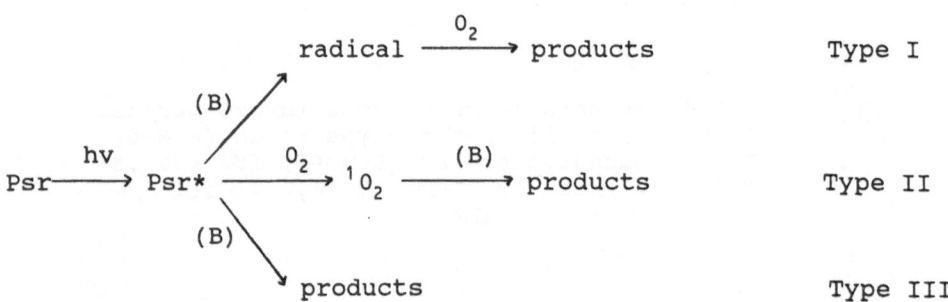

162

proteins and enzymes (oxidation of aminoacids) and also nucleic (guanosine is susceptible to oxidation). Pathway III involves photoreactions between furocoumarins and a substrate not involving oxygen: the main important events are C_4-photocycloadditions of furocourmarins to pyrimidine bases and to unsaturated fatty acids.

On the other hand from a biological point of view, furocoumarins seem to be able to interact with living cells at various levels as shown by the following Scheme II.

PHOTODAMAGE TO MEMBRANE CONSTITUENTS

In the last few years various researchers have turned their attention toward the photodamages that furcoumarins may induce at the level of the cell membrane.

The research done in this field has evidenced that the photodamage to cell membrane constituents involves both the oxygen-dependent photodynamic pathway and the anoxic pathway (See Scheme III).

While the photodynamic pathway mainly leads to lipid peroxidation (Potapenko and Sukhorukov, 1984) and to formation of cross-linking in the ghost protein (Horniceck et al., 1985), the anoxic pathway leads to a photocycloaddition reaction between furocoumarins and unsaturated fatty acids (Kittler and Lober, 1984; Caffieri et al., 1987; Specht et al., 1988a,b).

Photohemolysis of erythrocytes by furocoumarins

Erythrocytes are a cell model useful in revealing membrane damage. They, in fact, in the presence of damage show increased permeability to cations and subsequently undergo hemolysis. In this connection, the photohemolysis of erythrocytes by furocoumarins has been studied.

In early studies of the mechanism of action of furocoumarins, Musajo and Rodighiero (1962) studied the possible photohemolytic effect of furocoumarins, although at a concentration of 10^{-5}M they did not observe any photohemolysis.

Scheme II   Effects of Furocoumarins on Living Cells

Receptors   dark- and photinteraction with membrane
                and cytoplasmic receptors

Cell membrane   lipid peroxidation, C_4-cycloaddition to
                unsaturated lipids, formation of cross
                links in membrane proteins.

Cytoplasm   photoreactions with proteins, inactivation
                of enzymes, photoreaction with DNA,
                inactivation of ribosomes.

Nucleus   phootreaction with DNA and chromatin.

163

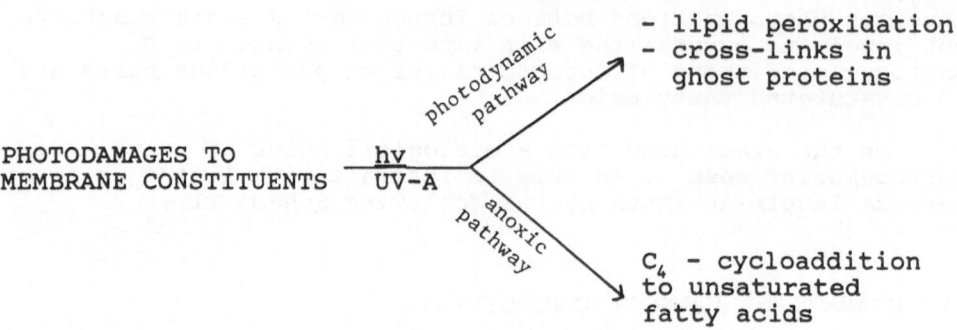

Scheme III

PHOTODAMAGES TO
MEMBRANE CONSTITUENTS $\frac{hv}{UV-A}$

photodynamic pathway
- lipid peroxidation
- cross-links in ghost proteins

anoxic pathway
C_4 - cycloaddition to unsaturated fatty acids

More recently, Potapenko et al., (1986a) studying the photosensitized modification of erythrocyte membrane induced by psoralen (10^{-4}M), were able to provoke photohemolysis in red blood cells under special conditions. After irradiation of red blood cells in the presence of psoralen, which should induce a prehemolytic stage, these authors were able to induce hemolysis by thermal incubation at 37°C.

Vedaldi et al., (1988) extended the above study to a large group of furocoumarins, having both linear and angular structure. Psoralen and 3-CPs proved to be the most active while the other compounds examined showed poor or no activity.

The above authors also showed a hemolytic effect in the dark, although at relatively high concentrations of furocoumarins (4.6 x 10^{-4} M).

Recently Vedaldi (unpublished results) studied the increase of membrane permeability to cations in terms of potassium leakage. A marked increase in potassium release is observed when erythrocytes are irradiated in the presence of psoralen. The effect is partly concentration-dependent. It should be pointed out that this effect can be observed without any thermal incubation, further supporting the observation that psoralen is able to photoinduce lesions to the erythrocyte membrane.

This toxic effect has also been studied in vivo by topical application of psoralen to the skin of albino guinea-pigs and then evaluating the damage induced in circulating red blood cells (Akimov et al., 1988). The results obtained show that topical PUVA on guinea-pig skin induces erythrocyte damage in terms of increased osmotic fragility, partial hemolysis and lipid peroxidation. This toxic effect should therefore be taken into account when topical photochemotherapy with psoralen is used.

Post-irradiation dark hemolysis (PIDH)

Potapenko et al (1986b) pre-irradiated an ethanol solution of psoralen, added it to a suspension of erythrocytes, and observed evident hemolysis (after thermal incubation) of red blood cells. This effect has been defined as post irradiation dark hemolysis (Vedaldi et al., 1988). This study was extended by the above authors to various furocoumarins, and showed the ability of these compounds to induce PIDH, although to a markedly lower extent.

Potapenko et al., (1986b) suggest a mechanism for this behaviour: in oxic conditions under UV-A irradiation, psoralen is oxidized forming a dimeric form of oxygen (O_2 - psoralen) which can be formed in ethanol or organic solvents but which is disrupted in water. The destruction of this photooxidized form causes oxidation of unsaturated fatty acids and may explain PIDH.

Taking into account that furocoumarins when irradiated with UV-A in ethanol, aqueous-ethanol or aqueous solution, undergo photomodification (Dall'Acqua and Caffieri, 1988), Vedaldi et al. (1988) suggested that the same products of photolysis may be responsible for PIDH. In fact they irradiated psoralen in saline physiologic solution and then added it to a suspension of erythrocytes. Psoralen and 3-carbethoxypsoralen (3-CPs) showed marked effects, while various other furocoumarins examined showed practically no effect.

Comparison between PUVA, PIDH in ethanol and saline

Recently Vedaldi (unpublished results) compared the effect of PUVA and of ethanol or saline solutions of psoralen pre-irradiated on erythrocyte suspensions (PIDH). Figure 2 shows the results obtained in these three conditions. In ethanol PIDH is very strong and is also present at the lowest concentrations of psoralen (1.4×10^{-5}M), while the other two conditions show lower (PIDH in saline) or no effect (PUVA). It should be noted that, in these conditions the hemolysis appeared without thermal incubation, suggesting that a different type of lesion or a similar but more pronounced lesion is probably involved in PIDH in ethanol.

Membrane permeability has also been studied in terms of potassium leakage. In this case too, the strongest effect is observed in the case of PIDH in ethanol, while the effects in the case of PIDH in saline and of PUVA are gradually lower.

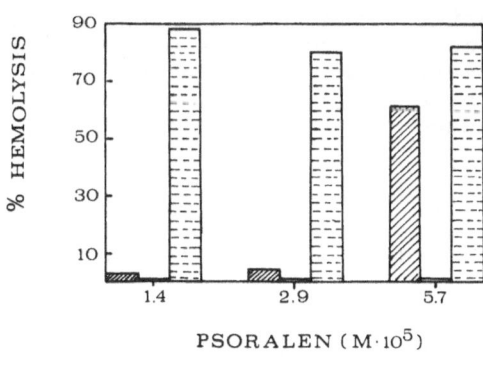

PSORALEN ($M \cdot 10^5$)

▨ Irradiated (PUVA)
▤ Pre-irradiated in ethanol
☐ Pre-irradiated in saline

Fig. 2   Hemolysis of erythrocytes by psoralen in various conditions.

Photoreactions (UV-A) between furocoumarins and unsaturated fatty acids

The first evidence of a photoreaction between unsatured fatty acids and TMP was reported by Kittler and Löber (1983, 1984).

Specht et al. (1988a and b) studied the photoreaction (360 nm) between TMP and oleic acid methyl ester (OAME), by irradiating a methanol solution of the compounds. The resulting photoproducts were separated by reverse phase HPLC. Four products with retention times longer than the parent fatty acid methyl ester were detected. On the basis of Mass and NMR spectrometry, UV absorption spectra and fluorescence properties, the photocompounds were shown to be cyclybutane adducts derived from [2+2] photocyloadditions, in which the 3,4 double bond of the TMP and the double bond of oleic acids were involved.

On the other hand, Caffieri et al., (1987) studied the photoreaction between four furocoumarins (psoralen, 8-MOP, TMA and 3-CPs) and linolenic acid. The photoadduct between psoralen and linolenic acid was isolated by reverse phase HPLC. When labelled psoralen was used, the isolated photocompounds proved to be radioactive, thus confirming the presence of the furocoumarinic moiety. Moreover, the photocompound psoralen-linolenic acid irradiated at 254 nm underwent photoreversion, forming psoralen and fatty acid. On the basis of its spectroscopic properties and ability to undergo photoreversion, the above authors suggest that the photocompound is derived from a C_4-cycloaddition between the 3,4-double bond of psoralen and a double bond of the fatty acid.

Caffieri et al (1988a, b) continued to study the photoreaction between psoralen and the methyl ester of linolenic acid (LAME). Various photocompounds are formed in the photoreactions. Two of them were isolated and characterized. They are C_4-cycloadducts between the pyrone double bond of the furocoumarin and the central (12-13) double bond of LAME (see Fig. 3).

Once it is established that only one psoralen molecule is linked per fatty acid molecule, the question arises of the regio- and stereo-arrangement of the two moieties in respect to cyclobutane: there are, in fact, eight possible isomers. Caffieri (unpublished results) studied the stereochemistry of

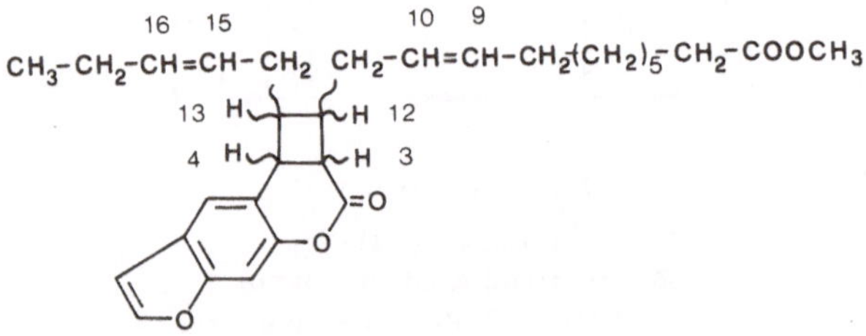

Fig. 3   General structure of the cycloadducts between psoralen and LAME.

one of the two isolated compounds; the two remaining parts of the linolenic acid are in trans with respect to the cyclobutane ring. Caffieri et al. (1988b) studied the photoreactions between psoralen and various unsaturated fatty acids. The adducts formed were isolated and characterized.

Caffieri et al (1988b), also studied the photoreaction between an angular furocoumarin and LAME. Also in this case, a C_4-cycloaddition reaction takes place between the pyroneside double bond of the angelicin and the central double bond of LAME. Caffieri et al. (1989a) also isolated and characterized the photoadducts formed in the photoreaction between TMA and oleic and linoleic acid methyl esters. These photoadducts show the same molecular structure of the previously described ones.

With the aim of verifying whether the photoreactions between furocoumarins and unsaturated fatty acids which take place in vitro can also occur in vivo, the following experiment was carried out. This was performed in similar conditions as described by Beijersbergen van Henegouwen et al. (1989). H-TMA was applied on the shaven skin of Wistar rats: the skin was irradiated ($I=20W/m^2$) after 15 minutes incubation, the animal was sacrificed, and the epidermis separated from the dermis. The epidermis was then submitted to dialysis under sink conditions to remove unbound ^{3}H-TMA and to selective extraction to obtain three phases: the lipid, DNA/RNA and the protein phases. The lipid phase was submitted to methanolysis to transform the natural unsaturated fatty acid esters and the photocompounds in the corresponding methyl esters. The transesterified lipid phase was examined by HPLC; in the chromatogram apart from linoleic and oleic acid methyl esters some peaks having retention times strictly similar to those shown by the adducts prepared in vitro were also present indicating that the photocycloaddition between furcoumarins and unsaturated fatty acids not only taks place in vitro but also in vivo (Caffieri et al., 1989a).

It is reasonable to expect that these photocycloaddition reactions to unsaturated fatty acids have some biological consequences. According to Midden (1988), a furocoumarin fatty-acid adduct may inhibit a phospholipase and therefore prevent the activation of protein kinase C or other regulatory enzymes. Inhibition of protein kinase C may account for the antiproliferative effect of PUVA therapy, as well as the inhibition of DNA synthesis. Moreover, because of the turnover of proteins and lipids, the temporary nature of the therapeutic benefit of PUVA in the treatment of psoriasis may be more easily explained by the furocoumarin fatty-acid adducts than by photoadducts induced in DNA.

Possible biological role of furocoumarin receptors

Laskin et al. (1985) have identified specific, saturable, high-affinity binding sites for 8-MOP on HeLa cells. These authors have also detected specific binding of 8-MOP to four other human cell lines and five mouse cell lines. In HeLa cells, binding is reversible and independent of the ability of the furocoumarin to form a molecular complex with DNA. Scatchard analysis indicates the presence in HeLa cells of two classes of furocoumarin binding sites: high affinity and low affinity sites. The high affinity binding sites become covalently modified by the psoralen molecule following UV-A exposure. In particular the psoralen receptor, labelled with H^3-8-MOP, was visualized in the cytoplasmic and plasma membrane. The receptor had an apparent molecular weight of about 22,000 daltons and has been shown to be sensitive to protease but not to nuclease treatment. Covalent binding of the tritiated 8-MOP to the receptor protein is inhibited by excess of unlabelled 8-MOP, indicating the covalent furocoumarin-receptor binding is saturable (Yurkov and Laskin, 1987).

Laskin and Laskin (1988) proposed a model in which furocoumarin receptors are localised in the cell membrane close to the receptor of the epidermal growth factor (EGF). Exposure to UV-A induces psoralen receptor activation which initiates intracellular signals leading to biological responses. These signals involve interaction with normal growth factor receptors such as EGF, decreasing the affinity of the receptors toward growth factors. This modulates protein kinase activity, thus also modulating the cell's response to a growth factor stimulus.

CONCLUSIONS

The extensive studies carried out on the photochemical and photophysical properties of furocoumarins have shown that both oxygen-dependent (types I and II) pathways and the oxygen independent pathway (type III, see Scheme I) are involved in the photosensitizing effects of these compounds. Moreover photobiological research performed in the last decade has shown that several mechanisms are involved in the photosensitizing, toxic and therapeutic activity of furocoumarins. Under UV-A activation these compounds are in fact able to induce various types of damage to the cell membrane, to photoreact with various cytoplasm components and to photobind to nuclear DNA and chromatin.

Better knowledge of the mechanisms of action of furocoumarins at the molecular level will allow a better understanding of their activity, suggest further possible applications or be of valid help in the drug design of new furocoumarins with reduced toxicity.

REFERENCES

Akimov, V.G., Potapenko, A.Ya., Lashmanova, A.P., Khorseva, N.I., Polyanskaya, N.P., and Bezdetnaya, L.N., 1988, Erythrocyte damage and induction of lipid peroxidation as a result of UV-irradiation or PUVA-treatment of skin of guinea-pigs, Studia Biophys., 124: 239.

Anders, A., Poppe, W., Herkt-Maetzky, C., Niemann, E.G., and Hofer, E., 1983, Investigations on the mechanism of photodynamic action of different psoralens with DNA, Biophys. Struct. Mech., 10: 11.

Beaumont, P.C., Parson, B.J., Navaratnam, S., Phillips, G.O., and Allen, J.C., 1980, The Reactivities of furcoumarin excited states with DNA in solution. A laser flash

photolysis and fluorescence study, <u>Biochim. Biophys.</u>
<u>Acta.</u>, 608: 259.

Beijersbergen van Henegouwen, G.M.J., Wijn, E.T.,
Schoonderwoerd, S.A., and Dall'Acqua, F., 1989, Method
for the determination of <u>in vivo</u> irreversible binding of
8-methoxypsoralen (8-MOP) to epidermal lipids, proteins
and DNA/RNA of rats after PUVA treatment, <u>J. Photochem.</u>
<u>Photobiol. B:Biol.</u>, 3: 631.

Bensasson, R.V., Land, E.J., and Salet, C., 1978, Triplet
excited state of furocoumarins: reaction with nucleic
acid bases and amino acids, <u>Photochem. Photobiol.</u>, 27:
273.

Cadet, J., Decarroz, C., Voituriez, L., Gaboriau, F., and
Vigny, P., 1984, Sensitized photoreactions of purine and
pyrimidine 2'-deoxyribonucleosides by 8-methoxypsoralen
and 3-carbethoxypsoralen, <u>in</u>: "Oxygen radicals in
chemistry and biology", W. Bors, M. Saran and D. Tait,
eds., Walter de Grujter and Co., Berlin and New York.

Caffieri, S., Tamborrino, G., and Dall'Acqua, F., 1987,
Formation of photoadducts between unsaturated fatty acids
and furocoumarins, <u>Med. Biol. Environ.</u>, 15: 11.

Caffieri, S., Daga, A., Vedaldi, D., and Dall'Acqua, F.,
1988a, Spectroscopic studies on the C_4-cycloadducts
between psoralen and unsaturated fatty acid methyl
esters, <u>Photomed. Photobiol.</u>, 10: 111.

Caffieri, S., Daga, A., Vedaldi, D., and Dall'Acqua, F.,
1988b, Photoaddition of angelicin to linolenic acid
methyl ester, <u>J. Photochem. Photobiol., B:Biol.</u>, 2: 515.

Caffieri, S., Schoonderwoerd, S.A., Daga, A., Dall'Acqua, F.,
and Beijersbergen van Henegouwen, G.M.J., 1989a,
Photoaddition of 4,6,4'-trimethylangelicin to unsaturated
fatty acids: possible <u>in vivo</u> occurrence, <u>Med. Biol.</u>
<u>Environ.</u>, in press.

Caffieri, S., Vedaldi, D., Daga, A., and Dall'Acqua, F.,
1989b, Photosensitizing furocoumarins: photocycloaddition
to unsaturated fatty acids, <u>in</u>: "Psoralens in 1988.
Past, present and future", T.B. Fitzpatrick, P. Forlot,
M.A. Pathak and F. Urbach eds., John Libbey Eurotext,
Montrouge (France).

Cimino, G.P., Gamper, H.B., Isaacs, S.T., and Hearst, J.E.,
1985, Psoralens as photoactive probes of nucleic acid
structure and function. Organic chemistry,
photochemistry and biochemistry. <u>Ann. Rev. Biochem.</u>, 54:
1151.

Craw, M., Bensasson, R.V., Ronfard-Haret, J.C., Sa e Melo,
M.T., and Truscott, T.G., 1983, Some photophysical
properties of 3-carbethoxypsoralen, 8-methoxypsoralen and
5-methoxypsoralen triplet states, <u>Photochem. Photobiol.</u>,
37: 611.

Craw., M., Chedekel, M.R., Truscott, T.G., and Land, E.J.,
1984, The photochemical interaction between the triplet
state of 8-methoxypsoralen and the melanin precurser L-
3,4-dihydroxyphenylalanine, <u>Photochem. Photobiol.</u>, 39:
155.

Dall'Acqua, F., and Caffieri, S., 1988, Recent and selected
aspects of furocoumarin photochemistry and photobiology,
<u>Photomed. Photobiol.</u>, 10: 1.

El Mofty, A.M., 1948, A preliminary clinical report on the
treatment of leukoderma with <u>Ammi majus</u>, <u>J. R. Egypt.</u>
<u>Med. Ass.</u>, 31: 651.

Horniceck, F.J., Malinin, G.I., Glew, W.B., Awret, U., Garcia, J.V., and Nigra, T.P., 1985, Photochemical cross-linking of erythrocyte ghost proteins in the presence of 8-methoxy and trimethylpsoralen. <u>Photobiochem. Photobiophys.</u>, 9: 263.

Kittler, L., and Löber, G., 1983, Furocoumarins: biophysical investigations on their modes of action, <u>Studia Biophys.</u>, 97: 61.

Kittler, L., and Löber, G., 1984, Photoreactions of furocoumarins with membrane constituents. Results with fatty acids and artificial bilayers, <u>Studia Biophys.</u>, 101: 69.

Land., E.J., and Truscott, T.G., 1979, Triplet excited state of coumarin and 4',5'-dihydropsoralen: reaction with nucleic acid bases and amino acids, <u>Photochem. Photobiol.</u>, 29: 861.

Laskin, J.D., Lee, E., Yurkov, E.J., Laskin, D.L., and Gallo, M.A., 1985, A possible mechanism of psoralen phototoxicity not involving direct interaction with DNA, <u>Proc. Natl. Acad. Sci.</u>, 82: 6158.

Laskin, J.D., and Laskin, D.L., 1988, Role of psoralen receptors in cell growth regulation, <u>in</u>: "Psoralen DNA photobiology", Vol. II, F.P. Gasparro ed., CRC Press, Boca Raton, Florida.

Laustriat, G., 1986, Molecular mechanism of photo-sensitization, <u>Biochimie</u>, 68: 771.

Midden, W.R., 1988, Chemical mechanisms of the bioeffects of furocoumarins: the role of reactions with proteins, lipids, and other cellular constituents, <u>in</u>: "Psoralen DNA photobiology", vol. II, F.P. Gasparro ed., CRC Press, Boca Raton, Florida.

Murray, R.D.H., Mendez, J., and Brown, S.A., 1982, The natural coumarins, John Wiley and Son Ltd, New York.

Musajo, L., and Rodighiero, G., 1962, The skin-photosensitizing furocoumarins, <u>Experientia</u>, 18: 153.

Parrish, J.A., Fitzpatrick, T.B., Tanenbaun, L., and Pathak, M.A., 1974, Photochemotherapy of psoriasis with oral methoxalen and long-wave ultraviolet light, <u>New Eng. J. Med.</u>, 291: 1207.

Parrish, J.A., Stern, R.S., Pathak, M.A., and Fitzpatrick, T.B., 1982, Photochemotherapy of skin diseases, <u>in</u>: "The science of photomedicine", J.D. Regan and J.A. Parrish eds., Plenum Publishing Co. New York.

Potapenko, A.Ya., and Sukhorukov, V.L., 1984, Photooxidative reactions of psoralens, <u>Studia Biophys.</u>, 101: 89.

Potapenko, A.Ya., Bezdetnaya, L.N., Lysenko, E.P., Sukhorukov, V.L., Remisov, A.N., and Vladimirov, Y.A., 1986a, Mechanisms of furocoumarin sensitized damage to biological membranes, <u>Studia Biophys.</u>, 114: 159.

Potapenko, A.Ya., Wunderlich, S., Pliquett, F., Bezdetnaya, L.N. and Sukhorukov, V.L., 1986b, Photosensitized modification of erythrocyte membrane induced by furocoumarins, <u>Photobiochem. Photobiophys.</u>, 10: 175

Rodighiero, G., Dall'Acqua, F. and Averbeck, D., 1988, New psoralen and angelicin derivatives, <u>in</u>: "Psoralen DNA photobiology", vol. I. F.P. Gasparro ed., CRC Press, Boca Raton, Florida.

Scott, B.R., Pathak, M.A., and Mohn, G.R., 1976, Molecular and genetic basis of furocoumarin reactions, <u>Mutat. Research</u>. 31: 29.

Song, P.-S., and Tapley, K.J., 1979, Photochemistry and photobiology of psoralens, Photochem. Photobiol., 29: 1177.

Specht, K.G., Bhan, P., Chedekel, M.R., and Midden, R.W., 1988a, Furocoumarin photosensitized reactions with fatty acids, in: "Photosensitization. Molecular, cellular and medical aspects", G. Moreno, R.H. Pottier and T.G. Truscott eds., NATO ASI Series, Springer Verlag, Berlin and Heidelberg.

Specht, K.G., Kittler, L., and Midden, R.W., 1988b, A new biological target of furocoumarins: photochemical formation of covalent adducts with unsaturated fatty acids, Photochem. Photobiol., 47: 537.

Vedaldi, D., Caffieri, S., Miolo, G., Dall'Acqua, F., and Arslan, P., 1988, Dark and photochemolysis of erythrocytes by furocoumarins, Z. Naturforsch., 43c: 888.

Yurkov, E.J., and Laskin, J.D., 1987, Characterization of a photoalkylated psoralen receptor in HeLa cells, J. Biol. Chem., 262: 8439.

STRUCTURE OF FUROCOUMARIN-DNA PHOTOADDUCTS

Paul Vigny[1], Annie Moysan[1], Jean Cadet[2], and
Lucienne Voituriez[2]

[1]Laboratoire de Physique et Chimie Biomoleculaire
(CNRS UA 198) Institut Curie & Universite Paris VI
11 rue P. et M. Curie 75231 Paris Cedex 05, France
[2]Laboratoires de Chimie, Centre d'Etudes
Nucleaires, DRF, 85 X, 38041 Grenoble Cedex, France

INTRODUCTION

Linear Furocoumarins (psoralens) as well as angular
furocoumarins (angelicins) give rise to specific
photoreactions with nucleic bases within DNA, when excited in
the UV-A domain. If the excitation wavelength is higher than
300 nm, DNA itself does not absorb the incident radiation and
the initial photoreactions occur via the excited singlet and
triplet states of the furocoumarin derivatives. Since the
early times of research on psoralen-DNA interactions, a
multistep model has been proposed to describe these
photoreactions which occur in the absence of oxygen (Musajo et
al., 1965, 1974): the first step, which happens independently
of any irradiation, is a non-covalent interaction of the
intercalation-type. This interaction involves mainly
hydrophobic forces which tend to maximize the overlap between
the psoralen ring and the adjacent base pairs allowing
favourable configurations for further photoreactions. The
second step is the formation of monoadducts under irradiation.
These usually involve the 5,6 double bond of a pyrimidine and
either the 4',5' bond of the psoralen yielding the so-called

---

*cis or trans : the cis stereochemistry denotes that both the
furocoumarin and pyrimidine are located to the same side of
the cyclobutyl ring. The same notation is used as for the
cyclobutadipyrimidines.
syn or anti : for a furan-side monoadduct to thymine or
cytosine the syn regiochemistry involves a convalent bond
between pyrimidine C(6) and furan C(5') whereas in the anti
regiochemistry these carbons are diagonally opposite with the
cyclobutane ring.
For a pyrone-side adduct, syn defines adducts in which
C(2) of the pyrone and N(1) of the pyrimidine are bonded to
adjacent corners of the cyclobutane ring whereas in the anti
regiochemistry these atoms are bound to opposite corners of
the cyclobutane ring.

furan-side monoadduct or the 3,4 bond giving rise to the pyrone-side monoadduct. In a third step the furan-side monoadduct can also be excited by the incident radiation. If the geometrical arrangement is favourable, it can undergo a second photoreaction with an adjacent pyrimidine yielding a diadduct thus forming an interstrand DNA cross-link. It is satisfactory to observe that this general scheme, based on simple geometrical and photochemical considerations, has been essentially confirmed during the present decade. This demonstration appeared as a necessity in view of the rapid development of the use of psoralens for biomedical applications. It has been greatly facilitated by the improvements in the physical and physico-chemical techniques of separation (high performance liquid chromatogrpahy) and of identification (high field mono- and two-dimensional NMR, soft-ionization mass spectrometries...). It is the aim of the present chapter to summarize both the expected and also unexpected results obtained recently in the structural approach of psoralen-DNA photoadducts and to point out some future lines of research in this area.

RECENT TRENDS IN THE CHEMICAL SYNTHESIS OF FUROCOUMARIN DERIVATIVES

The main clinical application of psoralens is their use, in association with UV-A, in the treatment of skin diseases such as psoriasis. In such treatments, the patient receives the psoralen derivatives (8-methoxypsoralen or 8-MOP, 5-methoxypsoralen or 5-MOP, 4,5',8-trimethylpsoralen or TMP) by topical applications or per os, before being locally exposed to the UV-A radiation delivered at increasing doses in order to avoid toxic reactions. Good results are obtained by this technique (Fitzpatrick et al., 1974) although the specific molecular events responsible for the medical effects are still not elucidated, psoralens being also able to interact in vivo with specific cellular macromolecules other than DNA (Laskin et al., 1985) and being distributed over the whole cell compartments (Amirand-Perchard et al., 1988). The main line of research in the chemical synthesis of new psoralens has consisted of looking for new derivatives with potential clinical activity but devoid of some weaknesses exhibited by the more common psoralens.

Water Soluble Psoralens

Psoralens show a poor-water solubility, ranging from less than 1μg/ml for TMP to about 20 μg/ml for 8-MOP. The use of intermediate organic solvents is thus required in biological experiments. In some cases their use has been shown to yield unexpected metabolic effects (Prognon et al., 1984). The goals in looking for new psoralen derivatives with higher water solubility were (i) to overcome these difficulties, (ii) to obtain compounds with higher DNA affinity, and finally (iii) to influence the pharmacokinetics of the drugs within the cells. In order to preserve the lipophilic property of the furocoumarin skeleton needed for DNA intercalative properties, the increased solubility was obtained by adding hydrophilic groups - mostly cationic - at the end of an aliphatic chain bound to the furocoumarin. The increase in water solubility and DNA affinity is particularly illustrated

174

in the case of the 4'-aminomethyl-4,5'-8-trimethylpsoralen or (AMT) which exhibits a water solubility of $10^4 \mu g/ml$ and a DNA association constant $K_a \sim 1.5 \times 10^5$ l.mol^{-1} (Isaacs et al., 1977, 1982). Several other derivatives were obtained by the same group (Isaacs et al., 1982) as well as 5- and 8- psoralen derivatives with an aminoalkoxy group (Antonello et al., 1979) and 5- and 8-Trimethylaminopropyloxy psoralens (Hansen et al., 1985). Interesting applications of nucleic acid photochemistry obtained with these water-soluble psoralens has been reviewed recently (Hearst, 1989).

Monofunctional Furocoumarins

DNA interstrand cross-links induced by the common psoralens are considered as responsible for the long term induction of cutaneous tumors after psoralen plus UV-A treatments. Another important line of research concerning psoralens has thus been the design of the new psoralens that are able to undergo the formation of monoadducts with DNA but are unable to induce the formation of interstrand cross-links (so-called "monofunctional psoralens"). The goal in obtaining compounds with antiproliferative activies, being effective in the photochemotherapy of psoriasis but with lower side-effects, was partially reached by using angular furocoumarins or specifically chemically-substituted linear furocoumarins. For geometrical reasons, the monoadducts formed with angelicins within DNA cannot be further converted to diadducts (Bordin et al., 1975). Several new monomethylangelicins (MA), dimethylangelicins (DMA) and trimethylangelicins (TMA) with high DNA photoaffinity have been synthesized by the Italian group in Padova, some of them exhibiting very interesting activities (Dall'Acqua et al., 1981, 1983; Baccichetti et al., 1981, 1984).

Following the same goal but by a different route, a program has been developed in France to block chemically the pyrone 3,4 double bond by chemical substitutions. Amongst the new compounds obtained, 3-carbethoxypsoralen (Queval and Bisagni, 1974) has been extensively studied from a sturctural point of view (Vigny et al., 1979; Averbeck et al., 1979; Gaboriau et al., 1981; Ronfard-Haret et al., 1982; Cadet et al., 1983; Amirand-Perchard et al., 1988; Moysan et al., 1988; Dardalhon et al., 1988; Gaboriau et al., 1989). It can therefore be considered as a reference monofunctional psoralen for further comparative mechanistic investigations. More recently, new psoralen derivatives with a fused pyridine ring on the 3,4 or 4',5' sites have been synthesized by the same group (Moron et al., 1983). Amongst them pyrido (3,4-c)psoralen (Pyps) and its methylated derivatives (MePyPs) have been shown to be of great clinical potential (Dubertret et al., 1985). Their molecular behaviour is beginning to be understood (Blais et al., 1984; Ronfard-Haret et al., 1987; Blais et al., 1987).

STRUCTURE OF THE MAJOR FUROCOUMARIN-DNA PHOTOADDUCTS

Pyrimidine bases, and thymine in particular, represent the preferential sites of furocoumarin photoadditions within DNA and with nucleosides (Hearst et al., 1984; Vigny et al., 1985; Cadet and Vigny, 1989). The chemical and stereochemical

structures of some of these photoadducts have been obtained
either by directly irradiating DNA-psoralen complexes or by
irradiating mixtures of psoralen and nucleic acid components
under conditions where the interactions between the
chromophores are enhanced, i.e. in the dry state or in frozen
solutions. The use of model photoreactions allows
photoadducts to be prepared in larger quantities and careful
checking of their chemical and photochemical stabilities. It
however also permits interactions to occur which are prevented
in psoralen-DNA complexes and thus yields a larger number of
chemical species. In this section we shall concentrate mainly
on the species produced and identified within native DNA.

Bifunctional Psoralens

Diadducts. The chemical structures of thymine-psoralen-
thymine diadducts produced with DNA have been determined for
8-MOP, TMP, HMT and psoralen itself. For each psoralen
studied, a single pair of diastereoisomers have been observed
in a cis-syn stereochemistry* and accounting for more than 90%
of the diadducts formed (Kanne et al., 1982a). This observed
stereochemistry is the direct result of the formation of an
intercalation complex and confirms the importance of the
geometrical parameters governing the non-covalent psoralen-DNA
interaction.

Furan-side monoadducts. In the case of 8-methoxypsoralen
(8-MOP) and of 4,5',8-trimethylpsoralen (TMP), three furan-
side monoadducts have been shown to be produced within DNA,
all showing a cis-syn stereochemistry (Kanne et al., 1982b).
The two major ones are a pair of thymidine diastereoisomers
whereas the minor one is a 2'-deoxyuridine adduct derived by
hydrolytic deamination of an initially formed 2'-
deoxycytidine. The crystal structure of the 8-MOP major
monoadduct isolated from calf thymus DNA after acidic
hydrolysis has been obtained (Peckler et al., 1982). It
indicates that both thymine and psoralen moieties remain
planar after the photoreaction but that the angle between the
two planes ranges between 44° and 53° when the cyclobutane is
formed. Such a distortion suggests that the photocrosslinking
may introduce a substantial kink in the DNA structure. This
point is however still the object of controversy, conflicting
conclusions being reached by various experimental and
theoretical approaches (Sinden & Hagerman, 1984; Tomic et al.,
1987; Perlman et al., 1985; Zhen et al., 1989). The
photochemical behaviour of the synthetic 4',hydroxymethyl-
4,5'8-trimethylpsoralen (HMT) towards DNA is similar to that
of the natural TMP. Four furan-side monoadducts have been
shown to be produced within DNA, all showing a cis-syn
stereochemistry (Straub et al., 1981). The two major ones,
which represent respectively 42% and 25% of the isolated
monoadducts, are a pair thymidine diastereoisomers whereas the
two minor ones, which together account for 28% of the isolated
monoadducts, are the two 2'-deoxyuridine monoadducts.

Pyrone-side monoadducts. In all bifunctional psoralen
studies (8-MOP, TMP and HMT) a cis-syn pyrone-side monoadduct
has been shown to be formed within DNA. This type of
monoadduct is less abundant than the corresponding furan-side
ones and only represents 19%, 3% and 2% of the covalently
bound derivative in 8-MOP, TMP and HMT respectively (Kanne et

al., 1982b; Straub et al., 1981). In the case of 8-MOP, this contrasts strongly with the results obtained with model photoreactions where they are obtained in larger quantities than the furan-side monoadducts (Shim & Kim, 1983; Joshi et al., 1984: Cadet et al., 1984). Obviously, the constraints imposed by the DNA helix on the geometry of the non-covalent DNA-psoralen complex are important. The fact that pyrone-side adducts of 8-MOP to thymidine are very unstable in aqueous solutions (Cadet et al., 1984) should also be taken into account to explain this observation.

Monofunctional Furocoumarins

Angelicin and allopsoralen derivatives. These compounds have a monofunctional behaviour due to their angular geometry. It is interesting to notice that, contrary to linear furocoumarins, only furan-side monoadducts have been observed within DNA with angelicins until recently (Dall'Acqua et al., 1981, 1984). In a recent study concerning 4'-methylangelicin, however, a pyrone-side monoadduct showing a cis-anti structure has been characterised in addition to the two cis-syn furan-side monoadducts to thymine (Caffieri et al., 1988). In the case of the other angular derivatives 4,7,5'-trimethylallopsoralen and 4,7,4'-thrimethylallopsoralen, both furan-side and pyrone-side monoadducts have been observed within DNA and a cis-syn orientation has been proposed for all of them (Caffieri et al., 1985).

Psoralen derivatives. The two main photoadducts formed by 3-carbethoxypsoralen (3-CPs) within DNA have been characterized as a pair of furan-side cis-syn diastereoisomers formed with thymine (Moysan et al., 1986; Gaboriau et al., 1987). The same two monoadducts have been characterized in yeast and mammalian cells and a method has been proposed for their identification and quantitiative detection in cells at levels as low as 1 monoadduct per 10,000 base pairs (Moysan et al., 1988) which has been used for repair kinetics studies (Dardalhon et al., 1988). Based on fluorescence spectroscopy, it has been shown that the new generation of linear monofunctional psoralens with a fused pyridine ring, namely pyrido [3,4-c]psoralen or PyPs, and 7-methylpyrido[3,4-c]psoralen or MePyPs, also give rise to furan-side monoadducts within DNA (Blais et al., 1984). The same conclusion has been reached for its isomer 7-methylpyrido[4,3-c]psoralen (Blais et al., 1987). More recently the isolation of two major 7-MePyPs-DNA adducts has been performed. These are furan-side monoadducts formed with thymidine. Several spectroscopic investigations indicate that they are also a pair of diastereoisomers (Moysan, 1987; Moysan et al., unpublished) this being very similar to what is observed with 3-CPs.

NEW PHOTOCHEMICAL AND PHOTOPHYSICAL EVENTS

Adducts Formed with Purines

Little attention has been given to the photoreaction of psoralen derivatives with purines. A few years ago, preliminary results concerning the isolation of several types of photoadducts of 8-MOP or of 3-CPs with 2'-deoxyadenosine when irradiating mixtures of these compounds in the dry state have been reported (Vigny et al., 1983; Cadet et al., 1984, 1986), with partial assignment of the 3-CPs adducts. It was later shown that the covalent bond formation occurs between the pyrone ring of 8-MOP at carbon C(4) and the furanose ring of the nucleoside either at the 1' or the 5' position (Cadet et al., 1988). These new types of furocoumarin-nucleic acid components, which appear to be specific to 2'-deoxyadenosine, were shown to result from recombination of the 3,4-dihydropyro-4-yl radical of 8-MOP with 2'deoxyadenosyl radical.

Cyclobutane Thymine Dimer Photosentization by Pyridopsoralens

In addition to the formation of a pair of diastereoisomeric furan-side monoadducts described above, pyridopsoralens have been recently and unexpectedly shown to photosensitize the formation of other photoproducts when excited at 365 nm. These have been rapidly identified as cyclobutane thymine dimers (Moysan et al., 1989 and to be published) with a rate of formation of 1.5 Thy<>Thy per 1000 nucleosides in a DNA sample containing 7 furan-side monoadducts of MePyPs per 1000 nucleosides. The induction of thymine dimers has been shown to be sequence dependent and to arise at strong sites of photoaddition i.e. AT rich sites. It is also dependent on the chemical structure of the pyridopsoralen derivative with a higher yield for pyrido[3,4-c]psoralen itself. This phenomenon has not been observed with other psoralens such as 8-MOP, 5-MOP, TMP or 3-CPs. Spectroscopic experiments and CNDO/S quantum chemistry calculations have been used to determine the excited T_1 state levels of the various psoralens and to correlate them with the observed thymine dimer efficiencies. A triplet-triplet energy transfer mechanism from pyridopsoralens to thymine is proposed to explain this observation (Costalat et al., 1989).

CONCLUSION

Significant progress was made in the characterization of the structures of the main furocoumarin-DNA photoadducts, with the aim to better understand molecular mechanisms of interaction of these photosensitizers with DNA. This has been largely facilitated by the impressive improvements of the analytical tools used in the purification and identification of the modified DNA components. It is clear that unexpected routes of interaction of furocoumarins, such as those briefly reported in the last section, are still to be discovered. However, the main challenge probably lies in the ability to identify such lesions and to detect them in a quantitative manner in cellular DNA. As pointed out recently, various physical and biochemical methods have become available for this purpose (Vigny and Cadet, 1987; cadet and Vigny, 1988).

REFERENCES

Amirand-Perchard, C., Nocentini, S., Vigny, P., Angiboust, J.F. and Manfait, M., 1988, A UV-microspectrophotofluorometric study of the uptake and photoreaction of furocoumarins in human cultured cells, in: "Spectroscopy of Biological Molecules - New

Advances", E.D. Schmid, F.W. Schneider and F. Siebert eds., John Wiley and Sons, Chichester & New York, pp. 455-458.

Amirand-Perchard, C., Vigny, P., Moysan, A., Ballini, J.P., Angiboust, J.F and Manfait, M., 1988, Improvements in the detection of very weak fluorescences from biological molecules by the use of microspectrophotofluorometry, in: "Spectroscopy of Biological Molecules - New Advances", E.D. Schmid, F.W. Schneider and F. Siebert eds., John Wiley & Sons, Chichester & New York, pp 459-462.

Antonello, C., Marciani, M., Gia, O., Carlassare, F., Baccichetti, F. and Bordin, F., 1979, Diethylaminoalkyloxy-coumarin and -Furocoumarin derivatives, Farmaco Ed. Sci., 34:139.

Averbeck, D., Bisagni, E., Marquet, J.P., Vigny, P. and Gaboriau, F., 1979, Photobiological activity in yeast of derivatives of psoralen substituted at the 3,4 and/or the 4',5' reaction site, Photochem. Photobiol., 30:547-555.

Baccichetti, F., Bordin, F., Carlassare, F., Peron, M., Guiotto, A., Rodighiero, P., Dall'Acqua, F. and Tamaro, M., 1981, 4,4'-dimethylangelicin, a monofunctional furocourmarin showing high photosensitizing activity, Photochem. Photobiol., 34:649-651.

Baccichetti, F., Carlassare, F., Bordin, F., Guiotto, A., Rodighiero, P., Vedaldi, D., Tamaro, M. and Dall'Acqua, F., 1984, 4,4',6-trimethylangelicin, a very photoreactive and non skin phototoxic monofunctional furocoumarin, Photochem. Photobiol., 39:525-529.

Blais, J., Averbeck, D., Moron, J., Bisagni, E. and Vigny, P., 1987, Effect of molecular structure on the photophysical properties, the photoreactivity with DNA and the photobiological activity of monofunctional pyridopsoralens, Photochem. Photobiol., 45:465-472.

Blais, J., Vigny, P., Moron, J. and Bisagni, E., 1984, Spectroscopic properties and photoreactivity with DNA of new monofunctional pyrido-psoralens, Photochem. Photobiol., 39:145-150.

Bordin, F., Marciani, S., Baccichetti, F., Dall'Acqua, F. and Rodighiero, G., 1975, Studies on the photosensitizing properties of angelicin, an angular furocoumarin forming only monofunctional adducts with the pyrimidine bases of DNA, Ital. J. Biochem., 24:258-267.

Cadet, J., Voituriez, L., Gaboriau, F., Vigny, P. and Della-Negra, S., 1983, Characterization of photocycloaddition products from reaction between thymidine and the monofunctional 3-carbethoxypsoralen, Photochem. Photobiol., 37:363-371.

Cadet, J. and Vigny, P., 1990, Photochemistry of Nucleic Acids, in: "Photobiochemistry of Nucleic Acids", H. Morrison ed., John Wiley and Sons, New York, pp 1-272.

Cadet, J., Voituriez, L., Ulrich, J., Joshi, P.C. and Wang, S.Y., 1984, Isolation and characterization of the monoheterodimers of 8-methoxypsoralen and thymidine involving the pyrone moiety, Photochem. Photobiol., 8:35-49.

Cadet, J., Voituriez, L., Gaboriau, F. and Vigny, P., 1984, Isolation and characterization of photoaddition products of 3-carbethoxypsoralen and 8-methoxypsoralen to 2'-deoxyadenosine, Photochem. Photobiol., 39:78S.

Cadet, J., Voituriez, L., Nardin, R., Viari, A. and Vigny, P., 1988, A new class of psoralen photoadducts to DNA components: isolation and characterization of 8-MOP adducts to the osidic moiety of 2'-deoxyadenosine, J. Photochem. Photobiol., B: Biology 2:321-339.

Cadet, J., Voituriez, L., Gaboriau, F. and Vigny, P., 1986, Isolation and characterization of psoralen photoadducts to DNA and related compounds in: "The role of cyclic nucleic acid adducts in carcinogenesis and mutagenesis, B. Singer and H. Bartsch eds., IARC Scientific Publication No. 70, pp. 247-251.

Cadet, J. and Vigny, P., 1988, Biochemical and chemical assays for monitoring the formation of DNA base photolesions, J. Photochem. Photobiol., 2:282-286.

Caffieri, S., Lucchini, V., Rodighiero, P., Miolo, G. and Dall'Acqua, F., 1988, 3,4 and 4',5' photocycloadducts between 4'-methylangelicin and thymine from DNA, Photochem. Photobiol., 48:573-577.

Caffieri, S., Rodighiero, P., Vedaldi, D. and Dall'Acqua, F., 1985, Methylallopsoralen-thymine 3,4 and 4'-5'-monoadducts formed in the photoreactions with DNA, Photochem. Photobiol., 42:361-366.

Costalat, R., Blais, J., Ballini, J.P., Moysan, A., Cadet, J., Chalvet, O. and Vigny, P., 1989, Formation of cyclobutyl thymine dimers photosensitized by pyridopsoralens: a triplet-triplet energy transfer mechanism, Photochem. Photobiol., 51:255-262.

Costalat, R., Blais, J., Ballini, J.P., Moysan, A., Cadet, J., Chalvet, O. and Vigny, P., 1989, The formation of thymine dimers photosensitized by pyridopsoralens: a possible mechanism. Proceed. 3rd Europ. Congress Photobiol., Budapest, Aug. 27-Sept. 2.

Dall'Acqua, F., Caffieri, S., Vedaldi, D., Guitto, A. and Rodighiero, G., 1981, Monofunctional 4',5'-photocycloadduct betwe4en 4',5'-dimethylangelicin and thymine, Photochem. Photobiol., 33:261-264.

Dall'Acqua, F., Vedaldi, D., Guiotto, A., Rodighiero, P., Carlassare, F., Baccichetti, F. and Bordin, F., 1981, Methylangelicins: New potential agents for the photochemotherapy of psoriasis. Structure-activity study on the dark and photochemical interactions with DNA, J. Med. Chem., 24:806-811.

Dall'Acqua, F., Caffieri, S., Vedaldi, D., Guitto, A. and Rodighiero, P., 1983, Monofunctional 3,4- and 4',5'-photocycloadducts between 4',5'-dimenthylangelicin and thymine, Photochem. Photobiol., 37:373-379.

Dall'Acqua, F., Vedaldi, D., Bordin, F., Baccichetti, F., Carlassare, F., Tamaro, M., Rodighiero, P., Pastorini, G., Guiotto, A., Recchia, G. and Cristofolini, M., 1983, 4'-methylangelicins: a new series of potential photochemotherapeutic agents for treatments of psoriasis, J. Med. Chem., 26:870-876.

Dall'Acqua, F., Vedaldi, D., Caffieri, S., Guiotto, A., Bordin, F. and Rodighiero, G., 1984, Chemical basis of the photosensitizing activity of angelicins, Natl. Cancer Inst. Monogr., 66:55-61.

Dardalhon, M., Moysan, A., Averbeck, D., Vigny, P., Cadet, J. and Voituriez, L., 1988, Repair of the two cis-syn diastereoisomers formed between 3-carbethoxypsoralen and thymidine in yeast cells, followed by a chemical method, J. Photochem. Photobiol. (B:Biology), 2:389-394.

Fitzpatrick, T.B., Pathak, M.A., Harber, L.C., Seiji, M. and Kukita, A., 1974, An introduction to the problem of

normal and abnormal responses of man's skin to solar
radiation, in: "Sunlight and Man", M.A. Pathak, L.C.
Harber, M. Seiji and A. Kubita eds, University of Tokyo,
Tokyo, pp 3-14.

Gaboriau, F., Vigny, P., Averbeck, D. and Bisagni, E., 1981,
Spectroscopic study of the dark interaction and of the
photoreaction between a new monofunctional psoralen: 3-
carbethoxypsoralen and DNA, Biochimine, 63:899-905.

Gaboriau, F., Vigny, P. and Moron, J., 1989 (in press),
Secondary structure of DNA modified by monofunctional
psoralen derivatives, Biochemsitry.

Gaboriau, F., Vigny, P., Cadet, J., Voituriez, L. and Bisagni,
E., 1987, Photoreaction of monofunctional 3-
carbethoxypsoralen with DNA; identification and
conformational study of the predominant cis-syn furan-
side monoadduct to thymine, Photochem. Photobiol.,
45:199-207.

Hansen, J.B., Bjerring, P., Buchardt, O., Ebbesen, P.,
Kaushup, A., Karup, G., Knudsen, P.H., Nielsen, P.E.,
Norden, B. and Ygge, B., 1985, Psoraleamines. 3.
Synthesis, pharmacological behavior and DNA binding of 5-
(aminomethyl)-8-Methoxy-,5[3-aminopropy)oxy methyl]-, and
8-[(3-aminopropyl)oxy]psoralen derivatives, J. Med.
Chem., 28:1001-1010.

Hearst, J.E., 1989 (in press), Use of psoralens as probes of
nucleic acid structure, in: "Photobiochemistry of Nucleic
Acids", H. Morrison ed., J. Wiley & Sons, New York.

Hearst, J.E., Isaacs, S.T., Kanne, D., Rapoport, H. and
Straub, K., 1984, The reaction of psoralen with
deoxyribonucleic acid, Quarterly Review of Biophysics,
17:1-44.

Isaacs, S.T., Shen, C.K., Hearst, J.E. and Rapoport, H., 1977,
Synthesis and characterization of new psoralen
derivatives with superior photoreactivity with DNA and
RNA, Biochemistry, 16:1058-1064.

Isaacs, S.T., Chun, C., Hyde, J.E., Rapoport, H. and Hearst,
J.E., 1982, A Photochemical characterization of reactions
of psoralen derivatives with DNA, in: "Trends in
Photobiology", C. Helene, M. Charlier, Th. Montenay-
Garestier and G. Laustriat, eds., Plenum Press, New York,
pp. 279-294.

Joshi, P.C., Wang, S.Y., Midden, W.R., Voituriez, L. and
Cadet, J., 1984, Heterodimers of 8-methoxypsoralen and
thymine, Photobiochem. Photobiophys., 8:51-60.

Kanne, D., Straub, K., Hearst, J.E. and Rapoport, H., 1982-a,
Isolation and characterization of pyrimidine-psoralen-
pyrimidine photoadducts from DNA, J. Am. Chem. Soc.,
104:6754-6764.

Kanne, D., Straub, K., Rapoport, H. and Hearst, J.E., 1982-b,
Psoralen deoxyribonucleic acid photoreaction.
Characterization of the monoaddition products from 8-
methoxypsoralen and 4,5',8-trimethoxypsoralen,
Biochemistry, 21:861-871.

Laskin, J.D., Lee, E., Yurkow, E.J., Laskin, D.L. and Gallo,
M.A., 1985, A possible mechanism of psoralen
phototoxicity not involving direct interaction with DNA,
Proc. Natl. Acad. Sci. USA, 82:6158-6162.

Moron, J., N'Guyen, C.H. and Bisagni, E., 1983, Synthesis of
5H-furo(3',2':6,7) (1)benzopyrano (3,4-c)pyridin-5-ones
and 8H-pyrano(3',3':5,6) benzo-furo(3,2-c)pyridin-8-ones

(pyridopsoralens), <u>J. Chem. Soc.</u>, Perkins Trans., 1:225-229.

Moysan, A., Vigny, P., Dardalhon, M., Averbeck, D., Voituriez, L. and Cadet. J., 1988, 3-carbethoxypsoralen-DNA photolesions: identification and quantitative detection in yeast and mammalian cells of the two <u>cis-syn</u> diastereoisomers with thymidine, <u>Photochem. Photobiol.</u>, 47:803-808.

Moysan, A., Gaboriau, F., Vigny, P., Voituriez, L. and Cadet, J., 1986, Chemical structure of 3-carbethoxypsoralen-DNA photoadducts, <u>Biochimie</u>, 68:787-795.

Moysan, A., Cadet, J., Moustacchi, E., Sage, E., Viari, A., Vigny, P. and Voituriez, L., 1989, Pyridopsoralens are able to photosensitize the formation of cyclobutane thymine dimers, Proceed. 3rd Europ. Congress Photobiol., Budapest, Aug. 27-Sept. 2.

Moysan, A., Cadet, J., Moustacchi, E., Sage, E., Viari, A., Vigny, P. and Voituriez, L., in preparation, Pyridopsoralens are able to photosensitize the formation of cyclobutyl thymine dimers with DNA upon UV-A irradiation.

Musajo, L., Rodighiero, G. and Dall'Acqua, F., 1965, Evidence of a photoreaction of the photosentizing furocoumarin with DNA and with pyrimidine nucleosides and nucleotides, <u>Experientia</u>, 21:22-24.

Musajo, L., Rodighiero, G., Caporale, G., Dall'Acqua, F., Marciani, S., Bordin, F., Baccichetti, F. and Bevilacqua, R., 1974, Photoreactions between skin-photosensitizing furocoumarins and nucleic acids, <u>in</u>: "Sunlight and Man", M.A. Pathak, L.C. Harber, N. Seiji and A. Kukita eds., University of Tokyo Press. Tokyo.

Pearlman, D.A., Holbrook, S.R., Pirkle, D.H. and Kim, S.H., 1985, Molecular models for DNA damaged by photoreaction, <u>Science</u>, 227:1304-1308.

Peckler, S., Graves, B., Kanne, D., Rapoport, H., Hearst, J.E. and Kim, S.H., 1982, Structure of a psoralen-thymine monoadduct formed in photoreaction with DNA, <u>J. Mol. Biol.</u>, 162:157-172.

Prognon, P., Blais, J., Averbeck, D., Averbeck, S., Vigny, P. and Gond, A., 1984, The metabolism of 8-methoxypsoralen by <u>Saccaromyces cerevisae</u>. Evidence for an inducing effect of ethanol, <u>II Farmaco</u>, 9:739-751.

Queval, P. and Bisagni, E., 1974, New synthesis of psoralen and related compunds, <u>Eur. J. Med. Chem.</u>, 9:335-340.

Ronfard-Haret, J.C., Averbeck, D., Bensasson, R.V., Bisagni, E. and Land, E.J., 1982, Some properties of the triplet excited state of the photosensitizing furocoumarin: 3-carbethoxypsoralen, <u>Photochem. Photobiol.</u>, 35:479-489.

Ronfard-Haret, J.C., Averbeck, D., Bensasson, R.V., Bisagni, E., Land, E.J. and Moron, J., 1987, Correlation between the triplet photophysical properties and the photobiological action in yeast of the two monofunctional pyridopsoralens, <u>Photochem. Photobiol.</u>, 45:235-239.

Shim, S.C. and Kim, Y.Z., 1983, Photoreaction of 8-methoxypsoralen with thymidine, <u>Photochem. Photobiol.</u>, 38:265-271.

Straub, K., Kanne, D., Hearst, J.E. and Rapoport, H., 1981, Isolation and characterization of the pyrimidine-psoralen-pyrimidine photoadducts from DNA, <u>J. Am. Chem. Soc.</u>, 103:2347-2355.

Tomic, M.T., Wemmer, D.E. and Kim, S.H., 1987, Structure of a psoralen cross-linked DNA in solution by nuclear magnetic resonance, Science, 238:1722-1725.

Vigny, P., Gaboriau, F., Duquesne, M., Bisagni, E. and Averbeck, D., 1979, Spectroscopic properties of psoralen derivatives substituted by carbethoxy groups at the 3,4 and/or 4',5' reaction site, Photochem. Photobiol., 30:557-564.

Vigny, P., Gaboriau, F., Voituriez, L. and Cadet, J., 1985, Chemical structure of psoralen-nucleic acid photoadducts Biochimie, 67:317-325.

Vigny, P., Spiro, M., Gaboriau, F., Lebeyec, Y., Della-Negra, S., Cadet, J. and Voituriez, L., 1983, 252Cf-Plasma Desorption Mass Spectrometry of covalently bound nucleic acid adducts: psoralen-nucleoside photoadducts, Int. J. Mass. Spectr. Ion Phys., 53:64-83.

Vigny, P. and Cadet, J., 1987, Determination of DNA lesions, in: "From Photophysics to Photobiology", A. Favre, R. Tyrell, J. Cadet eds., Elsevier, pp 123-129.

Zhen, W.P., Dahl, O., Buchardt, O. and Nielsen, P.E., 1988, On the DNA bending by psoralen interstrand crosslinking. A Gel electrophoretic study, Photochem. Photobiol., 48:643-646.

MONOFUNCTIONAL FUROCOURMARINS: PHOTOCHEMICAL, PHOTOBIOLOGICAL
AND PHOTOTHERAPEUTIC PROPERTIES OF PYRIDOPSORALENS

Louis Dubertret[1] and Dietrich Averbeck[2]

[1]Department of Dermatology, Inserm U, 312
Hôpital Henri Mondor, 94010 Créteil
[2]Institut Curie, CNRS U.A. 1292, 26, Rue d'Ulm
Paris, France

INTRODUCTION

New linear derivatives of psoralen, the pyridopsoralens,
pyrido (3,4-c) psoralen (PyPs) and 7-methylpyrido (3,4-c)
psoralen (MePyPs), were synthesized (Moron et al., 1983).

The aim was to develop new monofunctional furocourmarins
as photochemotherapeutic drugs which exhibit a high
potoreactivity towards DNA but possibly a reduced
photomutagenic activity on eukaryotic cells in comparison to
the bifunctional furocoumarins 8-methoxypsoralen (8-MOP) and
5-methoxypsoralen (5-MOP) actually in use for treating
psoriasis.

PHOTOPHYSICAL AND PHOTOCHEMICAL PROPERTIES OF PYRIDOPSORALENS

The introduction of a bulky chemical group, the pyrido
group, at the 3,4 reaction site of the psoralen, has been
perfomed to allow the production of only 4',5' (furan side)
monoadducts with pyrimidine bases in DNA. As a matter of
fact, studies in vitro demonstrated that the pyridopsoralens
photoreact, as expected, with DNA by forming only monoadducts
(Blais et al., 1984).

Furthermore, the presence of the pyrido group increases
the photoreactivity of the molecule by conferring, as in the
case of ellipticines (Tourbez-Perrin et al., 1980), a higher
degree of complexing to DNA which is an important preliminary
step for psoralen-DNA photoreactions (Dall'Acqua 1977). The
photoreactivity of MePyPs to DNA in vitro was found to be more
than ten fold higher than that of 8-MOP (Blais et al., 1984).
In eukaryotic cells the photobinding capacity of MePyPs was
about 5 to 20 times higher than that of the 8-MOP depending on
the cell type treated, i.e. yeast (Averbeck 1985, Magaña-
Schwencke and Moustacchi 1985) or mammalian cells in culture
(Papadopoulo et al., 1986, Nocentini, 1986). For example,

from data obtained in normal human fibroblasts, it can be estimated that one molecule of PyPs is photobound per 7.5×10^4 base pairs per kJ m^{-2} of UVA using a concentration of 1 μM and 1 kJ m^{-2} of UVA (13 molecules per 10^5 base pairs for MePvPs and 1.2 per 10^5 base pairs for 8-MOP) (Nocentini 1986).

In contrast, with 3-carbethoxypsoralen (3-CPs), a monofunctional furocoumarin previously developed by our group (Dubertret et al., 1979), the pyridopsoralens were quite stable under UVA radiation (Blais et al., 1984).

It appeared of major therapeutic importance to analyse the possible involvement of oxygen in the photoreactions of pyridopsoralens because oxygen dependent reactions have been inferred in furocoumarin induced unwanted clinical side effects such as erythema and photoallergic reactions (Pathak and Joshi, 1984, Dubertret et al., 1981, Kimura et al., 1985). Consistent with the fact that the quantum yields of singlet oxygen production under UVA radiation were lower for the pyridopsoralens than for 8-MOP (Ronfard-Haret et al., 1987) it was found in experiments performed in yeast that the inhibitory effect of pyridopsoralens on the colony forming ability (survival) of yeast cells was independent of the presence of oxygen (Averbeck 1984, Averbeck et al., 1985, Ronfard-Haret et al., 1987).

PHOTOBIOLOGICAL PROPERTIES OF PYRIDOPSORALENS

The photobiological activity of pyridopsoralens, i.e. PyPs and MePyPs, was first recognized in the haploid yeast Saccharomyces cerevisiae (Averbeck et al., 1982). Indeed, although they showed no significant effects in the absence of light activation, in the presence of UVA (365 nm radiation) PyPs and MePyPs were found to be 2 and 4 fold, respectively, more effective for the inhibition of colony forming ability, i.e. cell survival, than 8-MOP, and thus, far more photoactive than the monofunctional furocoumarin 3-CPs. The two pyridopsoralens photoinduced much more efficiently (10 - 20 x) cytoplasmic "petite" mutations, i.e. mitochondrial DNA damage, than 8-MOP. For the induction of nuclear mutations such as reverse mutations (his$^+$) or forward mutations (canR) in haploid yeast, PyPS and MePyPs were clearly less active than 8-MOP per unit dose of UVA radiation. As a function of survival, the mutagenic effects photoinduced by PyPs and MePyPs were very much lower than that of 8-MOP, indicating that the inhibitory effects on cell survival prevailed over the mutagenic effects (Averbeck et al., 1982, Averbeck et al., 1983, Averbeck et al., 1984, Averbeck 1984).

Repair of Pyridopsoralen-DNA photoadducts

Studies on the repair of MePyPs induced lesions in yeast (Moustacchi et al., 1983, Magaña-Schwencke and Moustacchi 1985) and in human fibroblasts (Nocentini 1986), using tritium labelled furocoumarins, showed the removal of MePyPs DNA adducts during post-treatment incubation of treated cells.

Evidence was obtained for the accumulation of DNA strand breaks during post-treatment incubation. In contrast, after treatment with 8-MOP and 3CPs, DNA strand breaks occurred only

transiently (Nocentini, 1986). In yeast, during the repair of
MePyPs induced lesions, single strand breaks were formed which
remained unsealed during 6 hours of post treatment incubation
(Moustacchi et al., 1983, Magaña-Schwencke and Moustacchi
1985). Also, in normal human fibroblasts, DNA strand breaks
appear to accumulate during incubation after treatment with
MePyPs (Nocentini 1986). The difficulty in rejoining the
breaks occurring during the repair of MePyPs induced lesions
may be due to the particular nature of the lesions induced, as
well as to their presence in high proportion. In this
connection, it should be noted that pyridopsoralens are able
to photosensitize the formation of DNA cyclobutyl thymine
dimers (Moysan et al., 1990). The mechanism involves a
triplet-triplet energy transfer. A correlation between the
relative quantum yield of thymine dimer production and the
lowest energy of the triplet states of the pyridopsoralens has
been observed (Costalat et al., 1990). Thus, it seems
possible that the pyridopsoralens induced pyrimidine dimers
contribute to the photobiological effects of these psoralen
derivatives.

Antiproliferative effects

A decreased efficiency for the repair of MePyPs versus 8-
MOP induced lesions is consistent with the observation that
normal human fibroblasts recovered semi conservative DNA
synthesis less rapidly following MePyPs than 8-MOP plus UVA
treatment (Nocentini 1986, Papadopoulo et al., 1986). This
finding together with the high inhibitory effect of MePyPs and
PyPs on the cloning capacity of yeast cells, V-79 Chinese
hamster cells and normal human fibroblasts (Averbeck et al.,
1983), when compared to that of 8-MOP, provided clear evidence
for a high antiproliferative activity of the pyridopsoralens.
This feature appears to be especially advantageous for their
prospective use in the treatment of proliferative skin
diseases such as psoriasis (Anderson and Voorhees 1980).

Mutagenic effects

Further studies on the mutagenicity of the
pyridopsoralens showed that, in diploid yeast cells, the
pyridopsoralens were only slightly less photomutagenic as a
function of UVA doses than 8-MOP but clearly less
photomutagenic at equal survival (Averbeck 1985). In V-79
Chinese hamster cells, however, pyridopsoralens were more
photomutagenic than 8-MOP as a function of UVA dose and
equally genotoxic as 8-MOP at equal survival levels
(Papadopoulo et al., 1986). At the same number of photobound
furocoumarin molecules, MePyPs induced lesions are clearly
more effective than 8-MOP induced lesions in the induction of
mutations in V-79 Chinese hamster cells (Papadopoulo et al.,
1986), whereas, in diploid yeast cells, MePyPs-DNA
photoadducts were less effective than 8-MOP-DNA photoadducts
(Averbeck 1985). This result is likely to be related to the
limited capacity of V-79 Chinese hamster cells to remove such
lesions by excision-repair, an error-free repair mechanism.
Post-replicational repair functions involving an error-prone
component may be responsible for the production of mutants in
hamster cells. Such an interpretation is supported by recent
data on the mutagenicity of the two furocoumarins in excision-

deficient human fibroblasts (see Papadopoulo et al., 1986; Papadopoulo, in preparation).

Additional lines of evidence exist demonstrating a decreased genotoxicity of the pyridopsoralens MePyPs and PyPs in comparison to 8-MOP. At the maximal level of the induction of sister-chromatid exchanges (SCE) in human fibroblasts, pyridopsoralens are found to be less effective than 8-MOP (Billardon et al., 1984). Furthermore, preliminary data obtained on the transformation of the C3H 10 T 1/2 mouse cells show that, despite a high efficiency of MePyPs in comparison with 8-MOP for the induction of transformed foci per unit dose of UVA, cell transformation in vitro by MePyPs appears to be clearly less efficient as a function of survival than that by 8-MOP (Averbeck et al., 1983; Papadopoulo, in preparation).

PHOTOCARCINOGENIC ACTIVITY OF PYRIDOPSORALENS IN MICE

The carcinogenic effects of the pyridopsoralens combined with UVA irradiation were investigated in two complementary series of experiments. They were carried out on albino mice, aged 12-14 weeks, belonging to the homozygous strain XVIIn c/Z. This strain of mice has already been used in previous experiments of photocarcinogenesis (Dubertret et al., 1979, Zajdela and Bisagni 1981). Groups of forty mice (males and females) were compared in each type of experiment.

Forty microlitres of an acetone solution of MePyPs, PyPs or 8-MOP (approx. 15 μg/cm^2) were applied to each ear 15 min. before irradiation. Irradiations were performed using Philips HPW 125 lamps, a Pyrex glass filter to cut off light under 340 nm and a water filter. At the level of mice ears, the irradiance for 365 nm was 28 Wm^{-2}.

The animals were treated five times a week. The standard irradiation dose was 1.68x10^4 Jm^{-2} (10 min. every time), the standard number of treatments was 115. The mice were observed until the end of their life, which was after 21 months.

A first series of experiments (Dubertret et al., 1985) was performed to determine the eventual photocarcinogenic activity of PyPs and MePyPs in comparison with 8-MOP. Twice as many treatment sessions and twice the ultraviolet doses at each session were necessary to obtain, with PyPs and MePyPs, the same kinetics of tumor appearance and the same number of tumors.

Under the conditions of irradiation used, examination of the ears showed no phototoxic side effect with the two pyridopsoralens, whereas mice receiving local treatment with 8-MOP and only half as many UVA sessions with half the radiation dose at each session, showed strong phototoxic reactions and virtual destruction of the ears at the end of the study.

A second series of experiments (Dubertret et al., 1985) was carried out with the pyridopsoralens to determine whether skin cancers could be induced by treating the ears of mice with the same number of sessions and the same UVA dose as with 8-MOP.

Since we used strictly comparable experimental conditions and homozygous albino animals, the carcinogenic properties of the different psoralens was compared using the Iball carcinogenic index: ICI (Iball 1939):

ICI = % of mice with tumors at the end of the experiment x 100
          average tumoral latency time

The Iball index for the different psoralens was: psoralen = 36; 8-MOP = 22.7; 5-MOP = 17.3; MePyPs = 11; PyPs = 8.

Despite their high biological activity under UVA radiation the pyridopsoralens are three to four times less carcinogenic than the bifunctional psoralens (Zajdela, in preparation). Particularly important from the therapeutic point of view, it should be noted that there was a considerable delay in the appearance of tumors with pyridopsoralens (14 months) as compared to 8-MOP or 5-MOP (10 and 12 months, respectively).

ACUTE SIDE EFFECTS AND THERAPEUTIC ACTIVITY OF THE PYRIDOPSORALENS

A cream containing 10^{-2} M of MePyPs, PyPs and 8-MOP finely ground in a substitute of lanolin (Hydrocerine Roc) heated at 70°C was prepared.

Acute side effects

Two hours after topical application of pyridopsoralens (10 $\mu g/cm^2$) on normal human skin (forearms of L. D.) the MED, three days after UVA radiation (Waldman 180), was 20 J/cm^2 (Dubertret et al., 1985). 10 volunteers (skin types 2 and 3) received increasing doses of UVA (2 to 10J/cm^2) two hours after 8-MOP, PyPs and MePyPs application. Two days after a strong erythema was observed on five subjects and skin blisters on the five others with 8-MOP+10 J/cm^2 of UVA. With PyPs and MePyPs +10 J/cm^2 no skin reaction was observed. Using the same topical preparations nine out of 12 japanese volunteers did not develop erythema following MePyPs application and UVA exposure to a dose as high as 15 J/cm^2. The other three subjects showed an MED at 8.4 and 12.6 J/cm^2 (Takashima et al., 1988). Using 8-MOP at the same molar concentration in the same base, the MED ranged from 0.6 to 1.7 J/cm^2, thus 8-MOP is 10 times more erythematogenic than MePyPs (Takashima et al., 1988). Furthermore, in our study, the pigmentogenic activity of pyrodiopsoralens (10^{-2}M) was unobservable one week after a single irradiation at 10 J/cm^2 of UVA, 2 hours after topical application, on skin type I to III. However, MePyPs was pigmentogenic on skin type IV, but 8.7 times less than 8-MOP (Takashima et al., 1988).

Therapeutic efficiency

Nine psoriatic patients were treated with pyridopsoralens (four patients, 10^{-2}M, treatment three times a week, topical application two hours before irradiation, 10 J/cm^2 of UVA) or, for comparison, by 8-MOP and pyridopsoralens on symmetrical lesions (five patients, 10^{-2}M, treatment three times a week, application two hours before irradiation, increasing doses of UVA from 1 J/cm^2). Psoralens were applied topically on area of 4 cm diameter into a psoriatic plaque.

In the four patients treated by topical applications of pyridopsoralens followed by UVA radiation at 10 J/cm^2, 10 sessions were necessary to achieve clearing. In the five patients treated by comparison with 8-MOP, the beginning of clearing was faster with 8-MOP but, at the end of the treatment, i.e. at high UVA doses, pyridopsoralens were equally, or perhaps slightly more, efficient than 8-MOP. In one case the relapse was studied. It was slower in the pyridopsoralen treated plaque than in the 8-MOP treated plaque, two weeks after stopping treatment. At the end of treatment, the pigmentation observed was nearly the same with pyridopsoralens and with 8-MOP. With this small number of patients it was not possible to show a difference between MePyPs and PyPs.

A Japanese study, with MePyPs and 8-MOP at the same concentration (10^{-2}M), in the same base (Hydrocerine), was performed by topical application on large skin surface areas on 6 psoriatics (Takashima et al., 1988). Each patient was treated by MePyPs plus a large dose of UVA (7.5 or 10 J/cm^2) and by 8-MOP with the first irradiation at the MED for 8-MOP with an increase of 1/2 J at each session up to 3MED. Three of these patients have been treated on an additional skin area by MePyPs but using the same protocol of UVA irradiation as with 8-MOP.

In the six patients, there were no significant differences in the number of treatment sessions needed for the beginning of resolution or for its completion with either MePyPs plus large doses of UVA (7.5 or 10 J/cm^2) or 8-MOP plus increasing doses of UVA. However, the cumulative doses of UVA necessary were quite different with the two drugs (mean UVA cumulative does for complete clearance : 101 J/cm^2 with MePyPs and 24 J/cm^2 with 8-MOP). When using the same protocol of UVA irradiation with MePyPs as with 8-MOP, MePyPs was also efficient, but required more than the double of sessions to achieve the same effect as 8-MOP.

In contrast to accidents of photosensitization, often severe, observed in 6 out of ten patients (Dubertret et al., 1979) and in 4 out of nine patients (Kimura et al., 1985) with another monofunctional furocoumarin, 3-CPs, such side effects were not observed with pyridopsoralens (Takashima et al., 1988).

In conclusion, pyridopsoralens, new monofunctional compounds of the psoralen family, represent a clear progress in photochemotherapy. In fact, in therapeutic conditions these compounds are as efficient as 8-MOP. They are devoid of acute side effects, and their genotoxicity is strongly reduced. Further encouraging studies on the therapeutic properties of these compounds are in progress in the Honigsmann group.

REFERENCES

Anderson, T.F. and Voorhees, J.J., 1980, Psoralen photochemotherapy of cutaneous disorders, Ann. Rev. Pharmacol. Toxicol., 20:235-257.

Averbeck, D., Dubertret, L., Bisagni, E. and Moron, J., 1982, Towards more safety in PUVA-therapy with new monofunctional psoralens, Int. Workshops on Investigative Dermatology, Kyoto, May 31-June 1, Abstr. W2-18.

Averbeck, D., Dubertret, L., Bisagni, E., Moron, J., Papadopoulo, D., Nocentini, S., Blais, J. and Zajdela, F., 1983, Photobiological and phototherapeutic properties of new monofunctional pyridopsoralens, J. Invest. Dermatol. 80:306.

Averbeck, D., 1984, Photochemistry and photobiology of psoralens, Proc. Jpn. Soc. Invest. Dermatol., 8:52-73.

Averbeck, D., Averbeck, S., Bisagni, E. and Moron, J., 1985, Lethal and mutagenic effects photoinduced in haploid yeast (Saccaromyces cerevisiae) by two monofunctional pyridopsoralens compared to 3-carbethoxypsoralen and 8-methoxypsoralen. Mutation Res., 148:47-57.

Averbeck, D., 1985, Relationship between lesions photoinduced by mono and bifunctional furocourmarins in DNA and genotoxic effects of diploid yeast, Mutation Res. 151:217-233.

Billardon, C., Levy, S. and Moustacchi, E., 1984, Induction in human skin fibroblasts of sister-chromatid exchanges (SCE) by photoaddition in two new monofunctional pyridopsoralens in comparison to 3-carbethoxypsoralen and 8-methoxypsoralen, Mutation Res. 138:63-70.

Blais, J., Vigny, P., Moron, J. and Bisagni, E., 1984, Spectroscopic properties and photoreactivity with DNA of new monofunctional pyridopsoralens Photochem. Photobiol., 39:145-156.

Costalat, R., Blais, J., Ballini, J.P., Moysan, A., Cadet, J., Chalvet, O. and Vigny, P., 1990, Formation of cyclobutane thymine dimers photosensitized by pyridopsoralens: a triplet-triplet energy transfer mechanism. Photochem. Photobiol., 51:255-262.

Dall'Acqua, F., 1977, New chemical aspects of the photoreaction between psoralen and DNA, in: Research in Photobiology, A. Castellani (Ed.), Plenum, New york, pp. 245-255.

Dubertret, L., Averbeck, D., Zajdela, F., Bisagni, E., Moustacchi, E., Touraine, R. and Laterjet, R., 1979, Photochemotherapy (PUVA) of psoriasis using 3-carbethoxypsoralen, a compound non carcinogenic in mice, Br. J. Dermatol., 101:379-389.

Dubertret, L., Averbeck, D., Bensasson, R., Bisagni, E., Gaboriau, F., Land, E.J., Nocentini, S., Macedo Sa E Melo, M.T., Moustacchi, E., Morliere, P., Ronfard-Haret, J.C., Santus, R., Vigny, P., Zajdela, F. and Latarjet, R., 1981, Photophysical, photochemical, photobiological and phototherapeutic properties of 3-carbethoxypsoralen, in: G Cahn, B.P. Forlot, C. Grupper, A.E. Meybeck and F. Urbach (Eds), Psoralens in Cosmetics and Dermatology, Pergamon, New York, pp. 245-256.

Dubertret, L., Averbeck, D., Bisagni, E., Moron, J.,
    Moustacchi, E., Billardon, C., Papadopoulo, D.,
    Nocentini, S., Vigny, P., Blais, J., Bensasson, R.,
    Ronfard-Haret, J.C., Land, E.J., Zajdela, F. and
    Latarjet, R., 1985, Chemotherapy using pyridopsoralens,
    Biochimie, 67:417-422.
Iball, J., 1939, The relative potency of carcinogenic
    compounds. Am. J. Cancer, 35:188-190.
Kimura, S., Mizuno, N., Hirano, S., Yoshikawa, K., 1985,
    Topical application of 3-carbetoxypsoralen plus UVA in
    the treatment of psoriasis. J. Dermatol (Tokyo), 12:251-
    257.
Magana-Schwencke, N. and Moustacchi, E., 1985, A new
    monofunctional pyridopsoralen: photoreactivity and repair
    in yeast, Photochem. Photobiol., 42:43-49.
Moustacchi, E., Cassier, C., Chanet, R., Magana-Schwencke, N.,
    Saeki, T. and Henriques, J.A.P., 1983, Biological role of
    photoinduced cross-lonks and monoadducts in yeast DNA:
    genetic control and steps involved in their repair. In:
    Cellular Responses to DNA damage (Eds.: E. C. Friedberg
    and B. A. Bridges) Alan Liss, New York, pp. 87-106.
Moron, J., Nguyen, C.H. and Bisagni, E., 1983, Synthesis of
    5H-Furol (3',2':6,7) (1) benzopyrano (3,4-c)pyrin-5-ones
    and 8H-pyrano (3',2':5,6) benzo-furo(3,2-c)pyridin-8-ones
    (Pyridopsoralens). J. Chem. Soc. Perkin Trans., 1:225-
    229.
Moysan, A., Cadet, J., Moustacchi, E., Sage, E., Viari, A.,
    Vigny, P. and Voituriez, L., 1990, Pyridopsoralens are
    able to photosensitize the formation of DNA cyclobutyl
    thymine dimers. Biochemistry, submitted.
Nocentini, S., 1987, DNA photobinding of 7-methyl pyrido (3,4-
    c) psoralen and 8-methoxypsoralen. Effects on
    macromolecular synthesis, repair and survival in cultured
    human cells. Mutation Res., 161:181-192.
Papadopoulo, D., Averbeck, D. and Moustacchi E., 1986,
    Mutagenic effects photoinduced in mammalian cells in
    vitro by two monofunctional pyridopsoralens. Photochem.
    Photobiol., 44:31-39.
Pathak, M.A. and Joshi, P.C., 1984, Production of active
    oxygen species (102 and 02-) by psoralens and ultraviolet
    radiation (320-340). Biochim. Biophys. Acta 798:115-126.
Ronfard-Haret, J.C., Averbeck, D., Bensasson, R., Bisagni, E.,
    Land, E.J. and Moron, J., 1987, Correlation between the
    triplet photophysical properties and the photobiological
    action in yeast of two monofunctional pyridopsoralens.
    Photochem. Photobiol., 45:235-239.
Takashima, A., Sunohara, A., Mizuno, N., 1988, Comparison of
    the relative therapeutic efficacy of 7-methyl
    pyridopsoralen and 8-methoxypsoralen in photochemetherapy
    in psoriasis treatment. The J. of Dermatol., 15:195-201.
Tourbez-Perrin, M.F., Pochon, F., Ducrocq, C., Rivalle, C. and
    Bisagni, E., 1980, Intercalative binding to DNA of new
    antitumoral agents: dipyrido (4,3-b) (3,4-f) indoles,
    Bull. Cancer (Paris), 67:9-13.
Zajdela, F. and Bisagni, E., 1981, 5-methoxypsoralen, the
    melanogenic additive to suntan preparations is
    tumorigenic in mice exposed to 365 nm UV radiation.
    Carcinogenesis, 2:121-127.

DAMAGE OF CHROMOSOMES OF HEALTHY PERSONS AND PSORIASIS PATIENTS UNDER PUVA CONDITIONS

L Kittler[1], V. Beensen[2], H. Schaarschmidt[3], B. Knopf[3], and G. Löber[1]

[1]Academy of Sciences of the GDR, Central Institute of Microbiology and Experimental Therapy, DDR-6900 Jena
[2]Friedrich Schiller University of Jena, Institute of Anthropology and Human Genetics, DDR-6900 Jena
[3]Department of Dermatology, Friedrich Schiller University of Jena, DDR-6900 Jena

SUMMARY

The effect of 8-methoxypsoralen (8-MOP), near UV-light (365 nm) and the combination of both (PUVA treatment) were studied on lymphocytes _in vitro_ taken from healthy persons and patients with psoriasis vulgaris and psoriasis arthritis. Chromosomes isolated from cell nuclei were visualized by means of the Giemsa staining technique and analyzed for induction of chromosomal defects, i.e. premature centromer division (PCD), major coiling (MC), and formation of gaps and fragile sites. Exposure of nonpsoriatic lymphocytes to 8-MOP, UVA or PUVA increased the rate of PCD or MC generation. In experiments with psoriatic lymphocytes a much weaker effect was found with a moderate increase of PCD and MC after UV-light or PUVA treatment in the case of psoriasis vulgaris, and of MC after UV-light treatment of psoriasis arthritis. No indication was obtained for the preference of certain chromosome groups or the appearance of "fragile sites", especially after methotrexate treatment. Our findings suggest PCD and MC investigations as possible sensitive tools for diagnosing latent psoriasis and for refined analysis of psoriatic cells or chromosomes.

INTRODUCTION

It is generally accepted that furocoumarin derivatives, e.g. 8-MOP in combination with UVA light produce in nucleic acids both thymine monoadducts and cross-links (Cole, 1971). The same technique applied in psoriatic patients has been found effective in attaining clinical remissions in psoriasis (Parrish et al., 1974). There is, however, no unambiguous evidence that the anti-psoriatic action is related to nucleic acid modifications, since proteins (Engel and Wulf, 1982; Lerman et al., 1982; Meffert et al., 1976), and membranes (Gast et al., 1982; Kittler and Löber 1983; Römer et al., 1983) were also suggested to act as cellular targets. Various papers are devoted to the nature and molecular basis of cutaneous photosensitivity reactions with attempts to judge the importance of the different cellular receptor sites (Kittler and Löber 1983; Römer et al., 1983; Pathak and Joshi, 1983) including photoimmunological effects (Morison et al., 1979). A long-term risk of developing skin for photochemotherapy has been reported from large cooperative studies (Roenigk and Caro, 1981; Stern et al., 1984). Irrespective of whether PUVA induced structural modifications of the cellular nucleic acids, in particular DNA, are the primary and essential events for the clearing process of psoriatic lesions, they are probably responsible for the occurrence of mutagenic effects in a variety of systems (Abel and Schimmer, 1981; Bridges et al., 1981; Igali et al., 1970; Kirkland et al., 1983) and for the induction of damage in eukaryotic chromosomes (Cassel and Latt, 1980; Hook et al., 1983; Linnainmaa and Wolff, 1982). Most frequently SCE was observed under PUVA treatment in vitro (Cassel and Latt, 1980; Linnainmaa and Wolff, 1982; Waksvik et al., 1977; Wulf, 1978) while there was no effect observable in vivo (Brogger et al., 1978; Lambert et al., 1978) and only a small effect when patients were PUVA treated over years. In the past, studies of chromosomal defects focused mainly on SCE, induction of constrictions, gaps and breaks. Less attention has been paid to deviations in the morphology of the chromosomes expressed in modifications of the centromer separation termed "premature centromer division" (PCD) (Vig, 1981) and in the appearance of screw-like coiled chromatid arrangements termed "major coiling" (MC) (Jorgensen and Bak, 1982). In a preliminary note it was shown that PUVA treatment of human lymphocytes in vitro, when taken from healthy persons, increased the number of metaphase plates displaying chromosomes with PCD and MC defects (Löber et al., 1982). This will be reported now in more detail. Our further interest deals with studies of chromosomes taken from lymphocytes of patients with psoriasis vulgaris and psoriasis arthritis to find out whether "psoriatic" chromosomes have latent defects which may be amplified under PUVA conditions.

MATERIAL AND METHODS

Lymphocytes gained from peripheral blood of healthy probands and patients with psoriasis vulgaris and psoriasis arthritis were prepared in 72h cultures according to standard procedures (Moorhead et al., 1960). 8-MOP, UVA or PUVA treatment was performed 24h before the preparation of chromosomes. One ml of the culture medium contained 5×10^{-5} M of 8-MOP which is more than one order of magnitude higher than found in the serum of PUVA treated patients (Thune, 1978). Irradiation of the lymphocytes by means of a high pressure mercury lamp, HBO 500, was started immediately after the drug addition. Reference experiments with MTX according to the method described by Schmidt and Passarge (1981) were done. Lymphocyte cultures were supplied with 10 μg/ml MTX 24 h before chromosome preparation. Again, this concentration is one order of magnitude higher than those determined in the sera of patients (Noble et al., 1975). Preparation of chromosomes and Giemsa staining was done as described elsewhere (Beensen, 1986). The controls and all treated probes contained the same amount of colchicine for harvesting of metaphase chromosomes. The preparation of control cultures and those for special treatment were taken from the same blood probe.

Patients

Five patients with psoriasis vulgaris and 5 patients with psoriasis arthritis (7 male, 3 female), with ages ranging between 19 and 74 years with an average of 43 years were investigated. In 3 cases MTX treatment preceded the PUVA experiments. PUVA effects on lymphocytes of psoriasis patients _in vivo_: The age of patients (6 male, 2 female) ranged between 27 and 57 years with an average age of 44.8 years. The average time of psoriasis manifestation was 16.6 years. Prior to our test all patients were treated with MTX and 6 with UVB. No patients with manifest or anamnestic tumor diseases are involved. Similarly, patients pretreated with arsen or X-rays in therapeutic doses were excluded. The PUVA therapy was performed according to recommendations given by Barth et al. (1982). Irradiation in PUVA therapy was achieved with a metal halogen lamp (NARVA Berlin, GDR) using a rasotherm glass filter in order to reduce the amount of UVB light (Sutherland and Leonhard, 1979).

RESULTS AND DISCUSSION

In vitro experiments

Metaphase chromosomes were prepared from peripheral lymphocytes of two healthy male probands, of five patients with psoriasis vulgaris and five patients with psoriasis arthritis. The frequences of detection of PCD and MC defects were quantified according to an all-or-none counting technique. If there is, for example, no detectable connection between the sister chromatids of a chromosome having a distance greater than the chromatid width, the metaphase plate was counted as PCD positive, independent of the number of chromosomes showing PCD modification. Analogously, if there are one or more chromosomes displaying helix-like chromatide, the metaphase plate was counted as MC positive, again independent of the number of chromosomes showing MC modification. Those metaphase plates on which none of the chromosomes has PCD or MC were counted as PCD or MFC negative, respectively. Each metaphase plate was analysed simultaneously for appearance of PCD and MC effects. PCD and MC can often be observed on the same metaphase plate while MC could be found without coincidental occurence of PCD. Quantitative differences of PCD and MC defects in the chromosomes of 8-MOP, UVA or PUVA treated "normal" lymphocytes and the untreated control were observed. 8-MOP alone induced the highest amount of PCD, while the effect of UVA and PUVA was clearly lower (Tab. 1). On the contrary various kinds of treatment did not show significant differences in MC, although all gave clearly higher numbers of metaphase plates with MC than the control. Actually, addition of 8-MOP to the

Table 1 Changes in the number of metaphase plates
showing premature centromer division (PCD)
and major coiling (MC) upon treatment with
8-MOP, UVA and in combination with both
(PUVA treatment).

PCD	No. of metaphase			P_l - P_u (%)	With/Total (%)
	Total	Without PCD	With PCD		
None	400	329	71	13.6 - 21.6	17.8
8-MOP	335	135	200	55.4 - 66.0	59.7
UVA	400	233	167	37.6 - 47.5	41.8
8-MOP + UVA	400	196	204	45.0 - 55.0	49.0

MC	Total	Without MC	With MC	P_l - P_u (%)	With/Total (%)
None	300	216	84	21.8 - 32.1	28.0
8-MOP	235	97	138	55.4 - 67.9	58.7
UVA	300	124	176	54.1 - 65.6	58.7
8-MOP + UVA	300	111	189	57.6 - 68.7	63.0

P_l and P_u are the lower and upper limits of
the 95% confidence interval, respectively.
The lymphocytes of two healthy probands were
treated 24 h before chromosome preparation.

PCD: Control<UVA<8-MOP + UVA<8-MOP
MC :   Control<8-MOP - UVA<8-MOP + UVA

The number of chromosomes per metaphase showing PCD ranged
from 6 - 9 and showed relatively small deviations between the
different kinds of treatment, and even from the control.

lymphocytes 24 h before the preparation and repeated
irradiation with doses of 1.5 J/cm^2 48 h before colchicine
addition yielded no measurable increase of the number of PCD
or MC metaphase plates. It is known that 5-bromodeoxyuridine
(Sutherland and Leonhard, 1979) or folic acid antagonists
(Noble et al., 1975) produce fragile sites which are
chromosome specific (Sutherland and Leonhard, 1979). This is
obviously not the case after PUVA, 8-MOP and UVA treatment,
where fragile sites were not found.

Most of the chromosome groups underwent comparable
enhancements in PCD, when the data are related to the number
of chromosomes belonging to each group. The centromeres of
different chromosomes separate probably in a non random,
apparently genetically controlled sequence (Vig, 1981). This
sequence cannot be influenced by the different pretreatment
conditions. The data obtained with peripheral lymphocytes
taken from patients with psoriasis vulgaris and psoriasis
arthritis are presented in Figs. 1 and 2. The in vitro
effects of 8-MOP, UVA PUVA and MTX on the occurrence of
metaphases with PCD and MC chromosomes is presented. The

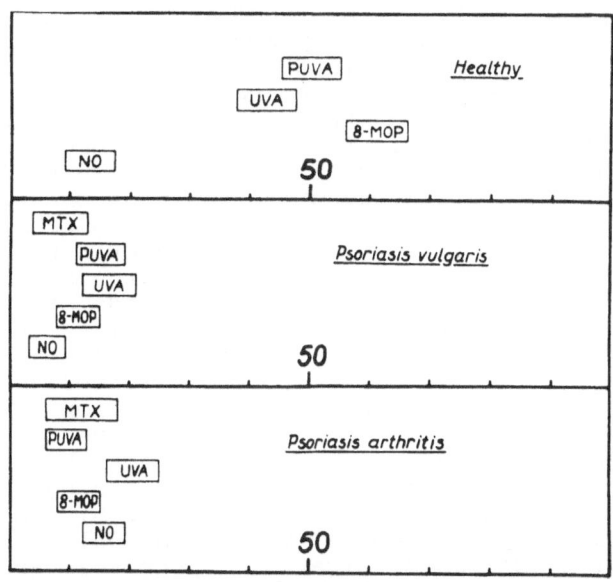

Fig. 1   Premature centromer division (PCD) of chromosomes from healthy persons and patients with
psoriasis vulgaris or psoriasis arthritis
(psoriasis arthropathica) after PUVA treatment _in vitro_ (with confidence intervals).

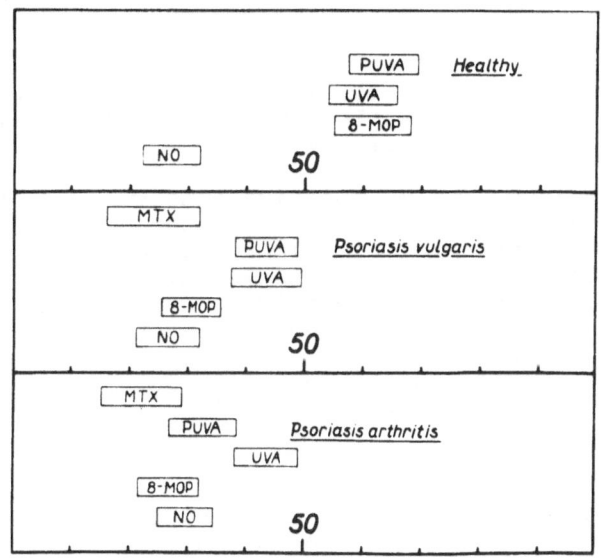

Fig. 2   Major coiling (MC) of chromosomes from
healthy persons and patients with
psoriasis vulgaris or psoriasis arthritis
(psoriasis arthropathica) after PUVA treatment _in vitro_ (with confidence intervals).

following features can easily be recognized: (i) the number of metaphase plates containing MC chromosomes is generally higher than the number of metaphase plates with PCD chromosomes, (ii) the number of metaphase plates from healthy persons with PCD or MC defects is higher after the treatment when compared with those of patients with psoriasis vulgaris and psoriasis arthritis, (iii) the antimetabolite MTX is known to inhibit competitively dehydrofolate reductase and thus DNA and RNA synthesis and to produce breaks in fragile chromosome segments (Schmidt and Passarge, 1981). Under our experimental conditions MTX did not give a measurable degree of metaphase plates with PCD or MC chromosomes in both kinds of psoriatic lymphocytes. The sets of 300-400 metaphase plates per proband and treatment were additionally analysed with respect to the occurence of other chromosomal aberrations (insertions, deletions, inversions, translocations, gaps) upon the phase of drug, UVA or PUVA administration. From the five types of chromosome aberration mentioned above (SCE not checked) only gaps were observed, whose number in no case exceeded the level of their spontaneous appearance. In conclusion, the occurrence of up to 400 metaphase plates in healthy probands and patients with psoriasis vulgaris or psoriasis arthritis investigaged in order to quantify chromosomal modifications by 8-MOP, UVA or PUVA treatment showed the sensitive response to chromosome defects like PCD or MC. This finding suggests that these techniques might be suitable tools for diagnosing latent psoriatic diseases, and for the refined analysis of psoriatic cells or chromosomes. No hint to a preferred attack of possible fragile sites could be obtained.

In vivo investigations. Immediately after PUVA therapy occurred chromosomes taken from peripheral lymphocytes of the psoriatic patients were analyzed. The found types of chromosomal aberrations and their frequencies correspond to their spontaneous appearance. This relates to chromatid and isochromatid gaps and, more rare, double minutes and deletions (observed hyperploid or hypoploid metaphase plates are probably methodological artefacts).

The analyzed PCD and MC modifications yielded increased frequencies only for MC. (Since the number of patients was small no care was taken for discrimination between psoriasis vulgaris and psoriasis arthritis patients). Moreover, the frequency of metaphase plates with MC modifications is with 50.9% clearly higher than the frequencies observed in in vitro experiments, i.e., when the isolated peripheral lymphocytes were PUVA treated (compare with data in Fig. 2).

REFERENCES

Abel, G. and Schimmer, O., 1981, Mutagenicity and toxicity of furocoumarins comparative investigations in two test systems, Mutat. Res., 90:451.

Barth, J., Heilmann, S., Meffert, H. and Metz, D., 1982, Arbeitsempfehlungen PUVA, Dermatol. Monatsschr., 168:53.

Beensen, V., 1986, Ein Beitrag zur Optimierung der Identifikation menschlicher Chromosomen, Thesis, Friedrich Schiller, University Jena.

Bredberg, A., Lambert, B., Lindblatt, A., Gunnar-Swanbeck, B.S. and Wennersten, G., 1983, Studies of DNA and chomosome damage in skin fibroblasts and blood lymphocytes from psoriasis patients treated with 8-methoxypsoralen and UVA irradiation, J. Invest. Dermatol., 81:93.

Bridges, B.A., Greaves, M., Polani, P.E. and Wald, I., 1981, Do treatments avaialble for psoraiasis patients carry a genetic or carcinogenic risk?, Mutat. Res., 86:279.

Brogger, A., Waksvik, H. and Thune, P., 1978, Psoralen/UVA treatment of chromosomes. II. Analysis of psoriasis patients, Arch. Dermatol. Res., 261:287.

Cassel, D.M. and Latt, S.A., 1980, Relationships between DNA adduct formation and sister chromatid exchange induction by ^{3}H 8-methoxypsoralen in Chinese hamster ovary cells, Exp. Cell. Res., 128:15.

Cole, R.S., 1971, Psoralen mono-adducts and interstrand cross-links in DNA, Biochim. Biophys. Acta. 254:30.

Engel, P.F. and Wulf, H.C., 1982, Localization of radioactivity in rat organs after oral administration of tritiated 8-methoxypsoralen in therapeutic doses, Arch. Dermatol. Res., 273:71.

Gast, W., Barth, J. and Rytter, M., 1982, PUVA-Wirkungen auf Lymphozyten, Dermatol. Monatsschr., 273:71.

Hook, G.J., Heddle, J.A. and Marshall, R.R., 1983, On the types of chromosomal aberrations induced by 8-methoxypsoralen, Cytogenet. Cell Genet., 35:100.

Igali, S., Bridges, B.A., Ashwood-Smith, M.J. and Scott, B.R., 1970, Mutagenesis in Escherichia coli IV. Photosensitization to near ultraviolet light by 8-methoxypsoralen, Mutat. Res., 9:21.

Jorgensen, A.L. and Bak, A.L., 1982, The last order of coiling in human chromosomes, Exp. Cell Res., 139:447.

Kirkland, D.J., Creed, K.L. and Mannisto, P., 1983, Comparative bacterial mutagenicity studies with 8-methoxypsoralen and 4,5',8-trimethylpsoralen in the presence of near-ultraviolet light and in the dark, Mutat. Res., 116:73.

Kittler, L. and Löber, G., 1983, Furocoumarins, biophysical investigations on their modes of action, Stud. Biophys., 97:61.

Lambert, G., Morad, M., Bredberg, A., Swanbeck, G. and Thyresson-Hok, M., 1978, Sister chromatid exchanges in lymphocytes from psoriasis patients treated with 8-methoxypsoralen and longwave ultraviolet light, Acta Dermato-Venerol., (Stockholm) 58:13.

Lerman, S., Megaw, J. and Gardner, K., 1982, Psoralen-longwave ultraviolet therapy and human cataractogenesis, Invest. Ophthalmol. Visual Sci., (St. Louis) 23:80.

Linnainmaa, K. and Wolff, S., 1982, Sister chromatid exchange induced by short-lived monoadducts produced by the bifunctional agents mitomycin C and 8-methoxypsoralen, Environ. Mutagen., 4:239.

Löber, G., Beensen, V. and Kittler, L., 1982, Effect of 8-methoxypsoralen (8-MOP) and UVA on the structure of human metaphase chromosomes, Stud. Biophys., 87:203.

Meffert, H., Diezel, W., Gunther, W. and Sönnichsen, N., 1976, Fotochemotherapie der Psoriasis mit 8-Methoxypsoralen und UVA. II. Bindung des Fotosensibilisators an Protein, Dermatol. Monatsschrift, 162:887.

Moorhead, P.S., Nowell, P.C., Mellman, W.J., Batipps, D.M. and Hungerford, D.A., 1960, Chromosome preparations of leucocytes cultures from human pericherical blood, _Exptl._ _Cell Res._, 20:613.

Morison, W.L. Parrish, J.A. and Epstein, J.A., 1979, Photoimmunology, _Arch. Dermatol._, 115:350.

Noble, W.C., Path, M.R.C., White, P.M. and Baker, H., 1975, Assay of therapeutic doses of methotrexate in body fluids of patients with psoriasis, _J. Invest. Dermatol._, 64:69.

Parrish, J.A., Fitzpatrick, T.B., Tanenbaum, L. and Pathak, M.A., 1974, Photochemotherapy of psoriasis with oral methoxalen and longwave ultraviolet light, _New Engl. J. Med._, 291:1207.

Pathak, M.A. and Joshi, P.C., 1983, The nature and molecular basis of cutaneous photosensitivity reactions to psoralens and coal tar, _J. Invest. Dermatol._, 80:66.

Potapenko, A. Ya. and Sukhorukov, V.L., 1984, Photooxidative reactions of psoralens, _Stud. Biophys._, 101:89.

Roenigk, H.H. and Caro, W.A., 1981, Skin cancer in the PUVA - 48 cooperative study, _J. Amer. Acad. Dermatol._, 4:319.

Römer, W., Kittler, L. and Löber, G., 1983, Inhibitory effects of furocoumarins plus near UV light (365 nm) on cAMP phosphodiesterase activity, _Stud. Biophys._, 94:33.

Schmidt, A. and Passarge, E., 1981, X-chromosomal erblicher Schwachsinn und brüchige Stelle am X-Chromosom, _Dtsch. Med. Wochenschr._, 106:460.

Stern, R.S., Laird, N., Melski, J., Parrish, J.A., Fitzpatrick, T.B. and Bleich, H.L., 1984, Cytaneous squamos cell carcinoma in patients treated with PUVA, _New. Engl. J. Med._, 310:1156.

Sutherland, G.R. and Leonhard, P., 1979, Hereditary recessives on human chromosomes. III. Detection of Fra (x) (972) in male with x linked mental retardation in their female relatives, _Hum. Genet._, 53:29.

Thune, P., 1978, Plasma levels of 8-methoxypsoralen and phototoxicity studies during PUVA treatment of psoriasis with meladinin tablets, _Acta Dermato-Venerol._, (Stockholm) 58:149.

Vig, E.K., 1981, Sequence of centromers separation: Analysis of mitotic chromosomes in man, _Hum. Genet._, 57:247.

Watksvik, H., Brogger, A. and Stene, J., 1977, Psoralen/UVA treatment and chromosomes. I. Aberrations and sister chromatid exchanges in human lymphocytes _in vitro_ and synergism with caffeine, _Hum. Genet._, 38:195.

Wulf, H.C., 1978, Acute effect of 8-methoxypsoralen and ultraviolet light on sister chromatid exchange, _Arch. Dermatol. Res._, 263:37.

5-MOP INDUCED PROTECTION AGAINST EPIDERMAL DNA DAMAGE BY
ULTRAVIOLET RADIATION IN HUMAN SKIN: THE ROLE OF THE SKIN
TYPE

Antony R. Young[1], Christopher S. Potten[2],
Caroline A. Chadwick[2], Gillian M. Murphy[1]
and A. Jeffrey Cohen[3]

[1]Photobiology Unit, Institute of Dermatology
United Medical and Dental Schools of Guy's
and St. Thomas's Hospitals, University of
London, UK

[2]Epithelial Biology Department, Paterson
Institute for Cancer Research, Christie Hospital
and Holt Radium Institute, Manchester, UK

[3]Toxicology Advisory Servies, Sutton, Surrey, UK

INTRODUCTION

It is now well recognised by the dermatological community
that solar exposre results in an increased risk of skin
cancer, especially in individuals of skin types I and II who
sunburn easily and have a limited ability to tan (MacKie et
al., 1987). In the last few years this information has been
widely publicised in the press and other media. Despite this,
a suntan is still a prized social asset and epidemiological
data indicate a continued rise in the incidence of skin
cancers. It also seems reasonable to suppose that the
possible depletion of the stratospheric ozone layer, over
populated areas of the earth's surface, would result in a
further long-term increase of skin cancer (van der Leun,
1988).

Ultraviolet radiation (UVR) induced damage of the skin
results in tanning and, in some cases, stratum corneum
thickening which are generally believed to offer subsequent
photoprotection from such damage. This view has been
supported by studies which show that a tan induced by UVB
(280-315 nm) or UVA (315-400 nm) protects against erythema
and/or DNA damage (Gange et al., 1985). However, such studies
do not necessarily represent what happens when a tan is
induced by solar UVR. It has also been demonstrated that a
tan acquired with UVR plus 8-methoxypsoralen (8-MOP) or 5-
methoxypsoralen (5-MOP) results in photoprotection from damage
such as erythema and sunburn cells formation by subsequent UVR

exposure (Cripps, 1981; Gschnait et al., 1978; Sambuco et al., 1987).

An important, and as yet unanswered, question is whether a tan acquired by solar exposure offers any protection from skin cancer from UVR. It is difficult, if not impossible, to answer this question directly, at least in humans. One approach would be to study protection against UVR-induced DNA damage, which is believed to be an important marker of malignancy.

We report on human studies which were designed to investigate the relative protective effects of different tans from DNA damage induced by solar simulated radiation (SSR). The damage was estimated from the levels of unscheduled DNA systhesis (UDS) in epidermal cells, assuming that levels of DNA repair reflect levels of DNA damage. The tans were acquired by repeated exposure of normal human skin to sub-erythemogenic doses of SSR in the presence or absence of 5-MOP preparations containing UVB sunscreens. 5-MOP is present in natural citrus oils, especially Citrus bergamia, that have been added to some commercial sunscreen preparations to enhance tanning.

The data presented here are an extension of previously published studies (Young et al., 1988; Potten et al., 1989) in which we showed, in individuals of skin type II, that a tan induced by 5-MOP containing sunscreens plus SSR offered superior protection against SSR induced DNA damage when compared with a tan induced by SSR alone. We have now compared the responses of individuals of skin types I and II (poor tanning ability) with those of types III, IV and V (progressively good tanning ability).

MATERIALS AND METHODS

Sunscreen Preparations

The sunscreen lotions were supplied by Laboratoires Bergaderm SA (France). One contained final concentrations of 30 ppm 5-MOP in citrus oils and the UVB filter, Parsol MCX (2-ethylhexyl 4'-methoxycinnamate) at 5% (w/w). The other contained the UVB filter only, at the same concentration. The concentrations of 5-MOP and UVB sunscreens in the test preparations were confirmed by Laboratoires Bergaderm.

SSR Source

The SSR source was a 2.5 kW Kratos (Westwood, N.J.) solar simulator with an air mass 1 filter (sun at 90° elevation). UVA irradiance was monitored with an International Light IL442A radiometer calibrated at 366 nm and was generally 11.25 mW.cm^{-2}. UVB irradiance was not determined.

Experimental Procedure

The test sites were previously untanned buttock skin of normal healthy subjects aged between 20-30, all of whom gave informed consent. Prior to the tanning protocol, the 24 hour SSR minimal eythema dose (MED) (just perceptible erythema) was determined for each subject using a $\sqrt{2}$ based incremental series of exposure times. Skin type was also assessed, based on skin coloration and tanning history. Details of skin type and MED are given in Table 1.

The tanning protocol was as follows. Test preparations were applied to the buttock skin and 30 minutes later, the sites were exposed to 0.70 MED (barely perceptible) of SSR. This procedure was repeated 5 days/week (Mon-Fri) for 2 consecutive weeks. The preparations were spread over 1 cm^2 sites, defined with an adhesive foil template, with a gloved finger so that there was no excess of preparation. It was estimated (by tests over larger areas) that the amount of preparations applied was of the order of $2\mu l/cm^2$. It can be assumed that any variation would be randomized over the two-week treatment. After a lapse of one week (day 19 after the beginning of the treatments), control and treatment sites were challenged with 2 MED SSR. No test preparations were applied immediately before this challenge.

The degree of erythema and tan on each site was noted on a daily basis, and on the day of challenge.

Tissue Handling

After treatment, biopsies were taken under local anaesthesia using a 3 mm full thickness dermal punch. Excess tissue was removed with a scalpel blade and the remaining circular piece of epidermal tissue sliced into approximately 1 mm thick strips. Within 10-15 minutes of SSR challenge, the biopsy samples were placed in 5-10 ml balanced salt solution containing 370 kBq (10 μCi)/ml of methyl ^{3}H-thymidine (^{3}HTdR) (sp act 185 GBq (5 Ci)/mM). The samples were left to incubate for 1 hour at 37°C with manual agitation every 5 minutes and were then placed in Carnoy's fixative for not less than 20 minutes before being placed in 70% ethanol for storage.

Sectioning and Autoradiography

A single 1 mm strip was then carefully embedded in paraffin and sectioned at right angles to the surface (5 μm). The sections were dewaxed, hydrated, and dipped in K5 liquid emulsion (Ilford Ltd) diluted 1:1 with water. After air drying at not more than 20°C, the slides were exposed for two weeks at 4°C after which they were developed and stained with hematoxylin and eosin.

Microscopy and Counting of UDS

The slides were analysed using a Zeiss planapo x 40 oil immersion lens. Six randomly selected areas of epidermis were used on each section. The number of silver grains overlaying each nucleus was counted and recorded. If S phase cells had more than 50 grains over their nuclei the number was not counted. Counts were kept separate for the basal layer (which contained most of the S cells) and the suprabasal layers of which there were generally more than 3-4. Generally about 15 basal cells were observed per field and about 50 suprabasal cells which provided about a total of about 100 basal cells and about 300 suprabasal cells per section. Generally, only one section per biopsy was scored. The average grain count for each volunteer was obtained excluding the S phase nuclei which were defined as having more than 30 grains per nucleus.

Scoring of Melanin and Stratum Corneum Thickness

Additional slides were stained using the Masson-Fontana silver reaction for melanin which also resulted in swelling of the stratum corneum. In this swollen state the number of cell layers was counted at 10 points in the section. The level of pigmentation was assessed by three independent assessors using an arbitary scale ranging from 0 (no detectable melanin at all) through to 6 (very high levels of pigmentation). The melanin levels in the basal layer and the suprabasal layers were determined separately.

RESULTS

The type I and some type II volunteers showed low grade erythema with confluent borders on some of the SSR only and some of the 5-MOP sunscreen sites during the treatments. In the type II subjects, erythema was usually restricted to the first week of treatment. Erythema was not seen on the sunscreen without 5-MOP sites. In general the intensity and duration of erythema for the 5-MOP containing sunscreen was similar to that of SSR alone. Some of the type I subjects showed erythema on the challenge day. This was not seen with type II, III, IV or V subjects. Transient erythema was very occasionally seen in some of the type III, IV, and V subjects during the treatment (SSR alone or SSR plus 5-MOP sunscreen) period.

Visual tanning was absent or barely perceptible on the challenge day in type I subjects. In general, the type II subjects started to show evidence of tanning on the SSR alone and 5-MOP sunscreen sites at the beginning of the second week. Type III, IV and V subjects started to show evidence of tanning after 1 to 3 days. Detailed results of visual tan on challenge day are not given here as, in general, they correlated very well with histologically observed tans indicated in Table 2. The tans obtained in skin types II, III, IV, and V with the 5-MOP sunscreen were similar to those obtained with SSR alone. Skin type IIs tanned minimally with the sunscreen only preparations, but skin types III, IV, and V showed some tanning with this treatment. Stratum corneum thickening is shown in Table 2.

The main results are presented in Table 1. These are expressed as protective indices (PI) of each individual of a given skin type for a given treatment. The PI for each treatment for each subject was calculated by dividing the post-challenge UDS score for the said treatment by the score obtained from the site challenged without prior treatment in the same volunteer. The latter (denominator) was usually in the region of 4-6 grains per nucleus. PIs < 1.00 show protection and scores > 1.00 show enhanced damage. The data in Table 1 are derived from counts in the basal layer. In general, the counts from the suprabasal layers showed the same trends. Analysis of data from completely untreated sites showed individual background UDS values (grains/basal cell) ranging from 0.2 to 1.0.

Table 1. Effect of Skin Type

MED (mins)	Skin Type	SSR only	S+SSR	5-MOP+S+SSR
	Details of subjects		UDS Ratio (PI)	
			PRE-TREATMENT	
14	I	1.43	1.07	0.70
14	I	1.00	0.80	0.33
14	I	0.89	1.23	0.93
14	I	1.12	0.98	0.39
14	I	1.07	1.39	0.27
mean±SD		1.10±0.20	1.03±0.29	0.52±0.28
14	II	0.81	0.88	0.61
14	II	1.27	1.03	0.73
14	II	1.34	1.38	0.60
14	II	1.14	0.80	0.33
mean±SD		1.14±0.23	1.02±0.26	0.56±0.17
20	III	0.41	1.15	0.40
20	III	0.62	0.91	0.41
20	III	0.58	0.92	0.23
20	III	0.90	1.03	0.48
mean±SD		0.63±0.20	1.00±0.11	0.38±0.11
*14(19)	IV	0.49	0.59	0.62
*28(37)	IV	0.73	1.18	0.39
*28(37)	IV	0.57	0.96	0.48
>40	IV	0.54	0.84	0.42
mean±SD		0.58±0.10	0.89±0.25	0.48±0.10
*40(53)	V	0.43	0.70	0.24

*In these subjects a higher irradiance (x 1.33) was used.
Value in parenthesis indicates "equivalent MED".

Table 1 shows the influence of skin type on MED and UDS
ratios with pretreatment with SSR alone, UVB sunscreen (S)
plus SSR, and 5-MOP containing sunscreen plus SSR. The
results show that pretreatment with SSR only shows no
protection in skin types I and II but gives protection with
skin types III plus. The application of the UVB sunscreen
blocked the protection shown with skin type III and reduced
that shown with skin types IV and V. Types I and II showed
protection after treatment with 5-MOP containing sunscreen and
improved protection, compared with SSR only, is seen with skin
types III plus.

Table 2 shows the effect of the various pre-treatments on
pigmentation and stratum corneum thickening compared with
sites that had no treatment at all.

Table 2. Effect of Pre-treatments on pigmentation and stratum corneum thickening.

| Skin Type | End Point | PRE-TREATMENTS | | | |
| | | None | SSR only | S+SSR | 5-MOP+S+SSR |
		Mean Score ± SD			
I	melanin	0.3±0.3	1.9±1.2	0.5±0.5	1.6±0.4
	sc layers	16.8±1.8	21.1±1.4	19.6±2.3	21.9±2.3
II	melanin	0.8±0.2	2.1±0.1	0.9±0.3	2.6±0.9
	sc layers	14.4±2.1	17.4±2.1	14.8±1.6	20.0±1.9
III	melanin	2.0±1.1	5.2±0.4	3.2±0.6	4.6±0.3
	sc layers	13.0±1.5	19.8±4.4	15.9±2.2	21.4±4.5
IV	melanin	4.8±0.9	6.0±0.8	5.3±1.6	6.1±1.1
	sc layers	14.6±2.4	23.2±4.5	20.0±4.5	21.0±2.7
V	melanin	5.7	6.0	6.0	6.0
	sc layers	12.2	13.2	12.3	13.0

DISCUSSION

UVB-induced pyrimidine dimers have been shown to persist in human skin for at least 24 hours (D'Ambrosio et al., 1981). Hönigsmann et al., (1987) have shown that UDS is still apparent at this time after a 2 MED exposure. We have delivered the challenge dose of 2 MED SSR 7 days after the last tanning exposure in order to eliminate the occurence of UDS as a consequence of the tanning protocol itself. No UDS was seen at 7 days in sites that were pretreated but not challenged (data not shown). Almost all of the basal cells would have undergone a cell division in 7 days (Potten, 1987) under normal steady-state conditions. The increase in the number of layers of the stratum corneum indicates a shorter turnover time.

The skin type I and II data show that pretreatment with a 5-MOP containing UVB sunscreen preparation plus SSR offers good protection from DNA damage (about a 50% reduction in UDS) caused by a challenge dose of 2 MED SSR. Associated with this photoprotection was an increase in the number of layers of the stratum corneum and an increase in pigmentation.

Pretreatment with a UVB sunscreen without 5-MOP plus SSR did not offer any photoprotection in skin types I and II. This is not surprising as this protocol resulted in minimal increases in skin pigmentation and number of stratum corneum layers. However, pretreatment with SSR alone offered no photoprotection despite increases of skin pigmentation and number of stratum corneum layers which, in general, were comparable to those that were obtained with the 5-MOP protocols.

These data indicate qualitative differences between the SSR alone and the 5-MOP plus SSR groups, despite the morphological similarities in tanning and stratum corneum thickening that were observed within skin types I and II. These differences may be in melaninization itself or other unrelated biochemical processes that may be important in photoprotection. The photoprotection obtained with 5-MOP in the 5 skin type I volunteers was the same as that observed in the skin type II subjects. The increase in pigmentation seen in this group (Type I) after 5-MOP pretreatment was minimal (other unpublished studies have shown 5-MOP induced tanning in skin type I under difference conditions). Thus, the type I data, along with the discrepancy between pigmentation (with SSR alone and 5-MOP plus SSR) and photoprotection in skin type II suggest that induced pigmentation per se may not be the determining factor in photoprotection. The skin type III, IV and V subjects also showed good photoprotection with the 5-MOP pretreatments but unlike the other skin types also showed photoprotection with SSR alone pretreatment.

Gange et al. (1985) reported that both a UVA- and a UVB-induced tan offered about a 45% reduction in UVB-induced (1 MED with well defined margins) DNA damage as measured by endonuclease sensitive sites. The authors did not define skin type but stated that the subjects were Caucasian with good ability to tan. We conclude that these subjects were probably skin type III (or more), and that these data are in good agreement with our skin type III plus subjects whose tan was obtained by SSR alone.

Any discussion on the use of 5-MOP containing sunscreens must acknowledge that, in the presence of UVA, 5-MOP is a potent mutagen (Ashwood-Smith et al., 1980) and a carcinogen in albino mouse skin (Zajdela and Bisagni, 1981; Young et al., 1983). Based on these studies, the use of 5-MOP in sunscreens has been taken to present a potentially enhanced risk of cancer in human skin. However, photocarcinogenicity of 5-MOP is significantly inhibited if UVB sunscreens are also present (Young et al., 1987). Thus, the risk associated with the use of 5-MOP in sunscreen preparations containing UVB filters is considerably lower than was previously estimated on data obtained with 5-MOP without UVB filters.

If a tan is the desired objective of a holiday it may be achieved by several means. Sunbathing without any sunscreen protection carries the risk of a sunburn and substantial DNA damage, especially in skin types I and II. The use of low SPF UVB sunscreens will allow a longer period in the sun but may not, unless carefully used, prevent sunburn or DNA damage. The use of effective high SPF (eg > 15) UVB sunscreens will prevent sunburn and, in all probability, reduce DNA damage but the desired tan may not be achieved in a reasonable time. Any tan obtained by such sunscreens is likely to have been partly induced by solar UVA. Recent data have shown that UVA is a skin carcinogen in mice and it has been noted that the UVB-UVA action spectra for human tanning and mouse skin photo-carcinogenesis are similar (Roza et al., 1989). Thus a UVA tan, whether obtained by sunlight or a sunbed, is likely to carry a risk similar to that of a UVB tan. Our data indicate that the judicious use of 5-MOP-containing sunscreens is a further option. This approach has the advantage of offering protection against DNA damage induced by sunlight which was not noted with the other options except in skin type III plus. Our data also indicate that after use of 5-MOP containing sunscreens, type II skin behaves more like type III plus in response to solar ultraviolet radiation.

It is likely that obtaining a tan by whatever means enhances the risk of skin cancer. Current knowledge does not allow us to predict with assurance the safest way of obtaining a tan. We believe that our data show that the use of 5-MOP containing sunscreens offers potential benefits which may outweigh potential risks. Our data also indicate that the use of sunscreens with UVB filters alone is not necessarily safer than the use of the UVB sunscreens containing 5-MOP. However, quantitative analysis cannot be undertaken without a considerable amount of further research.

ACKNOWLEDGEMENTS

This work was supported by a grant from Labortoires Bergaderm SA, France. We thank John Havlin, Pam Elliot and Judith Kinley for excellent technical assistance.

REFERENCES

Ashwood-Smith, M.J., Poulton, G.A., Barker, M., and Mildenberger, M., 1980, 5-methoxypsoralen, an ingredient in several suntan preparations, has lethal, mutagenic and clastogenic properties, Nature, 285:407.

Cripps, D.J., 1981, Natural and artificial photoprotection, J. Invest. Derm., 76:154.

D'Ambrosio, S.M., Slazinski, L., Whetstone, J.W., and Lowney, E., 1981, Excision repair of UV-induced pyrimidine dimers in human skin in vivo, J. Invest. Derm., 77:311.

Gange, R.W., Blackett, A.D., Matzinger,E.A. Sutherland, B.M., and Kochevar, I.E., 1985, Comparative protection efficiency of UVA-and UVB-induced tans against erythema and formation of endonuclease-sensitive sites in DNA by UVB in human skin, J. Invest. Derm., 85:352.

Gschnait, F., Brenner, W., and Wolf, K., 1978, Photoprotective effect of a psoralen-UVA-induced tan, Arch. Dermatol. Res., 263:181.

Hönigsmann, H., Brenner, W., Tanew, A., and Ortel, B., 1987, UV-induced unscheduled DNA synthesis in human skin: Dose response, correlation with erythema, time course, and split dose exposure in vivo, J. Photochem. Photobiol. (B: Biol.) 1:33.

MacKie, R.M., Elwood, J.M., and Hawk, J.L.M, 1987, Links between exposure to ultraviolet radiation and skin cancer, A Report of the Royal College of Physicians, J. Royal Coll. Phys., 21(2):1.

Potten, C.S., 1987, Possible defects in the proliferative organization and control mechanisms in psoriasis, in: "Proceedings of 4th International Symposium on Psoriasis, 1986 at Stanford University, California", Elsevier, New York.

Potten, C.S., Chadwick, C.A., Young, A.R., Murphy, G.M. and Cohen, A.J., 1989, A 5-methoxypsoralen-induced tan protects against DNA damage from a subsequent exposure to solar simulated radiation in human skin, in: "Psoralens; Photochemeprotection and Other Biological Activities", T.B. Fitzpatrick, P. Forlot, M.A. Pathak, and F. Urbach, eds., John Libbey Eurotext, Paris.

Roza, L., Baan, R.A., van der Leun, J.C., and Kligman, L., 1989, UVA hazards in skin associated with the use of tanning equipment, J. Photochem. Photobiol. (B: Biol.), 3:281.

Sambuco, C.P., Forbes, P.D., Davies, D.E., and Urbach, F., 1987, Protective value of skin tanning induced by ultraviolet radiation plus a sunscreen containing bergamot oil, J. Soc. Cosmet. Chem., 38:11.

van der Leun, J.C., 1988, Ozone depletion and skin cancer, J. Photochem. Photobiol. (B: Biol.), 1:493.

Young, A.R., Magnus, I.A., Davies, A.C., and Smith, N.P., 1983, A comparison of the phototumorigenic potential of 8-MOP and 5-MOP simulated radiation, Br. J. Derm., 108:507.

Young, A.R., Gibbs, N.K., and Magnus, I.A., 1987, Modification of 5-methoxypsoralen phototumorigenesis by UVB sunscreens: A statistical and histologic study in the hairless albino mouse. J. Invest. Dermatol., 89:611.

Young, A.R., Potten, C.S., Chadwick, C.A., Murphy, G.M., and Cohen, A.J., 1988, Inhibition of UV radiation-induced DNA damage by a 5-methoxypsoralen tan in human skin, Pig. Cell. Res., 1:350.

Zajelda, R., and Bisagni, E., 1981, 5-methoxypsoralen, the melanogenic additive in sun-tan preparations, is tumorigenic in mice exposed to 365 nm u.v. radiation, Carcinogenesis, 2:121.

QUANTITATIVE CHARACTERIZATION OF PHOTOSENSITIZER-NUCLEOPROTEIN
INTERACTIONS: A COMPARISON OF 4,6,4'-TRIMETHYLANGELICIN AND
4'-AMINOMETHYL-4,5'8-TRIMETHYLPSORALEN

Katalin Toth, Gabriella Csik and Györgyi Rontó

Institute of Biophysics, Semmelweis Medical Univ.
P O Box 263, H-1444 Budapest, Hungary

INTRODUCTION

The 4,6,4'-trimethylangelicin (TMA), synthesized in the
early 80s (Baccichetti et al., 1982) presents some
photochemotherapeutically advantageous features such as good
antiproliferative effects, inhibition of tumour transfer, and
lack of erythema induction (Baccichetti et al., 1984; Guiotto
et al., 1984). Its effectivity and side effects are usually
compared with those of 8-methoxypsoralen (8-MOP). Most of
these comparisons favour TMA, however there are reports of
controversary opinion concerning TMA (Santamaria et al., 1987)
as well as other monofunctional derivatives (Dall'Acqua 1988).
Investigations of side effects, studied on different systems
from bacterium to mice, focused on light induced effects. The
mutagenicity of TMA in the dark was found to be negligible on
Salmonella typhimurium (Guiotto et al., 1984). Generally very
little attention is paid to the possible deleterious side
effects caused by pre-irradiated furocoumarins. Recently we
have reported a significant increase of the dark genotoxicity
on phages in the case of pre-irradiated 4'-aminomethyl-4,5',8-
trimethylpsoralen (AMT) while no biological activity was
observed for pre-irradiated 8-MOP, 8-methyl-3-
carbethoxypsoralen and their thio-substituted derivatives
(Toth et al., 1988). In the case of TMA photodimerization is
known to occur (Caffieri et al., 1987) but its biological
importance has not yet been investigated further.

Due to differences in the test object sensitivities and
experimental conditions (concentrations, solvents, etc.) used
in the measurements it is hard to compare quantitatively the
beneficial and side effects of a given furocoumarin and so to
choose the best compound. Recently we proposed a simple but
very sensitive test system i.e. bacteriophages for a certain
quantification (Toth et al., 1988). Dark and light induced
inactivation as well as the role of pre-irradiation can be
studied on them. Moreover, in the case of DNA phages the
structural consequences of these interactions can be followed

as was the case for AMT (Toth et al., 1990). Phages differ from highly organized biological systems. Thus we suggest using them as models for testing:

-   the antiproliferative, beneficial effect as the light-induced phage-inactivation in the presence of furocoumarins,

-   the side-effects as inactivation on the dark either by original or by pre-irradiation drugs.

In the present paper all of these effects are quantified for TMA in terms of light or chemical doses leading to the same degree of inactivation of the T7 phage. AMT is used for comparison because of its easier handling and greater physico-chemical and photophysical similarities to TMA than to 8-MOP.

MATERIALS AND METHODS

Furocoumarins

4'-Aminomethyl-4,5'8-trimethylpsoralen (AMT) was purchased from H.R.I. Associates Inc., 4,6,4'-trimethylangelicin (TMA) was kindly donated by Prof. F. Dall'Acqua, University of Padua, Italy. Both AMT and TMA were dissolved in ethanol and then diluted by the buffer containing 50 mmol/L NaCl, 20 mmol/L Tris and I mmol/L EDTA, adjusted to pH = 7,5. The furocoumarin concentrations were determined from their absorption spectra measured on a Perkin-Elmer Lambda 15 spectrophotomer. Molar extinction coefficients at 300 nm are 10^4 L/mol.cm and $0,86.10^4$ L/mol.cm for AMT and TMA respectively (taken from Isaacs et al., 1977; Baccichetti et al., 1983). Concentrations between 1 and 100 μmol/L were applied. Absorption and fluorescence spectra were used to control furocoumarins in their dark and light mediated reactions.

Nucleoprotein

Bacteriophage T7 containing double stranded DNA (host cell E.coli B/r) was cultivated and purified as described earlier (Gaspar et al., 1981). The phage concentration was determined by the extinction coefficient at 260 nm (7300 L/mol(base).cm) corrected for light scattering (Toth 1981). A phage concentration of 10^{12}/mL (150μmol(base)/L) was used.

Irradiation

Furocoumarins in buffered solution and their mixtures with bacteriophages were irradiated by a 2,5 kW, Hg-Xenon lamp equiped with a Jobin-Yvon type grating monochromator (1400 grooves/mm). Irradiation wavelength varied between 340 and 380 nm, with a half-band width of 4 nm. The fluence rate (I_o) incident on the 1 cm silica cell was determined by a thermopile and was in the order of 5-30 W.m^{-2}. Samples were stirred during illumination time (t).

The absorbed dose (DA) is calculated taking into account the optical density (OD) of the sample measured at the irradiation wavelength:

$$DA = I_o t(1-10^{-OD})$$

and it is expressed in the averaged number of absorbed photons per mL. Irradiation of phages without sensitization was always performed for the same times, as a reference.

Inactivation

Phage inactivation was calculated either from the decrease in plaque-forming ability on Petri-dishes or by an automatic test method based on the light-scattering changes of bacteria caused by their lysis (Rontó et al., 1988). In both cases survival rates were evaluated as $\ln(N/N_o)$ where N_o and N correspond to the number of active phages before and after the treatment, respectively.

In the case of incubation in the dark, inactivation indices (MI values) were calculated from the slopes of the kinetic curves according to Rontó et al., (1986a). In the photoreaction a similar parameter was defined, the absorbed dose of 340 nm light, leading to 37% survival (DA_{340}). The ratio of MI and DA_{340} gives the quality factor (F) (Toth et al., 1988).

Thermal denaturation

Thermal denaturation of bacteriophage T7 was followed through the measurement of absorption changes at 260 nm as described previously (Toth and Rontó 1987) where identification of the different denaturation steps is given as well. Melting temperatures of the H-bond opening transition and that of the strand separation - coiling are now used to characterize the thermal stability of the nucleoprotein.

RESULTS AND DISCUSSION

Inactivation in the dark

Figure 1 presents the phage inactivation due to dark complex formation with the original and the pre-irradiated furocoumarins. For comparison, the inactivating effect of two well known genotoxic drugs: N-acetyl-ethylenimine and Lycurim are shown (Rontó et al., 1988). One can observe that the genotoxicity of both furocoumarins is of the same order of magnitude as the references are. The inactivation indices expressed in minute.mol/L units are in decreasing order: 3.10^{-2} (N-acetyl-ethyleneimine) $> 2.10^{-2}$ (AMT) $> 4,5.10^{-3}$ (TMA pre-irrad.) $> 2,2.10^{-3}$ (TMA) $> 10^{-3}$ (AMT pre-irrad.) $> 0.74.10^{-3}$ (Lycurim). The lower MI values correspond to higher genotoxic activity, i.e. for the standardized inactivation a lower product of incubation time and the used concentration is needed. Such significant genotoxicity of other furocoumarins was reported previously on the same phage (Rontó et al., 1986/b) as well as on an RNA phage MS2 (Toth et al., 1988).

The pre-irradiation of furocoumarins with UVA light may lead to some degradation, which in our cases results in changes in the optical as well as genotoxic characteristics. Figure 2 shows the absorption (a) and fluorescence (b) spectra

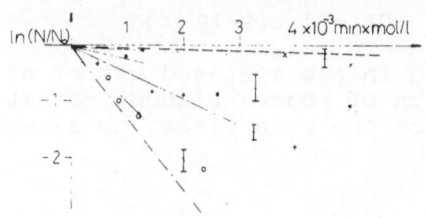

Fig. 1 Inactivation kinetics of phage T7 in the
dark with normal AMT (x) pre-irrad. AMT (o),
normal TMA (+) pre-irrad. TMA (o),
N-acetyl-ethyleneimine (- -) and Lycurim
(1,4-methylsulphonyl-oxyethylamino-1,4-
dideoxyerithrioldimethanesulphonate) (-·-).

of both furocoumarins before and after irradiating with 340 nm
light. The absorption band around 300 nm vanishes for both
drugs, whereas their fluorescence spectra change differently:
a significant decrease is observed for AMT, while only slight
red shifting is seen for TMA. Incubation of the phages with
the until saturation illuminated furocoumarins results in
inactivation as well: in the case of TMA the genotoxic
activity decreases to about half, while a fivefold increase is
observed for AMT. This behaviour of AMT seems to be unique
among the other furocoumarins investigated, whose genotoxicity
decreased or vanished upon pre-irradiation (Toth et al.,
1988). The decrease only to its half in the case of TMA may
imply that some of its photoproducts are biologically active.
The chemical structure of the photoproducts have not yet been
determined. However, in the case of TMA several products,
among them dimers, were isolated under similar irradiation
conditions and identified by NMR spectroscopy (Caffieri et
al., 1987).

The dark complexation of AMT was shown to result in a
decrease of the DNA thermal stability: one intercalated drug
per 200 base pairs represents about a 2°C decrease in strand
separation temperature (Toth et al., 1990). In the case

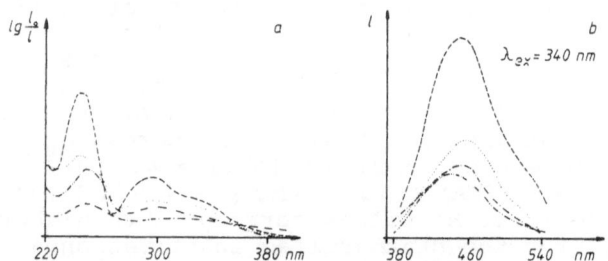

Fig. 2 Absorption (a) and fluorescence (b) spectra
of normal AMT (- -), pre-irradiated AMT
(····), normal TMA (-··-) and pre-irradiated
TMA (-·-). The pre-irradiation was carried
out at 340 nm, until the spectra did not
change any more (ca. 50 kJm^{-2} incident
dose). Furocoumarin concentrations were
60 μmol/L.

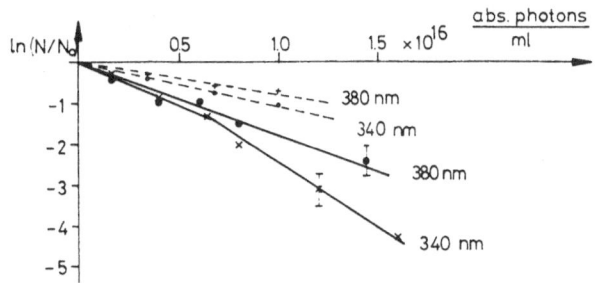

Fig. 3   Phage inactivation with photosensitizers
          AMT (——) and TMA (- -) with UVA light of
          340 nm and 380 nm. Drug concentrations
          were 1.5 μmol/L.

of TMA a similar tendency was observed, although the exact
intercalation rate is not yet available. With TMA we could
not obtain as strong destabilization as for AMT either because
the intercalation affinity is weaker or because of its
significantly lower solubility. The dark interaction of the
pre-irradiated drugs with the phages did not change the
thermal stability, suggesting a different type of interaction.

Photoreaction

     The photosensitizing effect of TMA was investigated with
narrow band UVA irradiation in order to separate possible
wavelength dependent actions. The phage inactivation in
photoreation is represented in Figure 3, as a function of the
absorbed photons at 340 nm and 380 nm for both furocoumarins.
The survival ratios are corrected both for the inactivation in
the dark and for the inactivation by UVA alone. The
photosensitizing ability of both furocoumarins is in the same
order of magnitude, AMT seems to be slightly more effective at
both wavelengths. The inactivation efficiencies of the short
and long wavelength photons are very similar to each other.

     The light induced genotoxicity was characterized by the
absorbed dose at 340 nm (DA_{340}) leading to 37% survival. This
value is 4.10^{15} photons/mL and 1.10^{16} photons/mL for AMT, and
TMA respectively.

     The similarity shown in the biological responses has not
been found in the structural consequences. The thermal
denaturation was measured, as a sensitive probe of the DNA
damage, through the amplifying effect of the cooperativity of
the melting. The melting process of the whole phage-
nucleoprotein takes place in several steps identified
previously (Toth and Rontó 1987). The photoreaction
influences only the step of highest temperature: the
separation and coiling of the DNA strands. Figure 4 presents
the changes of temperature of this transition as a function of
the absorbed dose.

     The photoreaction of the TMA does not significantly
change the thermal stability of the phage at any of the
wavelengths in the studied dose-range. It confirms the

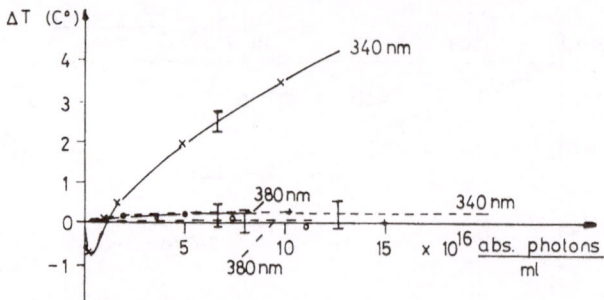

Fig. 4   Changes of the nucleoprotein melting
          temperature in photoreaction with AMT (——)
          and TMA (- -).   ΔT is expressed as the
          difference from the non-treated case.

monofunctional behaviour of TMA and the absence of cross-
links.   These are, however, formed with AMT at higher doses of
340 nm light.

Dark versus light effectivities

     As shown above, phages can be inactivated both in the
dark and by photoreaction.   We proposed to calculate the ratio
(F) of the quantitative characteristics (MI and DA_{340}) of the
two processes (Toth et al., 1988).   For standardisation the
incubation time of dark reaction is fixed at one hour while
(in the photoreaction) the concentration and wavelength are
fixed at 1,5 μmol/L and 340 nm respectively.   F is expressed
as the number of drug molecules for which the biological
consequence of a one hour incubation is equivalent to the
absorption of one UVA photon by the mixture.   The F values for
the TMA and AMT are 2±0,5 and 50±0,5, respectively,
representing a much higher risk of dark side effect for TMA.

CONCLUSIONS

     The quantitative characterization of the furocoumarin-
nucleoprotein interactions in the case of TMA and AMT lead us
to the following conclusions.

a)   Both furocoumarins are genotoxic on the phage T7 in the
      dark, TMA is about ten times more active.

b)   Pre-irradiation with 340 nm light influences differently
      the dark genotoxicities, a fivefold increase is found for
      AMT while it decreases to a half for TMA.

c)   The photosensitized inactivation of the phages is twice
      as strong with AMT than with TMA.

d)   A wavelength dependence of the photoreaction was only
      found for AMT, with which cross-links are formed at 340
      nm.

e)   The relative risk of side effects due to dark interaction
      is twenty five times higher for TMA than for AMT.

REFERENCES

Baccichetti, F., Bordin, F., Monti-Bragadin, C., Carlassare, F., Cristofolini, M., Dall'Acqua, F., Guiotto, A., Pastorini, G., Recchia, G., Rodighiero, G., Rodighiero, P. and Vedaldi, D., 1982, Italian Patent application No. 84148-A/82.

Baccichetti, F., Carlassare, F., Bordin, F., Guiotto, A., Rodighiero, P., Vedaldi, D., M.Tamaro and Dall'Acqua, F., 1984, 4,4',6-trimethylangelicin. A new very photoreactive and non skin-phototoxic monofunctional furocoumarin, Photochem. Photobiol., 39: 525-529.

Caffieri, S., Beijersbergen van Henegouven, G.M.J., Erkelens, C., de Bruijn C. and Dall'Acqua F., 1987, Photodimerization of 4,6,4'-trimethylangelicin, and 6,5'-dimethylangelicin, J. Photochem. Photobiol., B.1: 213-221.

Dall'Acqua, F. 1988, Dark and photochemical interactions between monofunctional furocoumarins and DNA, Biochem. Pharmacology. 37: 1793-1794.

Gaspar, S., Modos, K. and Rontó, Gy., 1980, Complex method for the determination of the physiological parameters of bacterium-phage systems in: "Advances in Physiological Science 34: Mathematical and Computational Methods in Physiology", L. Fedina, B. Kanyar and M. Kollai eds., Pergamon-Akademial Kiado, Budapest.

Guiotto, A., Rodighiero, P., Manzini, P., Pastorini, G., Bordin, F., Baccichetti, F., Carlassare, F., Vedaldi, D., Dall'Acqua, F., Tamaro, M., Recchia, G. and Cristofolini, M., 1984, 6-methylangelicins: a new series of potential photochemotherapeutic agents for the treatment of psoriasis. Journal of Medicinal Chemistry, 27: 959-967.

Isaacs, S.T., Shen, Sh.J., Hearst, J.E., and Rapoport, H. 1977, Synthesis and characterization of new psoralen derivatives with superior photoreactivity with DNA and RNA, Biochemistry, 16: 1058-1064.

Rontó, Gy., Tarjan, I., and Gaspar, S., 1986a, Phage T7 inactivation test. A possibility of quantitative mutagenicity screening, Physiol. Chem. Phys. Med. NMR, 18: 275-285.

Rontó, Gy., Fidy, J., Fekete, A., Toth, K., Csik,G. and Gaspar, S., 1986b, Sturctural and functional changes of bacteriophage-nucleoproteins in dark and photoreaction with furocoumarins, Stud. Biophys., 112: 63-70.

Rontó, Gy., Derka, I., Gaspar, S. and Modos, K., 1988, Quantitative test of physical and chemical effects on microbiological modes of chromosomes, cells and viruses using physical measurements, in: "I.E.E.E. Engineering in Medicine and Biology Society 10th Annual International Conference", New Orleans.

Santamaria, I., Bianchi, A., Amabold, A., Bianchi, L., Rizzi, R., Pizzalay, R., Andreoni, L., Susini, W. and Dall'Acqua, F., 1987, Photocarcinogenesis in mice and photomutagenesis in Salmonella tryphimurium TA102, Saccharomyces D7 and in UDS test induced by 4,4',6-trimethylangelicin (TMA) in: "Abstracts of 2nd Congress of the European Society for Photobiology, Padova", 94.

Toth, K., 1981, Etudes des bacteriophages T7, MS2 et OX-174 par dichroisme circulatire et l'absorption UV, Thesis, Paris.

Toth, K. and Rontó, Gy., 1987, Salt effects on bacteriophage

T7-I., <u>Physiol. Chem. Phys. and Med. NMR.</u>, 19: 59-66.

Toth, K., Csik, G. and Averbeck, D., 1988) Characterization of new furocoumarin derivatives by their dark and light-mediated action on RNA bacteriophage MS2, <u>J. Photochem. Photobiol. B.</u>, 2: 209-220.

Toth, K., Csik, G. and Rontó, Gy., 1990, Dark and photoreactivity of 4'-aminomethyl-4,5',8-trimethylpsoralen with phage T7, <u>J. Photochem. Photobiol. B.</u>, 5:167-178.

COMPARATIVE STUDIES ON THE PHOTOSENSITIZING POTENCY OF 5-MOP AND 8-MOP

Joachim Barth

Department of Dermatology, Medical Academy "Carl Gustav Carus" Dresden, GDR

INTRODUCTION

5-Methoxypsoralen (5-MOP) is an alternative photosensitizer in PUVA-therapy which has some advantages (Hönigsmann et al., 1979) compared to the commonly used 8-methoxypsoralen (8-MOP):

- lower potency to induce erythema, and, therefore,
- application of higher UV-doses possible,
- fewer side effects after systemic administration,
- faster pigmentation.

To compare and clarify the phototoxic potencies of these two psoralens we have studied them in different test systems.

MATERIALS AND METHODS

The following methods for evaluating phototoxicity have been used:

- photo-hemolysis according to Kahn and Fleischaker (1971) in a modification described by us (Barth et al., 1977),
- killing rate of human PMN (polymorphonuclear neutrophils) according to Gast et al., (1984),
- killing rate of human lymphocytes according to Gast et al., (1981),
- time course of the killing of the protozoans _Tetrahymena_ _pyriformis_ and _Paramecium caudatum_ according to Barth and Arnold (1976),
- inhibition of the growth of _Candida albicans_ according to Lohrisch et al., (1980),
- inhibition of the lymphocyte transformation rate according to Gast et al., (1983),
- induction and progress of mouse ear edema according to Barth (1978),
- induction and progress of erythema in human skin according to Barth et al., (1981).

The purity of the psoralens was authenticatecd by thin-layer chromatography, absorption spectroscopy and mass spectrometry. A fluorescent lamp, UVS 40-2 (Manufacturer: VEB Narva, 9231, Brand-Erbisdorf/GDR), was used as the UV-source. This lamp has an emission spectrum ranging from 306 to 425 nm with a maximal output at 345 nm. It is used in PUVA-therapy in the GDR (Barth et al., 1987). The UV-output was measured by means of a calibrated caesium-antimony photocell.

RESULTS

Photohemolysis

We could not induce any photohemolysis of human erythrocytes with 5-MOP or 8-MOP at the different concentrations used.

Killing rate of human PMN

By using the trypan-blue method we could demonstrate that under comparable conditions, 5-MOP-photosensitization causes a significantly higher lethality rate of PMN than 8-MOP-photosensitization at concentrations of 0.92 μM (p $\leq$ 0.01) but not at concentrations of 4.6. μM and 23.0 μM (Fig. 1).

Killing rate of human lymphocytes

In our test system we could, by utilizing the trypan-blue method, demonstrate a significantly stronger cytotoxicity of 5-MOP-photosensitization (p $\leq$ 0.01) compared to that of 8-MOP (Fig. 2) at all culture times assessed.

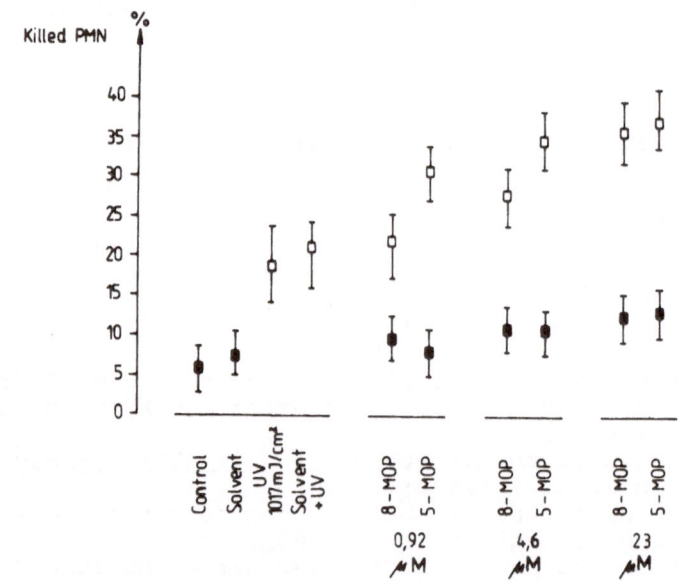

Fig. 1   Killing rate of PMN ± SD without (dark square) and with (light square) UV-radiation (10.17 kJm⁻²).

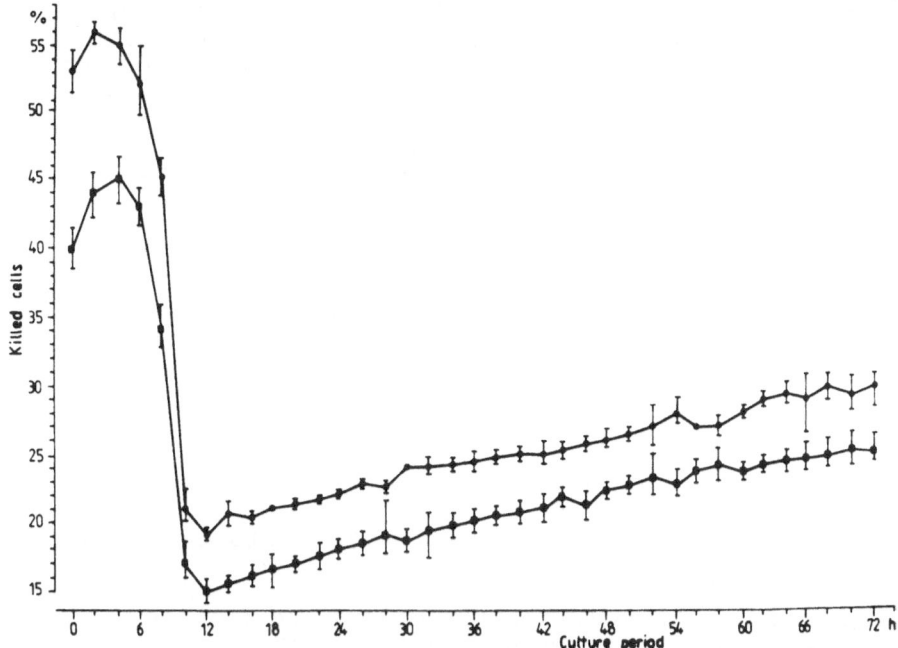

Fig. 2   Killing rate of lymphocytes (n = 5) ± SD at
          different culture periods after photo-
          sensitization with 4.6 μM 5-MOP (dark circle)
          and 8-MOP (dark square).   UV-dose 10.17 kJm^{-2}.

Time course of the killing of Tetrahymena pyriformis and Paramecium caudatum

        Protozoan lethality starts earlier and reaches its
endpoint sooner after 5-MOP-photosensitization compared to
that of 8-MOP-photosensitization (Fig. 3).

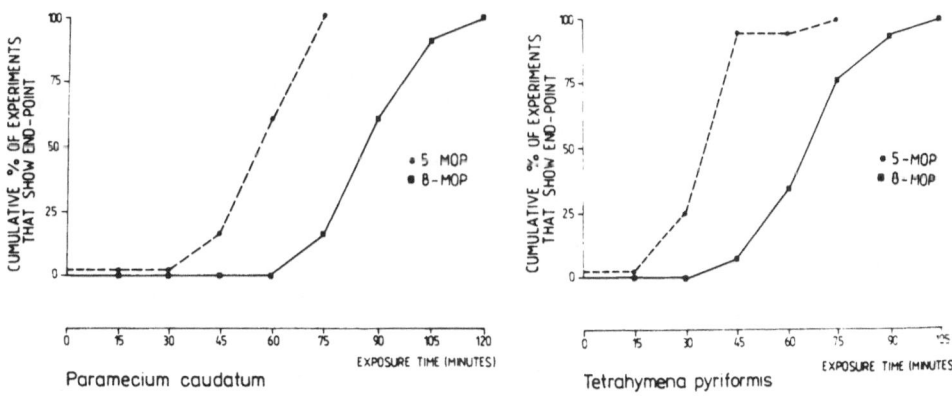

Fig. 3   Time course of the killing of Paramecium
          caudatum (n = 12) and Tetrahymena
          pyriformis (n = 21).   Fluence rate of UV-
          radiation 23 Wm^{-2}.   Final concentration of
          the photosensitizers 23.0 μM.

Table 1 Mean growth inhibition areas ± SD after
Candida albicans photosensitization with
1.74×10^{-3} M 5-MOP and 8-MOP at different
UV-doses (n = 30 each).

	20 kJm^{-2}	40 kJm^{-2}	60 kJm^{-2}	80 kJm^{-2}
5-MOP	4.8±0.5	5.7±0.6	5.7±0.5	7.8±0.7
8-MOP	4.3±0.7	5.1±0.7	5.3±0.7	6.1±0.7

Inhibition of the growth of Candida albicans

5-MOP-photosensitization inhibits the growth of Candida
albicans to a significantly higher degree ($p \leq 0.01$) at UV-
doses of 20.0 and 80.0 kJm^{-2} than 8-MOP-photosensitization
(Table 1).

Moreover, in monochromator studies we found the action
spectrum maximum of 5-MOP-photosensitization at 340 nm whilst
the action spectrum maximum of 8-MOP-photosensitization ranges
from 320 to 340 nm (Young and Barth, 1982). At 340 nm 5-MOP-
photosensitization causes the same inhibitory effect on
Candida growth as 8-MOP with less than half the UV-dose.

Inhibition of the lymphocyte transformation rate

The two compounds also inhibit lymphocyte transformation
without UV-radiation (dark effect) depending on concentration.
With UV-doses of 2.16 kJm^{-2} and higher, 8-MOP causes
significantly greater inhibition ($p \leq 0.01$) of the lymphocyte
transformation rate (Fig. 4) at the lower concentration (0.92
μM), but 5-MOP is more effective at the higher concentration
(23.0 μM).

Induction and progress of mouse ear edema

After subcutaneous injection, 8-MOP induces a more
intensive mouse ear edema than 5-MOP which can be demonstrated
by the significantly higher increase ($p \leq 0.01$) of mouse ear
swelling 72 hours after irradiation (Fig. 5).

Induction and progress of erythema in human skin

The erythema caused by topical photosensitization with
1.74×10^{-3} M solution of 8-MOP is most pronounced 48 hours
after irradiation, whilst the erythema caused by 5-MOP-
photosensitization shows no further increase 24 hours after
irradiation. This can also be confirmed by recording skin
temperature with liquid crystals (Fig. 6).

Pigmentation through 5-MOP-photosensitization also starts
without preceding erythema and shows a more reddish-brown
colour compared to the greyish-brown of the pigmentation
resulting from 8-MOP-photosensitization.

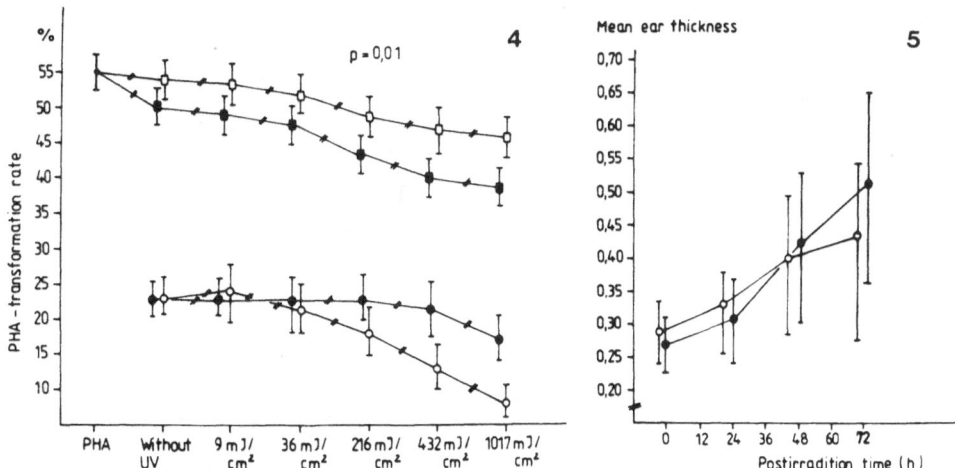

Fig. 4   Inhibition of the lymphocyte transformation
          rate ± SD after different UV-doses and
          different concentrations of the photo-
          sensitizers (n = 10)

          Light square   5-MOP 0.92 μM
          Light circle   5-MOP 23.0 μM
          Dark square   8-MOP 0.92 μM
          Dark circle   8-MOP 23.0 μM

Fig. 5   Mouse ear thickness ± SD after photo-
          sensitization with 5-MOP (light circle) and
          8-MOP (dark circle) with 2.5 x 10^{-3} mg/g
          body weight. UV-dose 48 kJm^{-2} (n = 25).

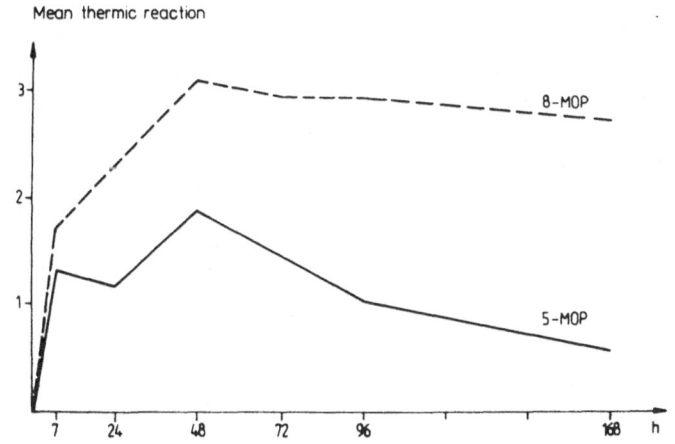

Fig. 6   Mean thermic reaction after topical photo-
          sensitization with 1.85 x 10^{-3} M solution
          of 5-MOP and 8-MOP (n = 16). UV-doses
          16.0 kJm^{-2}.

DISCUSSION

In those test systems which primarily demonstrate cytotoxic effects of photosensitization, 5-MOP proved to be the more potent photosensitizer compared to 8-MOP. This refers to the killing of PMN, lymphocytes, protozoans and the growth inhibition of <u>Candida albicans</u>. Due to the fact that all these test objects have well developed cell organelles it is difficult to identify the primary target of photosensitization. We failed to demonstrate photohemolysis in human erythrocytes, and this points to the importance of nuclear and other intracellular structures for photosensitized cell killing.

Indeed, it has been reported that 5-MOP has a higher DNA-binding capacity (Artuc et al., 1981) and a stronger ability for cross-link-formation with DNA (Dall'Acqua et al., 1971). 5-MOP also has a higher affinity for protein dark binding <u>in vitro</u> than 8-MOP (Artuc et al., 1979). Moreover, Canistaro and van de Vorst (1977) have found that, compared to 8-MOP, 5-MOP, has a higher potency for singlet oxygen production; on the other hand, the formation of free radicals with thymine and thymine derivatives is higher after 8-MOP-photosensitization than after 5-MOP-photosensitization. Therefore, it is most likely that 5-MOP and 8-MOP partly differ in their effectiveness on subcellular structures. Furthermore, our studies confirm the higher inflammative potency of 8-MOP compared to that of 5-MOP, something that is also seen after systemic photosensitization in humans (Hönigsmann et al., 1979).

According to the results presented here, 5-MOP seems to be more appropriate for photochemotherapy of those dermatoses in which cytotoxic effects are desired like, T-cell lymphomas and others. However, one should bear in mind the possibility that mutagenic effects could be induced to a higher degree than by 8-MOP-photosensitization.

REFERENCES

Artuc, M., Stüttgen, G., Schalla, W., Schäfer, H. and Gazith, J., 1979, Reversible binding of 5- and 8-methoxypsoralen to human serum proteins (albumin) and to epidermis <u>in vitro</u>, <u>Br. J. Dermatol.</u>, 101:669.

Artuc, M., Schalla, G., French, S. and Ramshad, M., 1981, Binding of psoralens to DNA by laser irradiation of different wavelengths, <u>Arch. Dermatol. Res.</u>, 270:234.

Barth, J., 1978, Mouse screening test for evaluating protection to longwave ultraviolet radiation, <u>Br. J. Dermatol.</u>, 99:357.

Barth, J. and Arnold, Th., 1976, Zur Aussagekraft des Paramäzientestes für die Erfassung von Fotosensibilisatoren, <u>Dermatol. Mon. schr.</u>, 162:900.

Barth, J., Hofmann, Ch. and Fickweiler, E., 1977, Untersuchungen zur fotohämolytischen Potenz von Pharmaka und Industriesubstanzen, <u>Dermatol. Mon. schr.</u>, 163:613.

Barth, J., Herold, W. and Wiegel, D., 1981, Der Einsatz der Flüssigkristallthermographie zur Differenzierung von Fotosensibilisationsreaktionen, <u>Medicamentum</u>, 22:87.

Barth, J., Meffert, H., Schiller, F. and Sönnichsen, N., 1987, 10 Jahre PUVA-Therapie in der DDR - Analyse zum Langzeitrisiko, Z. klin. Med., 42:889.

Canistaro, S. and van de Vorst, A., 1977, ESR and optical absorption evidence for free radical involvement in the photosensitizing action of furocoumarin derivatives and for their singlet oxygen production, Biochim. et Biophys. Acta, 476:166.

Dall'Acqua, F., Marciani, S., Ciavata, L. and Rodighiero, G., 1971, Formation of inter-strand cross-linking in the photoreactions between furocoumarins and DNA, Z. Naturforsch., 266:561.

Gast, W., Barth, J. and Rytter, M., 1981, Untersuchungen zum letalen Fotosensibilisationseffekt von 5- bzw. 8-Methoxypsoralen auf menschliche Kulturlymphozyten, Dermatol. Mon. schr., 167:182.

Gast, W., Rytter, M. and Barth, J., 1983, Zur Wirkung von 5-Methoxypsoralen auf die Phytohamagglutinin-stimulierte Lymphozytentransformationsrate in vitro, Dermatol. Mon. schr., 169:195.

Gast, W., Walther, Th., Rytter, M., Barth, J. and Haustein, U.-F., 1984, Vitalitätsverlust polymorphkerniger neutrophiler Granulozyten durch 5- bzw. 8-Methoxypsoralen und UV-Strahlung, Dermatol. Mon. schr., 170:719.

Hönigsmann, H., Jaschke, E., Gschnait, F., Brenner, W., Fritsch, P. and Wolff, K., 1979, 5-Methoxypsoralen (bergapten) in photochemotherapy of psoriasis, Br. J. Dermatol., 101:369.

Khan, G. and Fleischaker, B., 1971, I. Red blood cell hemolysis by photosensitizing compounds, J. Invest. Dermatol., 56:85.

Lohrisch, I., Müller, R., Barth, J. and Schönborn, Ch., 1980, Fototoxizität von Akridin, ausgewählten Derivaten und Komplex-verbindungen im Kandidahemmtest, Dermatol. Mon. schr., 166:156.

Young, A.R. and Barth, J., 1982, Comparative studies on the photosensitizing potency of 5-methoxypsoralen as measured by cytolysis in Paramecium caudatum and Tetrahymena pyriformis, and growth inhibition and survival in Candida albicans, Photochem. Photobiol., 35:83.

ANALYZING 8-MOP PHARMACOKINETICS

Rik Roelandts

Department of Photodermatology
University of Leuven
Leuven, Belgium

INTRODUCTION

To obtain an optimal therapeutic effect in PUVA treatment (combining total body UV-A irradiation with the oral intake of 8-MOP), it is necessary that the UV-A irradiation be applied when the 8-MOP concentration in the skin is at its maximum. In addition, it has always been considered advisable that this concentration be as high as possible.

As an indicator of the 8-MOP skin concentration, the 8-MOP plasma levels are taken. This is usually done by blood sampling at the time of 8-MOP intake and at various times thereafter.

When comparing the 8-MOP dose taken orally with the 8-MOP plasma concentration, only a small amount of the 8-MOP is found in the plasma, which indicates bad resorption. This could be due to the lipophilic character of 8-MOP and its low water solubility. In addition, large differences in 8-MOP plasma levels are found between individuals and also in the same individual from one day to another. This illustrates the importance of the pharmacokinetics of 8-MOP.

THE RELATION BETWEEN 8-MOP SKIN AND PLASMA LEVELS

A poor response to PUVA therapy may be due to low 8-MOP plasma levels at the time of the UV-A irradiation (Wagner et al., 1979). Nevertheless, good results have been reported in some cases with very low plasma levels (Thune and Volden, 1977). This might be explained by the lack of a positive correlation between the blood and the skin concentrations. This raises the question of whether the 8-MOP plasma level is a good indicator of the skin concentration.

A few years ago, the pharmacokinetics of 8-MOP in the skin were compared to the plasma levels (Roelandts et al., 1983a). This study was conducted on animals because of the large number of biopsies that had to be taken and because

radioactive 8-MOP was used. By means of skin and blood
samples from eighty rats at specific intervals after oral
administration of [3H]-8-MOP, it could be shown that 8-MOP
plasma levels seem to be a good indicator of the skin levels.

DIETARY INFLUENCES ON 8-MOP PLASMA LEVELS

     Perhaps the most important point is not to have the
highest possible plasma levels but to have them as consistent
as possible. This means that the 8-MOP levels have to peak
every day at the same time after oral intake in as many
patients as possible. It is, therefore important to analyze
possible influences such as diet, the crystal size of the 8-
MOP, and try to find out if differences in resorption and
metabolism could be due to the galenic form in which the 8-MOP
is applied.

     To study the effect of diet on 8-MOP plasma levels, these
levels were measured three times in 20 psoriasis patients:
while fasting, after a low-fat breakfast, and after a fat-rich
breakfast (Roelandts et al., 1981a).

     When the subject fasts and after a low-fat breakfast, the
maximum 8-MOP plasma level was reached most often in two
hours, whereby the difference between the two values was
statistically insignificant (p > 0.8). This indicates that a
low-fat breakfast will not have an important influence on the
8-MOP resorption compared with a fasting condition.

     After a fat-rich breakfast, the 8-MOP plasma levels
differ, the mean after two hours being much lower. The
maximum is reached after 3 to 4 hours and is higher than after
the two previous diets.

     Eating meals before drug intake usually prolongs the time
the drug remains in the stomach and therefore delays
absorption, which usually occurs with the highest efficacy
from the small intestine (Brickl et al., 1984). Because
lipids as well as emulsified lipids can reduce
gastrointestinal tract motility, the 8-MOP has more time to
dissolve in the region of the small intestine. This could
explain why a fat-rich breakfast can not only delay the
resorption of the 8-MOP but also seems to give higher peak
levels. The higher peak values after a high-fat breakfast
could also be explained by the fact that 8-MOP is a lipophilic
substance.

     Ideally, the drug should be given in a dietary
standardized way. However, standardizing the diet is, in
practice, not an easy thing to do. The reason why diet can be
important is the poor resorption of the 8-MOP. An easier and
better way of standardizing 8-MOP plasma levels is by using an
8-MOP formulation with better resorption. Uniform 8-MOP
absorption from the gastrointestinal tract is necessary to
avoid day-to-day variations in 8-MOP plasma levels in the same
patient. By increasing the resorption, additional factors,
such as food, decline in importance. This could explain the
different results reported in the literature concerning
dietary influences on 8-MOP plasma levels (Levins et al.,
1984; Bonnot et al., 1985).

INCREASING THE RESORPTION OF THE 8-MOP

One way of increasing the resorption of 8-MOP is by changing the particle size.

The influence of the particle size and the crystal form of the 8-MOP is illustrated in a study comparing the 8-MOP plasma levels after oral administration of three different forms: gelatin capsules containing 8-MOP with a mean particle size of 200 μm, gelatin capsules containing 8-MOP in microcrystalline form with particle size between 20 μm and 30 μm, and a formulation containing the same microcrystalline 8-MOP but in an emulsion base (Roelandts et al., 1983b; Van Boven et al., 1985a).

The reduction of the particle size of 8-MOP in capsule form enhances bioavailability since the drug dissolves more rapidly when its surface area is increased.

Although reducing the particle size significantly enhances the 8-MOP plasma concentration, interindividual differences are still quite important. However, when the micronized 8-MOP is used in an oil-in-water emulsion base, these differences are markedly less. The emulsion gives a high uniform absorption resulting in better and thus more standardized plasma levels as compared to the capsule forms. One may assume that many patients are not getting the most out of their PUVA treatments because of the form of 8-MOP being used. Several studies have demonstrated that the 8-MOP plasma levels depend in large measure on the brand administered and the galenical form (Thune, 1978; Andersen et al., 1980; Ljunggren et al., 1980; Stolk et al., 1980; Menne et al., 1981; Nitsche et al., 1981). More standardized levels also means more predictable levels.

In all of these studies, major differences were found in the maximum plasma levels as well as in the time when the maximum concentrations were obtained. The formulations dissolved in a liquid seem to give higher plasma levels and reach maximum concentration earlier than most of those with 8-MOP in solid form.

It would be optimal to try to combine the advantages of the improved resorption of a liquid form and the practicality of a solid form. Therefore, five new formulations were made, all with oil as the vehicle as 8-MOP is lipophilic. Two of these five were emulsion forms, and three were capsule forms (Roelandts et al., 1984; Van Boven et al., 1985b). The object was to prove that the capsules would be as good as the emulsions and much better than what was commonly being used. The five self-prepared formulations were compared with seven brands widely used in Europe. The experiment was performed on dogs and not on human subjects because each of the twelve preparations had to be given to the same individuals, which means that for each individual 108 blood samples were taken. The formulations that show early peaking times (60 minutes or less) and high peak concentrations are formulations in which the 8-MOP is present in dissolved or partially dissolved form which obviously provides for good and rapid resorption.

The use of liquid formulations would certainly be a large step toward dose standardization (Roelandts et al., 1981b; Roelandts et al., 1983b; Van Boven et al., 1985a), although oral liquid formulations are less practical for common use. These drawbacks can be avoided by using liquid 8-MOP in soft gelatin capsules. With the soft gelatin capsules containing the 8-MOP in solution higher 8-MOP plasma levels are obtained than with micronized 8-MOP powder in hard gelatin capsules, the maximum peak levels being reached earlier and the relative bioavailability being higher for the soft capsules. Therefore, such formulations should be used.

DISCUSSION

Although the dose of 8-MOP usually given is a constant dose of 0.6 mg/kg body weight two hours prior to UV-A irradiation, major variations in the 8-MOP plasma levels have been observed (Steiner et al., 1977; Thune and Volden, 1977; Ehrsson et al., 1978; Herfst et al., 1978; Steiner et al., 1978; Thune, 1978; Swanbeck, 1979; Wagner et al., 1979; Andersen et al., 1980; Ljunggren et al., 1980; Smyth et al., 1980; Launis and Wilen, 1981; Menne et al., 1981; Stevenson et al., 1981; Herfst, 1982; Herfst and De Wolff, 1982a, 1982b; Stolk, 1982; Jansen et al., 1983; Monbaliu, 1983; Reymond et al., 1988). The occurrence of variations in 8-MOP plama levels in patients treated in a similar way and with the same dose of 0-MOP considerably complicates PUVA therapy. Very likely, varying 8-MOP plasma levels are at least partly due to individual differences in pharmacokinetics that probably arise from differences in first-pass metabolism in the liver, but this does not explain significant day-to-day variations in the same patient.

Because 8-MOP is a phototoxic drug, it is much better to have a short steep high peak value than a broader plateau value lasting for several hours. On the other hand, it is also important that this maximum 8-MOP level always be reached at the same time, since the irradiation in clinical practice is always administered at a well-defined time after oral intake of the 8-MOP.

With good drug formulations, the 8-MOP plasma level has to peak within 1 hour and will be close to zero after 4 to 6 hours. It is advisable to use doses well above the dose necessary to have a saturation of the first-pass effect, because inter- and intraindividual differences in 8-MOP plasma levels tend to be large at doses in the range of the saturation process (Brickl et al., 1984). This could explain the large inter- and intraindividual variability in 8-MOP plasma and serum levels reported in the literature (Steiner et al., 1977; Thune and Volden, 1977; Ehrsson et al., 1978; Herfst et al., 1978; Steiner et al., 1978; Thune, 1978; Swanbeck, 1979; Wagner et al., 1979; Andersen et al., 1980; Ljunggren et al., 1980; Smyth et al., 1980; Launis and Wilén, 1981; Menne et al., 1981; Stevenson et al., 1981; Herfst, 1982; Herfst and De Wolff, 1982a 1982b; Stolk, 1982; Jansén et al., 1983; Monbaliu, 1983; Reymond et al., 1988).

It is not of prime importance that the 8-MOP plasma levels be the highest possible after administration. In fact, if they are too low, the 8-MOP dose can always be increased to achieve satisfactory plasma levels. In addition, the subjective side effects will often increase when formulations are used that give higher plasma levels. Compensating unsatisfactory plasma levels by increasing the UV-A dose is less advisable since the importance of the first-pass effect is higher for low plasma levels. Because of inter- and intraindividual variability, this could result in unexpected overdosage with erythema or an unsatisfactory clinical response. It is, therefore, better to use formulations that give predictable plasma levels, peaking as much as possible at the same time so that the UV-A irradiation is always administered at the optimal moment.

Liquid formulations and liquid capsules and their variants give more predictable plasma levels than other formulations, the influence of the first-pass effect being diminished because of their good and rapid resorption.

That there is not always a clear-cut relationship between the 8-MOP plasma level and the therapeutic response to PUVA treatment (Thune and Volden, 1977; Wagner et al., 1979; Andrew et al., 1981; Stevenson et al., 1981) can be explained by the fact that solitary 8-MOP serum measurements yield only limited information due to wide intraindividual variations. Repeated measurements are, therefore, more appropriate to optimize 8-MOP plasma levels.

Adjusting the UV-A irradiation to coincide with the maximum plasma level (Wagner et al., 1979; Hönigsmann et al., 1982) or increasing the 8-MOP dose (Thune and Volden, 1977; Andrew et al., 1981) or changing to soft gelatin capsules (Langner and Wolska, 1981; Hönigsmann et al., 1982; Siddiqui et al., 1984; Lowe et al., 1987; Stolk and Siddiqui, 1988) can lead to an improvement in the therapeutic result for problem patients.

REFERENCES

Andersen, K. E., Menne, T., Gammeltoft, M., Hjorth, N., Larsen, E., and Solgaard, P., 1980 Pharmacokinetic and clinical comparison of two 8-methoxypsoralen brands, Arch Dermatol Res 268: 23-29.

Andrew, E., Nilsen, A., Thune, P., and Wiik, I., 1981, Photechemotherapy in psoriasis-clinical response and 8-MOP plasma concentrations at two dose levels, Clin Exp Dermatol 6: 591-600.

Bonnot, D., Beani, J. C., Beriel, H., Boitard, M., Amblard, P., and Reymond, J. L., 1985, Influence of the diet on the plasma kinetics of 8-methoxypsoralen, Dermatologica 171: 442-445.

Brickl, R., Schmid, J., and Koss, F. W., 1984, Clinical pharmacology of oral psoralen drugs, Photodermatology 1: 174-186.

Ehrsson, H., Eksborg, S., and Wallin, I., 1978, Metabolism of 8-methoxypsoralen in man: Identification and quantification of 8-hydroxypsoralen, Eur J Drug Metab Pharmacokinet 3: 125-128.

Herfst, M. J., Koot-Gronsveld, E. A. M., and De Wolff, F. A.,
1978, Serum levels of 8-methoxypsoralen in psoriasis
patients using a new fluorodensitometric method, <u>Arch
Dermatol Res</u> 262: 1-6.

Herfst, M. J. and De Wolff, F. A., 1982a, Influence of food on
the kinetics of 8-MOP in serum and suction-blister fluid
in psoriatic patients, <u>Eur J Clin Pharmacol</u> 23: 75-80.

Herfst, M. J. and De Wolff, F. A., 1982b, Difference in
bioavailability between two brands of 8-methoxypsoralen
and its impact on the clinical response in psoriatic
patients, <u>Br J Clin Pharmacol</u> 13: 519-522.

Herfst, M. J., 1982, Pharmacological aspects of PUVA therapy.
Thesis, Leiden.

Hönigsmann, H., Jaschke, E., Nitsche, V., Brenner, W.,
Rauschmeier, W., and Wolff, K., 1982, Serum levels of 8-
methoxypsoralen in two different drug preparations:
Correlation with photosensitivity and UV-A dose
requirements for photochemotherapy, <u>J Invest Dermatol</u> 79:
233-236.

Jansén, C. T., Wilén, G., Ylitalo, P., and Malmiharju, T.,
1983, Interindividual and intraindividual variations in
serum methoxsalen levels during repeated oral exposure,
<u>Curr Ther Res</u> 33: 258-264.

Langner, A. and Wolska, H., 1981, New galenical form of 8-
methoxypsoralen in photochemotherapy of psoriasis, <u>Arch
Dermatol Res</u> 271: 461-462.

Launis, J. and Wilén, G., 1981, Bioavailability of three 8-
methoxypsoralen brands in ten psoriatic patients,
<u>Diagnosis</u> 6: 693-696.

Levins, P. C., Gange, R. W., Momtaz, T. K., Parrish, J. A.,
and Fitzpatrick, T. B., 1984, A new liquid formulation of
8-methoxypsoralen: bioactivity and effect of diet, <u>J
Invest Dermatol</u> 82: 185-187.

Ljunggren, B., Carter, D. M., Albert, J., and Reid, T., 1980,
Plasma levels of 8-methoxypsoralen determined by High-
Pressure Liquid Chromatography in psoriatic patients
ingesting drug from two manufacturers, <u>J Invest Dermatol</u>
74: 59-62.

Lowe, H. J., Urbach, F., Bailin, P., and Weingarten, D. P.,
1987, Comparative efficacy of two dosage forms of oral
methoxsalen in psoralens plus ultraviolet A therapy of
psoriasis, <u>J Am Acad Dermatol</u> 16: 994-998.

Menne, T., Andersen, K. E., Larsen, E., and Solgaard, P.,
1981, Pharmacokinetic comparison of seven 8-
methoxypsoralen brands, <u>Acta Derm Venereol (Stockh)</u> 61:
137-140.

Monbaliu, J., 1983, Intra-individual variation in disposition
of 8-methoxypsoralen, <u>Pharm Weekbl [Sci]</u> 5: 39.

Nitsche, V., Raff, M., and Bardach, H., 1981, 8-
Methoxypsoralen (8-MOP) in neuer galenischer Zubereitung
und seine Beziehung zum 8-MOP-Serumspiegel, <u>Arch Dermatol
Res</u> 271: 11-17.

Reymond, J. L., Beani, J. C., Racinet, H., Bonnot, D., Beriel,
H., and Amblard, P., 1988, Comparative pharmacokinetics
of 8-MOP in serum and in suction blister fluid,
<u>Photodermatology</u> 5: 51-52.

Roelandts, R., Van Boven, M., and Adriaens, P., 1983b,
Verdient toediening van 8-methoxypsoraleen in de vorm van
meladininetabletten de voorkeur bij PUVA-therapie? <u>Ned
Tijdschr Geneeskd</u> 127: 311-312.

Roelandts, R., Van Boven, M., Adriaens, P., De Schryver, F.,
Degreef, H., 1983a, The relation between 8-
methoxpysoralen skin and blood levels, J Invest Dermatol
81: 331-333.
Roelandts, R., Van Boven, M., Deheyn, T., Vanderstichele, G.,
Degreef, H., and Daenens, P., 1981a, Dietary influences
on 8-MOP plasma levels in PUVA patients with psoriasis,
Br J Dermatol 105: 569-572.
Roelandts, R., Van Boven, M., and Adriaens, P., 1981b,
Methoxsalen serum level variations in psoralen and
ultraviolet A (PUVA) therapy, Arch Dermatol 117: 758.
Roelandts, R., Van Boven, M., Degreef, H., Kinget, R.,
Adriaens, P., and Daenens, P., 1984, Variations in plasma
levels with 12 different forms of methoxsalen, Arch
Dermatol 120: 1281-1282.
Siddiqui, A. H., Stolk, L., and Cormane, R. H., 1984,
Comparison of serum levels and clinical results of PUVA
therapy with three different dosage forms of 8-
methoxypsoralen, Arch Dermatol Res 276: 343-345.
Smyth, R. D., Van Harken, D. R., Pfeffer, M., Nardella, P.A.,
Vasiljev, M., Pinto, J. S., and Hottendorf, G. H., 1980,
Biological disposition of 8-methoxsalen in rat and man,
Arzneim Forsch 30: 1725-1730.
Steiner, I., Prey, T., Gschnait, F., Washüttl, J., and
Greiter, F., 1977, Serum level profiles of 8-
methoxypsoralen after oral administration, Arch Dermatol
Res 259: 299-301.
Steiner, I., Prey, T., Gschnait, F., Washüttl, J., and
Greiter, F., 1978, Serum levels of 8-methoxypsoralen 2
hours after oral administration, Acta Derm Venereol
(Stockh) 58: 185-186.
Stevenson, I. H., Kenicer, K. J. A, Johnson, B. E., and Frain-
Bell, W., 1981, Plasma 8-methoxypsoralen concentrations
in photochemotherapy of psoriasis, Br J Dermatol 104: 47-
51.
Stolk L. M. L., 1982, Farmacokinetische en biofarmaceutische
aspecten van enige in de dermatologie gebruikte
psoraleenderivaten, Thesis, Amsterdam.
Stolk, L., Kammeyer, A., Cormane, R. H., and Van Zwieten, P.
A., 1980, Serum levels of 8-methoxypsoralen: difference
between two oral methods of administration, Br J Dermatol
103: 417-420.
Stolk, L. M. L. and Siddiqui, A. H., 1988, Biopharmaceutics,
pharmacokinetics and pharmacology of psoralens, Gen
Pharmac 19: 649-653.
Swanbeck, G., Ehrsson, H, Ehrnebo, M., Wallin, I., and
Jonsson, L., 1979, Serum concentration and phototoxic
effect of methoxsalen in patients with psoriasis, Clin
Pharmacol Ther 25: 478-480.
Thune, P., 1978, Plasma levels of 8-methoxypsoralen and
phototoxicity studies during PUVA treatment of psoriasis
with meladinin tablets, Acta Derm Venereol (Stockh) 58:
149-151.
Thune, P. and Volden, G., 1977, Photochemotherapy of psoriasis
with relevance to 8-methoxypsoralen plasma level and low
intensity irradiation, Acta Derm Venereol (Stockh) 57:
351-355.
Van Boven, M., Roelandts, R., Adriaens, P., Daenens, P.,
Degreef, H., and Kinget, R., 1985a, Standardizing 8-MOP
plasma profiles by using an emulsion form, J Am Acad
Dermatol 12: 822-827.

Van Boven, M., Roelandts, R., Degreef, H., Kinget, R., Adriaens, P., and Daenens P, 1985b, A pharmacokinetic comparison in dogs of seven brands of 8-MOP and five new formulations, <u>Photodermatology</u> 2: 27-31.

Wagner, G., Hofmann, C., Busch, V., Schmid, J., and Plewig, G., 1979, 8-MOP plasma levels in PUVA problem cases with psoriasis, <u>Br J Dermatol</u> 101: 285-292.

MUTAGENESIS PHOTOINDUCED BY PSORALENS IN YEAST

Dietrich Averbeck

Institut Curie-Biologie, CNRS UA 1292, rue d'Ulm
F-75231 Paris Cedex 05, France

INTRODUCTION

The use of furocoumarins (psoralens) in the photochemotherapy of several human skin diseases (Parrish et al., 1982) and their presence in cosmetic and skin tanning preparations have led photobiologists to investigate the genotoxic activities of these photosensitizing compounds and the mechanisms involved (see for review Averbeck 1984, Ben-Hur and Song, 1984). Furocoumarins have been shown to photoreact not only with nucleic acids but also with proteins and lipids (for review see, Midden 1988). Nevertheless, cellular DNA has been recognised as one of the major targets of psoralen photoreactions (Ben-Hur and Song 1984). Following complexation to DNA in the dark psoralens produce in the presence of near-ultraviolet light (UVA, 315-400 nm) C_4-cyclobutane adducts to pyrimidine bases in DNA (for review see, Cimino et al., 1985; Musajo and Rodighiero 1972) consisting of furan-side (MA_f) and pyrone-side (MA_p) monoadditions as well as of diadducts (DNA interstrand cross-links, CL). Different types of photoadducts in DNA have been characterized (Vigny et al., 1985), and their distribution and sequence specificity were determined (Sage et al., 1987; Boyer et al., 1988). Because of their specific photoreactivity with DNA psoralens have become sensitive probes for elucidating the structure of DNA (Cimino et al., 1985; Wollenzien 1988), the processing of adducts by genetically controlled repair systems (for review see, Smith 1988) and the genetic consequence of adduct formation (Averbeck 1985, Averbeck et al., 1990a).

The furocoumarins 8-methoxypsoralen (8-MOP), 5-methoxypsoralen (5-MOP) and 4,5',8-trimethylpsoralen (TMP) used in photochemotherapy are bifunctional compounds photoinducing mono- and diadducts in DNA (see Fig. 1). Other, more recently developed compounds of photochemotherapeutic interest are linear and angular psoralens of the monofunctional type photoinducing monoadducts in DNA (Fig. 1). These include 3-carbethoxypsoralen (3-CPs), the pyridopsoralens, for example,

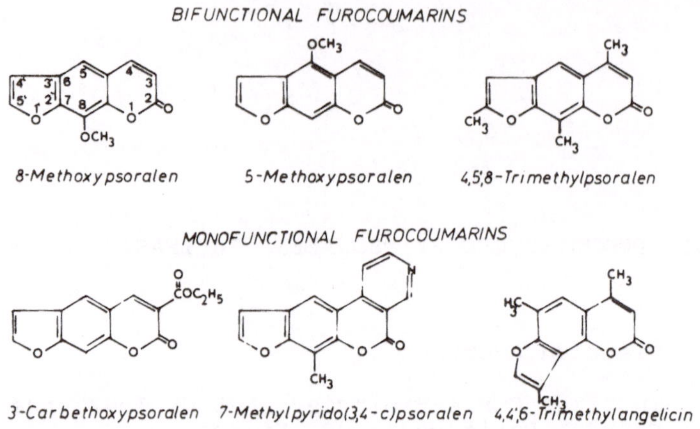

BIFUNCTIONAL FUROCOUMARINS

8-Methoxypsoralen   5-Methoxypsoralen   4,5',8-Trimethylpsoralen

MONOFUNCTIONAL FUROCOUMARINS

3-Carbethoxypsoralen   7-Methylpyrido(3,4-c)psoralen   4,4',6-Trimethylangelicin

Fig. 1   Molecular structures of mono- and bi-
          functional furocoumarins of photochemothera-
          peutic interest.

7-methylpyrido (3,4-c) psoralen (MPP) and methylated
derivatives of angelicin, for example, 4,4',6-
trimethylangelicin (TMA) (for review see, Rodighiero et al.,
1988).

     Psoralens have been found to exert photomutagenic effects
in pro- and eukaryotic cell systems (Averbeck 1984; Ben-Hur
and Song 1984; Cassier-Chauvat and Averbeck 1988; IARC 1986;
Rodighiero et al., 1988; Scott et al., 1976). Valuable
information has been gained from studies in bacteria, however,
because of the more complex (chromosomal) organisation and
the presence of different repair systems, investigations
performed on eukaryotic cells are thought to be important in
view of genotoxic effects encountered after human exposure
(Stern et al., 1988).

     As a typical eukaryote the yeast Saccharomyces cerevisiae
has been useful for studying the relationship between the
induction and the repair of psoralen induced damage and the
induction of mutations. The present overview summarizes
recent data focussing on known mono- and bifunctional
furocoumarins.

MATERIALS AND METHODS

     The furocoumarins used were chromatographically pure 8-
methoxypsoralen (8-MOP), 5-methoxypsoralen (5-MOP), 3-
carbethoxypsoralen (3-CPs) and 7-methylpyrido (3,4-c) psoralen
(MPP) (Averbeck 1985; Moron et al., 1983) and the
pyranocoumarin 8,8-desmethylxanthyletine or homopsoralen (HPs)
(Faulques et al., 1983; Averbeck et al., 1990a).

     The biological material consisted of the diploid strain
D7 of the yeast Saccharomyces cerevisiae (Zimmermann et al.,
1975). Cell survival and the induction of reverse mutations
(ILV+) were determined as previously described (Averbeck
1985). As a source of UVA radiation an HPW 125 Philips lamp
emitting mainly at 365 nm (UVA) (Averbeck 1985) was used and

for exposures to defined wavelengths (365 nm or 405 nm) a 2.5 kW Xenon lamp with a Kratos GM252 high intensity grating monochromator (Kratos Analytical Instruments, Ramsay, USA) was used (Averbeck 1985; Averbeck et al., 1987). Psoralen photoadducts were detected by using radioactively labeled psoralens and alkaline elution analysis according to Averbeck et al., 1987.

RESULTS

Induction of Mutations by Bifunctional Psoralens and Single Exposure to UVA

In yeast, the photochemotherapeutically active bifunctional furocoumarins 8-MOP, 5-MOP and TMP have been found, as a function of UVA dose, highly mutagenic in comparison to monofunctional furocoumarins such as 3-CPs, MPP and angelicin (Averbeck 1985; Averbeck and Moustacchi 1980; Averbeck et al., 1981). In an attempt to define the role of psoralen induced MA and CL in the mutagenic effects observed with bifunctional furocoumarins, the induction of ILV⁺ reverse mutations has been determined in diploid yeast (D7) in parallel to the photobinding to DNA after treatments with 8-MOP and 5-MOP. The results (Fig. 2) show that at an equal number of total photoadducts, 5-MOP induced lesions are more effective than 8-MOP induced lesions. It has been estimated from alkaline step elution experiments in yeast (Dardalhon et al., unpublished data), in accord with data obtained with mammalian cells in culture using alkaline elution or hydroxylapatite chromatography (Papadopoulo and Averbeck 1985; Dardalhon et al., 1988), that 5-MOP is approximately as

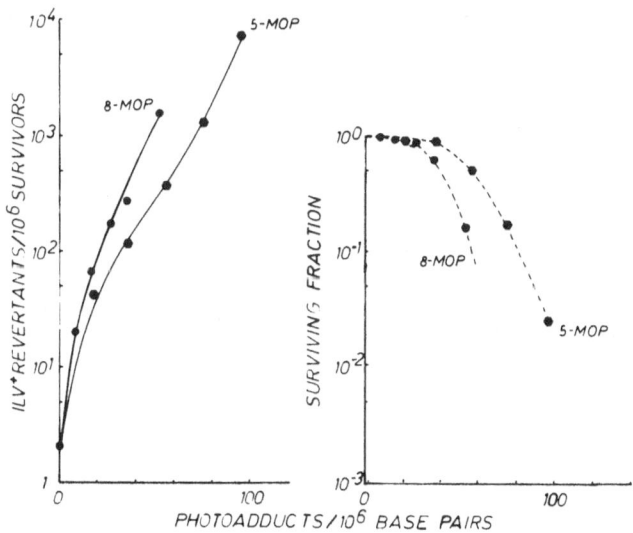

Fig. 2 Induction of ILV⁺-revertants (left panel) and lethal effects (right panel, broken lines) in diploid yeast by the bifunctional furocoumarins 8-MOP and 5-MOP as a function of total photoadducts.

effective as 8-MOP for the induction of CL. Thus, the higher
mutagenicity of 5-MOP in comparison to 8-MOP may be explained
by the relatively higher capacity of 5-MOP to produce more MA
than CL at identical adduct levels. In other words, the ratio
of mono-over diadducts induced is important and MA contributes
to the induction of mutations by bifunctional furocoumarins.

Induction of Mutations by Bifunctional Furocoumarins and Exposures to Monochromatic 365 nm and 405 nm Radiation

Recently, by the use of different activating wavelengths,
it has become possible to analyse the role of monoadducts
alone and mixtures of mono- and diadducts induced by the same
bifunctional furocoumarins. Based on previous work on DNA in
vitro (Chatterjee and Cantor 1978; Tessman et al., 1985) it
has been shown for 8-MOP by photobinding and alkaline elution
experiments in yeast that at 405 nm mainly DNA monoadducts are
induced whereas at 365 nm mixtures of MA and CL are induced
(Averbeck et al., 1987; Cundari and Averbeck 1988).
Furthermore, in a reirradiation regimen with 8-MOP consisting
of a first exposure to 405 nm radiation (induction of MA_f and
MA_p) followed, after washing out of unbound psoralen
molecules, by a second expsoure to 365 nm radiation (leading
to the conversion of part of the MA_f into CL), the level of CL
could be substantially increased without changing the number
of total photoadducts (Cundari and Averbeck 1988).

The induction of $\underline{IVL}^+$ revertants by 8-MOP and 5-MOP was
measured in the diploid yeast strain D7 following exposures to
365 nm or 405 nm radiation. Fig. 3 shows that, as a function
of survival, treatments at 365 nm (inducing a mixture of MA
and CL) are clearly more mutagenic than treatments at 405 nm

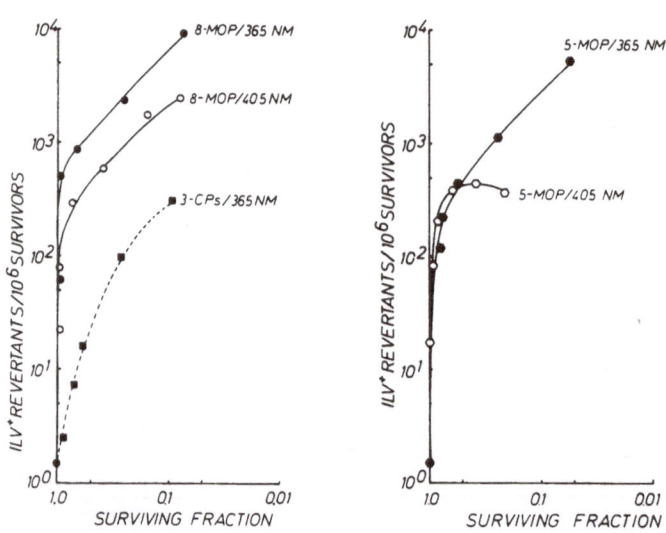

Fig. 3   Induction of $\underline{ILV}^+$-revertants in diploid
yeast by 8-MOP (left panel) and 5-MOP
(right panel) at different activating wavelengths as a function of survival. Data
obtained with the monofunctional 3-CPs
(dotted lines) are included.

238

(inducing only MA$_p$ and MA$_f$). The latter treatment is more mutagenic than 3-CPs plus 365 nm inducing only MA$_f$. In general, at equal survival, monofunctional furocoumarins were significantly less mutagenic than bifunctional furocoumarins (Averbeck 1985), suggesting that the relationship between lethal and mutagenic events is highly dependent on CL.

Interestingly, the differences in mutagenicity of the two types of exposure are already striking at high survival levels in the case of 8-MOP, but are significantly different in the case of 5-MOP only at survival levels below 37%. This is probably due to the fact that 5-MOP in combination with 365 nm radiation induces for a given survival, MA at higher relative amounts than 8-MOP. As reported previously for 8-MOP (Averbeck et al., 1987; Averbeck et al., 1990b), as a function of total photoadducts, 5-MOP plus 365 nm radiation is clearly more effective in the induction of mutations than 5-MOP plus 405 nm radiation (Fig. 4). Thus, the presence of CL appears to play an important role for the induction of mutations. Also, the effectiveness of certain chemical filters (UVA) in decreasing the genotoxic effects of preparations containing 5-MOP (Averbeck et al., 1990c) may be in part attributed to a shift towards longer activating wavelengths (above 390 nm) showing a decreased crosslinking potential.

Furthermore, the effects of 8-MOP plus single exposures (at 405 nm or 365 nm) and reirradiation regimens consisting of first exposures to 365 or 405 nm light followed after washing out of unbound psoralen molecules by a second exposure to 365 nm radiation, are illustrated in Fig. 5. The results show that at a constant level of total photoadducts induced (20

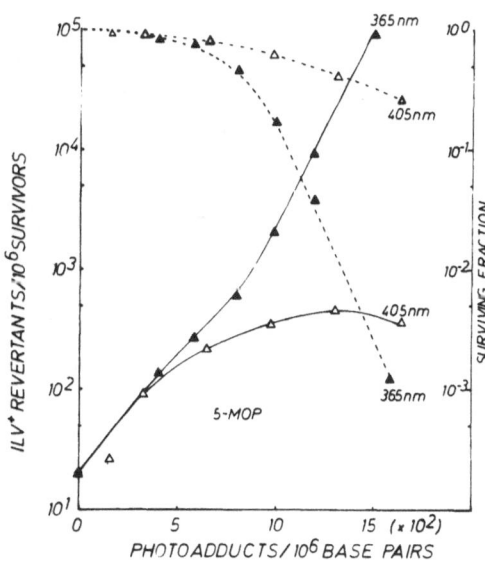

Fig. 4   Induction of ILV$^+$-revertants and lethal effects in diploid yeast by the bifunctional furocoumarin 5-MOP at different activating wavelengths (365 nm and 405 nm) as a function of total photoadducts.

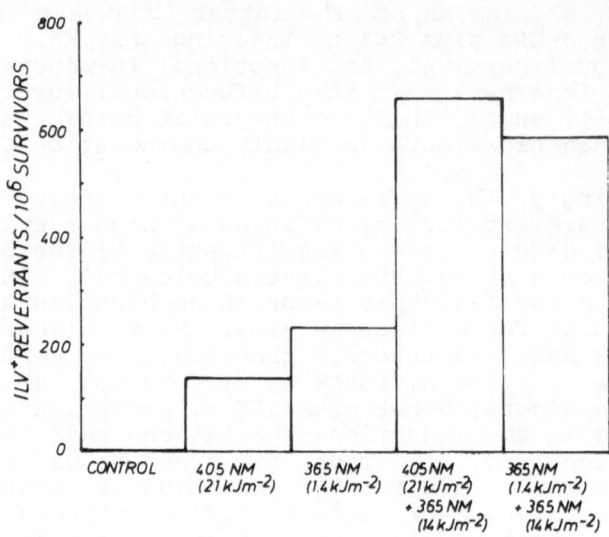

Fig. 5   Effects of single and double exposures.
         Mutagenicity of 8-MOP in diploid yeast
         after single exposures to 405 nm or 365 nm
         radiation and 405 nm/365 nm or 365 nm/365
         nm reirradiation regimens.

adducts/10^6 nucleotides) (Averbeck et al., 1987), a single
exposure to 405 nm light is much less mutagenic than a single
exposure to 365 nm light. The mutation frequency increases
strikingly when applying the reirradiation protocols 405
nm/365 nm or 365 nm/ 365 nm due to the increase in the level
of crosslinking brought about by the conversion of MA_f into CL
by the second exposure at 365 nm. The frequency of 8-MOP
induced mutations is higher after 405 nm/365 nm than after 365
nm/365 nm reirradiation. This is probably related to the
previous observation (Averbeck et al., 1987) that more cross-
linkable monoadducts (MA_f) are produced by 8-MOP with 405 nm
than with 365 nm radiation as a first exposure.

Mutagenic Effectiveness of Monoadducts Photoinduced by the Bifunctional Furocoumarin 8-MOP or the Monofunctional Furocoumarins 3-CPs and MPP

Since mixtures of MA and CL were found more mutagenic
than MA alone, the question arises whether the MA induced by
mono- and bifunctional furocoumarins are comparable with
respect to the mutagenicity induced in diploid yeast. Table 1
shows how many photoadducts are needed to produce 50 IVL[+]
mutants/10^6 survivors in the case of MA induced by 8-MOP plus
405 nm light, 3-CPs and MPP plus 365 nm light. For
comparison, the amount of photoadducts yielding the same
mutation frequency after treatment with 8-MOP and 365 nm light
is included.

8-MOP plus 405 nm light induced MA (MA_f and MA_p) are much
more mutagenic than those induced by 3-CPs and MPP consisting
exclusively of MA_f. Assuming that the effects are solely due
to the induction of MA, the difference in mutagenicity could
indicate a role for MA_p in 8-MOP induced mutagenesis.

Table 1 Amount of photoadducts needed for the
induction of 50 $\underline{ILV}^+$ revertants/10^6
survivors in diploid yeast

8-MOP plus 405 nm light:	1.6 adducts/10^6 base pairs
3-CPS plus 365 nm light:	7.5 adducts/10^6 base pairs
MPP plus 365 nm light:	10.0 adducts/10^6 base pairs
8-MOP plus 365 nm light:	0.6 adducts/10^6 base pairs

In addition, the apparent difference in mutagenicity of
3-CPs and MPP could indicate the presence of other,
concomitantly induced, mutagenic lesion(s). In this
connection, it is noteworthy that for MPP the formation of
pyrimidine dimers by energy transfer from photoexited MPP via
the triplet excited state has been recently demonstrated in
vitro (Costalat et al., 1990) and thymine dimers have been
detected in yeast cell DNA following treatment with MPP and
UVA (M. Dardalhon, A. Moysan, L. Andreu, P. Vigny, unpublished
results). Cyclobutane dimers are well known mutagenic lesions
(see Brash 1988). Thus, it seems possible that thymine dimers
induced by MPP and UVA contribute to the overall
photomutagenic effect of MPP.

No data are as yet available showing the mutagenic
effectiveness of pyrone-side monoadducts induced by psoralens.
In an attempt to approach this question we determined the
mutagenicity of pyridopsoralens derivatives (Moron et al.,
1983) able to either photoinduce furan-side (MPP) or pyrone-
side monoadducts (pyridomethylpsoralen, PMP, or methyl-8-H-
pyrano (3',2':5,6) benzo-furo(3,2-c)pyridin-8-one). At the
same survival level, the MA_p inducing compound (PMP) appeared
to be more mutagenic than the MA_f inducing compound MPP (data
not shown). The apparent higher mutagenic effectiveness of
MA_p needs to be confirmed by comparisons taking into account
the number and distribution of lesions induced.

On the Possible Involvement of Oxygen Mediated Damage in Psoralen Induced Mutagenesis

In recent years the capacity of furocoumarins to undergo
oxygen mediated reactions has been established (De Mol et al.,
1981; Vedaldi et al., 1983; Joshi and Pathak 1983), and the
possible involvement of such reactions in mutation induction
has been suggested (De Mol et al., 1981). In order to analyse
this further, experiments were performed in diploid yeast
using treatments with 3-CPs and UVA, since 3-CPs is well-known
for its high capacity for singlet oxygen production (Vedaldi
et al., 1983). A dose-dependent oxygen effect on mutation
induction was observed (Averbeck 1988).

As a function of UVA, mutations were apparently more
effectively induced by 3-CPs in the presence rather than in
the absence of oxygen (Averbeck 1988). Addition of 10 mM N-
acetyl cysteine partially suppressed the oxygen effect due to

chemical quenching. However, when plotting the data as a function of survival, mutation induction by 3-CPs was the same in the presence and in the absence of oxygen, indicating that the relationship between lesions and lethal and mutagenic events induced, and thus the mechanism of mutation induction, remained unaltered by oxygen. Similar results were obtained in E. coli after treatments with 8-MOP (De Mol et al., 1986) In accord with these results, Sage et al. (1989) demonstrated that oxygen-dependent damage induced by 3-CPs can be repaired by repair enzymes with high fidelity.

With bifunctional furocoumarins such as 5-MOP and 8-MOP no oxygen effect has been observed on cell survival in yeast despite their capacity to generate singlet oxygen (Averbeck 1984). Recently, no oxygen effect on the mutagenic response was seen in experiments performed with 5-MOP and solar simulated radiation. The same was true for bergamot oil containing 5-MOP (Averbeck et al., 1990). Thus, psoralen induced oxygen mediated damage appears to be repairable.

On the Importance of Monoadducts and Cross-Links for the Induction of Mutagenic Effects by Bifunctional Derivatives

Using radioactively labeled compounds we were recently able to show that 8-MOP is as effective per unit dose of UVA as the pyranocoumarin HPs (homopsoralen) in photobinding to DNA. However, the two compounds differed clearly in their mutagenic effectiveness, 8-MOP being more effective than HPs (Averbeck et al., 1990a). Alkaline elution analysis of treated yeast cells revealed that 8-MOP was also more effective than HPs in the induction of CL. This supports the notion that increased levels of CL can be associated with high mutagenicity. The results reported in Table 2 appear to substantiate this further.

Table 2 Relationship of CL to mutations ($\underline{ILV}^+$) photoinduced by 8-MOP and HPs in diploid yeast

Treatment Compound	UVA dose (kJm^{-2}) 1st	2nd exposure	CL/genome[1]	$\underline{ILV}^+/10^6$ survivors
HPs	–	–	–	40
HPs	2.4	–	12	558
HPs	2.4	30	49	3225
8-MOP	–	–	–	27
8-MOP	2.4	–	40	2735

[1] according to alkaline elution analysis

When in a 365 nm/365 nm reirradiation regimen with HPs the level of CL/genome is increased up to the same level as that induced by 8-MOP plus a single UVA exposure, the mutation frequently increases accordingly. Consistent with this, the removal of photoadducts induced by HPs plus 365 nm/ 365 nm reirradiation followed the same kinetics as the removal of 8-MOP plus UVA induced adducts (Averbeck et al., 1990a). The increase in mutagenicity of HPs in the reirradiation protocol is probably due to the increase in the relative amount of CL. However, it may be kept in mind that the relative amount of MA_p increases when the amount of MA_f decreases by the conversion into CL.

On the Possible Involvement of an Inducible Component in Psoralen Induced Mutagenesis

When analysing mutation frequency curves Haynes et al. (1985) and Eckardt et al. (1981) recognized non-linear dose response patterns which after mathematical transformation in a so-called "apparent survival test" can infer the possible existence of inducible components of the mutagenic response in the yeast Saccharomyces cerevisiae. The test consists of calculating the survival of the mutants induced (Haynes et al., 1985). The apparent survival is given, as a function of dose, by the ratio of the mutant yield to the corresponding linear component of mutation frequency. If the apparent survival rises to values greater than unity before declining at high doses, this indicates the existence of a positive two-hit or higher order component in the mathematical process. Since after UV exposure this component was found to be cycloheximide sensitive (Haynes et al., 1985), its presence is likely to reflect an inducible component.

When the curves for the induction of HIS^+ revertants by 8-MOP plus UVA in haploid yeast were analysed in the same way, evidence was obtained for a rise in the apparent survival of mutant cells which is in favor of the presence of an inducible component in psoralen induced mutagenesis (Averbeck and Eckardt, unpublished data).

In yeast, several genes (for example, RAD2 (Robinson et al., 1986) and RAD54 (Cole et al., 1987)) involved in the repair of specific DNA lesions have been found inducible by a variety of DNA damaging agents. In order to find out whether the inducing signal might differ depending on the presence of specific types of DNA lesions induced by psoralens, we tested the effects of the mono- and bifunctional furocoumarins 3-CPs and 8-MOP, respectively, at a comparable level of total photoadducts on the inducible response of the RAD54 gene using a RAD54-lacZ gene fusion described previously by Cole et al. 1987. Colorimetric beta-galactosidase assays were performed with the colorimetric substrate o-nitrophenyl-beta-D-galactopyranoside (Guarente 1983) after treatment of diploid yeast cells in logarithmic growth phase with 3-CPs, 8-MOP and UVA. When plotted as a function of total photoadducts induced (Averbeck 1985), the level of beta-galactosidase activity, correlated with the amount of fusion gene product present in the cells, is clearly 2 to 4 fold higher in the presence of 8-MOP induced MA and CL than in the presence of 3-CPs induced MA (data not shown).

From this, it appears that mixtures of MA and CL constitute a stronger inducing signal than MA alone. Interestingly, this parallels the results obtained on the role of these lesions in psoralen induced mutagenesis and adds some support to the involvement of an inducible component in the mutational process. Whether the inducing signal is related to repair intermediates, such as excision-repair related single strand breaks in the case of MA, and/or to the occurrence of double strand breaks during the repair of CL in yeast, remains an open question that needs further clarification by more detailed molecular studies.

DISCUSSION

In the first period of psoralen research, psoralens were tested for their antiproliferative and genotoxic activity in view of their use in photochemotherapy. The UVA dose dependence of mutagenic effects was established for the different psoralens. The yeast <u>Saccharomyces cerevisiae</u> as a typical eukaryotic organism was found to be a useful system. It was shown (Averbeck et al., 1981; Averbeck 1985; Cassier-Chauvat and Averbeck 1988; Bianchi et al., 1990) that in equimolar conditions the order of mutagenic activity in yeast was the following:

TMP > 5-MOP > TMA > 8-MOP > MPP > 3-CPs > angelicin

Interestingly, monofunctional furocoumarins were always less mutagenic than bifunctional furocoumarins at equitoxic levels, suggesting that the presence of CL has an enhancing effect on mutation induction. At equal solubility levels and equimolar concentrations, 5-MOP was always found more mutagenic in yeast than 8-MOP (Averbeck 1984), and similar results were obtained in other eukaryotic cell systems (Schimmer 1981; Loveday and Donahue 1984; Papadopoulo and Averbeck 1985). This type of approach was helpful for photochemotherapeutic considerations and genotoxic risk evaluations (IARC 1986). Of course, the order of activity did not take into account differences in penetration, UV absorption, photostability and DNA photobinding activity.

Classical monofunctional furocoumarins like 3-CPs and angelicin were always found less active (per unit dose of UVA as well as a function of survival) than the bifunctional furo-coumarins 8-MOP, 5-MOP and TMP. This was apparently valid for the induction of reverse and forward mutations in yeast (Averbeck 1984; Averbeck 1985; Cassier-Chauvat and Averbeck 1988) and suggested that MA alone induced by psoralens are more easily repairable than mixtures of MA and CL.

Studies with repair deficient mutants in yeast revealed that different pathways are involved in the repair of psoralen induced lesions (for reviews see, Moustacchi 1987; Friedberg 1988). The repair of MA appears to involve repair pathways such as excision repair (complementation group <u>RAD3</u>) and mutagenic repair (complementation group <u>RAD6</u>) which also operate on UV (254 nm) induced lesions. However, the processing of DNA containing psoralen induced CL (together with MA) is found to rely on a number of different repair pathways including those already mentioned: excision repair

(RAD3), mutagenic repair (RAD6), double strand break and
recombinational repair (RAD50) and specific repair pathways
for DNA CL (PSO2, SNM1) (for review see, Friedberg 1988) and
recombination (PS04) (Andrade et al., 1989; Henriques et al.,
1989). As in the repair of CL in E. coli (Cole 1973; Van
Houten et al. 1986) it can be deduced from repair studies in
yeast (for review see, Moustacchi 1987, Friedberg 1988) that
also in eukaryotic cells the repair of CL occurs by a
sequential excision-recombination mechanism.

In E. coli the incision of CL by UvrABC excinuclease
appears to be a step in the path to psoralen induced
mutagenesis (Sladek et al., 1989) rendering CL recombinogenic.
This may well also be the case in eukaryotic cells since the
frequency of recombinogenic events increases in yeast with
increasing levels of CL induced (Cassier et al., 1984;
Averbeck et al., 1987; Cundari and Averbeck 1988).

With regard to the mutagenicity of psoralen induced
monoadducts in eukaryotic cells a similar model as proposed in
E. coli for UV induced leasions (Sedgwick 1976) and psoralen
MA (Sladek et al., 1989) may apply, assuming MA to become
mutagenic due to a blockage of DNA synthesis via a particular
high local density or an interaction of lesions leading to
overlapping gaps in opposite nascent DNA strands.

Cassier et al. (1985) and Cassier and Moustacchi (1988)
were able to show that furocoumarin plus UVA induced
mutagenesis in yeast depends on the function of specific genes
controlling error-prone repair.

The development of new, very photoreactive monofunctional
psoralen derivatives for photochemotherapeutic use, such as
pyridopsoralens (Moron et al., 1983) and methylangelicins
(Rodighiero et al., 1988), threw a new light on mutational
studies since some of the new monofunctional compounds, for
example, MPP (Averbeck 1985) and TMA (Bianchi et al., 1990),
were respectively slightly less or even more photomutagenic
(per unit dose of UVA) than known bifunctional furocoumarins
such as 8-MOP. A relatively high mutagenicity of 4,5'-
dimethylangelicin was observed in yeast (Averbeck 1984;
Averbeck et al., 1981) and mammalian cells (Swart et al.,
1985). Obviously, the DNA photobinding activity of each
compound constitutes an important factor. In order to
determine the role of monoadducts and cross-links in the
mutagenic process, comparisons had to be made on the basis of
equal numbers of DNA lesions induced.

A comparison of the mutagenicity of two bifunctional
furocoumarins 5-MOP and 8-MOP indicates that not only the
presence of CL, i.e. the concomitant induction of CL with MA,
is responsible for the relatively high mutagenicity of these
compounds but also the relative amount of MA, i.e. the ratio
of MA over CL, is important. Experiments performed at
different wavelengths and reirradiation experiments appear to
confirm these conclusions. They are valid for reverse
mutations as shown here and for the induction of forward
mutations in yeast (Cassier et al., 1984: Cassier-Chauvat and
Averbeck 1988), green algae (Schimmer 1983) and V-79 Chinese
hamster cells (Babudri et al., 1981). Also, experiments with
8-MOP and the bifunctional pyranocoumarin HPs showed that the

final level of CL induced is a determinant factor for the repair of photoadducts and for mutagenesis (Averbeck et al., 1990a).

In reirradiation experiments the conversion of MA_f into CL is likely to be the most important factor. However, the decrease in MA_f due to conversion drastically changes also the ratio of MA_p over CL and thus may provide a role for pyrone-side adducts.

Unfortunately, little is known about the mutagenic effects of MA_p. At the same survival level, certain psoralen derivatives such as the methylated pyridopsoralen PMP (Moron et al., 1983) known to induce exclusively MA_p, are apparently more mutagenic than compounds like MPP which induces only MA_f in yeast. 8-MOP plus 405 nm induced MA (MA_p and MA_f) are more efficient in mutation induction than MA_f alone induced by 3-CPs plus 365 nm (Averbeck et al., 1987).

Interestingly, the two monofunctional furocoumarins 3-CPs and MPP do not exhibit the same photomutagenic effects (Averbeck 1985) although both are known to induce the same type of photoadduct, i.e. MA_f. From the data presented above, it is unlikely that oxygen mediated reactions contribute significantly to the mutagenic effects of 3-CPs since, at least _in vitro_, the oxygen-mediated DNA damage appears to be repaired with high fidelity (Sage et al., 1989). The comparatively high photomutagenicity of MPP may rely not only on the induction of MA_f but, surprisingly enough for a psoralen derivative, on the concomitant induction of pyrimidine dimers via energy transfer from the triplet excited state of MPP (Costalat et al., 1990). The details of the interaction of MA_f and pyrimidine dimers are under investigation. Such an interaction may be implicated in the cytotoxic and mutagenic activity of MPP in yeast (Averbeck 1985) and mammalian cells (Papadopoulo et al., 1986) and may be related to its high therapeutic efficiency (Dubertret et al., 1985).

It has to be kept in mind that psoralen induced mutagenesis in eukaryotic cells represents a rather complex situation due to the fact that in normal exposure conditions with bifunctional furocoumarins at least three types of different adducts are induced. Bredberg and Nachmansson (1987), using a shuttle vector plasmid propogated in primate cells, were able to show high mutagenicity of DNA cross-links. In _E. coli_, Zhen et al. (1986) showed that a plasmid containing a single CL (without any MA) upon transformation into excision proficient _E. coli_ cells yielded mutations at a very high frequency. Thus, it is clear that CL alone are mutagenic and not just lethal in accord with previous findings (Saffran and Cantor, 1984; and Piette et al., 1985) indicating that base changes in psoralen-containing vectors occur predominantly at favorable sites for cross-linking.

At present, even in _E. coli_, the mechanism by which psoralen cross-links lead to mutations is unknown, but it appears that damage inducible functions (SOS functions) are involved (Saffran and Cantor 1984; Sladek et al., 1989). Interestingly, in yeast the signal for inducible functions, for example, the expression of the inducible _RAD54_ gene, is much stronger in the presence of CL than in the presence of MA. In parallel, also the induction of mutations appears to be strongly influenced by the presence of CL. Thus, it seems likely that in yeast as in E. coli inducible functions are involved in the processing of CL and psoralen induced mutagenesis.

ACKNOWLEDGEMENTS

Support by the CNRS UA1292, the CEA (Saclay), la Ligue Nationale Francaise Contre le Cancer and contracts intersection of the Institut Curie are gratefully acknowledged. Many thanks are also due to Madame S. Averbeck for excellent technical assistance.

REFERENCES

Andrade, H.H.R., Marques, E.K., Schenberg, A.C.G. and Henriques, J.A.P., 1989, The PSO4 gene is responsible for an error-prone recombinational DNA repair pathway in Saccharomyces cerevisiae. Mol. Gen. Genet., 217, 419-426.

Averbeck, D., 1984, Photochemistry and photobiology of psoralens, Proc. Jpn. Soc. Invest. Dermatol., 8, 52-73.

Averbeck, D., 1985, Relationship between lesions photoinduced by mono- and bifunctional furocoumarins in DNA and genotoxic effects in diploid yeast, Mutat. Res., 151, 217-233.

Averbeck, D. and Moustacchi, E., 1980, Decreased photo-induced mutagenicity of monofunctional as opposed to bifunctional furocoumarins in yeast, Photochem. Photobiol., 31, 475-478.

Averbeck, D., Magana-Schwencke, N. and Moustacchi, E., 1981, Genetic effects and repair in yeast of DNA lesions induced by 3-carbethoxypsoralen and other photoreactive furocoumarins of therapeutic interest. In: "Psoralens in Cosmetics and Dermatology", J. Cahn, P. Forlot, C. Grupper, A. Meybeck, F. Urbach, eds., Pergamon Press, New York.

Averbeck, D., Averbeck, S. and Cundari, E., 1987, Mutagenic and recombinogenic action of DNA monoadducts photoinduced by the bifunctional furocoumarin 8-methoxypsoralen in yeast (Saccharomyces cerevisiae). Photochem. Photobiol., 45, 371-379.

Averbeck, D., 1988, Photmutagenicity induced by psoralens: modulation of the mutagenic response in eukaryotes, Arch. Toxicol. Suppl., 12, 35-46.

Averbeck, D., Cundari, E., Dardalhon, M., Dall'Acqua, F. and Vedaldi, D., 1990a, Genetic effects and repair of DNA photoadducts induced by 8-methoxypsoralen and homopsoralen (pyranocoumarin) in diploid yeast. J. Photochem. Photobiol., B. 5, 179-195.

Averbeck, D., Dardalhon, M., and Magana-Schwencke, N., 1990b, Mutagenic effects of psoralen-induced photoadducts and their repair in eukaryotic cells. Proc. 10th Int. Congr. Photobiol. Jerusalem 20 Oct - 5 Nov 1988, in press.

Averbeck, D., Averbeck, S., Dubertret, L., Young, A. and Morliere, P., 1990c, Genotoxicity of bergapten and bergamot oil in Saccharomyces cerevisiae. J. Photochem. Photobiol., B., in press.

Badudri, N., Pani, B., Venturini, S., Tamaro, M., Monti-Bragadin, C. and Bordin, F., 1981, Mutation induction and killing of V-79 Chinese hamster cells by 8-methoxypsoralen plus near ultraviolet light: relative effects of monoadducts and crosslinks. Mutat. Res., 91, 391-394.

Ben-Hur, E. and Song, P.S., 1984, The photochemistry and photobiology of furocoumarins (psoralens), Adv. Radiat. Biol., 11, 131-171.

Bianchi, L., Bianchi, A., Dall'Acqua, F. and Santamaria, L., 1990, Photobiological effects in Saccharomyces cerevisiae induced by the monofunctional furocoumarin 4,4',6-trimethylangelicin (TMA) and the bifunctional furocoumarin 8-methoxypsoralen (8-MOP), Mutat. Res., 235, 1-7.

Boyer, V., Moustacchi, E. and Sage, E., 1988, Sequence specificity in photoreactions of various psoralen derivatives with DNA: role in biological activity. Biochemistry, 27, 3011-3018.

Brash, D.E., 1988, UV mutagenic photoproducts in Escherichia coli and human cells: a molecular genetics perspective on human skin cancer, Photochem. Photobiol., 48, 59-66.

Bredberg, A. and Nachmansson, N., 1987, Psoralen adducts in a shuttle vector pasmid propagated in primate cells: high mutagenicity of DNA cross-links. Carcinogenesis, 8, 1923-1928.

Cassier, C., Chanet, R. and Moustacchi, E., 1984, Mutagenic and recombinogenic effects of DNA cross-links induced in yeast by 8-methoxypsoralen photoaddition. Photochem. Photobiol., 39, 799-803.

Cassier, C., Chanet, R. and Moustacchi, E., 1985, Repair of 8-methoxypsoralen photoinduced crosslinks and mutagenesis: role of different repair pathways in yeast, Photochem. Photobiol., 41, 289-294.

Cassier, C. and Moustacchi, E., 1988, Allelism between psol-1 and rev3-1 mutants and between pso2-1 and snml mutants in Saccharomyces cerevisiae, Curr. Genet., 13, 37-40.

Cassier-Chauvat, C. and Averbeck, D., 1988, Photomutagenic effects induced by psoralen derivatives in the yeast Saccharomyces cerevisiae. In: "Psoralens 1988, Past, Present and Future of photochemoprotection and other biological activities", Fitzpatrick, T.B., Forlot, P., Pathak, M.A. and Urbach, F., eds., John Libbey Eurotext, Montrouge, France, pp. 329-335.

Chatterjee, P.K. and Cantor, C.R., 1978, Photochemical production of psoralen-DNA monoadducts capable of subsequent photo-crosslinking, Nucleic Acids Res., 5, 3619-3633.

Cimino, G.D., Gamper, H.B., Isaacs, S.T. and Hearst, J.E., 1985, Psoralens as photoactive probes of nucleic acid structure and function: organic chemistry, photochemistry and biochemistry. Ann. Rev. Biochem., 54, 1151-1193.

Cole, R.S., 1973, Repair of DNA containing interstrand cross-links in Escherichia coli: sequential excision and recombination, Proc. Natl. Acad. Sci. USA., 70, 1064-1068.

Cole, G.M., Schild, D., Lovett, S.T. and Mortimer, R., 1987, Regulation of RAD54-and RAD52-lacZ gene fusion in Saccharomyces cerevisiae in response to DNA damage, Mol. Cell. Biol., 7, 1078-1084.

Costalat, R., Blais, J., Ballini, J.P., Moysan, A., Chalvet, O. and Vigny, P., 1990, Formation of thymine dimers photosensitized by pyridopsoralens: a triplet-triplet energy transfer mechanism, Photochem. Photobiol., 51, 255-262.

Cundari, E. and Averbeck, D., 1988, 8-methoxypsoralen photoinduced DNA cross-links as determined in yeast by alkaline step elution under different reirradiation conditions. Relation with genetic effects, Photochem. Photobiol., 48, 315-320.

Dardalhon, M. and Averbeck, D., 1988, Induction and removal of DNA interstrand cross-links in V-79 Chinese hamster cells masured by hydroxylapatitie chromatography after treatment with bifunctional furocoumarins, Int. J. Radiat. Biol., 54, 1007-1020.

De Mol, N.J., Beijersbergen van Henegouwen, G.M.J., Mohn, G.R., Glickmann, B.W. and Van Kleef, P.M., 1981, On the involvement of singlet oxygen in mutation induction by 8-methoxypsoralen and UVA irradiation in Escherichia coli K-12, Mutat. Res., 82, 23-30.

De Mol, N.J., Beijersbergen van Henegouwen, G.M.J., Weeda, B., Knox, C.N. and Truscott, T.G., 1986, Photobinding of psoralens to bacterial macromolecules in situ and induction of genetic effects in a bacterial test system. Effects of singlet oxygen diagnostic aids D_2O and DABCO, Photochem. Photobiol., 44, 747-751.

Dubertret, L., Averbeck, D., Bisagni, E., Moron, J., Moustacchi, E., Billardon, C., Papadopoulo, D., Nocentini, S., Vigny, P., Blais, J., Bensasson, R.V., Ronfard-haret, J.C., Land, E.J., Zajdela, F. and Latarjet, R., 1985, Photochemotherapy using pyridopsoralens, Biochimie, 67, 417-422.

Faulques, M., René, L., Royer, R., Averbeck, D. and Moradi, M., 1983, Synthèse at propriétés biologiques photoinduities des dérivés déméthylés des pyranocoumarines naturelles, xanthylétine et séséline, Eur. J. Med. Chem., 18, 9-14.

Friedberg, E.C., 1988, Deoxyribonucleic acid repair in the yeast Saccharomyces cerevisiae, Microbiol. Rev., 52, 70-102.

Eckardt, F., Kunz, B.A. and Haynes, R.H., 1981, UV induced mutation and recombination in yeast: involvement of an inducible component of repair, Radiat. Environ. Biophys., 19, 292 (abstract).

Guarente, L., 1983, Yeast promoters and Lac Z fusions designed to study expression of cloned genes in yeast, Methods in Enzymol., 101, 181-191.

Haynes, R.H., Eckardt, F. and Kunz, B.A., 1985, Analysis of non-linearities in mutation frequency curves, Mutation Res., 150, 51-59.

Henriques, J.A.P., Vincente, E.J., Da Silva, K.V.C.L. and Schenberg, A.C.G., 1989, PSO4: a novel gene involved in error-prone repair in Saccharomyces cerevisiae, Mutation Res., 218, 111-124.

IARC, 1986, In: "IARC Monographs on the Evaluation of the Carcinogenic Risk of Chemicals to Humans", Vol. 40, Lyon, France.

Joshi, P.C. and Pathak, M.A., 1983, Production of singlet oxygen and superoxide radicals by psoralens and their biological and significance, Biochem. Biophys. Res. Commun., 112, 638-546.

Loveday, K.S. and Donahue, B.A., 1984, Induction of sister chromatid exchanges and gene mutations in Chinese hamster ovary cells by psoralens, Natl. Cancer Inst. Monogr., 66, 146-155.

Midden, W.R., 1988, Chemical mechanisms of the bioeffects of furocoumarins: the role of reactions with proteins, lipids and other cellular constituents. In: "Psoralen DNA Photobiology", Vol. II, Gasparro, F.P., ed., CRC Press, Inc. Boca Raton, Florida.

Moustacchi, E., 1987, DNA repair in yeast: genetic control and biological consequences. In: "Advances in Radiation Biology", Lett, J., ed., Academic press, New York.

Moron, J., Nguyen, C.H. and Bisangni, E., 1983, Synthesis of 5H-furo(3',2':6,7)(1)benzopyrano (3,4-c)pyridin-5-ones and 8H- pyrano (3',2':5,6)benzo-furo(3,2-c)pyridin-8-ones (pyridopsoralens). J. Chem., Perkin Trans., 1, 225-229.

Musajo, L. and Rodighiero, G., 1972, Mode of photosensitizing action of furocoumarins. In: "Photobiology" Vol. VII, Giese, A.C. ed., Academic Press, New York.

Papadopoulo, D. and Averbeck, D., 1985, Genotoxic effects and DNA photoadducts induced in Chinese hamster V-79 cells by 5-methoxypsoralen and 8-methoxypsoralen. Mutat. Res., 151, 281-291.

Papadopoulo, D., Averbeck, D. and Moustacchi, E., 1986, Mutagenic effects photoinduced in mammalian cells in vitro by two monofunctional pyridopsoralens. Photochem. Photobiol., 44, 31-39.

Parrish, J.A., Stern, R.S., Pathak, M.A. and Fitzpatrick, T.B., 1982, Photochemotherapy of skin diseases. In: "The Science of Photomedicine", Regan, J.D. and Parrish, J.A., eds., Plenum Press, New York.

Piette, J., Decuyper-Debergh, D. and Gamper, G., 1985, Mutagenesis of the lac promoter region in M13mp10 phage DNA by 4'-hydroxymethyl4,5',8-trimethylpsoralen, Proc. Natl. Acad. Sci. USA., 82, 7355-7359.

Robinson, G., Nicolet, C., Kaloinov, D. and Friedberg, E., 1986, A yeast excision-repair gene is inducible by DNA damaging agents, Proc. Natl. Acad. Sci. USA, 83, 1843-1846.

Rodighiero, G., Dall'Acqua, F. and Averbeck, D., 1988, New psoralen and angelicin derivaties. In: "Psoralen DNA Photobiology", Vol I, Gasparro, F.P., ed., CRC Press, Inc., Boca Raton, Florida.

Saffran, W.A. and Cantor, C.R., 1984, Mutagenic SOS repair of site-specific psoralen damage in plasmid pBR322, J. Mol. Biol., 178, 595-609.

Sage, E. and Moustacchi, E., 1987, Sequence context effects on 8-methoxypsoralen photobinding to defined DNA fragments, Biochemistry, 26, 3307-3314.

Sage, E., Le Doan, T., Boyer, V., Helland, D.E., Kittler, L., Helene, C. and Moustacchi, E., 1989, Oxidative DNA damages photoinduced by 3-carbethoxypsoralen and other furocoumarins. Mechanism of photo-oxidation and recognition by repair enzymes. J. Mol. Biol., 209, 297-314.

Schimmer, O., 1981, Vergleich der photomutagenen Wirkungen von 5-MOP (Begapten) und 8-MOP (Xanthotoxin) in Chlamydomonas reinhardii., Mutation Res., 89, 283-296.

Schimmer, O., 1983, Effect of re-irradiation with UVA on irradiation and mutation induction in cells of

Chlamydomonas reinhardii pretreated with furocoumarins plus UVA., _Mutat. Res._, 109, 195-205.

Scott, B.R., Pathak, M.A. and Mohn, G.R., 1976, Molecular and genetic basis of furocoumarin reactions, _Mutat. Res._, 39, 29-74.

Sladek, F.M., Melian, A. and Howard-Flanders, P., 1989, Incision by UvrABC excinuclease is a step in the path to mutagenesis by psoralen crosslinks in _Escherichia coli._, _Proc. Natl. Acad. Sci. USA_, 86, 3982-3986.

Smith, C.A., 1988, Repair of DNA containing furocoumarins adducts. _In_: "Psoralen DNA Photobiology", Vol. II, Gasparro, F.P., ed., CRC Press, Inc., Boca Raton, Florida.

Stern, R.S., Lange, R. and members of the photochemotherapy follow-up study, 1988, Non-melanoma skin cancer occurring in patients treated with PUVA five to ten years after first treatment, _J. Invest. Dermatol._, 91, 120-124.

Swart, R.N., Beckers, M.A. and Schothorst, A.A., 1983, Phototoxicity and mutagenicity of 4,5'-dimethylangelicin and long-wave ultraviolet irradiation in Chinese hamster cells and human fibroblasts, _Mutat. Res._, 124, 271-279.

Tessman, J.W., Isaacs, S.T. and Hearst, J.E., 1985, Photochemistry of the furan-side 8-methoxypsoralen-thymidine monoadduct inside the DNA helix. Conversion to diadduct and to pyrone-side monoadduct. _Biochemistry_, 24, 1669-1676.

Van Houten, B., Gamper, H., Holbrook, S.R., Hearst, J.E. and Sancar, A., 1986, Action mechanism of ABC excision nuclease on a substrate containing a psoralen crosslink at a defined position, _Proc. Natl. Acad. Sci. USA_, 83, 8077-8081.

Vedaldi, D., Dall'Acqua, F., Gennaro, A. and Rodighiero, G., 1983, Photosensitized effects of furocoumarins: the possible role of singlet oxygen, _Z. Naturforsch. C._, 38, 866-869.

Vigny, P., Gaboriau, F., Voituriez, L. and Cadet, J., 1985, Chemical structure of psoralen-nucleic acid photoadducts. _Biochimie_ 67, 317-325.

Wollenzien, P.L., 1988, Psoralens as probes of nucleic acid structure and function. _In_: "Psoralen DNA Photobiology", Vol. II, Gasparro, F.P., ed., CRC Press, Inc., Boca Raton, Florida.

Zhen, W.-P., Jeppesen, C., and Nielsen, P.E., 1986, Repair in _Escherichia coli_ of a psoralen-DNA interstrand crosslink site specifically introduced into $T_{410}A_{411}$ of the plasmid pUC19, _Photochem. Photobiol._, 44, 47-51.

Zimmerman, F.K., Kern, R. and Rasenberger, M., 1975, A yeast strain for simultaneous detection of induced mitotic crossing-over, mitotic gene conversion and reverse mutation, _Mutat. Res._, 28, 381-388.

SECOND GENERATION PHOTOSENSITIZERS FOR THE PHOTODYNAMIC
THERAPY OF TUMOURS

Giulio Jori and Elena Reddi

Department of Biology, University of Padova, Italy

INTRODUCTION

It is now generally accepted that photodynamic therapy
(PDT) represents a promising approach to cancer treatment.
The technique involves the systemic administration of a
photosensitizer displaying a preferential affinity for tumours
and absorbing light wavelengths longer than 600 nm (Wilson and
Jeeves, 1987; Manyak et al., 1988); such wavelengths are not
absorbed by the endogeneous chromphores of animal tissues and
are endowed with a maximal penetration power into most
biological tissues. After a suitable time interval following
administration of the drug, the neoplastic area is illuminated
with light specifically activating the photosensitizer, thus
inducing selective damage of the tumour. At present, most
clinical PDT protocols involve the use of a chemical
derivative of hematoporphyrin (HpD or its partially purified
form Photofrin II, see Kessel and Cheng, 1985) or
hematoporphyrin itself (Tomio et al., 1984) as tumour-
localizing and photosensitizing agents. Until now, a few
thousand patients have been treated by PDT and a vast majority
of them have objectively benefitted from the treatment (Manyak
et al., 1988). Table 1 summarizes the main indications
emerging from PDT studies performed with selected tumour
types. In spite of the favourable results obtained,
especially in the treatment of superficial and early-stage
neoplasias, some important factors still limit the efficacy of
PDT, including the hetergeneous and partially unknown chemical
composition of Photofrin II, its low extinction coefficient in
the red spectral region and the relatively poor selectivity of
tumour targetting as compared with most peritumoural tissues
or some normal tissues (e.g. skin). Several ongoing research
lines are exploring (i) potential approaches for enhancing the
specificity of photosensitizer localization in the tumour (for
a recent review see Jori and Reddi, 1989), and (ii) the
possible use of new tumour-photosensitizing agents in order to
overcome or minimize the present drawbacks of PDT. Desirable
features of a photosensitizer to be used for therapeutical
applications include; chemical purity, stability under

Table 1. Indications and Results Obtained in Clinical Applications of the Photodynamic Therapy of Tumors

District	Treated tumours	Results	Complications
Cutaneous and subcutaneous malignancies	Malignant melanoma; basal cell carcinoma; squamous cell carcinoma; Kaposi's sarcoma; metastatic breast carcinoma	Good response esp. for basaliomas and small-sized tumours	General skin photosensitivity
Head and neck cancer	Recurrent squamous cell carcinoma	Substantial reduction in tumour size; often recovery of impaired function (e.g. tongue)	High percentages of recurrencies; oro-cutaneous fistulae and chronic ulcerations
Bladder carcinoma	Transitional cell carcinoma	Good response esp. for superficial lesions, only partial for invasive papillary tumours	Bladder volume contraction; transient dysuria
Endobronchial and esophageal tumours	Early and obstructive carcinoma	Good response for early-stage squamous cell lung carcinoma; significant palliation for obstructive tumours	Few cases of hemoptysis; excessive mucosal secretions and sloughing
Intraocular malignancies	Choroidal melanoma; retinoblastoma	Reduction in tumour mass, a few complete tumour destructions	Chemosis, lid swelling, exudative retinal detachment (with subsequent recovery)
Gynaecological malignancies	Vaginal/epidermal recurrencies	Partial or complete response in all the treated cases	
Brain tumours	Glioma, astrocytoma	Partial response with improved survival, esp. if applied in combination with surgery	Cerebral edema (seldom); originating recurrencies from peritumoural regions

physiological conditions, lack of cytoxicity in the dark, and superior spectroscopic properties as compared with Photofrin.

It is the purpose of the present paper to critically discuss the physico-chemical characteristics of second generation photosensitizers in the ground and electronically excited states which could allow an adequate fulfilment of the above outlined criteria, based on presently available information.

PHYSICO-CHEMICAL PROPERTIES IN THE GROUND-STATE

The requirement for efficient light absorption in the 600-900 nm range restricts the choice of PDT agents to photosensitizers characterized by extended electronic delocalization. This goal can be achieved by a rational design of the chemical structure of the photosensitizer, althouh one must keep in mind that the structural features can profoundly influence both the hydrophobic/hydrophilic properties of the dye (hence its affinity for tumour tissues and its distribution among the intratumoural compartments) and the degree of dye aggregation in a variety of media (hence photosensitizing activity) (Dougherty, 1987). The optimal compromise is being sought through the interplay of selected factors, including the expansion of the conjugated electron system, the presence and type of the metal ion coordinated with the porphyrin-type macrocycle, and the nature of the functional groups possibly replacing the peripheral hydrogen atoms.

Porphyrin Analogs with Enhanced Long Wavelength Absorption

Three possibilities for increasing the efficiency of red light absorption by porphyrins through a manipulation of the electron conjugation have been proposed. In the first place, it is well known that chlorins (i.e. porphyrins in which one pyrrole ring is partially reduced) possess an intense absorption band in the 650-700 nm range (extinction coefficient above 10^5 $M^{-1}cm^{-1}$). Both pheophorbide derivatives (Bommer and Burnham, 1986; Roder, 1986) and purpurins, i.e. chlorin analogs with extended conjugation through one of the meso carbon atoms (Morgan et al., 1987), have been prepared and shown to display promising tumour-localising and -photosensitizing properties. In particular, the hydrophilic monoaspartyl-chlorin e_6 (a derivative bearing one acetyl group in a meso position and conjugated to an aspartyl residue via a peptide linkage) shows very attractive properties in vivo (Oseroff et al., 1986). In a detailed study of the structural features which enhance the extinction coefficient of chlorins in the red region, Morgan et al. (1987) pointed out the need for the simultaneous presence of a five- or six-membered exocyclic ring and a meso-substituent with π-electrons conjugated to the aromatic π-electron cloud. In this case, molar absorptivity values at 670-690 nm about ten-fold larger than that typical of HpD at 630 nm are obtained (See Table 2). It is also worth mentioning that bacteriochlorins are a class of efficient photosensitizers with remarkable absorbance at 780-800 nm (Borland et al., 1987), although no detailed investigation on their possible use in PDT has been performed until now.

Table 2   Absorption Properties of Potential Photo-
            sensitizers for the Photodynamic Therapy
            of Tumours

Photosensitizer class	Absorption range in the red (nm)	Extinction coefficient $(M^{-1}cm^{-1})$
Hematoporphyrin derivative, Photofrin II	630	3,500-4,000
Chlorins/Purpurins with one exocyclic ring	680-700	40,000
Bacteriochlorins	780-800	150,000
Benzoporphyrins	690-700	13,000
Phthalocyanines	680-720	200,000
Naphthalocyanines	780-820	350,000
22 π-porphyrin	470	1,100,000
26 π-porphyrin	550 780	910,000 30,000
34 π-porphyrin	660 995	370,000 25,000

A modest increase (ca. four-fold) of the extinction
coefficient in the red is observed (Table 2) upon preparation
of monobenzoporphyrin compounds, through the condensation of a
phenyl ring with either pyrrole ring A or pyrrole ring B of
protoporphyrin (Pangka et al., 1986). A substantially more
pronounced improvement of the spectroscopic properties can be
obtained with tetraaza-tetrabenzoporphyrins (phthalocyanines)
and tetraaza-tetranaphthaloporphyrins (naphthalocyanines), as
is exemplified in Table 2. Both water-insolubale
phthalocyanines (Ricchelli et al., 1987) and naphthalocyanines
(Firey and Rodgers, 1987), as well as their water-soluble
sulfonated derivatives (Spikes, 1986; McCubbin and Phillips,
1986), have been characterised and extensively tested in
cellular and animal models. The aza-nitrogen inter-ring bonds
and the extended conjugated pathway along the macrocyclic
skeleton cause a large bathochromicity and an enhanced
probability of photoexcitation.

These features are also found for porphyrinoid structures
obtained by insertion of conjugated double bonds into the ring
system. Through ingenious chemical synthesis, Franck et al.
(1988) succeeded in modifying the porphyrin structure so that
the aromatic 18 π-electron cloud is extended to 22 π-, 26π-,
or even 34 π-electron systems. The expanded porphyrins are
characterized by Soret bands in the blue-green spectral region

whose extinction coefficient is around 10^6 M^{-1}cm^{-1}, i.e. the highest observed for organic pigments. Moreover, these porphyrin analogs have a large molor absorptivity also in the red (640-670 nm), while the 34 π-porphyrin exhibits significant absorbance even at 990-1000 nm (Table 2).

All the above mentioned dyes show a significant tendency to undergo aggregation especially in aqueous media owing to the onset of hydrophobic and $\pi-\pi$ intermolecular forces which are aimed at minimizing the interaction betweeen the planar aromatic macrocycle and the polar environment. Aggregation usually results in a broadening of the absorption band, hypochromicity and a drop in the photosensitizing activity (Jori and Spikes, 1984; Spikes, 1986). The formation of aggregated species is prevented by the presence of electric charges or bulky substituents in the dye molecule, while disruption of the aggregates can be achieved by incorporation of the dye into protein matrices or apolar systems, such as surfactant micelles and liposomes.

Lastly, some lipophilic cationic dyes derived from phenoxazine and phenothiazine (Cincotta et al., 1987), triarylmethane (Wadwa et al., 1988), and Kryptocyanines (Ara et al., 1987), have been proposed as PDT agents, since they are characterized by intense red-absorption bands and noticeable specificity of localization in the mitochondria of neoplastic cells.

Role of the Central Metal Ion

Porphyrins and their analogs can coordinate a large number of metal ions with the four pyrrole nitrogens (Jori and Spikes, 1984; Moser and Thomas, 1982). It is not clear at present to what extent the presence of the metal ions influences the tumour-localizing ability of a photosensitizer; at least for Photofrin II the distribution in tumour-bearing animals is very similar with that of its Cu(II)-derivative (Wilson et al., 1988; Winkelman, 1967). On the other hand, the insertion of a metal drastically affects the spectroscopic, photochemical and photosensitizing properties of the dye. In the case of porphyrins, the four Q-band envelope above 500 nm is converted into a two-band system upon complexation of metal ions; in particular, there is no residual absorbance at wavelengths longer than 600 nm. This effect is less pronounced for chlorins, phthalocyanines and naphthalocyanines, although a limited spectral shift of the metal derivatives with respect to the parent compounds often takes place; at the same time, the metal ion generally induces a sharpening of the absorption bands and hyperchromicity.

Of particular importance is the presence of ions which can coordinate one or two axial ligands that are perpendicular to the ring plane. These ligands can decrease the tendency of a photosensitizer to undergo aggregation as a result of steric hindrance. Thus, under identical experimental conditions, Al(III)-phthalocyanine solutions contain a larger fraction of monomeric dye as compared with the corresponding Zn(II) and Mg(II) compounds, since the tervalent aluminium binds a chloride ion orthogonally to the phthalocyanine macrocycle. Moreover, aggregated Zn(II)-phthalocyanines can be readily monomerized by addition of pyridine or morpholine (Valduga et al., 1987; Ford et al., 1989), which give octahedral complexes through coordination of the amines to the metal ion. The two axial ligands can also be utilized for synthetically attaching organic residues designed to convey specific solubility properties to the photosensitizer, thus controlling its affinity for specific tumour compartments. For example, the hydrophobic Si(IV)-naphthalocyanine molecule can be made water-soluble by ligation of the silicon ion with two polyethylene glycol units; on the other hand, the degree of lipid solubility can be modulated by attaching branched alkyl chains of different length (Ford et al., 1988).

Role of the Peripheral Substituents

The nature of the substituents which are present in the 1-8 peripheral positions of porphyrins can affect the relative intensities of the Q_x and Q_y absorption bands (Jori and Spikes, 1984). In particular, the intensity of the band above 600 nm, which is most interesting from the point of view of PDT applications, undergoes a modest increase in the presence of electron donors. An about 20 nm-red shift of this band, as well as a somewhat more important enhancement of its intensity, are obtained by the introduction of phenyl groups in the methine carbons bridging the pyrrole rings; the extinction coefficient of meso-tetra(4-sulfonatophenyl)- and meso-tetra(hydroxyphenyl)-porphines at 650 nm is about 10,000 $M^{-1}cm^{-1}$. The hydroxyphenyl derivatives have been reported to posses a greater tumour selectivity than Photofrin II (Berenbaum et al., 1986; Peng et al., 1987). Analogously, in the case of phthalocyanines and naphthalocyanines, placing sulfonate or carboxylate groups on carbon atoms distal to the point of fusion between the benzene or naphthalene ring and the pyrrole moiety has no significant effect on the spectroscopic features. However, at least in the case of Zn(II)-phthalocyanine, the α-alkoxy regioisomers show a red shift in the Q band of 60-70 nm with only slight changes in the extinction coefficient (Ford et al., 1989). It is apparent that more detailed investigations on the relationship between chemical structure and spectroscopic properties of these macrocyclic compounds would be useful.

Once again, the polarity and bulkiness of the substituents can affect both the tendency of the photosensitizers to aggregate and their water/lipo-solubility. Thus, the octacarboxylic uroporphyrin, which has eight negative charges at neutral pH values, is almost completely monomeric in aqueous solution at 10 μM concentrations, while dicarboxylic porphyrins, such as hemato- or proto-porphyrin are heavily aggregated (Jori and Spikes, 1984). Similarly mono- and di-sulfonated phthalocyanines are sparingly soluble and extensively aggregated in aqueous media; only their tetra-sulfonated analogs are largeley in a monomeric state at micromolar concentrations (Spikes, 1986). In general, the hydrophilicity of macrocyclic compounds can be increased by the addition of sulfonated, hydroxyl and alkoxy functional groups (Leznoff et al., 1989).

The relative position of the substituents in the macrocycle also plays an important role. In a systematic study of the relative photodynamic efficiency of selectively sulfonated tetraphenyl-porphines, Kessel et al., (1987) showed that the di-sulfonic isomer having two groups on vicinal phenyl rings is accumulated by tumours in larger amounts than the isomer bearing the groups on opposite rings. Apparently, the amphiphilic character of the former isomer favours the crossing of the lipid-rich cell membranes while endowing the dye with sufficient water solubility to allow its transport in the bloodstream. Similar conclusions were reached for sulfonated phthalocyanines (Wagner et al., 1987).

PHOTOCHEMICAL PROPERTIES

Several porphyrins and chlorins, as well as their metal complexes, undergo irreversible modification upon irradiation with visible light (Jori and Spikes, 1984; Moan et al., 1988). The photoreactions are usually very complex and lead to a variety of photoproducts, whose relative yield is dependent on the chemical structure of the porphyrin and the experimental conditions. Thus, oxygen-dependent processes (possibly involving the intermediacy of singlet oxygen) result in ring cleavage (Morgan and Tertel, 1986) and/or alteration of peripheral substituents, such as the vinyl groups of protoporphyrin IX (Jori and Spikes, 1984). However, oxygen-independent processes also occur, especially in the presence of metal ions having a low redox potential; these reactions are likely to involve intermolecular electron transfer steps originating radical-type transients (Tanno et al., 1980). A large efficiency of dye photodegradation has been observed for polysulfonated derivatives of phthalocyanines and naphthalocyanines, whereas the related unsubstituted compounds display a remarkable photostability (Spikes and Bommer, 1987; Jori, G., unpublished observations).

In general, the overall photoprocess is accompanied by the disappearance of the visible absorption bands of the photosensitizer, hence it is often defined as photobleaching. The occurrence of photobleaching has also been observed during PDT in vivo (Mang et al., 1987); as a consequence, its importance in competition with photosensitization has profound implications for determining the outcome of the phototreatment. Actually, the removal of photosensitizer from skin and other normal tissues by this route has been proposed as one way to overcome the main side effect of PDT, namely the persistence of cutaneous photosensitivity for some weeks in several patients that have been injected with Photofrin II. On the other hand, it can be argued that photobleaching lowers the concentration of the photosensitizer in the tumour area, thus reducing its efficiency in inducing tumour necrosis; moreover, unknown photoproducts are generated, whose toxicity for tissues and clearance rate from the organism are not defined as yet. However, some photoproducts are very good photosensitizers with large quantum yields for cell inactivation (Moan, 1988).

PHOTOPHYSICAL AND PHOTOSENSITIZING PROPERTIES

In vivo photosensitization of tumour necrosis requires the presence of oxygen (Zhou, 1989; Moan and Sommer, 1985), which justifies the definition "photodynamic" for this type of phototherapy. Photodynamic effects can occur by two competing mechanisms; in type I processes an electron or hydrogen transfer reaction between the triplet photosensitizer and a substrate molecule takes place, whereas in type II processes ground state oxygen is promoted to the highly reactive $^1\Delta_g$ singlet state via energy transfer from the triplet photosensitizer (Foote, 1988). Consequently, in order to design an effective photosensitizer and to evaluate its photoactivity in vivo, it is necessary to assess the influence of the molecular structure and environmental factors (e.g. medium, pH) on the physico-chemical properties of the triplet state, including (i) the lifetime and quantum yield of its photogeneration, which determine the reactivity of triplet photosensitizer in diffusion-controlled processes; (ii) the redox potential, which controls the direction and efficiency of electron exchange with suitable substrates; and (iii) the energy level, which determines the efficiency of $^1O_2$ formation.

In vitro studies demonstrate that a large variety of red light-absorbing photosensitizers can promote both type I and type II photoprocesses (Spikes, 1986; Foote, 1988); however, no definite evidence has been obtained until now, as regards the occurrence and importance of either pathway in cells and tissues, although for porphyrins and phthalocyanines a type II process is generally invoked. Some photophysical properties of the triplet state of selected porphyrins, chlorins, phthalocyanines and naphthalocyanines are listed in Table 3. Clearly, the presence of aggregated species can drastically reduce the triplet lifetime, which is reflected by the diminished photosensitizing activity. Thus, the quantum yield of singlet oxygen generation sensitized by hematoporphyrin is lower in polar solutions where the dye is extensively aggregated, than in apolar media, where the porphyrin is almost exclusively in a monomeric state (Table 4). Photofrin II, which contains a large proportion of dimeric and oligomeric porphyrins, is an inefficient photosensitizer in vitro; its ability to photoinduce tumour necrosis in vivo is possibly a consequence of deaggregation processes taking place when the porphyrin is bound in specific loci of neoplastic cells (Dougherty, 1987). Analogously, metallo-porphyrins or -phthalocyanines bearing one or two axial ligands (which hinders aggregation) are more powerful photosensitizing agents than most tetra-coordinated metal complexes (Tables 3 and 4). It is also apparent that the presence of paramagnetic metal ions usually inhibits the photosensitization of biological substrates by porphyrins and other macrocyclic dyes; once again, this is due to a drastic shortening of the lifetime of triplet photosensitizers (Jori and Spikes, 1984). On the other hand, the diamagnetic metallo-porphyrins, -chlorins or -phthalocyanines often display a photosensitizing efficiency at least comparable with that typical of the corresponding free bases (Table 3 and 4).

In general, the triplet energies of these photosensitizers range between 25 and 36 kcal/mole, hence they can potentially generate $^1\Delta_g O_2$ (energy level 22.6 kcal/mole) with high efficiency. The situation is more complex for naphthalocyanine derivatives (Firey and Rodgers, 1987; Ford et al., 1989). The triplet energy of Si(IV)-naphthalocyanine is 22 kcal/mole in benzene solutions so that energy transfer to oxygen in order to give $^1O_2$ is slightly endoergonic. This

Table 3   Photophysical properties of the lowest
          excited state of red light-absorbing
          photosensitizers

Photosensitizer[a]	Medium	Triplet quantum yield
HpD	H_2O, pH 7.4	0.20
Hp 17 μM	H_2O, pH 7.4	0.63
4 μM	H_2O, ph 7.4	0.80
	90% Methanol	0.83
	Albumin-bound	0.66
	Micelle	0.87
Mg(II)-Hp	H_2O, pH 7.4	0.51
Zn(II)-Hp	H_2O, pH 7.4	0.60
	Micelle	1.10
Cu(II)-Hp	Methanol	-
Fe(III)-Hp	Methanol	0.83
Hemin	H_2O, pH 7.0	-
Octaethylpurpurin	Benzene	0.85
Zn(II)-etiopurpurin	Benzene	1.00
Bacteriochlorin	Methanol	0.54
Pc	Methanol	0.22
Pc-octabutoxy	Benzene	0.09
PcS_4	H_2O, pH 7.0	0.22
Mg(II)-Pc	DMSO	0.18
Zn(II)-Pc	Methanol	0.57
Zn(II)-Pc octabutoxy	Benzene	0.51
Zn(II)-PcS_4	H_2O, pH 7.4	0.46
Al(III)-Pc	DMSO	0.41
Al(III)-PcS_4	H_2O, pH 7.4	0.50
Cu(II)-Pc	DMSO	0.95
Cu(II)-PcS_4	DMSO	0.92
Si(IV)-Nc	Benzene	0.39
Al(III)-NcS_4	Methanol	0.45

(a)   Abbreviations: Hp, hematoporphyrin; Pc, PcS_4 =
      phthalocyanine and its tetrasulfonated derivative; Nc,
      NcS_4 = naphthalocyanine and its tetrasulfonated
      derivative.

Table 4 Quantum yield of singlet oxygen generation
by red light-absorbing photosensitizers

Photosensitizer[a]	Medium	Quantum Yield
HpD	H_2O, pH 7.4	0.11
	Liposome	0.23
Hp	H_2O, pH 7.4	0.32
	90% Methanol	0.53
	Micelle	0.56
	Albumin-bound	0.30
Cu(II)-Hp	H_2O, pH 7.4	0.00
	Methanol	0.00
Protoporphyrin dimethylester	Benzene	0.57
Hemin	Methanol	0.00
22 π-porphyrin	Benzene/Methanol	0.29
26 π-porphyrin	Benzene/Methanol	0.44
34 π-porphyrin	Benzene/Methanol	0.00
Octaethylpurpurin	Benzene	0.66
Zn(II)-etiopurpurin	Benzene	0.54
Bacteriochlorin	Methanol	0.20
Zn(II)-Pc	Ethanol	0.53
	Liposome	0.70
PcS_4	H_2O, pH 7.0	0.17
Al(III)-PcS_4	H_2O, pH 7.0	0.34
Zn(II)-PcS	H_2O, pH 7.0	0.35
Cr(IV)-PcS_4	H_2O, pH 7.0	0.50
Cu(II)-PcS_4	H_2O, pH 7.0	0.00
Fe(III)-PcS_4	H_2O, pH 7.0	0.00
Zn(II)-Pc octabutoxy	Benzene	0.45
Si(IV)-Nc	Benzene	0.19

(a) Abbreviations are defined in Table 3.

fact does not prevent the formation of $^1O_2$ via a reversible
energy transfer process; actually, Si(IV)-naphthalocyanine has
been shown to photosensitize tumor necrosis in experimental

animals (Ford et al., 1988). The picture can be altered by insertion of metal ions other than silicon into the naphthalocyanine macrocycle (Ford et al., 1989).

These considerations emphasize the importance of considering the energy gap between the lowest excited singlet and triplet states upon designing red light-absorbing photosensitizers for clinical use: the $^1O_2$ luminescence, corresponding to the decay of this excited species to the ground state, is observed at 1,270 nm; therefore, any dye absorbing at shorter wavelengths can in principle yield $^1O_2$. However, singlet-triplet gaps of the order of 10-15 kcal should be taken into consideration; this would limit the longest wavelength absorption band to 800-850 nm, if a type II ($^1O_2$-involving) photosensitization mechanism is to be promoted. Actually, as one can see in Table 4, the 34 π-porphyrin (Franck et al., 1988), which has a fluorescence emission at 1,295 nm, does not generate $^1O_2$. Of course, under suitable experimental conditions, these compounds could photosensitize biological damage by type I reactions. In this case, the redox potential rather than the energy level of the triplet state is the key factor. Only few experimental determinations of this parameter are reported in the literature. Evidence indicating the photodynamic activity of photosensitizers which are known to be inefficient generators of $^1O_2$ has recently been obtained (Wadwa et al., 1988).

REFERENCES

Ara, G., Aprille, G.R., Malis, C.D., Kane, S.B., Cincotta, L., Foley, J., Bonventre, J.V., and Oseroff, A.R., 1987, Mechanism of mitochondrial photosensitization by cationic dye, N,N'bis(2-ethyl-1,3-dioxylene) - kryptocyanine, Cancer Res., 47: 6580-6585.

Berenbaum, M.C., Akande, S.L., Bonnett, R., Kour, H., Ioannou, S., White, R.D., and Winfield, U.J., 1986, Meso-tetra(hydroxyphenyl)porphyrins, a new class of potent tumour photosensitizers with favourable selectivity, Br. J. Cancer, 54: 717-725.

Bommer, J.C., and Burnham, B.F., 1986, Tetrapyrrole compounds, Eur. Patent Application, EP 168, 831.

Borland, C.F., McGarvey, D.J., Truscott, T.G., Codgell, R.J., and Land, E.J., 1987, Photophysical studies of bacteriochlorophyll a and bacteriopheophytin a: singlet oxygen generation, J. Photochem. Photobiol., B: Biology, 1: 93-101.

Cincotta, L., Fowley, J.W., and Cincotta, A.H., 1987, Novel red absorbing benzo[a]phenoxazinium and benzo[a] phenothiazinium photosensitizers: in vitro-evaluation, Photochem. Photobiol., 46: 751-758.

Dougherty, T.J., 1987, Photosensitizers: therapy and detection of malignant tumours, Photochem. Photobiol., 45: 879-889.

Firey, P.A., and Rodgers, M.A.J., 1987, Photoproperties of silicon naphthalocyanine, a potential photosensitizer for photodynamic therapy, Photochem. Photobiol., 45: 535-538.

Foote, C.S., 1988, Mechanistic characterization of photosensitized reactions, in: "Photosensitization," G. Moreno, R.H. Pottier and T.G.Truscott, eds., NATO ASI Series in Cell Biology, Vol. 15, Springer Verlag, Berlin.

Ford, W.E., Firey, P.A., Sounik, J.R., Rihter, B., Kenney,

M.E., and Rodgers, M.A.J., 1988, Photoproperties of
        naphthalocyanines, in: "Advances in Photochemotherapy,"
        T. Hasan, ed., Proc. SPIE 997, Washington.
Ford, W.E., Richter, B.D., Kenney, M.E., and Rodgers, M.A.J.,
        1989, Photoproperties of alkoxy-substituted
        phthalocyanines with deep-red optical absorbance,
        Photochem. Photobiol., in press.
Franck, B., Fulling, G., Gosmann, M., Knubel, G., Mertes, H.,
        and Schroder, D., 1988, Chemical principles in the design
        of improved porphyrin photosensitizers, in: "Advances in
        Photochemotherapy," T. Hasan, ed., Proc. SPIE 997,
        Washington.
Jori, G., and Reddi, E., 1990, Strategies for tumor targeting
        by photodynamic sensitizers, in: "Photodynamic Therapy,"
        D. Kessel, ed., CRC Press, Boca Raton, Florida.
Jori, B., and Spikes, J.D., 1984, Photobiochemistry and
        porphyrins, in: "Topics in Photomedicine", K.C. Smith,
        ed., Plenum Press, New York.
Kessel, D., and Cheng, M., 1985, On the preparation and
        properties of dihematoporphyrin ether, the tumor-
        localizing component of HpD, Photochem. Photobiol., 41:
        277-282.
Kessel, D., Thompson, P., Saatio, K., and Nanturi, K.D., 1987,
        Tumour localization and photosensitization by sulfonated
        derivatives of tetraphenylporphine, Photochem.
        Photobiol., 45: 787-790.
Leznoff, C.C., Vigh, S., Suirskaya, P.I., Greenberg, S., Ben-
        Hur, E., and Rosenthal, I., 1989, Synthesis and
        photocytotoxicity of some new substituted
        phthalocyanines, Photochem. Photobiol., 49: 279-284.
Mang, T.S., Dougherty, T.J. Potter, W.R., Boyle, D.G., Somer,
        S., and Moan, J., 1987, Photobleaching of porphyrins used
        in photodynamic therapy and implications for therapy,
        Photochem. Photobiol., 45: 501-506.
Manyak, M.J., Russo, A., Smith, P.D., and Glatstein, E., 1988,
        Photodynamic therapy, J. Clin. Oncol, 6: 380-391.
McCubbin, I., and Phillips, D., 1986, The photophysics and
        photostability of zinc (II) and aluminum (III) sulfonated
        naphthalocyanines, J. Photochem., 34: 187-195.
Moan, J., 1988, A change in the quantum yield of
        photoinactivation of cells observed during photodynamic
        treatment, Lasers Med. Sci., 3: 93-97.
Moan, J., and Sommer, S., 1985, Oxygen dependence of the
        photosensitizing effect of hematoporphyrin derivative in
        NHIK 3025 cells, Cancer Res., 45: 1608-1610.
Moan, J., Western, A. and Rimington, C., 1988,
        Photomodification of porphyrins in biological systems,
        in: "Photosensitization", G. Moreno, R.H. Pottier and
        T.G. Truscott, eds., NATO ASI Series in Cell Biology,
        Vol. 15, Springer Verlag, Berlin.
Morgan, A.R., Nonis, S., and Reampersaud, A., 1987, New dyes
        for photodynamic therapy, in: "New Directions in
        Photodynamic Therapy," D.C. Neckers, ed., Proc. SPIE 847,
        Washington.
Morgan, A.R., and Tertel, N.C., 1986, Observations on the
        synthesis and spectroscopic characteristics of purpurins,
        J. Org. Chem., 51: 1347-1350.
Moser, F.H., and Thomas, A.L., 1982, "The Phthalocyanines,"
        Vol. 1, CRC Press, Boca Raton, Florida.
Oseroff, A.R., Ohuoha, D., Hasan, T., Bommer, J.C., and
        Yarmush, M.L., 1986, Antibody-targeted photolysis:

selective photodestruction of human T-celll leukemia cells using monoclonal antibody-chlorin e_6 conjugates, <u>Proc. Natl. Acad. Sci. USA,</u> 83: 8744-8748.

Pangka, J.S., Morgan, A.R., and Dolphin, D., 1986, Diels-Alder reactions of protoporphyrin IX dimethylester with electron-deficient alkynes, <u>J. Org. Chem.,</u> 51: 1094-1100.

Peng, Q.,Evensen, J.F., Rimington, C., and Moan, J., 1987, A comparison of different photosensitizing dyes with respect to uptake C3M-tumors and tissues of mice, <u>Cancer Lett.,</u> 36: 1-10.

Ricchelli, F., Biolo, R., Reddi, E., Tognon, G., and Jori, G., 1987, Liposomes as carriers of hydrophobic photosensitizers <u>in vivo</u>: increased selectivity of tumour targeting, <u>in</u>: "New Directions in Photodynamic Therapy", D. C. Neckers, ed., Proc. SPIE 847, Washington.

Roder, R., 1986, Pheophorbide a: a new photosensitizer for the photodynamic therapy of tumours, <u>Stud. Biophys.,</u> 114: 183-186.

Spikes, J.D., 1986, Phthalocyanines as photosensitizers in biological systems and photodynamic therapy, <u>Photochem. Photobiol.,</u> 43: 691-699.

Spikes, J.D., 1989, Quantum yields and kinetics of the photobleaching of chlorins and phthalocyanines: candidate sensitizers for the photodynamic therapy of tumours, <u>Photochem. Photobiol.,</u> 49: 84S.

Spikes, J.D., and Bommer, J.C., 1987, Effects of the degree of sulphonation on the photophysical and photosensitizing properties of aluminum and zinc phthalocyanine, <u>Photochem. Photobiol.,</u> 45: 79S.

Tanno, T., Wohrle, D., Kaneko, M., and Yamada, A., 1980, Rapid photoreduction of methyl viologen with visible light using metal phthalocyanines as photosensitizers, <u>Ber. Bunsen Phys. Chem.,</u> 84: 1032-1034.

Tomio, L., Calzauara, F., Zorat, P.L., and Corti, L., 1984, Photoradiation therapy of cutaneous and subcutaneous malignant tumours using hematoporphyrin, <u>in</u>: "Porphyrin Localization and Treatment of Tumors," D. Doiron and C.J. Gomer, eds., Alan R. Liss, New York.

Valduga, G., Reddi. E., and Jori, G., 1987, Spectroscopic studies on zinc (II)-phthalocyanine in homogeneous and microheterogeneous systems, <u>J. Inorg. Biochem.,</u> 29: 59-65.

Wadwa, K., Smith, S., and Oseroff, A.R., 1988, Cationic triarylmethane photosensitizers for selective photochemotherapy: Victoria blue 80, Victoria blue R and Malachite green, <u>in</u>: "Advances in Photochemotherapy," T. Hasan, ed., Proc. SPIE 997, Washington.

Wagner, J.R., Ali, H., Langlois, R., Brasseur, N., and Van Lier, J.E., 1987, Biological activities of phthalocyanines. VI. Photooxidation of L-tryptophan by selectively sulfonated gallium phthalocyanines: singlet oxygen yields and effect of aggregation, <u>Photochem. Photobiol.,</u> 45: 587-594.

Wilson, B.C., Firnau, G., Jeeves, W.P., Brown, K.L., and Burns-McCormick, D.M., 1988, Chromatographic analysis and tissue distribution of radiocopper-labelled haematoporphyrin derivative, <u>Lasers Med. Sci.,</u> 3: 71-80.

Wilson, B.C., and Jeeves, W.P., 1987, Photodynamic therapy of cancer, <u>in</u>: "Photomedicine," E. Ben-Hur and I. Rosenthal, eds., CRC Press, Boca Raton, Florida, Vol. 2.

Winkelman, J., 1967, Metabolic studies on the accumulation of
    TPPS in tumors, _Experientia_, 23: 949-950.
Zhou, C., 1989, Mechanisms of tumour necrosis induced by
    photodynamic therapy, _J. Photochem. Photobiol., B:
    Biology_, 3: 299-318.

PHOTOPHYSICAL STUDIES OF SYSTEMS RELATED TO PHOTODYNAMIC THERAPY AND PORPHYRIC DISEASE

T.G. Truscott

Department of Chemistry, University of Keele
Staffs ST5 5BG, U.K.

INTRODUCTION

Excited singlet molecular oxygen ($^1O_2$), formed by energy transfer from the lowest excited triplet state of a porphyrin (^3P), to the ground state of oxygen ($^3O_2$) has long been thought to be the mediator of photodamage in various photosensitised systems. Typical examples are the well-known experiment in which Dr. Meyer-Betz self-administered haematopophyrin (H_p) and the range of porphyric diseases, such as erythropoeitic protoporphyria (EPP), which lead to substantial skin photosensitivity. In the latter case carotenoids and particularly β-carotene (β-C) are effective in ameliorating the sensitivity due to either or both of the quenching reactions:

$$^3P + \beta\text{-C} \rightarrow P + {^3\beta}\text{-C}$$

$$^1O_2 + \beta\text{-C} \rightarrow {^3O_2} + {^3\beta}\text{-C}$$

with the $^3\beta$-C losing its excess energy by harmless thermal deactivation.

In photosynthesis the protective effect of β-carotene (or bacterial carotenoids in bacterial photosynthetic systems) is entirely analogous to the ^3P quenching in that the quenching of chlorophyll or bacteriochlorophyll triplet precludes the formation of the damaging $^1O_2$.

Photodynamic Therapy (PDT), using, as the sensitiser a porphyrin mixture known as haematoporphyrin derivative (HPD), is a relatively new method of cancer treatment and several thousands of people worldwide have now been treated by this therapy. The early work concerning the observation of the selective retention of HPD (or at least some components of HPD) in malignant tissues and the procedure for HPD preparation are due to Schwartz, Lipsom and others. However, the major initiative for the use of PDT as a cancer treatment is due to the reports of Dougherty and co-workers from about 1978 onwards (see, for example, Dougherty et al., 1978, 1985 and 1987).

As is well known the treatment consists of injection of 2-5 mg/kg body weight of HPD or the commercial product Photofrin II followed by laser (although coherent light is not necessary) irradiation about 48-72 hours later at a wavelength of about 630 nm.

Excitation of porphyrins with such red light leads to the formation of the first excited singlet state (S_1) and this electronic excited state is responsible for at least two competitive processes, namely fluorescence emission to the ground state (s_0) and intersystem crossing (ISC) to the lowest excited triplet state (T_1). As noted above, $^1O_2$ is then formed via energy transfer and it is generally believed that this is the cytotoxic species leading to the initial steps of tumour destruction in PDT. The well-known deleterious side effect of PDT is prolonged skin photosensitivity (i.e. an induced porphyria) and this is also due to $^1O_2$. Overall this so-called Type II mechanism can be written as:

$$P + hv \ (630 \ nm) \ \rightarrow P(S_1) \tag{1}$$

$$P(S_1) \ \rightarrow P + Fluorescence \ (red) \tag{2}$$

$$P(S_1) \ \rightarrow \ ^3P \qquad ISC \tag{3}$$

$$^3P + \ ^3O_2 \ \rightarrow P + \ ^1O_2 \tag{4}$$

$$^1O_2 + biological \ substrate \rightarrow oxidation \ products \tag{5}$$

Also, as noted above, the role of β-C in treating EPP and in photosynthesis is to quench the ^3P. However, to date β-C has not been successful in ameliorating the skin photosensitivity associated with PDT because it takes several weeks to build up to the required concentration in the skin and it cannot be administered prior to PDT or it might protect the malignancy.

In systems where the Type II mechanism is important it is worthwhile to determine the triplet quantum yields of formation (ΦT), the $^1O_2$ quantum yield ($\Phi \Delta$) and the reactivity of $^1O_2$ with other biological substrates. It should be noted that, under the conditions usually employed, $\Phi \Delta$ is related to ΦT by the equation

$$\Phi \Delta / \Phi T = S_\Delta \tag{6}$$

where S_Δ is the fraction of oxygen quenching interactions leading to the production of $^1O_2$.

Recently, attention has turned to the possibility of replacing HPD for PDT with a 'second-generation' drug which, amongst other properties, would absorb further to the red than HPD. The contenders for such use include chlorins and purpurins, bacteriochlorins, phthalocyanines and naphthalocyanines. In this paper we will report some photophysical studies of H_p and related material and some of

the possible alternatives such as the bacteriochlorins and
purpurins. Also, even though unsuccessful to date for use in
ameliorating the side effects of PDT, our recent measurements
of β-carotene photophysical parameters will be briefly
reported.

EXPERIMENTAL

In fluid solution at physiological or room temperatures
porphyrin triplet state, carotenoid triplet states and $^1O_2$ are
short-lived transient species which last typically 5-1000 μs
and it is therefore necessary to use time-resolved fast reaction techniques for their study. Such techniques often use a
pump-probe technique in which the excited state is generated
by a 'pump' laser operating in a Q-switched mode which can
generate pulses of intense monochromatic light in the
picosecond-nanosecond range or flash lamps for micro-second
study. Following generation of the excited species they are
monitored by a suitable fast detection system. An alternative
to the laser as an excited state 'pump' is a pulsed beam of
high energy ionising electrons - this technique is known as
pulse radiolysis. In this method the high energy radiation is
absorbed mainly by the solvent (at least in dilute solutions).
In non-polar environments charge recombination processes lead
to excited state production (singlets and triplets) whereas in
polar environments radicals are generated.

A wide range of techniques are available for the
detection of excited states and radicals. The major methods
used to date are measurement of the time-resolved absorption
of the probe or monitoring light sources (Bensasson et al.,
1972) or the measurement of the light emission as fluorescence
or phosphorescence. The important species, $^1O_2$, is now
almost routinely measured by time-resolved near infra-red
luminescence at - 1268 nm corresponding to the transition from
the $^1\Delta_g$ state of O_2 to the $^3\Sigma_g^-$ state (i.e. ground state
$^3O_2$, Rodgers, 1987). However, other monitoring techniques had
made useful contributions. These include time-resolved
thermal lensing (Redmond and Braslavsky, 1988 and Redmond et
al., 1987) in which the change in the local refractive index
of the solvent due to the radiationless processes of the
solute are measured - this is detected as a change in
intensity of a probe laser. The time-resolved resonance Raman
technique in which the probe laser is scattered - this
technique is particularly useful for structural studies of
excited triplet states and has been applied quite successfully
to the carotenoids (Wilbrandt and Jensen, 1983). Other
monitoring techniques are based on spin methods.

RESULTS AND DISCUSSION

No results are presented in this paper on excited singlet
lifetimes (see, for example, Andreoni and Cubeddu, 1984 and
Yamashita et al., 1988). In this study we are concerned with
the use of both laser flash photolysis and pulse radiolysis to
study a wide range of porphyrins and carotenoids in non-polar
solvents (see, for example, Pottier and Truscott, 1986). The
molecules studied include esters of haematoporphyrin and also
porphyrins associated with various porphyric diseases. In

addition the Φ_Δ values of such systems have been determined via time-resolved 1270 nm $^1O_2$ luminescence.

The Porphyrins

Typical Φ_Δ values for the porpyrins lie in the range 0.6-0.9 with Φ_Δ values in the range 0.5-0.8 so that virtually all non-metallo and diamagnetic porphyrins studied are good for $^1O_2$ generation. In such measurement it is worthwhile to monitor the effect of enhanced intersystem crossing of oxygen since this could lead to erroneously high Φ_Δ values if this parameter Φ_Δ is measured in air or oxygen saturated conditions.

Several workers have investigated the effect of the microenvironment on porphyrin triplet and $^1O_2$ yields (see, for example, Rodgers, 1985 and Craw et al., 1984). Typical examples involve detergents, proteins, liposomes and, more recently, cells. In detergents, proteins and liposomes high triplet and $^1O_2$ yields are again detected although, at least in liposomes, the kinetics of the triplet decay are not simple.

Our recent studies with Hp in transformed baby hamster kidney cells (Truscott et al., 1988) have shown that the triplet state of such molecules can be readily detected and that the lifetime of the triplet species is reduced in the presence of oxygen (similar behaviour was observed using the mixture HPD and also HPD from which all porphyrin monomers had been removed by gel chromatography). However, in no case could we detect the $^1O_2$ luminescence at 1270 nm. At this stage it must be assumed that the $^1O_2$ species is formed but reacts with some cell component so fast that its lifetime is less than the limit of our equipment for such luminescence ($\sim 1~\mu s$). Similar results have been obtained by ourselves with mouse myeloma cells and also by Firey et al.,(1988). Recently, (McLean and Truscott - Unpublished) a detailed kinetic study of the triplet state of Hp in BHK cells shows that the triplet decays by two first-order processes. Also, at high laser intensities, photo-ionisation of Hp occurs to produce a long-lived radical.

Table 1 gives a selection of ΦT and Φ_Δ values reported for a range of molecules being considered as 'second generation' photosensitisers for PDT together with several porphyrins relevant to HPD and porphyric disease.

As can be seen for virtually every sensitiser studied a substantial ΦT and Φ_Δ arises and the ratios S_Δ (= $\Phi_\Delta/\Phi T$) lie in the range 0.8±0.2, i.e. the fraction of triplets quenched by oxygen leading to $^1O_2$ is high. As an example of a contrasting situation the paramagnetic molecule copper phthalocyanine has $\Phi T \sim 1$ but, due to efficient enhanced intersystem crossing back to the ground state from T_1 the Φ_Δ value is near zero, so that the S_Δ value is of little interest.

The Carotenoids

It has been well established for over twenty years (Chessin et al., 1966) that the triplet yield of the non-

Table 1. Typical ΦT and $\Phi \Delta$ Values

Class	Sensitiser	Solvent	ΦT	$\Phi \Delta$
Porphyrin	HpDME	Benzene	0.72	0.57
	PpDME	Benzene	0.80	0.57
	Hp	Water	0.63	0,51
	Up	Water	0.93	0.52
	SnPp	Methanol	0.83	0.70
Meso-Substitute	TPP	Benzene	0.67	0.63
Porphyrin	o-THPP	Methanol	0.68	0.65
	p-THPP	Methanol	0.65	0.68
	m-THPP	Methanol	0.69	0.69
Bacteriochlorins	Bchl a	Benzene	0.32	0.32
	Bpheo a	Benzene	0.73	0.46
Chlorins and	PhotoPpDME(A)	Benzene	0.65	0.67
Purpurins	PhotoPpDME(B)	Benzene	0.65	0.66
	Octaethyl-purpurin (NT2)	Benzene	0.8	0.67
	Zn-ethiopurpurin (ZnEt2)	Benzene	0.83	0.56
Phthalocyanines	ZnPcTS	Methanol	0.47	0.43
and related	AlPcTS	Water	0.38	?
Molecules	Si-Naphthalo cyanine	Benzene	0.39	0.35

Pp = protophyrin: Up = uroporphyrin: DME = dimethylester

hetero atom containing carotenoids such as β-carotene (a C_{40} carotenoid) have a near zero ΦT and hence generate no $^1O_2$. Furthermore, the major role of such molecules in photobiology (photosynthesis) and photomedicine (treatment of EPP) is to protect from photodynamic damage by quenching the porphyrin triplet state, i.e.

$$^3P_p + \beta\text{-}c \rightarrow P_p + {}^3\beta\text{-}c$$

as discussed above. Despite the importance of such carotenoids in the above processes, the C_{20} polyenes in vision and the treatment of dermatological disorders, and the widespread use of β-carotene, β-apocarotenal and canthaxanthin as food colourants, relative large gaps exist in our understanding of their photochemistry. Thus, the lowest triplet energy level of the C_{40} cartenoids such as β-carotene is unknown, the mechanism of cis=trans isomerisation is still debated and the details of the $^1O_2$ -carotenoid interaction is not clear. We have recently studied both the triplet energy level (E_t) of β-carotene and also studied the $^1O_2/\beta$-carotene interaction (Conn et al., 1989), also we are attempting to study the $^3P_p/\beta$-carotene interaction in red blood cells taken from EPP sufferers on β-carotene therapy.

The method used in our attempt to obtain $E_T{}^{\beta\text{-}c}$ is via the kinetics of energy transfer from a series of donor triplets of known E_T - if $E_T^D > E_T{}^{\beta\text{-}c}$ a second -order rate constant of ¯$10^9/^{10}$ $M^{-1}s^{-1}$ is expected whereas if $E_T^D < E_T{}^{\beta\text{-}c}$ this value will fall by several orders of magnitude.

Table 2 gives typical data we have obtained together with some data on oxygen quenching for comparison.

Comparison of the rate constants for pentacene and SiNc-$(R)_2$ would appear to imply a $E_T^{\beta-c}$ - 23±2 kcal/mol, i.e. very near that of $^1O_2$ and which is similar to some previous estimates/predictions (Hertstroeter, 1975; Bensasson et al., 1976 and Gorman et al., 1988) However the situation is more complex than this due to two factors.

1. Back energy transfer from $^3\beta$-C to SiNc$(R)_2$ may be a possibility, i.e. $^3\beta$-C + ^{1}SiNc → $^1\beta$-C + ^{3}SiNc$(R)_2$, but we have studied such a process via pulse radiolysis (in which $^3\beta$-C can be generated without a donor triplet via charge recombination processes such as β-C$^{\cdot+}$ + β-C$^{\cdot-}$ → $^3\beta$-C*) thus allowing the study of such back transfer.

2. Steric effects, i.e. do the bulky axial alkyl side groups (R) on SiNc reduce the rate of energy transfer.

As far as possibility (1) is concerned our pulse radiolysis results [for which we are grateful to Dr. E. J. Land, Christie Hospital, Manchester, for assistance], imply back transfer is not important and is of little or no consequence in our estimation of $E_T^{\beta-c}$ by the energy transfer technique.

However, the situation is quite different as far as the steric factors are concerned. To investigate these we have compared the rate of energy transfer to β-C from two phthalocyanines (Pc) with known triplet energy levels (~ 26 kcal/mol), well above that of β-C, but with substantially different steric hindrance. These Pc's are zinc Pc (ZnPc) and SiPc$(R)_2$ with the same axial side groups as in SiNc$(R)_2$. The rate constants obtained for ZnPc and SiPc$(R)_2$ were found to be

Table 2.   Second-Order Rate Constants for the Quenching of Donor Triplets by All-Trans β-carotene.

Carotenoid	SiNc-(R)$_2$ 21.5	Penta-cene 23	SiPc-(R)$_2$ 26^{+2}	ZnPc 26	PPIXDME 36	Anthra-cene 42
All-trans β-C	3.3×10^8	3×10^9	5.3×10^8	1.2×10^9	3.5×10^9	7.4×10^9
9-cis β-carotene	2.2×10^8	-	-	-	-	-
15-cis β-carotene	1.8×10^8	-	-	-	-	-
All-trans lycopene	5.1×10^8	5×10^9	-	-	-	5×10^9
Decapreno-β-carotene	3.9×10^8	-	-	-	-	-

$^+$ assumed to be as for ZnPc

1.2 x 10^9 M^{-1}s^{-1} and 0.53 x 10^9 M^{-1}s^{-1} respectively. That is, a steric factor of 2.26 arises between these Pc's. Applying such a factor to the SiNc(R)$_2$ rate constant leads to a 'sterically corrected' trend with no sharp fall-off just a slow reduction in kq with increasing size of donor. Indeed, it may well be that if an additional steric factor is applied to correct for donor size there would be no reduction at all in the energy transfer rate down to SiPc(R)$_2$ with E$_T$ of 21.5 kcal/mol. Our preliminary conclusion at this stage is that E$_T^{\beta-C}$ is < 21.5 kcal/mol by at least 2 kcal/mol. Future experiments will use donors with still lower energy levels than SiNc (R)$_2$ in an attempt to measure E$_T^{\beta-C}$.

The $^1O_2/\beta$-carotene interaction

Briefly, we have studied the process

$$^1O_2 + \beta\text{-}C \rightarrow {}^3O_2 + {}^3\beta\text{-}C$$

under conditions where the $^1O_2$ is generated from the triplet state of a sensitiser (acridine) and little or none (< 10%) of the sensitiser triplet reacts with the β-C.

We have measured the amount of $^1O_2$ reacting from the intensity of the 1270 nm luminescence and the concentration of β-C triplet produced from laser flash photolysis monitoring at the maximum of the β-carotene triplet-triplet absorption ˜ 520 nm (Chessin et al., 1966). Unfortunately the oxygen (ground state, $^3O_2$) also quenches the β-C triplet

$$^3\beta\text{-}C + {}^3O_2 \rightarrow \beta\text{-}C + {}^3O_2$$

by enhanced intersystem crossing so that it is difficult to obtain precise data due to the large kinetic correction factor which must be applied in determining the concentration of $^3\beta$-C. Following such corrections we find that only about $^2/_3$ of the $^1O_2$ reacting with β-C leads to $^3\beta$-C. At this stage we have not invesitgated what constitutes the remainder of the $^1O_2/\beta$-C interaction - there is no spectroscopic evidence of significant photo-oxidation products. However, the kinetic correction factors we apply are large and our result must be regarded as tentative.

Our very preliminary laser flash photolysis studies on RBC from porphyria patients before β-carotene therapy indicates that the Pp triplet can be detected but not once carotene therapy has begun. Presumably the Pp triplet is being very efficienty quenched by the β-carotene - so far we have not attempted to detect the formation of β-carotene triplet or singlet oxygen in this situation.

CONCLUSIONS

1. Porphyrins related to PDT and porphyric disease give high yields of triplet state and $^1O_2$.

2. Related molecules for future PDT with enhanced red absorption also show high yields of triplet and $^1O_2$.

3.   Porphyrin triplet states are readily detected in cellular environments as is the reaction with other molecules of biological relevance such as oxygen and β-carotene.

4.   Singlet oxygen cannot be detected in cellular environments probably due to its extreme reactivity.

5.   The energy of the lowest triplet state of β-carotene (formed by energy transfer from a porphyrin triplet or $^1O_2$) is lower than predicted and probably $\leq$ 20 kcal/mol.

6.   The yield of triplet β-carotene from singlet oxygen may be less than expected on simple energy-transfer consideration.

ACKNOWLEDGEMENTS

The CRC, MRC and Hoffman-La Roche are thanked for financial support. Thanks are due to many collaborators including P. Conn, E.J. Land, D. J. McGarvey, J. G. McKeen, A. J. McLean, A. N. Macpherson, R. W. Redmond and W. Schalch.

REFERENCES

Andreoni, A., and Cubeddu, R., 1984, Fluorescence properties of HpD and its components, in: "Porphyrins in Tumour Phototherapy, " A. Andreoni and R. Cubeddu, eds., Plenun Press, New York.

Bensasson, R.V., Land, E.J. and Maudinas, B., 1972, Triplet states of carotenoids from photosynthetic bacteria studied by nanosecond ultraviolet and electron pulse irradiation, Photochem. Photobiol., 23: 189.

Chessin, M., Livingston, R and Truscott, G.G., 1966, Direct evidence for the sensitized formation of a metastable state of β-carotene, Trans. Faraday Soc., 62: 1519.

Craw, M., Redmond, R. and Truscott, T.G., 1984, Laser flash photolysis of haematoporphyrin in some homogeneous and heterogeneous environments, J.C.S., Faraday Trans. I, 80: 2293.

Dougherty, T.J., Kaufman, J.E., Goldfarb, A., Weishaupt, K.R., Boyle, D.G. and Mittelman, A., 1978, Photoradiation therapy for the treatment of malignant tumours, Cancer Res., 38: 2628.

Dougherty, T.J., 1985, Photodynamic therapy, in: "Photodynamic Therapy of Tumours and Other Diseases," Jori and Perria, eds., Libreria Progetto, Padova.

Dougherty, T.J., 1987, Photosensitizers: therapy and detection of malignant tumours, Photochem. Photobiol., 45: 879.

Firey, P.A., Jones, T.W., Jori, G. and Rodgers, A.J., 1988, Photexcitation of zinc phthalocyanine in mouse myeloma cells: the observation of triplet states but not of singlet oxygen, Photochem. Photobiol., 48: 357.

Gorman, A.A., Hamblett, I., Lambert, C., Spencer, B. and Standen, M.C., 1988, Identification of both preequilibrium and diffusion limits for reaction of singlet oxygen, O_2 $(^1\Delta_g)$, with both physical and chemical quenchers: variable-temperature, time-resolved infrared luminescence studies, J. Am. Chem. Soc., 110: 8053.

Herkstroeter, W.G., 1975, The triplet energies of azulene, β-carotene, and ferrocene, <u>J. Am. Chem. Soc.</u>, 97: 4161.

Pottier, R. and Truscott, T.G., 1986, The photochemistry of haematoporphyrin and related systems, <u>Int. J. Radiat. Biol.</u>, 50: 421.

Redmond, R.W., Heihoff, K., Braslavsky, S.E. and Truscott, T.G., 1987, Thermal-lensing and phosphorescence studies of the quantum yield and lifetime of singlet molecular oxygen ($^1\Delta_g$) sensitized by hematoporphyrin and related porphyrins in deuterated and non-deuterated ethanols, <u>Photochem. Photobiol.</u>, 45: 209.

Redmond, R.W. and Braslavsky, S.E., 1988, Absolute determination of quantum yields of photsensitisation by time resolved thermal lensing, <u>in</u>: "Photosensitisation: Molecular, Cellular and Medical Aspects, NATO ASI Series H: Cell Biology, Vol.15," G. Moreno, R.H. Pottier and T.G.Truscott, eds., Springer-Verlag.

Rodgers, M.A.J., 1985, The Photoproperties of Porphyrins in biological environments, <u>in</u>: "Photodynamic Therapy of Tumours and Other Diseases, "Jori and Perria, eds., Libreria Progetto, Padova.

Rodgers, M.A.J., 1987, Singlet oxygen quantum yields, <u>Amer. Chem. Soc. Symposium Series</u>, 339: 76.

Truscott, T.G., 1988, Detection of haematoporphyrin derivative and haematoporphyrin excited states in cell environments, <u>Cancer Letts.</u>, 41: 31.

Wilbrandt, R. and Jensen, N-H., 1983, Applications of time-resolved resonance Raman spectroscopy in radiation chemistry and photobiology: structure and chemistry of carotenoids in the excited triplet state, <u>in</u>: "Time-Resolved Vibrational Spectroscopy," Academic Press.

Yamashita, M., Tomono, T., Kobayashi, S., Torizuka, K., Aizawa, K. and Sato, T., 1988, Picosecond fluorescence spectroscopy on incorporation processes of haematoporphyrin derivative into malignant tumour cells <u>in vitro</u>, <u>Photochem. Photobiol.</u>, 47: 189.

THE EFFECT OF SOLUTION pH ON THE AGGREGATION DEGREE OF HEMATOPORPHYRIN DIACETATE

Ruta Kapočiute[1], Tatiana Szito[2], Ricardas Rotomskis[1] and Jurgita Rotomskiene[1]

[1]Vilnius University Laser Research Center
Sauletekio ave. 9, corp. 3, Vilnius, 232054
Lithuania
[2]Institute of Biophysics, Semmelweis Medical
University, Budapest VIII., Puskin u. 9, POB 263
H-1444, Hungary

INTRODUCTION

Recent interest in hemtoporphyrin derivative (HPD) and related compounds is due to their increasing applications in tumor therapy. For the comprehension of porphyrin sensitized tumor therapy, the interaction of light with HPD and its aggregates are of great importance. Previous investigations on one of the compounds of HPD, hematoporphyrin diacetate (HP-Diac) (Bonnet et al., 1980) revealed that light induced processes by the photosensitizer depend on light intensity (Gadonas et al., 1986) and that these processes vary in the picosecond and nanosecond time scales (Gadonas et al., 1987). The elucidation of the aggregation characteristics should help to clarify the physical, chemical and biological properties of porphyrin type photosensitizers used in tumor therapy.

Aggregation of porphyrins is known to be very sensitive to the nature of the solvent. It has been reported that the aggregation degree depends on drug concentration for solutions of hematoporphyrin (HP) (Brown et al., 1976), HPD (Andreoni et al., 1982) and Photofrin (Pottier et al., 1985), as well as on the solvent polarity (Redmond et al., 1985; Ricchelli and Grossweiner, 1984). It was shown that temperature also influences the aggregation of HPD (Bottiroli et al., 1984). A systematic investigation on the dependence of HP-Diac aggregation (Kapociute, 1989) reveals that in aqueous solution of a concentration (C) higher than ca. 10^{-5} M, HP-Diac forms highly aggregated species referred to as conglomerates. Changes in solution concentration, polarity and temperature induce alterations in the degree of HP-Diac aggregation similar to those reported for other porphyrins. A generalization of the dependence of conglomerates on environmental parameters are summarized in Fig. 1. Porphyrin aggregation is greatly effected by extremes of solvent pH (pH < 6 and pH > 8).

Light in Biology and Medicine. Volume 2 Edited by R.H. Douglas *et al.*
Plenum Press. New York, 1991

277

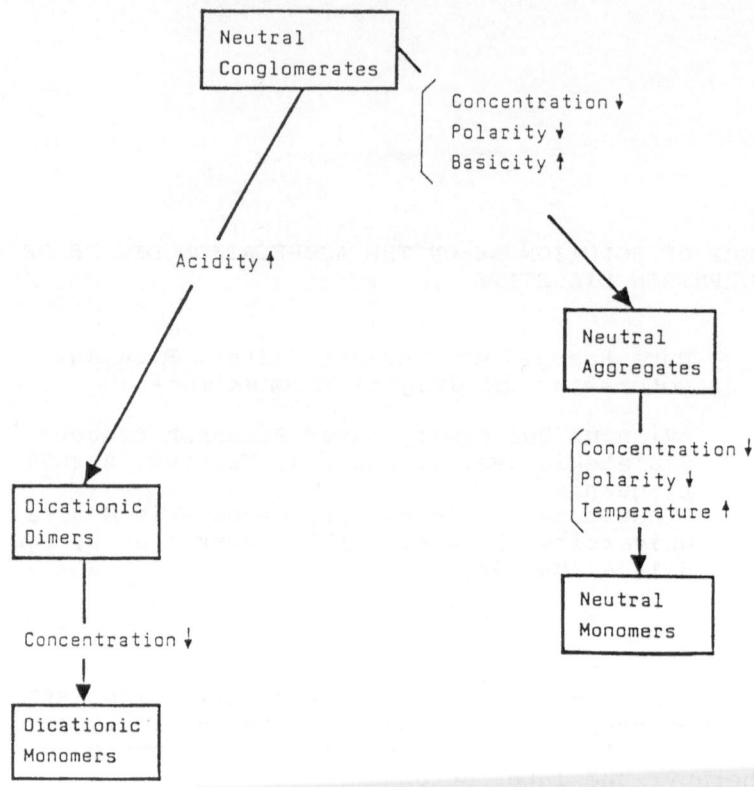

Fig. 1 Dissociation scheme of HP-Diac conglomerates.
The brackets indicate the synergy of the
environmental parameters. Δ increase,
↓ decrease.

From a biological point of view, the influence of extreme
solution pH on the aggregation of photosensitizer is of minor
importance. However, investigations at pH values close to
physiological condition may be useful for the better
understanding of porphyrin retention and accumulation in
tumors. The pH range 6-7 is of special significance since
many tumors have been reported to have pH values in this
vicinity (Gullino et al., 1965; Wike-Hooley et al., 1984). In
this paper we report the results of investigations on the
dependence of HP-Diac aggregation on solution pH with
emphassis on the intermediate pH range.

MATERIALS AND METHODS

HP-Diac was kindly supplied by Dr. A.N. Nizhnik (Moscow
Institute of Fine Chemical Technology) and was used as
received. Stock HP-Diac solution (1.5×10^{-3} M) was prepared
by solving 1 mg of solid HP-Diac in 1 ml 0.1 M NaOH.
Phosphate buffered solutions (PBS) of varying pH were used to
dilute the stock solution. pH values lower than 4.5 and
higher than 8.9 were adjusted by the addition of 1 M HCl or
0.1 M NaOH, respectively.

Solution pH was measured by a Radelkis precision digital pH meter OP-208/1, having an accuracy of ± 0.001.

Absorption sepctra were recorded by the use of a Carl Zeiss M-40 and a Perkin-Elmer 200 UV-vis spectrophotometers. Measurements were carried out on fresh solutions in dim light at room temperature. A 1 cm length cuvette was used for all measurements.

The data presented are average values obtained from three parallel experiments. The relative errors were 5% in absorbance and ± 20 cm^{-1} in full band width at half maximum ($\Delta\nu$).

RESULTS AND DISCUSSION

Fig. 2 shows the absorption spectra of 1.5 x 10^{-5} aqueous solutions of HP-Diac in neutral and acidic conditions. At pH 7.4 the Soret band has a maximum at 372 nm, a shoulder at about 390 nm and is similar to that of aggregated HP (Brown et al., 1976) and HPD (Andreoni et al., 1982), as well as Photofrin II (Andreoni and Cubeddu, 1984). Upon dilution, the Soret band maximum shifts to 395 nm and the absorbance is not linear with concentration above 10^{-8}M. This implies that HP-Diac is highly aggregated in aqueous solutions of 1.5 x 10^{-5} M and pH 7.4. A detailed analysis of the Soret band characteristics as a function of concentration reveals that aggregates of higher degree than dimers are also present at

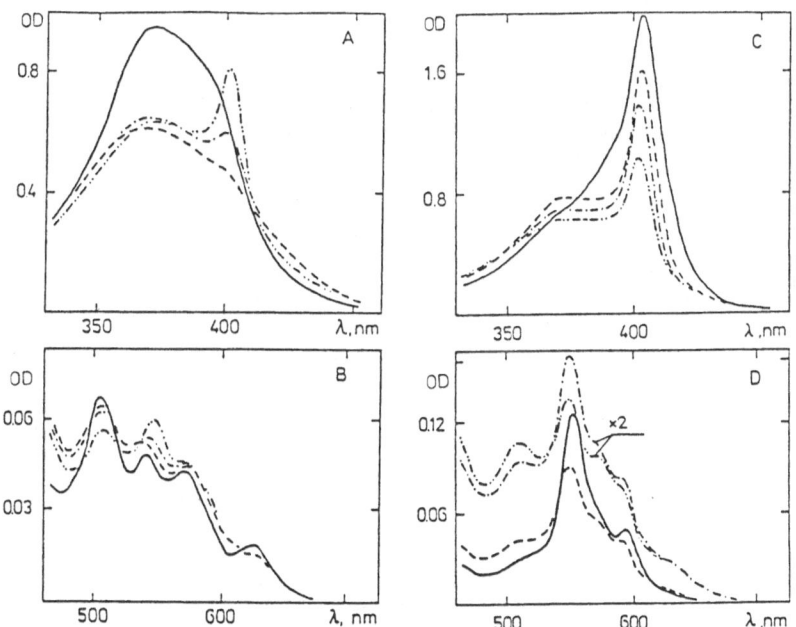

Fig. 2   HP-Diac absorption spectrum in PBS.   pH values:
A,B ——— 7.4, − − − 3.7, −·− 3.3 & −··− 2.9;
C,D −··− 2.7, − ·− 2.1, − − − 1.7 & ——— 1.2.
HP-Diac concentration 1.5 x 10^{-5} M.

high concentration values (Kapociute, 1989). The absorbance spectrum observed for aggregated HP-Diac is consistent with the formation of united electronic systems from π-orbitals of the individual porphyrin molecules, as described by Karns et al. (1979).

Fig. 2 shows that at pH less than ca.5 and higher than ca.3, no new absorbance bands are detected. In comparison, when the pH is decreased below 3, a new absorbance band centred at about 400 nm is observed along with intensity redistribution of the Q-bands in the visible spectral range. The observed absorption changes of HP-Diac at low pH are similar to those of Photofrin II (Pottier et al., 1985). At pH 1.2 (Fig. 2c and d) the Soret band is narrow and intense, with a maximum at 403 nm. In the visible region only two bands (Q(0 - 0) and Q(0 - 1)) are present. This absorption spectrum is characteristic of dicationic porphyrins (Smith, 1975) and is consistent with a change of symmetry of HP-Diac molecule from D_{2h} to a full square D_{4h} symmetry, in which the excited states become doubly degenerate (Gurinovich et al., 1961). The metal-free porphyrins are known to form dianions in highly alkali medium (Smith, 1975), but we could not observe significant changes of the HP-Diac absorbance with pH increase up to 12.6. This is consistent with the low deprotonation ability of the HP-Diac tetrapyrrole core.

Fig. 3a summarizes the Soret band characteristics of 1.5 x 10^{-5}M HP-Diac at various pH's. The absorbance at 370, 395 and 403 nm corresponds to the absorbance of neutral aggregates and monomers, and dicationic monomers, respectively. It is seen that in the pH range from about 5 to 3 the absorbance of the HP-Diac species is lowest and the Soret band Δy is highest (ca. 6.0 x 10^3 cm^{-1}). These spectral characteristics may be explained by an increase in the degree of aggregation via the monocationic HP-Diac species. The dependence of the Soret band characteristics on pH was found to be similar at the two HP-Diac concentrations studied (See Fig. 3a and b).

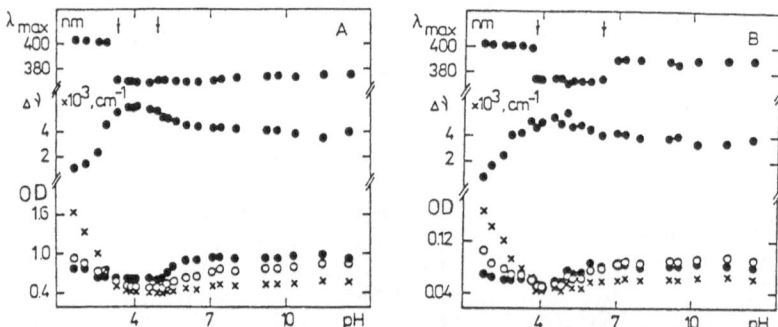

Fig. 3 Dependence of Soret band characteristics on solution pH. Absorbance is plotted at ● - 370, 0 - 395 and X - 403 nm, the wavelengths corresponding to absorption maximum of neutral aggregates and monomers, and dications, respectively. HP-Diac concentration: A- 1.5 x 10^{-5} M, B - 1.5 x 10^{-6} M. The arrows indicate the intermediate pH range.

The results presented in Fig. 3 indicate that HP-Diac aggregation is increased in the pH range 3-5. The pH range of high aggregation formation depends on the HP-Diac concentration, with lower concentration favouring the formation of aggregation at the higher solution pH. Δy for the Soret band is ca.6.0 x 10^3 cm^{-1} at 1.5 x 10^{-6}M and 5.5 x 10^3 cm^{-1} at 1.5 x 10^{-6} M. These values are higher than that observed for extremely aggregated HP-Diac in neutral solutions at a concentration 1.5 x 10^{-4} M, where Δy = 4.2 x 10^{-3} cm^{-1} (Kapociute, 1989). This discrepancy may in part be explained by assuming the existence of other HP-Diac species besides the neutral aggregates and monomers found in the intermediate pH range. It is reasonable to assume that monocations are formed in the pH range 3-5. The pK values for similar porphyrins reported by Pottier et al., (1985) and Bonnett et al., (1985) support this suggestion. Further, if the concentration of HP-Diac is decreased to < 10^{-7}, the Soret maximum is centred at 400 nm, characteristic of a monocationic monomer. Experimental evidence for the formation of monocations in roughly the same pH range has recently been presented for hematoporphyrin (Pottier et al., 1988). Thus it is reasonable to assume that in the pH region 3-5, one is monitoring the spectral characteristic of a mixture of the aggregated species of HP-Diac and the corresponding monocation monomeric species.

CONCLUSION

The efficiency of porphyrin sensitized tumor therapy is believed to be related to the aggregated species of the drug. Our results indicate that the high degree of aggregation observed at low concentrations of HP-Diac for pH values characteristic of tumors may be due to the interaction between neutral HP-Diac species and monocationic ones. The different ionic species found at low tumor pH may play an important role in both the selective biodistribution and photosensitizing ability of porphyrin type photochemotherapeutic agents.

ACKNOWLEDGEMENTS

Authors greatly appreciate close collaboration and very fruitful discussions with Dr. Katalin Toth.

REFERENCES

Andreoni, A., Cubeddu, R., De Silvestri, S., Laporta, P., Jori, G. and Reddi, E., 1982, Hematoporphyrin derivative: experimental evidence for aggregated species, Chem. Phys. Lett., 88:33.

Andreoni, A. and Cubeddu, R., 1984, Photophysical properties of Photofrin II in different solvents, Chem. Phys. Lett., 108:141.

Bonnett, R., Ridge, R.J., Scourides, P.A. and Berenbaum, M.C., 1980, Haematoporphyrin derivative, J. Chem. Soc. Chem. Cummun., 24:1198.

Bottiroli, G., Freitas, I., Docchio, F., Ramponi, R. and Sacchi, C.A., 1984, The time-dependent behaviour of hematoporphyrin-derivative in saline: a study of spectral modifications, Chem.-Biol. Interactions, 49:1.

Brown, S.B., Shillcock, M. and Jones, P., 1976, Equilibrium and kinetic studies of the aggregation of porphyrins in aqueous solution, Biochem. J., 153:279.

Gadonas, R., Danielius, R., Kapociute, R. and Piskarskas, A., 1986, Picosecond relaxation processes in biological molecular complexes and dyes aggregates. In: "Investigation of Structure, Physical Properties and Energetics of Biological Active Molecules," S.A. Achmanov and A.S. Piskarskas, eds., Mokslas, Vilnius.

Gadonas, R., Kapociute, R., Krasauskas, V. and Piskarskas, A., 1987, Picosecond electronic excitation deactivation process in monomers and aggregates of hematoporphyrin derivative, in: "Lasers and Optical Nonlinearity," P.A. Apanasevich, ed., BSSR University, Minsk.

Gullino, P.M., Grantham, F.H., Smith, S.H. and Haggerty, A.C., 1965, Modifications of the acid-base status of the internal milieu of tumors, J. Natl. Cancer. Inst., 34:857.

Gurinovich, G.P., Sevchenko, A.N. and Solovyev, K.N., 1961, Limited polarization of porphyrins fluorescence, Opt. Spektrosk., 10:750.

Kapociute, R., 1989, Candidate Thesis, Budapest.

Karns, G.A., Gallagher, W.A. and Elliot, W.B., 1979, Dimerization constants of water-soluble porphyrins in aqueous alkali, Bioorg. Chem., 8:69.

Pottier, R., Laplante, J.P., Chow, Y.-F. A. and Kennedy, J., 1985, Photofrins: a spectral study, Can. J. Chem., 63:1463.

Pottier, R.H., Kennedy, J.C., Chow, Y.-F.A. and Cheung, F., 1988, The pK/a values of hematoporphyrin IX as determined by absorbance and fluorescene spectroscopy, Can. J. Spectros., 33:57.

Redmond, R.W., Land, E.J. and Truscott, R.G., 1985, Aggregation effects on the photophysical properties of porphyrins in relation to mechanisms involved in photodynamic therapy, in: "Methods in Porphyrin Photosensitization," D. Kessel, ed., Plenum Press, New York and London.

Ricchelli, F. and Grossweiner, L.I., 1984, Properties of a new state of hematoporphyrin in dilute aqueous solution, Photochem. Photobiol., 40:599.

Smith, K.M., 1975, Porphyrins and metalloporphyrins, Elsevier Scientific Publishing Company, Amsterdam, Oxford, New York.

Wike-Hooley, J.L., Haveman, J. and Reinhold, H.S., 1984, The relevance of tumor pH to the treatment of malignant diseases, Radiother. Oncol., 2:366.

POSSIBLE EFFECTS OF INCREASED ULTRAVIOLET RADIATION ON HUMAN
SKIN

F. Urbach

Temple University School of Medicine, Philadelphia
PA 19140 USA

INTRODUCTION

Examination of the sun's role in the production of human
skin cancer does not lend itself to direct experimentation.
However, extensive astute observations have strongly suggested
the etiologic significance of ultraviolet radiation.

Human skin cancers, especially basal cell and squamous
cell carcinomas, are closely associated with chronic repeated
exposure of skin to solar ultraviolet radiation. Three types
of evidence indicate that the most effective wavelengths are
shorter than 320 nm: (1) extensive experiments in mice show
that wavelengths longer than 320 nm are relatively ineffective
for induction of skin cancer. (2) Wavelengths shorter than
320 nm are highly effective in inducing photochemical changes
in DNA and killing of cells in tissue culture. Furthermore,
damage to DNA is considered to be one of the events leading to
carcinogenesis, and a number of carcinogenic chemicals mimic
UV radiation damage to DNA. (3) The effective wavelengths for
human skin erythema production are mostly below 320 nm. Also,
individuals who sunburn easily and have high exposure to solar
UV radiation have a much higher incidence of non-melanoma skin
cancer than those who rarely sunburn and have little exposure
to the sun. (Blum, 1955).

This last observation has been used as a basis for
assuming that the human skin erythema action spectrum and the
skin carcinogenesis action spectrum are closely related.

RELATIONSHIP OF INCIDENCE OF SKIN CANCER IN MAN TO POTENTIAL
CHANGES IN STRATOSPHERIC OZONE

Of the total solar radiant energy that reaches earth,
approximately 5% consists of ultraviolet radiation (UVR).
Depending on sun angle (i.e. air mass, time of day) the
biologically most effective UVR comprises 10% or less of the

total ultraviolet radiation (UVB). Because ozone begins to absorb UVR significantly about 325 nm, the segment of the solar UVR spectrum that would be augmented in the event of diminution of atmospheric ozone concentration includes on the small waveband between 290 and 325 nm. Wavelengths longer than 325 nm are not appreciably influenced by changes in atmospheric ozone concentration. Although the total energy that would be added to the solar (and even solar UVR) insolation would be very small, this region has notable photochemical efficiency in biologic systems.

The balance between increasing biological effectiveness of decreasing wavelengths in the UVR, and increasing absorption by ozone of the shorter wavelengths becomes critical when considering the biological implications of solar UVR changes as a function of alteration of the stratospheric ozone layer.

For most biological action spectra, it is clear that changes in stratospheric ozone will result in increases in biologically effective UV irradiance disproportionately greater than might be indicated by simply integrating the increasing total UVR energy flux.

Extensive studies of the relative biological effectiveness of wavelengths in the UVB (i.e. action spectra) on such diverse biological materials as purified DNA, bacteria, plant and human cells in tissue culture and human skin and eye _in vivo_ have shown that wavelengths of 290 nm are one thousand to ten thousand times as effective in producing damage (Thymine dimer production, single and double strand breaks in DNA, production of mutations, cell killing, production of skin erythema and skin cancer) than wavelengths of 330 nm.

Based on these observations, it has been estimated that a 1% decrease in the total ozone column would increase the biologically effective (action spectrum weighted) UVB radiation

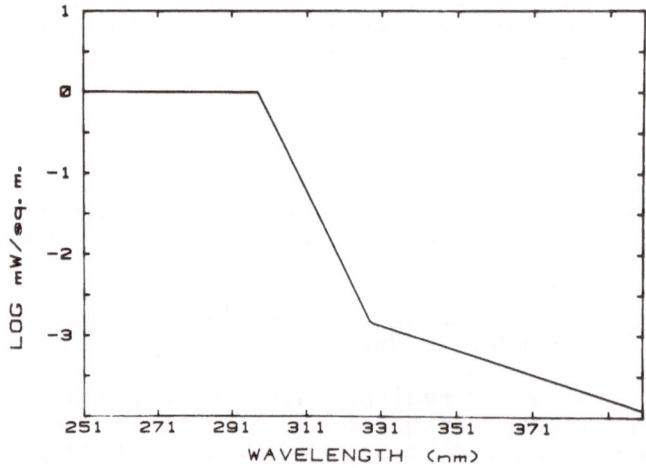

Fig. 1   Relative spectral effectiveness of UVR.
(From: McKinlay and Diffey, 1987).

effects by between 1.7 and 2% with use of the 1935 CIE action spectrum, and between 1.25 and 1.3 with use of the 1987 CIE action spectrum (McKinlay and Diffey, 1987) (Figure 1). This has been referred to as the "optical" amplification factor (AF_o).

A recent report of the Ozone Trends Panel (a joint effort of WHO, NASA, NOAA, EPA, CSIRO and various universities) has reported that in the 20 year period 1967-1986, total column ozone has decreased by 2 to 3 percent, depending on latitude. (NASA 1988) (Table 1).

Since the decreases were greater in the winter months (when very much less UVB reaches the ground) than in the summer, the annual average increases are somewhat deceptive as far as biologically effective UVB is concerned (Diffey et al., 1988).

From ground-based (biologically effective) UVB measurements performed with Robertson-Berger meters at Davos, Switzerland (Lat. 46.8°N) and Norrkiping, Sweden (Lat. 58.75°N) for 1984, it is possible to calculate the expected UVB increases, based on the WMO/NASA report.

Table 1   Coefficients of Multiple Regressions to Re-analysed Dobson Total Ozone Measurements collected into Band Averages (total percent changes for periods 1969-1986).

MONTH	LATITUDE BAND		
	53°-64°N	40°-52°N	30°-39°N
JAN	-8.3±2.2	-2.6±2.1	-2.2±1.5
FEB	-6.7±2.8	-5.0±2.2	-1.2±1.9
MAR	-4.0±1.4	-5.6±2.3	-3.5±1.9
APR	-2.0±1.4	-2.5±1.7	-1.7±1.3
MAY	-2.1±1.2	-1.3±1.1	-1.7±0.9
JUN	+1.1±0.9	-1.8±1.0	-3.3±1.0
JUL	+0.0±1.1	-2.2±1.0	-1.3±1.0
AUG	+0.2±1.2	-2.4±1.0	-1.0±1.0
SEP	+0.2±1.1	-2.9±1.0	-1.0±0.9
OCT	-1.1±1.2	-1.5±1.5	-0.9±0.8
NOV	+1.5±1.8	-2.4±1.3	-0.1±0.8
DEC	-5.8±2.3	-5.5±1.7	-2.1±1.1
ANNUAL AVERAGE	-2.3±0.7	-3.0±0.8	-1.7±0.7
WINTER AVERAGE	-6.2±1.5	-4.7±1.5	-2.3±1.3
SUMMER AVERAGE	+0.4±0.8	-2.1±0.7	-1.9±0.8

At Davos (Lat 46.8°N), assuming a decrease in column ozone of 3%, and an AF° of 1.8, the (biologically effective) UVB would have been expected to increase by 5.4%. Utilizing the monthly changes from that report, UVB would have increased 4.54%. At Norrkoping (Lat. 58.75°N) the increase in UVB, based on an average decrease of 2.3% of O_3 should have been 3.9%. Based on the monthly changes, it would have been only 0.92%. If the AF° were 1.25 (as in the McKinlay-Diffey action spectrum) the UVB increases would be +3.15% for Davos and +0.64% for Norrkoping. It is thus important that estimates of biologically effective UVB be based on monthly changes (greater in winter) than an annual averages. (Moan et al., 1989).

Finally we (Scotto et al., 1988) have reported that in the 11 year period (1974-1985) there has been no increase in ground level, biologically effective UVB as measured by Robertson-Berger meters, which have a response spectrum very similar to the biologic hazard spectrum of McKinlay and Diffey. Dahlback et al. (1989) also found no particular trend in ozone values between 1978 and 1988 over Oslo (60°N).

It is most likely that the lack of UVR increase measured by the R-B meters is due to metereological, climatic and environmental factors taking place in the troposphere, which attenuate UVB (Scotto et al., 1988; Dahlback et al., 1989; Henriksen et al., 1989).

In the preceding section, the importance of weighting solar UVB radiation with a biologic effectiveness action spectrum has been discussed. Extensive studies on action spectra for two biologic effects - acute skin erythema and skin carcinogenesis - have been performed in man (experimental and epidemiologic) and in animal models (hairless mouse) (Cole et al., 1986). Based on these studies it has been shown that there is a precipitous, practically semilogarithmic drop in biologic effectiveness of wavelengths from 290 to 330 nm.

A modern skin erythema action spectrum has been proposed by McKinlay and Diffey (1987) and accepted by both CIE and IEC. It has been found to predict accurately the erythemal effectiveness of several polychromatic light sources differing greatly in spectral composition (Urbach, 1987).

Review of extensive photocarcinogenesis experiments in animals, performed with light sources of differing UVB content by Cole et al (1986) suggest that, as long as a light source contains 2% or more UVB, radiation of wavelengths longer than 330 nm have no measurable effect on photocarcinogenesis. That action spectrum closely simulates the older erythema action spectra. However, detailed experimental work in animals by Sterenborg and van der Leun (1987) showed that the action spectrum for mouse skin cancer carcinogenesis is very similar to the new (McKinlay-Diffey) human erythema action spectrum, i.e. wavelength longer than 330 nm are measurably effective for skin cancer induction.

Calculations of biologically effective UVR for the summer solstice with these two (differing in UVA effect) action spectra, performed by Prof. John Frederick (1988) show that with the action spectrum used by Cole the optical amplification factor (AF°) is 1.8, and with that used by Sterenborg and van der Leun the optical AF° is 1.25, or a difference of approximately 30%.

Epidemiologic studies of human Non-Melanoma skin cancer (NMSC) in the US (Scotto et al., 1974; Vitaliano and Urbach 1980) have shown that there is indeed a progressive relationship of NMSC incidence with latitude and amount of outdoor sunlight exposure. Furthermore there is a different relationship between Basal Cell and Squamous Cell Carcinoma, the latter being much more influenced by solar exposure.

Based on an extensive epidemiologic study of Non-Melanoma skin cancer, Scotto et al. (1983) estimated the relative increase in incidence at Basal Cell (BCC) and Squamous Cell Carcinoma (SCC) of the skin in man associated with a 1% increase in UVB by geographic location in the United States. This leads to biological amplification factors (AF_B) of 3 at 35° N and 1.44 for latitudes greater than 47° N. For malignant Melanoma, Scotto and Fears (1987) found a biological amplification factor of 0.6.

In most previous reports by the US National Academy of Sciences an AF_0 of 2, and AF_B of 2 were used for estimation of skin cancer increases, potentially due to ozone decreases (at equilibrium).

Comparing these estimates with those based on new data, one can estimate that Non-Melanoma skin cancer should have increased in the 18 years from 1969 to 1986 by approximately 7% at 35° N and by 1% at 60° N. Malignant Melanoma should have increased by 2% at 35° N and 0.5% at 60° N.

It is interesting to compare these calculated increases in skin cancer incidence with reported epidemiologic data.

In the 20 year period 1965 to 1985, Malignant Melanoma has increased by at least 5% per year, the actual increase in some areas of the US has been 100% or more in this period. In the last decade, Malignant Melanoma incidence in the US has been 7% per year. In Denmark, Malignant Melanoma has increased by 83% in males and by 68% in females between 1968 and 1985 (Osterlind and Jensen, 1986).

The Australians believe that the increase in incidence of Malignant Melanoma began with people born at the end of the 19th century and has continued since. Since there is evidence that stratospheric ozone either remained stable or even increased prior to 1965, it is clear that factors other than increasing solar UVR must operate in the genesis of Malignant Melanoma.

In conclusion, I would like to most strongly emphasize that I and my colleagues are extremely concerned by the probability that stratospheric ozone will be decreased by man-made effluents. Although changes to date appear to be small and may not yet have caused major problems for living things on earth, the potential for an inexorable catastrophe is great. This would be particularly true if the Montreal protocol is not adhered to.

There is every reason to believe that the effects of increased biologically effective UVR on plants and plankton and other living materials at the beginning of the food chain may be much greater than that on people, who after all can take measures to protect themselves, while stationary living things cannot do so.

It is thus of the utmost importance that the Treaties now proposed be put into effect and even stregthened and expanded to include chemicals other than CFC's. Changes in the stratosphere develop slowly but can be reversed only in hundreds of years.

REFERENCES

Blum, H.F., 1959, Carcinogenesis by Ultraviolet Light, Princeton University Press, Princeton, New Jersey.

Cole, C.A., Forbes, P.D. and Davies, R.E., 1986, An action spectrum for UV photocarcinogenesis, Photochemistry and Photobiology, 43: 275-284.

Dahlback, A., Henriksen, T., Larsen, S.H.H. and Stamnes, K., 1989, Biological UV doses and the effect of an ozone layer depletion, Photochemistry and Photobiology, 49: 621-625.

Diffey, B.L., Meanwell, E.F., and Loftus M.T., 1988, Ambient ultraviolet radiation and skin cancer incidence. Photodermatology, 5: 175-178.9

Executive Summary of Ozone Trends Panel. NASA, Washington, D.C., March 1988.

Henriksen, T., Stamnes, K., Valden, C. and Folk,E.S., 1989, Ultraviolet radiation at high latitudes and the risk of skin cancer, Photodermatology, 6: 110-117.

McKinlay, A.F. and Diffey, B.J., 1987, A reference action spectrum for ultraviolet induced erythema in human skin, in: "Human Exposure to Ultraviolet Radiation: Risks and Regulations. Passchier, W.R. and Bosnajokovic, F.F.M. eds., Elsevier Science Publishers, Amsterdam.

Moan, T., Dahlback, A., Larsen, S., Henriksen, T. and Stamnes, K., 1989, Ozone depletion and its consequences for the fluence of carcinogenic sunlight. Cancer Research, 49: 4245-4250.

Osterlind, S. and Jensen, D. 1986, Trends in incidence of Malignant Melanoma of the skin in Denmark 1943-1982, in: "Recent Results in Cancer Research", 102: 8-17.

Scotto, J., Kopf, A.W. and Urbach, F., 1974, Non-melanoma skin cancer in four areas of the United States, Cancer, 34: 1333-1338.

Scotto, J., Fears, T. and Fraumeni, J.F., 1983, Incidence of non-melanoma skin cancer in United States. DHEW Publication NIH 83-2433, National Cancer Institute, Bethesda, MD.

Scotto, J. and Fears, T.R., 1987, The association of solar ultraviolet and skin melanoma incidence among caucasians in United States. Cancer Investigation, in press.

Scotto, J., Cotton, G., Urbach, F., Berger, D. and Fears, T., 1988, Biologically effective ultraviolet radiation: Surface measurements in the United States 1974 to 1985. Science, 239: 762-764.

Sterenborg, H.J.C.M. and van der Leun, 1987, Action spectra for tumorigenesis by ultraviolet radiation, in: "Human

Exposure to Ultraviolet Radiation: Risks and Regulations". Passchier, W. R. and Bosnajokovic, B.F.M., eds., Elsevier Science Publishers, Amsterdam.

Urbach, F., 1987, Man and Ultraviolet Radiation, in: "Human Exposure to Ultraviolet Radiation: Risks and Regulations. Passchier, W.R. and Bosnajokvic, B.F.M., eds., Elsevier Science Publishers, Amsterdam.

Vitaliano, P.P. and Uurbach, F., 1980, The relative importance of risk factors in non-melanoma carcinoma. Arch. Derm., 116: 454-456.

OZONE DEPLETION AND SKIN CANCER

Johan Moan[1], Arne Dahlback[2], Soren Larsen[2],
and Knut Magnus[3]

[1]Institute for Cancer Research, Montebello, Oslo 3
[2]Inst. of Physics, University of Oslo, Oslo 3
[3]The Cancer Registry of Norway, Montebello, Oslo 3

INTRODUCTION

It has been known for a long time that different kinds of
human activities, including supersonic transport, nuclear
explosions, and waste of chlorofluorocarbons might reduce the
ozone layer (Giese, 1976). The first reports of the Antarctic
ozone hole were therefore quite alarming (Farman et al., 1985;
Stolarski, 1988). A small negative trend was announced also
for the Northern Hemisphere (Lindley, 1988). However, the
ozone level fluctuates from year to year, due to climatic
factors and sun-spot activity, and the negative trend for the
Northern Hemisphere is probably not statistically significant.

In order to evaluate the consequences of an ozone
depletion for the incidence of skin cancer one needs to
determine two constants: the radiation amplification factor A_r
which gives the percent increase in annual exposure to
carcinogenic sunlight at a given location per percent
reduction of the ozone layer, and the biological amplification
factor A_b which gives the percent increase in skin cancer
incidence per percent increase in annual fluence of
carcinogenic sunlight. The product of these two factors, $A =
A_r A_b$ then gives the percent increase in skin cancer incidence
per percent reduction of the ozone layer.

In the evaluation of A_r, as well as of A_b, an action
spectrum for carcinogenesis is needed. Of course, no such
spectrum has been determined, so one has to use
approximations. In the present work we have used the CIE
"risk spectrum" which is an approximation to the action
spectrum for erythema induction (McKinley and Diffey, 1987).
We have shown that this spectrum is similar to the action
spectrum for carcinogenesis in mice as well as to the action
spectrum for mutation of cells in the basal layer of the skin
(Moan et al., 1989a, b).

In the present work we have considered only basal cell

Light in Biology and Medicine, Volume 2 Edited by R.H. Douglas et al.
Plenum Press, New York, 1991

291

carcinomas (BCCs) and squamous cell carcinomas (SCCs) since
melanoma induction is thought to be related to episodes of
sunburn and therefore not only to the accumulated exposure to
carcinogenic sunlight.

MATERIALS AND METHODS

O_3-measurements

The amount of O_3 in the atmosphere over Oslo and Tromsö
was measured with Dobson spectrophotometers as described
earlier (Moan et al., 1989b). The instruments have been
calibrated at NOAA's laboratory.

Annual exposure to carcinogenic sunlight

The fluence rate of carcinogenically effective solar
radiation is defined by the expression $E_c = \int E\phi_c d\lambda$, the
integration being performed over the wavelength region of the
solar spectrum. ϕ_c is the action spectrum for skin
carcinogenesis which is approximated by the CIE spectrum
(McKinlay and Diffey, 1987). The solar irradiance at the
earth's surface, E, was determined by using a discrete
ordinate algorithm to calculate the propagation of light in
vertically inhomogenous plane parallel media (Stamnes et al.,
1988). The model atmosphere used was the "US Standard
Atmosphere 1976" which was divided into 39 homogenous layers
with a thickness of 2 km. We used the extraterrestrial solar
radiation spectra as well as all orders of scattered light
(Rayleigh scattering) from the atmosphere. The ground albedo
was set equal to 0.2. The absorption spectrum of ozone was
taken from "World Meteorological Organization, Global ozone
research and monitoring project". Ref. No. 16. Atmospheric
ozone, Vol. 1, 355-358, 1985.

The annual exposure to carcinogenic radiation from the
sun is:

$D = \int E_c dt$, the integral being taken over one year. In
our calculations the integral was approximated by the
sum:

$D = \Sigma_t \Sigma_\lambda E_\lambda \phi_c \Delta\lambda \Delta t$, with $\Delta t = 1$ hour and $\Delta\lambda = 1$ nm. The
seasonal average ozone levels at different latitudes were
used.

Epidemiological data

Data for incidence of skin cancer were provided by the
Norwegian Cancer Registry. All the incidences were age
adjusted to the European standard population (Hill, 1971).

Norway was divided into six zones. The latitudes of the
centre of gravity of the populations in these zones were
$59.0°$, $60.9°$, $61.0°$, $63.1°$, $66.9°$ and $69.6°$, respectively. The
population of the regions ranged between 200,000 and 700,000.
The area around Oslo is densely populated compared with the
other regions and we have reasons to believe that differenct
practices of reporting skin cancer may apply in this region.
Therefore, it was excluded a priori. In all areas between 50

and 70 percent of the population lives in rural areas.
Practically all inhabitants are Caucasians and we have no
reason to believe that there is any difference between the
regions with respect to the distribution of persons with
different skin types. A small fraction of the total exposure
of the population to carcinogenic solar light may be acquired
during vacations to countries at low latitudes. Using
information about the vacation habits of Norwegians, we have
estimated that due to this factor, the biological
amplification factor may be underestimated by at most 10 per
cent.

RESULTS AND DISCUSSION

     Fig. 1 indicates how the ozone level at high northern
latitudes has fluctuated over the past 50 years. Deviations
from the average level by as much as 15 percent can be seen.
It is evident that it is impossible to detect a permanent
decrease in the level by a few percent superimposed on these
normal fluctuations. Based on Dobson measurements from the
ground, like those shown in figure 1, The Ozone Trend Panel
announced a small negative trend for the ozone level over the
northern Hemisphere for the period 1969-1986 (Lindley, 1988).
Such a decrease is in agreement with the data shown in figure
1 for the same period. However, if data for the time periods
before 1969 and after 1986 are included, this conclusion is no
longer warranted. A slow fluctuation may seem to take place,
showing minima in the periods 1969-1965 and 1983-1986,
respectively, and a maximum in the period 1970-1975, and
possibly also one in the period 1940-1943. This pattern does

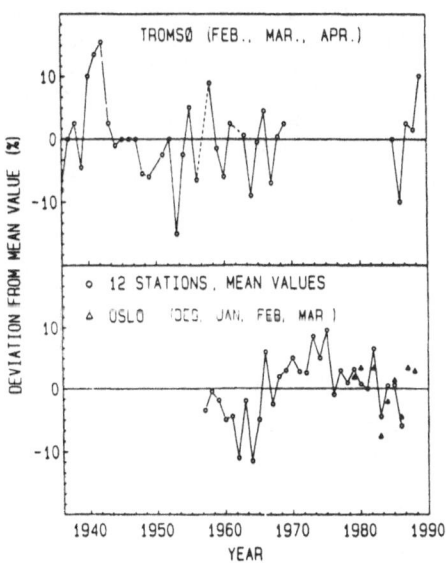

Fig. 1   Ozone levels measured from the ground.
         Each point represents an average value for
         a given year. Average values for 12
         stations north of 59° (Bojkov, 1988) are
         also given. These values are very close to
         those measured in Oslo.

not coincide with the sun-spot cycle (maxima in 1958, 1969, 1980) which has been proposed to have a slight influence on the ozone level (Giese, 1976).

Annual exposure (D) to carcinogenic sunlight

Using the CIE action spectrum as an approximation for the action spectrum of skin photocarcinogenesis, we have calculated the annual exposure to carcinogenic sunlight at different latitudes Fig. 2).

Fig. 3 shows how ozone depletion will lead to an increase in the annual exposure to carcinogenic sunlight in Oslo. If the ozone is completely removed the exposure will increase by more than a factor of 60. From Figs. 2 and 3 one can see that a 70 percent ozone depletion will have a similar effect with respect to increases in annual exposure to carcinogenic sunlight as moving from Oslo to the Equator. Correspondingly, a movement from Oslo to Budapest corresponds to a decrease of the ozone level above Oslo by about 30 percent.

The radiation amplification factor A_r

From Fig. 3 it is possible to determine the radiation amplification factor $A_r = (dD/d[O_3])([O_3]/D)$. This factor is found to increase from about 1.1 in the Nordic countries to about 1.2 at the Equator when the CIE spectrum is used in the calculation (Moan et al., 1989b).

The biological amplification factor A_b

The incidence of non-melanoma skin cancers in sun-exposed areas of the body (face and head) increases with increasing exposure to sunlight as shown in Figure 4. Sunlight is the main cause of these skin cancers (Moan et al., 1989a). Data from Norway and from Australia agree extremely well. For men basal cell carcinomas are about twice as frequent as squamous cell carcinomas. Squamous cell carcinomas are about twice as frequent among men as among women, while the incidence of basal cell carcinomas is similar for the two sexes.

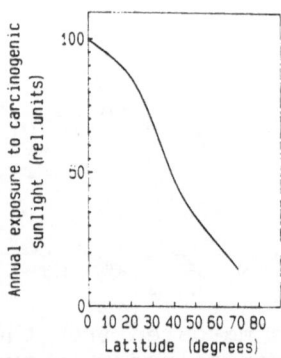

Fig. 2   Annual exposure to carcinogenic sunlight at different latitudes. Calculated by use of the CIE action spectrum and known ozone levels at different latitudes.

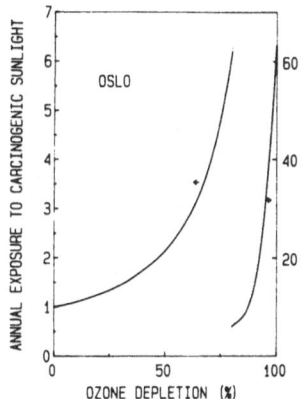

Fig. 3 Annual exposure to carcinogenic sunlight in
        Oslo as a function of the percent of ozone
        depletion from the present average level.

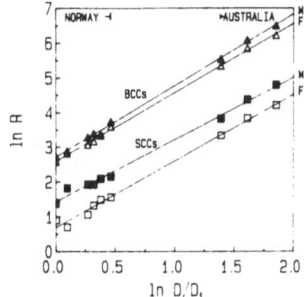

Fig. 4 The annual incidence R of non-melanoma skin
        cancer located in face and head as a
        function of annual expsoure to carcinogenic
        sunlight. The incidence R is given in
        cases per 100,000, age adjusted to the
        "European Standard Population" (Hill, 1971).
        Data for Australia: Giles et al., 1988.
        The curves for men and women are marked
        M and F, respectively.

     From the slopes of the curves in Figure 4 the biological
amplification factor can be determined directly: A_b =
(dR/dD)(D/R).   For BCCs A_b = 2.06 ± 0.03 for men and within
the error limits similar for women: A_b = 1.99±0.03.   For SCCs
the values are slightly lower: A_b = 1.800.05 for men and 1.92
for women.   These numbers are very close to those obtained
earlier on the basis of the Norwegian Data alone (Moan et al.,
1989a).

The overall amplification factor A

     The overall amplification factor for non-melanoma skin
cancer is defined as the ratio of the increment in skin cancer
production to the decrement in the ozone level: A =
(dR/d[O_3])([O_3]/R) = $A_r A_b$.   Taken together our data lead to an

amplification factor of about 2.2 in Nordic countries and of about 2.4 at the Equator. Thus, an ozone reduction by one percent will lead to an increase in the incidence of non-melanoma skin cancer by slightly more than two percent. When the absorption spectrum of DNA, corrected for the transmission through the epidermis, is used in the calculations, a significantly higher amplification factor is found. However, UVA is clearly more carcinogenic than indicated by such a spectrum.

Even though the amplification factor as calculated here is lower than has been assumed by authorities all over the world, one should not reduce the efforts to prevent an ozone depletion caused by human activities.

ACKNOWLEDGEMENT

The present work was supported by The Norwegian Council for Science and The Humanities (NAVF).

REFERENCES

Bojkov, R., 1988, Ozone variations in the northerns polar region, J. Meteorology and Atmospheric Physics, 38:117.

Farman, J.C., Gardiner, H. and Shanklin, J.D., 1985, Large losses of total ozone in Antarctica reveal seasonal $C10_x/NO_x$ interaction, Nature, 315:207.

Giese, A.C., 1976, "Living with our sun's ultraviolet rays". Chapter 9. Plenum Press, New York and London.

Giles, G., Marks, R. and Foley, P., 1988, Incidence of non-melanocytic skin cancer treated in Australia, Br. Med. J., 296:13.

Hill, A.B., 1971, "Principles of medical statistics", pp. 204-210, The Lancet Limited, London.

Lindley, D., 1988, CFCs cause part of global ozone decline, Nature, 323:293.

McKinlay, A.F. and Diffey, B.L., 1987, A reference action spectrum for ultraviolet induced erythema in human skin, CIE-Journal, 6:17.

Moan, J., Dahlback, A., Henriksen, T. and Magnus, K., 1989a, Biological amplification factor for sunlight-induced non-melanoma skin cancer at high latitudes, Cancer Res., 49:5207.

Moan, J., Dahlback, A., Larsen, S., Henriksen, T. and Stamnes K., 1989b, Ozone depletion and its consequences for the fluence of carinogenic sunlight, Cancer Res., 49:4247.

Pearl, D.D. and Scott, E.L., 1986, The anatomical distribution of skin cancers, in J. Epidemiol., 15:502.

Scotto, J., Cotton, G., Urbach, F., Berger, D. and Fears, T., 1988, Biologically effective ultraviolet radiation: Sunface measurements in the United States., Science, 231:762.

Stamnes, K., Tsay, S., Wiscombe, W. and Jayaweera, K., 1988, Numerically stable algorithm for discrete-ordinate-method radiative transfer in multiple scattering and emitting layered media, Applied Optics, 32:2502.

Stolarski, R.S., 1988, The antartic ozone hole, Sci. Am., 258:20.

RISK FACTORS FOR SKIN MELANOMA

Ernst G. Jung

Department of Dermatology
Mannheim Faculty of Clinical Medicine
Univeristy of Heidelberg, FRG

INTRODUCTION

Malignant melonomas of the skin are regarded as highly malignant and lead to death owing to early and unpredictable metastatic spreading. In recent decades, the incidence of melanoma has more than doubled and is now assumed to have a level of four to 20 new cases per 100,000 inhabitants a year. At the same time, it is striking that persons between 20 and 40 years old are developing melanomas with ever increasing frequency.

For prevention of melanomas and early diagnosis of melanoma precursors, it is necessary to explore the pathogenesis of these tumors. Three pathogenetic paradigms can be distinguished and described as risks of melanoma.

PIGMENT NEVI AS POTENTIAL MELANOMA PRECURSORS

Pigment nevi are laid down as extensive flat or nodular accumulations of pigment cells in the skin during the genesis and formation of the ectoderm and are either already present at birth (congenital nevi) or only become manifest in the course of the first two decades of life by growth and pigmentation. Every person develops some pigment nevi and many people have up to 200 of these pigment lesions of different size and appearance. Special attention is to be paid to congenital nevi, which are present at birth and as a rule grow with the children. In recent years, the following observations have been substantiated, (Jung, 1989).

Congenital nevi which are larger in diameter than 1.5 cm

These entail a raised risk of melanoma, which is manifested from the 18th year of life. Six percent of these congenital nevi will develop a melanoma in the course of adulthood. It is therefore urgently necessary to excise

congenital nevi which are larger than 1.5 cm in diameter
before the 18th year of life.

Large congenital nevi which have a diameter greater than 10 cm at birth

These bathing-trunk or giant nevi, such as occur in
neurocutaneous melanosis, entail a risk of melanoma of 10% to
20%, a large proportion of which are manifested in the first
10 years of life.

SOLAR RADIATION (UVB) AS A RISK OF MELANOMA

The epidemiological data, the multiplicity of
manifestation forms, and the different age of manifestation of
melanomas, indicate that endogenous and exogenous factors
contribute to the multifactorial genesis of melanomas. The
genetically fixed pigmentation of the human being plays an
important role here. Fair skinned people with type I skin
(Fitzpatrick 1988) develop substantially more melanomas than
light pigmented people of skin type III or IV. In addition, a
genetic disposition to melanoma can be detected on the basis
of model diseases. Furthermore, exogenous UV exposure of the
skin plays a crucial role in the genesis of melanomas, even if
this is to be considered in differentiated terms. This
increase in melanoma risk from UVB radiation of sunlight and
also from artificial light sources can be described in model
terms by two fundamentally different risk patterns.

Cumulative Total Lifetime Exposure

Photo-induced damage occurs by repeated exposures to UV
distributed throughout life and acting cumulatively. The
damage extends extensively on to the areas exposed to light
and becomes manifest relatively late in life with multiple
foci (Tab. 1). Besides actinic elastosis (heliosis), actinic
keratoses, basaliomas and spinaliomas are manifested as
precanceroses and carcinomas of the kertinocyte system. Of
the melanomas, only the lentigo-maligna melanomas (LMM) are
located here. In view of their age of manifestation,
morphodynamics, multilocular disposition and the strict light
localisation, they correspond to the photo-induced damage
mentioned, and account for 5% to 10% of cutaneous melanomas.

Table 1   Characteristics of melanomas induced by
          Cumulative total Lifetime Exposure

```
Damage on light exposed areas
Multiple foci
Late manifestation

Aktinic keratoses
Basal cell carcinoma (BCC)
Squamous cell carcinoma (SCC)
Lentigo maligna melanoma
Activation of dysplastic nevi
```

Experimental data on fibroblasts and kertinocytes show that the effectiveness of the cellular repair systems and their overtaxing play a major role. The excision repair, measured as unscheduled DNA synthesis (UDS) can repair a light-induced DNA damage in the first hours after irradiation without errors. This enzymatic repair system of cell nuclei is limited to 1/2 to 1 MED (minimal erythema doses) and is exhausted by irradiation repeated at short intervals. It can no longer repair all the DNA damage caused by UV, so that this damage persists, and other error-prone repair systems are activated. In this way, oncogenic point mutations and development of malignant cloni in the epidermis as well as in the melanocytes may occur (Hönigsman et al., 1987; Jung 1986 and 1988).

The autosomal-recessive disease Xeroderma pigmentosum (XP) is the genetic model for this risk group. This condition is rare, but very characteristic. With raised sensitivity to light, these patients suffer chronic photoinduced damage to the skin after a single or after very few UV exposures. Besides the pigment incontinence, the epidermal atrophy and the actinic elastosis, multiple benign and malignant tumors of the skin occur. Half of the XP patients develop one or several lentigo malignas and melanomas, exclusively of the lentigo-maligna melanoma type. It is shown that the light-dependent risk of melanoma is raised by a factor of 2,000 in xeroderma pigmentosum (Jung 1986 and 1988, Jung et al. 1986) and by far exceeds the other differences in risk due to racial pigmentation.

Particularly intense exposure in early life

As indicated by epidermiological observations, case control studies and investigations on patients who have undergone "migration" (Sober 1987) one or several (as a rule only a few) excessive exposures to UV in childhood and in youth which led to severe sunburns, which can be remembered by the paitents and mostly receives medical treatment, appear to characterise a further risk of melanoma (Tab. 2). In locations which do not have to be typical for a regular light exposure, nodular (NM) or superficially spreading melanomas (SSM) occur solitarily between the 20th and 50th year of life (i.e. earlier than LMM). However, excessive exposure to sunlight may also activate or provoke melanomas in dysplastic nevi (Tab. 3). This appears to lead to a reduction of the age of manifestation. Without being able to present exact figures on the risk of excessive UV exposure in youth, it is suspected that most NM and SSM of the skin arise from this risk type.

Table 2   Characteristics of melanomas induced by
          Particularly Intense Exposure in Early Life

```
Atypical localisation
Solitary
Early manifestation

Nodular and superficial spreading
Melanoma (NM, SSM)
Activation of dysplatic nevi
Activation of congenital nevi
```

Table 3 Characteristics of Dysplastic Nevi (DN)

```
Polycyclic, irregular borders ( >5 mm)
Polychromasia (black, brown, rose)
Regression and satellites
Surface flat and lichenoid
```

In this context, it is helpful to examine the congenital Dysplastic Nevus Syndrome (DNS). This is an autosomal polygenic syndrome with multiple dysplastic nevi which spread and proliferate in a bizzare, irregular way with change of form after adolescence (Tab. 4). Carriers of this syndrome develop solitary or multiple melanomas from dysplastic nevi with an age of manifestation between the 25th and 40th year of life. On extrapolation, at least one melanoma has been manifested in each carrier of DNS at the age of 70 (Kraemer et al. 1983). The activity and number of dysplastic nevi is substantially raised at skin sites with exposure to sunburn compared to covered skin (Kopf et al. 1985). Melanomas based on DNS probably account for 20% or more of NM and SSM. Excision repair is in the normal range in solitary melanomas as well as in the dysplastic nevus syndrome, (Thielmann et al. 1987) whereas increased chromosomal anomalies are found in melanocytes and fibroblasts in patients with DNS. Moreover, cytogenetic indications for hypermutability are regularly found under experimental UV exposure of cultivated cells and cell lines (Jung et al. 1986, Perera et al. 1986) which enable an approach to cytogenetic detection and description of the endogenous risk. The hypermutability is the same in patients with and without manifest melanoma, indicating that it is due to the syndrome and does not arise in the course of development of the disease (Jung et al. 1986). The relative risk of melanoma can be described in figures based on experience. In DNS, an excessive UV exposure in youth very probably leads to manifestation or to a decisive promotion of the risk. This appears to be the case to a lesser extent in large congentital nevi and neurocutaneous melanosis (NCM).

Table 4   Characteristics of Dysplastic Nevus
          Syndrome (DNS)

```
Dysplastic Nevi
- multiple (10 to > 100)
- buttocks, trunk, head
- increase in number and size (in adults)

Genetic
- autosomal
- dominant or polygenetic
- associated with fair complexion

Melanoma in DN
- early appearance (20-40 y)
- often multiple
- up to 100 % (> 70 y)
```

A first mutation of the genome can be assumed for the syndrome of dysplastic nevi. This shows autosomal inheritance and leads to chromosomal instability and to hypermutability. One or several excessive UV exposures early in life are to be considered as a second exogenously induced somatic mutation. This explains the reduced age manifestation, the familial occurrence and the multiplicity of the melanomas in DNS.

The risk of melanoma in human skin can thus be divided into three different risk groups. In two risks, the nature and degree of UV exposure play a major role. This must be reduced. The third risk is characterised by specific melanoma precursors. These are to be detected early and exactly. They can then be excised before manifestation of an invasive tumor.

REFERENCES

Fitzpatrick, T.B., 1988, The validity and practicality of sunreactive skin types I through VI Arch. Derm. 124:869.
Hönigsmann, H.. Brenner, W., Tanew A., Ortel, B., 1987, UV-induced unscheduled DNA synthesis in human skin. Dose response, correlation with erythema, time course and split dose exposure in vivo. J Photochem. Photobiol. B 1:33.
Jung, E.G., 1986, Xeroderma Pigmentosum. Int. J. Derm. 25:629.
Jung, E.G., Bohnert, E., Boonen, H., 1986, Dysplastic nevus syndrome. Ultraviolet hypermutability confirmed in vitro by elevated sister chromatid exchanges. Dermatologica (Basel) 173:297.
Jung, E.G., 1988, Ist das Melanom-Risiko kalkulierbar? Z. Haut- u. Geschl. - Kr. 63:559.
Jung, E.G., 1989, Wie kann man Melanome verhindern? Dtsch. med. Wschr. 114:393.
Kraemer, K.H., Greene, M.H., Tarone, R., Edler D.E., Clark, W.H. jr., Guerry, D., Dysplastic nevi and cutaneous melanoma risk. Lancet 1983/II, 1076.
Kopf, A.W., Lindsay, A.C., Rogers, G.S., Friedman, R.J., Rigel, D.S., Levenstein, M., 1985, Relationship of nevocytic nevi to sun exposure in dysplastic nevus syndrome. J. Amer. Acad. Derm. 12:656.
Perera, M.I., Um, K.I., Greene, M.H., Waters, H.L., Bredberg, A., Kraemer, K.H., 1986, Hereditary dysplastic nevus syndrome. Lymphoid cell ultraviolet hypermutability in association with increased melanoma susceptibility. Cancer Res. 46:1005.
Sober, A.J., 1987, Solar exposure in the etiology of cutaneous melanoma. Photodermatology 4:23.
Thielmann, H.W., Edler, L., Burkhardt, M.R., Jung, E.G., 1987, DNA repair synthesis in fibroblasts strains from patients with actinic keratosis, squamous cell carcinoma, basal cell carcinoma, or malignant melanoma after treatment with ultraviolet light, N-acetosy-2-acetyl-aminofluorene; methyl methanesulfonate, and N-methyl-N-nitrosocured. J. Cancer Res. Clin. Oncol. 113:171.

PREVENTION OF PHOTOCARCINOGENESIS BY SUNSCREENS

Per Thune and Henrik Flindt-Hansen

Department of Dermatology
Ullevål Hospital
0407 Oslo 4, Norway

INTRODUCTION

Sunscreens have been primarily used to protect skin against sunburn. The cosmetic benefits of sunscreens have been widely accepted with the recent indictment of photoaging as a major cause of skin wrinkles. It has been much more difficult to achieve the same public acceptance of sunscreen use to prevent the development of nonmelanoma skin cancer. Most human squamous cell, and approximately 2/3 of basal cell cancers are directly related to the total life-time exposure dose of solar radiation. The use of sunscreens thus serves two purposes: the first one which is apparent to everybody, is an immediate or short-term effect - namely the prevention of sunburn. The other purpose, the prevention of skin cancer, is not apparent to lay people since the development of skin cancer takes several years. It seems probably that altered sun habits are the main factors to be blamed for the large increase in both malignant melanoma and basal and squamous cell carcinoma of the skin which has been observed during the last decades. Epidemiologic studies indicate that almost 90% of these last two types of cancers can be attributed to expsoure of UVB (Slaper and van der Leun, 1987; Stern et al., 1986).

It is well documented that sunscreens protect against sunburn and actinic damage including vascular changes (Grove and Kaidbey 1986; Kligman et al., 1982; Sambuco et al., 1984). Animal studies have demonstrated that these agents also reduce the risk of UVB induced tumours (Knox et al., 1960; Wulf et al., 1982; Snyder and May, 1975). The tumour risk reduction observed in animals was correlated with the effectiveness of the sunscreen. Using a mathematical model based on epidemiological data, Stern et al., (1986) have estimated that regular use of a UVB absorbing sunscreen, with an effective sun protective factor of 7.5, for the first 18 years of life would reduce the cumulative incidence of nonmelanoma skin cancer by more than 75%.

During the last years the need for sunscreens with protection against UVA has been increasingly realized. UVA can induce erythema and pigmentation, dermal elastosis and possible skin cancer (van der Leun, 1987). Increased use of sunscreens which effectively only blocks UVB radiation and the warning sign of UVB erythema, could lead to an increased expsoure to solar UVA. Accordingly, increased attention has been paid to UVA as well as UVB blockers during the last few years. At the same time we have seen that the sun protection factor (SPF) has been raised and that better methods have been developed for the evaluation of the SPF (Sayre et al., 1979; Arase and Jung, 1986). At the present time there is no consensus to the action spectrum for UV carcinogenesis (van der Leun, 1987; Forbes, 1981; van der Leun, 1987b; Cole et al., 1983). Experimental results indicate that it is a fair approximation to the erythemal action spectrum, but real information on the carcinogenic effectiveness of the whole UVA spectrum, i.e., the various UVA wavelengths, is still lacking (van der Leun, 1987b; Cole et al., 1983).

How can sunscreens prevent cancer? There is good evidence that damage to DNA is an important factor in photocarcinogenesis of the skin (Peak and Peak, 1989). The appearance of sunburn cells and cyclobutyl pyrimidine dimers in the epidermis is evidence of UV-induced damage (Young, 1987; Freeman et al., 1988; Cesarini et al., 1989). Freeman et al., (1988) showed for the first time that a chemical sunscreen could prevent UV-induced pyrimidine dimers in human skin in situ. Since the induction of pyrimidine dimers in DNA is correlated with transformation and tumorigenesis, the experiments did support the view that sunscreens can prevent photocarcinogenesis. The action spectra for dimer formation, for the induction of unscheduled DNA synthesis (UDS) and for the inhibition of semiconservative DNA synthesis have been well elucidated and seem to be the same (Arase and Jung, 1986; Peak and Peak, 1989). Arase and Jung (1986) developed an in vitro method for the evaluation of the photoprotective efficacy of sunscreens using fibroblasts in culture. They measured quantitatively the protective effect on the UVB-induced UDS and on the inhibition of semiconservative DNA synthesis (SDS).

In our own studies we have been concentrating on the protective effect of PABA on UVB-induced carcinogenesis. We have studied the effect of a) full time application, b) part-time application and c) the application of a pre-irridiated solution containing degradation products of PABA (Flindt-Hansen et al., 1990a, b).

MATERIALS AND METHODS

PABA and the photodegradation product diazobenzoic acid (DABA), were tested for protected effect on photocarcinogenesis. Irradiated and non-irradiated PABA (5%) in 70% ethanol and 5% glycerol in water were applied to the backs of different groups of hairless mice (Table 1). The pre-irradiation dose of UVB before application of the PABA solution was $27J/cm^2$ corresponding to 3h and the degree of photodegradation has been estimated as 40% by mass spectroscopy and UV spectroscopy (Flindt-Hansen et al., 1988).

Table 1 Treatment schedule for 9 groups of mice

Group	topical treatment	UVR
A	-	-
B	-	+
C	PABA full-time	+
C'	PABA part-time [a]	+
D	PABA full-time	-
E	Vehicle	+
F	Vehicle	-
G	Pre-irrad. PABA [b]	-
H	Pre-irrad. PABA	+

[a]Part-time treatment with PABA from week 16 to week 26.
[b]The solution had been pre-irradiated with 27 J/cm^2.

270 lightly pigmented (hr/hr) hairless mice were randomized in to 9 groups of 30 each. One group had treatment with PABA from week 16 to week 26 and UVR from week one to week 30 (group C'). The treatment schedule is shown in Table 1. Immediately after painting, the mice were exposed to UVB from a Philips TL 40W/12 lamp in daily escalating doses from 155 mJ/cm^2 to 260 mJ/cm^2. This maximum exposure dose was reached after eights weeks of treatment. The total dose achieved after 30 weeks of irradiation was 49 J/cm^2. The development of skin tumours was recorded weekly for 40 weeks. All skin tumours were biopsied and all mice were autopsied. The skin tumours were classified into three classes ranging from hyperplasia without atypia, atypical hyperplasia to definite invasive squamous cell carcinoma. At the end of the study the mice were weighed and the dorsal skin was removed and weighed.

MED studies

A group of eight female, hairless lightly pigmented (hr/hr) mice 8-12 weeks old, were used for these experiments. The backs were covered with sections of Duoderm dressing bandages (Squibb) each with six punched out round holes (6 mm diameter) during anesthesia. The mice were exposed to UVB in doses ranging from 50 mJ/cm^2 to 300 mJ/cm^2 in increments of 50 mJ/cm^2. Observations of erythema and edema were made after 24 h and the presence of a circular border was used to indicate a positive reaction. The dose corresponding to the median rank was considered the MED as described by Cole et al., (1983). This was calculated to 175 mJ/cm^2.

RESULTS AND DISCUSSION

All animals developed tumours in the UVR-exposed non-protected groups. The application of PABA full-time (30 weeks) resulted in a significant ($p < 0.05$) retardation in tumour induction time and a significant ($p < 0.05$) increase in tumour free animals compared with non-protected UVR exposed groups. One animal in the PABA treated group developed a

Table 2   The number of squamous cell carcinomas per animal for nine groups of 30 mice

Group	Treatment	Number of tumours per animal						
		0	1	2	3	4	5	6
A	Control	30	0	0	0	0	0	0
B	UVR	3	3	4	7	9	2	0
C	UVR+PABA	26	1	0	0	0	0	0
C'	UVR+part.PABA[a]	17	4	5	2	0	0	0
D	PABA	29	0	0	0	0	0	0
E	UVR+vehicle	0	7	7	7	3	3	1
F	Vehicle	30	0	0	0	0	0	0
G	Preirr. PABA	29	0	0	0	0	0	0
H	UVR+preirr. PABA	28	0	0	0	0	0	0

[a]Part-time application of PABA (week 16 to week 26).

squamous cell carcinoma and 25 UVB exposed non-protected mice developed 78 squamous cell carcinomas (Table 2). The part-time application of PABA also resulted in a significant ($p < 0.05$) delay in tumour induction time (Fig. 1). Cessation of protection gave rise to an abrupt decline in the percentage of tumour free animals. The yield of carcinomas was 20 for the part-time protected group and this was significantly ($p < 0.05$) fewer than the UVR exposed non-protected group (Table 2). Likewise, non-protected mice had a significantly ($p < 0.05$) heavier mean weight of their dorsal skin compared with part-time protected mice and the latter did not differ from non-irradiated controls (Fig. 2). The percentage of tumours showing hyperplasia without atypia was significantly ($p < 0.05$) higher in the non-protected group of mice indicating a retardation in the development towards malignancy.

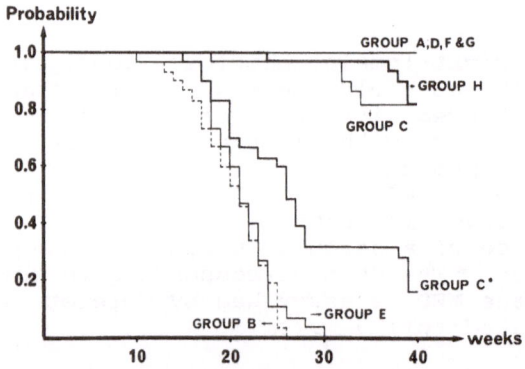

Fig. 1   The probability of tumour free animals in nine groups of hairless mice receiving no treatment (A), UVR (B), PABA+UVR (C), part-time protection with PABA (week 16 to 26) and full-time UVR (C'), PABA only (D), vehicle+UVR (E), vehicle only (F), pre-irradiated PABA (G) and UVR+pre-irradiated PABA (H).

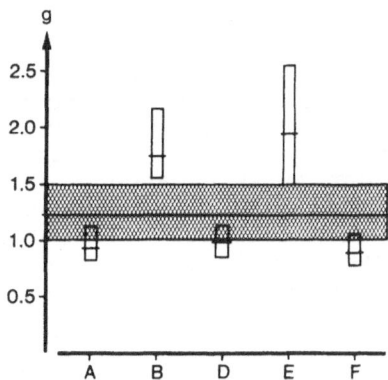

Fig. 2   The mean weight of the dorsal skin of six
         groups of mice receiving no treatment (A),
         UVR (B), PABA (D), vehicle and UVR (E) and
         vehicle only (F). The shaded area shows
         the confidence interval of the mean weight
         of the group treated part-time with PABA
         and full-time with UVB (C').

     Treatment with pre-irradiated PABA significantly (p<0.05)
retarded the tumour induction time (Fig. 1) and reduced
significantly (p<0.05) the number of squamous cell carcinomas
compared with non-protected, irradiated mice (Table 2).   This
tumour retarding ability did not differ significantly from the
effect achieved when using non-irradiated PABA (Fig. 1).   Also
the groups protected with pre-irradiated PABA showed a
significantly lower mean dorsal skin weight than UVR-exposed
non-protected mice (Fig. 3).   The group protected with the
pre-irradiated PABA did not differ from the groups which were
protected with unprepared PABA.

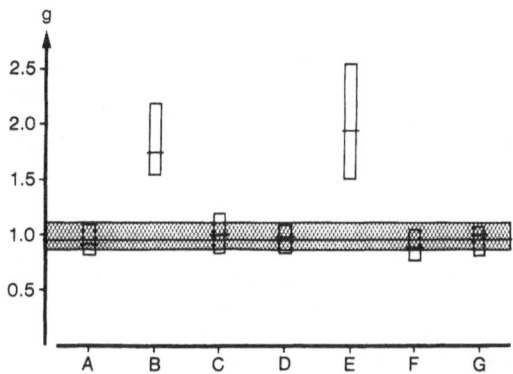

Fig. 3   The mean weight of the dorsal skin of
         eight groups of mice receiving no treatment
         (A), UVR (B), PABA+UVR (C), PABA only (D),
         vehicle and UVR (E), vehicle only (F),
         pre-irradiated PABA (G). The shaded area
         shows the confidence interval of the mean
         weight of group H treated with pre-
         irradiated PABA containing photoproducts,
         and UVR.

CONCLUSION

Both PABA and its photodegradation products provide a high degree of protection against UV-carcinogenesis. Part-time application of PABA on previously exposed skin, which mimics the conditions of human skin in vivo, significantly retards the development of skin cancer. No carcinogenic effect of pre-irradiated PABA, i.e. photodegradation products, could be demonstrated which opposes the suggestions that these products might be carcinogenic (Gasparro, 1986). In this study a very useful method for quantitative estimation of carcinogenesis, particularly when small and confluent tumours are developing making the count difficult. But more accurate data were obtained from calculating the probability curves as weighing of the dorsal skin did not show any significant difference for the part-time protected group compared to the full-time protected.

In this study the emission spectrum of the lamp differed markedly from the solar spectrum with regard to the longer wavelengths. PABA protects mainly against UVB and carcinogenic effects from UVA and possibly longer wavelengths can be expected with the use of only UVB blockers. It is therefore recommended that sunscreen testing should not only concern the MED which is mainly UVB dependent, but also various effects on epidermal and dermal cell constituents including the pigment cell system, which can be caused by wavelengths longer than UVB.

REFERENCES

Arase, S., Jung, E.G., 1986, In vitro evaluation of the photoprotective efficacy of sunscreens against DNA damage by UVA, Photodermatology, 3:56.

Cesarini, J.P., Chardon, A., Binet, O., Horseau, C. and Grollier, J.F., 1989, High protection sunscreen formulation prevents UVB-induced sunburn cell formation, Photodermatology, 6:20.

Cole, C.A., Davies, R.E., Forbes, P.D. and D'Aloisio, C.C., 1983, Comparison of the action spectra of acute cutaneous responses to ultraviolet radiation: man and albino hairless mice, Photochem. Photobiol., 37:623.

Flindt-Hansen, H., Nielsen, J.C. and Thune, P., 1988, Measurements of the photodegradation of PABA and some PABA derivatives, Photodermatology, 5:257.

Flindt-Hansen,, H., Thune, P. and Eeg-Larsen, T., 1990a, The effect of part-time and full-time application of PABA on photocarcinogenesis, Arch. Dermatol. Res., 282:38-41.

Flindt-Hansen, H., Thune, P. and Eeg-Larsen, T., 1990b, The effect of a partly photodegradated solution of PABA on photocarcinogenesis, Acta. Dermto-Venereol (Sth), in press.

Forbes, P.D., 1981, Photocarcinogenesis. An overview, J. Invest. Dermatol., 77:139.

Freeman, S.E., Ley, R.D. and Ley, K.D., 1988, Sunscreen protection against UV-induced pyrimidine dimers in situ, Photodermatology, 5:243.

Gasparro, F.P., 1986, PABA: Friend or Foe?, Photodermatology, 3:61.

Grove, G. and Kaidbey, K.H., 1980, Sunscreens prevent sunburn

cell formation in human skin, J. Invest. Dermatol.,
75:363.

Kligman, L., Atkin, F. and Kligman A.M., 1982, Prevention of
ultraviolet damage to the dermis of hairless mice by
sunscreens, J. Invest. Dermatol., 78:181.

Knox, J.M.., Griffin, A.C. and Hakim, R.E., 1960, Protection
from ultraviolet carcinogenesis, J. Invest. Dermatol.,
34:51.

van der Leun, J.C., 1987, Animal experiments in the study of
photocarcinogenesis, Review article, Photochem.
Photophysics. (suppl.), 353.

van der Leun, J.C., 1987, Interactions of different
wavelengths in effects of UV radiation on skin, Review
article, Photodermatology, 4:257.

Peak, M.J. and Peak, J., 1989, Solar-ultraviolet-induced
damage to DNA. Review article, Photodermatology, 6:1.

Sambuco, C.P., Forbes, P.D., Davies, P.E. and Urbach, F.,
1984, An animal model to determine sunscreen
protectiveness against both vascular injury and pediermal
damage, J. Am. Acad. Dermatol, 10:737.

Sayre, R.M., Agin, P.P., Le Vee, G.J. and Marlowe, E., 1979, A
comparison of in vivo and in vitro testing of
sunscreening formulas, Photochem. Photobiol., 29:559.

Slaper, H. and van der Leun, J.C., 1987, Human expsoure to
ultraviolet radiation: quantitative modelling of skin
cancer incidence, in: Human exposure to ultraviolet
radiation. Risks and regulations, W.F. Passchier, B.F.M.
Bosnjakovic, ed., Elsevier Science Publishers, Amsterdam.

Stern, R.S., Weinstein, M.C. and Baker, S.G., 1986, Risk
reduction for nonmelanoma skin cancer with childhood
sunscreen use, Arch. Dermatol, 122:537.

Snyder, D.S. and May, M., 1975, Ability of PABA to protect
mammalian skin from ultraviolet light-induced skin
tumours and actinic damage, J. Invest. Dermatol, 65:543.

Wulf, H.C., Poulsen, T., Brodthagen, H. and Hou-Jensen, K.,
1982, Sunscreens for delay of ultraviolet induction of
skin tumours, J. Am. Acad. Dermatol, 7:194.

Young, A., 1987, The sunburn cell. Review article,
Photodermatology, 4:127.

ADVANCES IN THE PHOTODERMATOSES

John Hawk

Photobiology Unit
St. Thomas' Hospital
London SE1 7EH

ACQUIRED IDIOPATHIC PHOTODERMATOSES

The term polymorphic light eruption (PLE) describes a persistent acquired abnormal reaction to sunlight of delayed onset and varied morphology affecting 10 - 20% of the population in temperate climates. The cause of PLE is unknown, but a delayed type hypersensivitiy (DTH) response has been postulated because of the delay in onset of lesions of up to 2 days after sun exposure. We have shown that exposure of previously affected sites to suberythemogenic doses of simulated radiation will induce typical PLE lesions in a significant proportion of patients. Immunochistological study (Norris et al., 1989a) of the evolution of such lesions has demonstrated changes consistent with DTH, including onset within 5 hours of perivascular infiltration predominantly of CD4+ (helper) T cells, increased Langerhans cell numbers in both the dermis and epidermis, and by 3 days increased expression of intercellular adhesion molecule-1 by keratinocytes, the latter probably secondary to release of gamma-interferon by immunologically activated lymphocytes in the underlying dermis.

Prophylactic prevention of PLE may be achieved in most patients by restriction of ultraviolet (UV) exposure and regular application of highly protective sunscreens, with more severe cases sometimes requiring phototherapy or psoralen photochemotherapy. Other empirical but unproven and usually not fully effective therapies include hydroxychloroquine, thalidomide, beta-carotene and nicotinamide. We have successfully treated two exceptionally severely affected PLE patients who were unable to tolerate PUVA with azathioprine, again consistent with DTH (Norris and Hawk, in press).

Chronic actinic dermatitis (CAD) includes the conditions of actinic reticuloid and photosensitive eczema. Photobiological, histological and immunohistological findings in 14 patients with CAD have been compared (Norris, et al., in

Light in Biology and Medicine. Volume 2 Edited by R.H. Douglas *et al.*
Plenum Press, New York, 1991

311

press). Dermal infiltrates consisted predominantly of T
lymphocytes, with a significant trend towards lower CD4+/CD8+
(cytotoxic/suppressor) ratios in patients with histological
appearances resembling cutaneous T cell lymphoma, also
reported in persistent allergic contact dermatitis (Kanerva et
al., 1988). Photosensitivity extending beyond 340 nm was
detected in 8 patients but width of the spectrum of
photobiological abnormality did not correlate with
histological severity of variation in T cell subsets in
lesional skin. Thus photosensitive eczema and actinic
reticuloid apparently represent extremes of a spectrum of
photosensitive dermatitis, encompassed by the term CAD, in
which (immuno)histological changes are not closely related to
the extent of photosensitivity but would be consistent with a
persistent DTH reaction, possibly to an endogenous UV-induced
antigen (Editorial 1988). A double-blind controlled trial has
confirmed the efficacy of azathioprine in CAD. In patients
unable to tolerate azathioprine or who fail to respond, we
have found cyclosporin an effective alternative (Norris et
al., 1989b).

GENETIC AND METABOLIC

In eyrthropoietic protoporphyria (EPP) the activity of
haem synthetase is deficient and free protoporphyrin
accumulates, causing marked photosensivity. We have developed
a novel assay for this enzyme based on the formation of zinc
protohaem, a technique easy to use and sensitive enough to
allow determination of haem synthetase activity in mononuclear
cells obtained from 10 mls of peripheral blood. Using this
assay to investigate the inheritance of EPP in 9 affected
families we have shown that EPP is not transmitted as a simple
autosomal dominant trait and that inheritance of more than one
gene may be required for disease expression (Norris et al.,
1989c).

There is evidence for defective DNA repair in xeroderma
pigmentosum (XP), Cockayne's syndrome (CS) and some patients
with trichothiodystrophy (TTD), but for increased cancer risk
only in XP. It has been suggested that impaired immune
surveillance might contribute to cancer susceptibility in XP,
and defects in adaptive T cell-mediated responses have been
reported previously in some but not all patients. We have
studied immune responses in XP, CS and TTD, and defective
natural killer (NK) cell activity was demonstrated in XP
(Norris et al., 1988) but not CS or TTD, whilst T cell
mediated responses were normal in all three conditions. Thus
defective immune surveillance particularly by NK cells may
contribute to the increased risk of skin neoplasma in XP.

REFERENCES

Kanerva, L., Estlander, T., and Jolarka, R., 1988,
    Immunohistochemistry of lymphocytes and Langerhans cells
    in long-lasting allergic patch tests, Acta. Derm.
    Venereol., 108: 116-22.
Norris, P. G., and Hawk, J.L.M., 1989, Successful treatment of
    severe polymorphic light eruption with azathioprine.
    Arch. Dermatol., 125:1377-1379.

Norris, P.G., Limb, G.A., Hamblin, A.S., and Hawk, J.L.M., 1988, Impairment of natural killer cell activity in xeroderma pigmentosum. <u>New England Journal of Medicine,</u> 319: 1668-9.

Norris, P.G., Morris, J., McGibbon, D., Chu, A.C. and Hawk, J.L.M., 1989a, Polymorphic light eruption: an immunopathological study of evolving lesions. <u>Br. J. Dermatol.</u>, 120: 173-84.

Norris, P.G. and Hawk, J.L.M., 1990 Chronic actinic dermatitis - a unifying concept, <u>Arch. Dermatol.</u>, 126:376-378.

Norris, P.G., Camp, R.D.R., and Hawk J.L.M., 1989b, Actinic reticuloid: response to cyclosporin. <u>J. Am. Acad. Dermatol.</u>, 21: 307-309.

Norris, P.G., Nunn, A.V., Hawk, J.L.M. and Cox, T.M., 1989c, Analysis of the genetic defect in erythropoietic protoporphyria by measurement of lymphocyte zince chelatase. <u>Br. J. Dermatol.</u>, 120: 295-6.

Norris, P.G., Morris, J., Smith N.P., Chu, A.C., and Hawk, J.L.M., 1989, Chronic actinic dermatitis: an immunohistologic and photobiologic study. <u>J. Am. Acad. Dermatol.</u>, 21:966-971.

PERSISTENT PHOTOSENSITIVITY INDUCED BY INGESTED DRUGS

William Frain-Bell

Department of Dermatology
University of Dundee
Dundee, Scotland

INTRODUCTION

The ability of an increasing number of drugs which are
commonly used in medical therapeutics, such as the
phenothiazines, the sulphonamides, the tetracyclines, and the
non-steroidal anti-inflammatory agents (NSAIDs) to cause
cutaneous photosensitivity as an unwanted side effect has led
to the study of the clinical features and the course of such
reactions, and to the identification by phototesting of the
responsible wavelengths of light. However, abnormal reactions
of the exposed skin are caused by a variety of environmental
factors, only one of which is 'light', and it is therefore
often necessary to confirm by phototesting the diagnosis of
such cutaneous photosensitivity suspected on clinical grounds.
The technique of phototesting allows for the exposure of the
skin to measured amounts of artificial irradiation with a
series of wavebands throughout the sunlight spectrum, the
resultant reactions being compared with those found in normal
non-photosensitive subjects similarly exposed. Such an
investigation also identifies the responsible wavelengths,
information which is useful not only in indicating the
specific photodermatosis, but in the case of it being drug
induced, which drugs are likely to be responsible.

The diagnosis of drug induced photosensitivity is based
on the recognition of the general clinical features which
suggest the presence of photosensitivity (Frain-Bell, 1986) to
be followed by the selection of the specific form of
photosensitivity present on the basis of morphological and
other features. Most commonly these morphological changes are
those of dusky erythema with variable oedema, scaling and even
blistering, leading to pigmentation which may be persistent.
With some photoactive drugs the symptoms of the
photosensitivity response, for example burning, pain, and
irritation, may overshadow the signs, which was the case with
benoxaprofen, and simulates that seen in erythropoietic
protoporphyria. In this way drug induced photosensitivity

rather than one of the other photodermatoses can be identified
with reasonable certainty. The improved understanding of the
various forms of abnormal photosensitivity in the human has
been mainly due to the development of reliable phototesting
techniques. These techniques have also facilitated the study
of the natural progression of a variety of drug
photosensitivity reactions both during the period of
administration of the drug and subsequently. We therefore
know that with some drugs, and in some individuals, the
photosensitivity may continue for a variable period after the
drug has been withdrawn but at most for a period of months and
not years (Tromovitch and Jacobs 1963; Ferguson, et al., 1986:
Addo et al., 1987; Ferguson et al., 1987). It has been
suggested, however, that one of the non-steroidal anti-
inflammatory durgs (NSAIDs), benoxaprofen, was different in
that it could not only cause cutaneous photosensitivity during
the period of administration of the drug but that in some
instances this photosensitivity could persist for years after
the drug had been withdrawn (Sneddon 1986), i.e. that a state
of persistent light reaction had developed which did not
require further exposure to the original causal drug for its
maintenance. Such a sequence of events had only been
previously noted when the causal photoactive substance had
reached the skin by external contact.

Thus, before considering the possiblity of persistent
photosensitivity following the withdrawal of a systemically
administered drug, it is useful to look at the case which has
been made in the past for this occurring when the exposure to
the photoactive substance has been through external skin
contact. It has been suggested, ever since the epidemic of
the early 1960's of contact and photocontact dermatitis to the
halogenated salicylanilides (Wilkinson 1961), that some of
those affected through external exposure to such substances
continued to react abnormally to light for many years
thereafter, i.e. that they had developed a state of persistent
photosensitivity although contact with the causal
photosensitiser had ceased and that a state of so-called
'persistent light reaction' now existed. The assumpton being
that the original salicylanilide induced photosensitivity
reaction had changed the skin in such a way that it (Kochevar
and Harber 1977; Kochevar 1979) continued to have the
potential for reacting abnormally to light without further
exposure to the primary photosensitiser. This hypothesis,
however, remains to be proved. There is the alternative
explanation that the persisting photosensitivity is due to
continued undetected exposure to the original photoactive
substance and/or allied to other photoactive substances and
that perhaps it was also more liable to occur in skin which is
affected by a long-term chronic inflammatory response at least
in part due to cell mediated contact allergy (Frain-Bell and
Johnson 1979). Over the years since the 1960s the substances
incriminated in this production of persistent photosensitivity
are known to be both contact allergens as well as
photosensitisers (Frain-Bell and Johnson 1979; Addo et al.,
1985), fragrance materials (Addo, et al., 1982), oak moss
(Thune, 1977), musk ambrette (Wojnarowska and Calnan 1985),
and the study of such patients with chronic photosensitivity
dermatitis shows positive patch test evidence for contact
allergic sensitivity to a variety of substances including
those suspected as being responsible also for the photocontact

reaction, i.e. for the photosensitivity. The identification of these substances as photosensitisers is based on the demonstration in the laboratory of their photoactivity (Johnson et al., 1986) and in the patient by positive photopatch test reactions (Addo, et al., 1985). Such a reaction may be mediated through either a phototoxic or a photoallergic mechanism. There is most evidence for the former.

It is perhaps not surprising that with the view being held by some that persistent photosensitivity of a number of years duration can be a sequel of photosensitivity induced as a result of external contact with a photosensitiser, that it might equally as well occur when the photoactive substance has reached the skin following ingestion as with the administration of a therapeutic drug such as benoxaprofen. It is recognised in some individuals, and with some drugs more than others, that cutaneous photosensitivity can persist after the ingested drug has been discontinued but that when it does occur it is only for a period of months at the most, the assessment being based on the continuation of the abnormal clinical features along with the abnormal action spectrum as demonstrated by phototesting (Addo et al., 1987, Ferguson et al., 1986, 1987). It is a reasonable assumption that the amounts of the drug sufficient to maintain the phototoxic response are present for longer in some individuals and with some drugs because of the way that a similar dose regime is individually handled as well as the pharmcokinetics of the drug. The two reports which have suggested otherwise, i.e. persistent photosensitivity of some years duration, have unfortunately been incomplete in that either phototesting (McKerrow and Greig 1986) and/or alternative forms of cutaneous photosensitivity such as photosensitivity dermatitis have not been satisfactorily ruled out (McKerrow and Greig 1986; Robinson et al., 1985).

The criteria for the diagnosis of persistent photosensitivity of the type previously reported as being due to external exposure to a photoactive substance was the repeated demonstration of an abnormal action spectrum on phototesting which usually involved a broad spectrum including the whole of the UV waveband and in some instances also into the visible light (Frain-Bell 1986; Ive et al., 1969; Frain-Bell, et al., 1974), also that this abnormal action spectrum remained over the years, as did the chronic inflammatory reaction of the skin often with acute and subacute episodes. Thus, one would expect such a persistent abnormal action spectrum to be also present when persistent photosensitivity has been postulated, this time however when the photoactive substance reached the skin following ingestion, i.e. after a systemically administered drug.

To establish whether such persisitent photosensitivity occurs after the withdrawal of an ingested photoactive drug requires the demonstration of the clinical features of cutaneous photosensitivity plus an abnormal action spectrum on phototesting. It is rarely possible to obtain such information in all of those originally diagnosed because of problems inherent in any long-term follow up study involving patients. I would like however to describe the experience of the Dundee unit over the last few years with regard to drugs such as the thiazides, quinine, tetracylines, phenothiazines, amiodarone, and in particular the NSAID benoxaprofen at one time used for the treatment of arthritis before it was withdrawn in 1982 because of unwanted side effects, the most common being that of cutaneous photosensitivity. With this latter drug it was possible to study a group of subjects first in 1982 while taking benoxaprofen, and again in 1985, i.e. 3 years after the withdrawal of the drug, and also a separate group which gave a retrospective history of continuing photosensitivity since 1982 but who had not been studied originally in 1982.

All of the subjects studied, with the exception of the benoxaprofen group, had been referred routinely to the unit in Dundee for investigation of photosensitivity suspected on clinical grounds, and were subsequently identified as suffering from drug induced photosensitivity based on history, clinical features, and results of phototesting. The follow up assessment consisted in the recording of the continuing presence or otherwise of evidence for abnormal reactions to light and/or abnormal phototest results. The benoxaprofen group of subjects differed in that some of them were first of all assessed in 1981 while under treatment with the drug and experiencing cutaneous photosensitivity as a side effect and later in 1985 (10 subjects), and a second separate group (42 subjects) who had not been assessed in either 1981 or 1985 but were studied in 1987 because of a retrospective history of continuing cutaneous photosensitivity of variable severity since the withdrawal of treatment with the drug in 1982; this latter group was unselected except that they were involved in litigation.

The techniques of phototesting used have been described elsewhere (Mackenzie and Frain-Bell 1973: Johnson and Mackenzie 1982; Frain-Bell 1986) and made use of both monochromator and solar simulator irradiation sources. They consisted of exposure of the back skin to a range of measured doses from a series of wavebands throughout the solar spectrum from 290 to 700 nm. Recordings were made of both symptoms (irritation, pricking, burning, pain, etc) and signs (erythema, urtication, etc) at the following time intervals after irradiation, 0, 5, 10, 15, 30 minutes and at 7, 24 and 48 hours. Any reactions were compared to those known to occur on the back skin of normal non-photosensitive subjects similarly exposed and in this way abnormality was detected, both for immediate and delayed reactions, and the responsible wavelengths identified. The programme for the benoxaprofen group was planned to follow closely that used for the phototesting of the original group of subjects studied in 1981 whilst under treatment with benoxaprofen which led to the definition of the action spectrum of the induced photosensitivity (Ferguson et al., 1982). It also covered the programmes routinely used in the unit for the diagnosis of suspect photosensitivity of all kinds (Frain-Bell 1986) including drug induced (Addo, 1987). It differed by the addition of a number of extra wavebands and a series of higher doses which had been shown to be relevant in the 1981 study (Table 1).

Table 1. Phototesting Programme

A Monochromator

Waveband ± 5 nm	305	310	315	320	325	330	340	360
Dose (mJ/cm²)	8.2	39*	100*	100*	100*	100*		
	22	100*	220*	220*	220*	220*		
	56*	220*	470*	470*	470*	470*	470*	470*
	150* **	470* **	1200* **	1200*	1200*	1200*	1200*	1200*
	390* **	820* **		2200* **	2200*	2200*	2200*	2200*
	820* **							

± 30 n Dose (mJ/cm²)	305	310	315	320
	335	365	400	430
	470	1200	4700	10000
	1200	2700	10000	22000
	3300	5600	22000	47000
	6800*	12000	47000	82000

B Solar Simulator

Waveband Dose (J/cm²)	290 – 700nm	320 – 700
	5	20
	10	40
	15	70
	20*	100
	25* **	170
	30* **	200*

Legend

The minimal response dose (MRD) for immediate and delayed erythema and for urtication or abnormal sensations was recorded on day 1, and on day 2 the interval between the minimal response dose and the neighbouring no response dose was filled in by a further range of doses so as to determine the MRD as accurately as possible. Also on day 2 if no abnormal immediate responses had been noted with the highest doses at wavebands 315, 320 and 325 nm, the skin was then irradiated with the following single doses at these wavebands – 315 nm (2500 mJ/cm²), 320 nm (2850 mJ/cm²), 325 nm (2500 mJ/cm²).

The wavebands and dosages marked with an asterisk were additional to those used in the routine diagnostic phototesting programme and were included for comparison with those which had been shown to produce abnormal reactions in the 1981 study(?). A few of the highest dosages, those marked **, were not used in all cases as explained in the text.

Filters

- WG305 used with wavebands 360, 335 and 365 nm
- WG345 used with wavebands 400, 430 nm
- Copper sulphate and WG305 used with solar simulator wavebands
- 290 – 700 nm and WG345 used with waveband 320 – 700 nm

RESULTS

Unfortunately, it was not possible, because of the limitations of long-term follow up in patients, who were often elderly and referred from other parts of the country, to obtain clear cut data as to continuing photosensitivity after the withdrawal of the original responsible drug. However, in those in whom such a long-term assessment was possible in none did it persit beyond a maximum of 6 months and was usually shorter, being a matter of a few months at the most (Ferguson, et al., 1986, 1987). Examples of such data can be seen in Table 2 which deals with the follow up of the 42 subjects with thiazide induced photosensitivity. However, the results of the benoxaprofen study of the subjects not previously assessed in 1981 and 1985, were more clear cut. Tables 3 and 4 detail the results of phototesting both for immediate and delayed responses. These showed that in the majority, 31 out of 41 subjects (No. 21 was not tested) the results were normal. Of the remaining 10 in whom some abnormality was detected, in 2 (No. 20 and 24) it was confined to some discomfort of the skin which was not reproducible, and No. 35 suffered from a separate photosensitivity dermatitis. Of the remaining 7 subjects abnormal immediate or delayed erythema was recorded at a single (No 7, 16, 33 and 42) or at multiple wavebands (No. 11, 19 and 29), however, all of these with the exception of case No. 11 and 19 were being treated at the time of testing with other potentially photoactive drugs such as fenbufen (case No. 7), hydrochlorthiazide (case no. 16), mefenamic acid (case No. 33) and case no. 42 was taking both frusemide and ibuprofen and case no. 29 both tiaprofenic acid and hydrochlorthiazide. Thus there was an explanation for the abnormal findings in 5 of these remaining 7 subjects the exceptions being case No. 11 and 19, further assessment of whom was not possible in 1987 for a variety of reasons.

DISCUSSION

Despite the limitations of any long-term follow up assessment, the experience of the unit in Dundee (2,723 patients studied during the period 1970-1989) has been that

Table 2.   Thiazide Photosensitivity

No. of subjects		42
Loss of follow up or otherwise incomplete assessment		15
Drug continued		
Clear	12	
Improved	2	(one subject was also taking ketoprofen and the other suffered from polymorphic light eruption)
Unchanged	7	(5 subjects also suffered from photosensitivity dermatitis and 2 from polymorphic light eruption)

Table 3. Minimal immediate erythema dose estimations (Benoxaprofen subjects)

Case No.

	Waveband 305 ±5nm	310 ±5nm	315 ±5nm	320 ±5nm	325 ±5nm	330 ±5nm	340 ±5nm	360 ±5nm	335 ±30nm	365 ±30nm	400 ±30nm	430 ±30nm	290-700	320-700
7	>47	NT	NT	>2200	NT	NT	NT	NT	3300	>12000	>47000	>82000	>10 J	>200 J
11	390	820	2500	>2850	>2500	>2200	>2200	>2200	6800	>12000	27000	82000	>30 J	>170 J
16	>47	NT	NT	NT	NT	NT	NT	NT	4700	>12000	>47000	>82000	NT	NT
19	820	>820	2500	>2850	>2500	>2200	>2200	>2200	4700	>12000	>47000	>82000	30 J	100 J
33	>56	>220	>470	>1200	>2200	>2200	>2200	>2200	>6800	>12000	27000	>82000	>20 J	>200 J
All others	>820	>820	>2500	>2850	>2500	>2200	>2200	>2200	>6800	>12000	>47000	>82000	>30 J	>200 J
Normal values (lowest normal)	820	>820	>1200	>2200	>2200	>2200	>2200	>4700	>6800	12000	47000	82000	30 J	47 J

NT = Not tested

Immediate erythema is defined as erythema lasting longer than 5 mintes

J = Joules, otherwise doses are in mJ/Cm^2

321

Table 4   Minimal delayed erythema dose estimations (Benoxaprofen subjects)

Doses are in millijoules (mJ/cm²) except for wavebands 290 – 700 and 300 – 700 nm where they are in Joules (J/cm²).

Case No.	305 ± 5nm	310 ± 5nm	315 ± 5nm	320 ± 5nm	325 ± 5nm	330 ± 5nm	340 ± 5 nm	360 ± 5 nm	335 ± 30 nm	365 ± 30 nm	400 ± 30 nm	430 ± 30 nm	290-700 nm	320-700 nm
1	150	·470	1200	2850	>2500	>2200	>2200	>2200	>6800	>12000	>47000	>82000	15 J	200 J
2	120	470	2500	>2850	>2500	>2200	>2200	>2200	>6800	>12000	>47000	>82000	20	>200
3	56	220	470	1200	2200	1200	>2200	>2200	>6800	>12000	>47000	>82000	10	200
4	56	470	1200	>2850	>2500	>2200	>2200	>2200	>6800	>12000	>47000	>82000	8	200
5	>56	180	>470	>1200	>2200	>2200	>2200	>2200	3300	>12000	>47000	>82000	20	>200
6	150	470	>2500	>2500	>2200	>2200	>2200	>2200	>6800	>12000	>47000	>82000	20	>200
7	47	NT	NT	>2200	NT	NT	NT	NT	4700	>12000	>47000	>82000	>10	>200
8	150	470	>1200	>1200	>2200	>2200	>2200	>2200	6800	>12000	>47000	>82000	20	>200
9	68	180	820	>2850	>2500	>2200	>2200	>2200	6800	>12000	>47000	>82000	15	>200
10	150	470	2500	>2850	>2500	>2200	>2200	>2200	>6800	>12000	>47000	>82000	24	>200
11	56	180	470	>2200	>2500	>2200	>2200	>2200	4700	>12000	>47000	>82000	15	>170
12	82	NT	NT	NT	NT	NT	NT	NT	6800	>12000	>47000	>82000	15	>200
13	56	220	>470	>1200	>2200	>2200	>2200	>2200	6800	>12000	>47000	>82000	10	>200
14	150	470	1200	2200	>2500	>2200	>2200	>2200	6800	>12000	>47000	>82000	15	>200
15	150	470	2500	>2850	>2500	>2200	>2200	>2200	6800	>12000	>47000	>82000	20	>200
16	>47	NT	NT	NT	NT	NT	NT	NT	4700	>12000	>47000	>82000	NT	NT
17	>56	>220	>470	>1200	>2200	>2200	>2200	>2200	>6800	>12000	>47000	>82000	>20	>200
18	33	180	390	2850	>2500	>2200	>2200	>2200	6800	>12000	>47000	>82000	15	>200
19	56	180	1200	2850	>2500	>2200	>2200	>2200	4700	8200	33000	>82000	10	200
20	56	>220	>470	>1200	>2500	>2200	>2200	>2200	6800	>12000	>47000	>82000	15	>200
21	NT	NT	NT	NT	NT	NT	NT	NT	NT	NT	NT	NT	NT	NT
22	150	82	2500	>2850	>2500	>2200	>2200	>2200	>6800	>12000	>47000	>82000	NT	NT
23	82	300	1200	2200	>2500	>2200	>2200	>2200	>6800	>12000	>47000	>82000	20	200
24	39	220	680	>2850	>2500	>2200	>2200	>2200	6800	>12000	47000	>82000	7	>200
25	150	470	1200	>2850	>2200	>2200	>2200	>2200	>6800	>12000	>47000	>82000	15	>200
26	150	220	1200	2850	>2500	>2200	>2200	>2200	2200	10000	>47000	>82000	30	>200
27	47	820	2500	2200	>2500	>2200	>2200	>2200	>8200	>12000	>47000	>82000	20	>200
28	56	470	1200	2200	>2500	>2200	>2200	>2200	3300	12000	>47000	>82000	15	>200
29	56	220	470	680	470	680	>2200	>2200	6800	>12000	>47000	>82000	>20	>200
30	56	>220	>470	>1200	>2200	>2200	>2200	>2200	6800	>12000	>47000	>82000	15	>200
31	150	470	1200	>2850	>2500	>2200	>2200	>2200	6800	>12000	>47000	>82000	15	>200
32	>47	>220	>470	>1200	>2200	>2200	>2200	>2200	>6800	>12000	>47000	>82000	>20	>200
33	>56	>220	470	>1200	>2200	>2200	>2200	>2200	>6800	>12000	>47000	>82000	>20	>200
34	100	270	1200	>2850	>2500	>2200	>2200	>2200	>6800	>12000	>47000	>82000	20	>200
35	2.7	39	100	220	220	470	2200	>2200	100	2700	>47000	>82000	1	>200
36	150	220	2500	2850	2500	>2200	2200	>2200	>6800	>12000	>47000	>82000	10	>200
37	100	330	1000	>2850	>2500	>2200	2200	>2200	6800	>12000	>47000	>82000	20	>200
38	>56	>220	470	>1200	>2200	>2200	2200	>2200	>6800	>12000	>47000	>82000	>20	>200
39	>56	220	>470	>1200	>1200	>1500	>1200	NT	>6800	>12000	>47000	>82000	15	70
40	>56	150	>470	>1200	2200	>2200	>2200	>2200	4700	>12000	>47000	>82000	15	170
41	47	180	>470	>1200	2200	>2200	>2200	>2200	6800	>12000	>47000	>82000	10	>200
42	56	>220	>470	>2200	>2200	>2200	>2200	>2200	6800	12000	>47000	>82000	15	>200
Normal Values Range (mJ/cm²)	47-82	180-330	390->680	>1200	>2200	>2000	>2200	>2200	3300-6800	12000-22000	47000->47000	>82000	8.2-15J/cm²	100-200 J/cm²
Lowest Dose (mJ/cm²)	27	150	330	1000	1200	1200	1500	>2200	1800	8200	47000	>82000	7J/cm²	79J/cm²

NT = Not Tested

The range of normal values is that within which 70% of the normal subjects tested responded

although photosensitivity may persist after the withdrawal of an ingested drug, this is for a maximum period of some months and not for years and a reasonable assumption is that when it does persist this is probably due to the pharmacokinetics of the drug and the way that it is individually handled by the subject so affected. In such assessment it is also important to identify the presence of the administration of other photoactive drugs and also other photodermatoses, e.g. we have noted the administration of thiazides at various times in the natural history of 40 of a group of 243 subjects with chronic photosensitivity dermatitis (Addo, et al., 1987).

The benoxaprofen study, on the other hand, was more rewarding in that it provided results at 3 and 5 year intervals after the withdrawal of the drug and in particular in subjects with a positive retrospective history of clinical photosensitivity over the years since discontinuing the drug in 1982. In this group normal results for phototesting were found (31 out of 41 - 76%) and where they were abnormal the abnormality was not that of a broad spectrum of response but usually restricted to a single waveband and/or associated with the current administration of a photoactive drug. We had also in 1985 the opportunity in a group of 10 subjects who had previously been phototested in 1982 whilst taking benoxaprofen, to show by a study of the reaction at the 320 nm waveband, which was the peak of the abnormal response (Ferguson, et al., 1982), that they had returned to normal by 1985 (Frian-Bell 1989). Thus, with the exception of the few cases in whom the continuing photosensitivity was due to the presence of another photodermatosis, e.g. photosensitivity dermatitis in case No. 35, it would seem that in the absence of confirmation of abnormal findings by phototesting in 1987 that the abnormal reactions to light which the patients described over the years since the withdrawal of the benoxaprofen in 1982 were episodic and related to the administration of other known photoactive drugs and in particular other NDSAIDs in that they had been taking over long periods of time since 1982, often for years, one or more known photoactive drugs (Frain-Bell 1989).

It is difficult to determine from a study of the literature the incidence of cutaneous photosensitivity induced by the NSAID group of drugs, although there are reports of clinical photosensitivity resulting from the administration of piroxicam (Stern 1983; Halasz 1987; Fjellner 1983; Serrano et al., 1984; Figueiredo et al., 1987), naproxen (Howard, et al., 1985; Judd et al., 1986; Mayou and Black 1986; Shelley et al., 1986; Burns 1987), azapropazone (Olsson, et al., 1985), and two studies which reported the induction of photosensitivity as demonstrated by abnormal phototest results after the administration of piroxicam and azapropazone (Diffey et al., 1983; Diffey et al., 1986). There is probably also a number of reasons why it is difficult to determine the frequency of drug induced photosensitivity in general and from the NSAIDs in particular, in that those affected, because of the similarity sometimes of the response of the skin to that of normal sunburn, may not seek medical advice and also with the reaction often being episodic may find that they can deal with it by avoiding excessive exposure to sunshine and using appropriate topical photoprotection. These drugs, including benoxaprofen, would appear to act through a phototoxic mechanism, the characteristic of which is that it is limited to a defined period of time when there is sufficient of the drug present in the skin. There is no evidence that such photosensitivity based on a phototoxic mechanism can lead to photosensitivity persisting for years after withdrawal of the drug in the same way as has been reported when the skin has been exposed to the photoactive substance by external contact as in the photocontact reaction and the subsequent development of a state of "persistent light reaction" as seen in the photosensitivity dermatitis and actinic reticuloid syndrome. In this latter situation involvement of immunological mechanisms has been proposed whereas there is little, if any, evidence for the involvement of such mechanisms, i.e. 'photoallergy', in the photosensitivity reactions in the human skin as the result of the systemic administration of drugs, although it has been shown that it is possible to induce photoallergy in mice with both sulphonamide and chlorpromazine (Guidici and Maquire 1985). It is probable therefore that in the subjects studied a phototoxic mechanism was once again involved, as was the case with the original benoxaprofen induced reactions and is the mechanism implicated in most of the recorded drug photosensitivity reactions. Although such a phototoxic reaction will occur in any person, provided there is present in the skin adequate amounts of the photoactive drug. This amount will vary in different subjects despite the same dosage regime depending on the way the dose is individually handled and between one drug and the next depending on the pharmacokinetics.

It would appear therefore that photosensitivity induced by ingested drugs may in certain circumstances persist for some months after the drug has been discontinued. However, that the state of persistent light reaction, i.e. persistent phototoxicity of some years duration established as a feature of photosensitivity induced by external exposure to a photoactive substance, does not occur when the drug has been administered systemically and later withdrawn; and that in situations where patients who have suffered from photosensitivity induced by a particular drug continue to have problems with abnormal reactions to light in subsequent years, alternative reasons, especially the taking of other photoactive drugs should be considered as the probable cause.

REFERENCES

Addo, H.A., Ferguson, J. and Frain-Bell, W., 1987, Thiazide-induced photosensitivity: a study of 33 subjects, Br. J. Dermatol., 116: 749.

Addo, H.A., Sharma, S.C., Ferguson, J., Johnson, B.E. and Frain-Bell, W., 1985, A study of Compositae plant extract reactions in photosensitivity dermatitis, Photodermatol., 2: 68.

Addo, H.A., Ferguson, J., Johnson, B.E. and Frain Bell, W., 1982, The relationship between exposure to fragrance materials and persistent light reaction in the photosensitivity dermatitis with actinic retinculoid syndrome, Br. J. Dermatol., 107: 261.

Addo, H.A., Hosie, J. and Frain-Bell, W., 1985, Unpublished data.

Burns, D.A., 1987, Naproxen pseudoporphyria in a patient with vitiligo. Clin. Exp. Dermatol., 12: 296.

Diffey, B.L., Daymond, T.J. and Fairgreaves, H., 1983, Phototoxic Reactions to Prioxicam, Naproxen and Tiaprofenic Acid, Br. J. Rheumatol., 22: 239.

Diffey, B.L., Pal, B. andRobson, J., 1986, Azapropazone therapy and photosensitivity, Photodermatol., 3: 304.

Ferguson, J., Addo, H.A., Jones, S., Johnson, B.E. and Frain-Bell, W., 1986, A study of cutaneous photosensitivity induced by amiodarone, Br. J. Dermatol., 113:537.

Ferguson, J., Addo, H.A., Johnson, B.E. and frain-Bell, W., 1987, Quinine induced photosensitivity: clinical and experimental studies, Br. J. Dermatol., 117: 631.

Ferguson, J., Addo, H.A., McGill, P.E., Woodcock, K.R., Johnson, B.E. and Frain-Bell, W., 1982, A study of benoxaprofen induced photosensitivity, Br. J. Dermatol., 107: 429.

Figueiredo, A., Ribeiro, C.A., Goncalo, S., Caldeira, M.M., Poiares-Baptista, A., Teixeira, F., 1987, Piroxicam-induced photosensitivity, Contact Dermatitis, 17 (2): 73.

Fjellner, B.O., 1983, Photosensitivity induced by piroxicam. Acta. Derm-venereol. (Stockholm), 63: 557.

Frain-Bell, W., 1986, Cutaneous Photobiology. Oxford University Press. Oxford.

Frain-Bell, W. and Johnson, B.E., 1979, Contact allergic sensitivity to plants and the photosensitivity dermatitis and actinic reticuloid syndrome, Br. J. Dermatol., 101: 503.

Frain-Bell, W., Lakshmipathi, T., Rogers, J., Wilcock, J, 1974, The syndrome of chronic photosensitivity dermatitis and actinic reticuloid. Br. J. Dermatol., 91: 617.

Frain-Bell, W., 1989, A study of persistent photosensitivity as a sequel of the prior administration of the drug benoxaprofen. Br. J. Dermatol., 121: 551.

Guidici, P.A. and Maguire, H.C., 1985, Experimental Photoallergy to Systemic Drugs, J. Invest. Dermatol., 85: 207.

Halasz, C.L., 1987, Photosensitivity to the Nonsteroidal Anti-inflammatory Drug Piroxicam, Cutis, 339: 37.

Howard, A.M., Dowling, J. and Varigos, G., 1985, Pseudoporphyria due to Naproxen, Lancet, ii: 819.

Ive, F.A., Magnus, I.A., Warin, R.P. and Wilson Jones, E., 1969, Actinic Reticuloid: a chroniic dermatosis associated with severe photosensitivity and the histological resemblence to lymphoma, Br. J. Dermatol., 81: 469.

Johnson, B.E. and Mackenzie, L.A., 1982, Techniques used in the study of the photodermatoses, Seminars Dermatol., 1: 217.

Johnson, B.E., Walker, E.M. and hetherington, A.M., 1986, In Vitro Models for Cutaneous Phototoxicity, in: "Skin Models", R. Marks andG. Plewig eds., Springer Verlag.

Judd, L.E., Henderson, D.W. and Hill, D.C., 1986, Naproxen-Induced Pseudoporphyria, Arch. Dermatol., 122: 451.

Kochevar, I. and Harber, L.C., 1977, Photoreactions of 3,3',4',5-tetrachlorosalicylanilide with proteins, J. Invest. Dermatol., 68: 151.

Kochevar, I., 1979, Photoallergic responses to chemicals, Photochem. Photobiol., 30: 437.

Mackenzie, L.A. and Frain-Bell, W. 1973, The construction and development of a grating monochromator and its application to the study of the reaction of the skin to light, Br. J. Dermatol., 89: 251.

McKerrow, K.J. and Greig, D.E., 1986, Piroxicam-induced photosensitive dermatitis, J. Am. Acad. Dermatol., 15: 1237.

Mayou, S. and Black, M.M., 1986, Pseudoporphyria due to naproxen, Br. J. Dermatol., 114: 519.

Olsson, S., Biriell, C. and Boman, G., 1985, Photosensitivity during treatment with azapropazone, Br. Med. J., 2291: 99939.

Robinson, H.N., Morison, W.L. and Hood, A.F., 1985, Thiaziade Diuretic Therapy and Chronic Photosensitivity, Arch. Dermatol., 121: 522.

Serrano, G., Bonillo, J., Aliaga, A., Gargallo, E. and Pelufo, C., 1984, Piroxicam-induced photosensitivity, _J. Am. Acad. Dermatol._, 11: 113.

Shelley, W.B., Elpern, D.J. and Shelley, E.D., 1986 Naproxen Photosensitization Demonstrated by Challenge, _Cutis_, 38:169.

Sneddon, I.B., 1986, Persistent photoxicity after benoxaprofen, _Br. J. Dermatol._, 115: 515.

Stern, R.S., 1983, Phototoxic reactions to prioxicam and other nonsteroidal antiinflammatory agents, _New Eng. J. Med._, 309: 186.

Tromovitch, T.W. and Jacobs, P.H., 1963, Photosensitivity to Oxytetracycline, _Ann. Int. Med._, 58: 529.

Thune, P., 1977, Contact allergy due to lichens in patients with a history of photosensitivity, _Contact Dermatitis_, 3: 267.

Wilkinson, D.S., 1961, Photodermatitis due to tetrachlorsalicylanilide, _Br. J. Dermatol._, 73: 213.

Wojnarowska, F. and Calnan, C.D., 1985, Contact and photocontact allergy to musk ambrette, _Br. J. Dermatol._, 114: 667.

DNA REPAIR AND LIGHT SENSITIVITY IN DERMATOLOGY

Irene Horkay, Laszlo Varga[1], Hans Altmann[2]
and Agnes Kosa

Department of Dermatology, Univ. Med. School
H-4012 Debrecen
[1]F. J. Curie Nat. Res. Inst. for Radiobiol. and
Radiohyg., H-Budapest 22, Hungary
[2]Inst. of Biol., Research Centre, A-2444
Seibersdorf, Austria

This paper presents a short review of the present
knowledge concerning the photodermatological aspects of DNA
damage and repair after UV radiation and of our own
investigations in this field.

UV energy may be absorbed by many chromophores at various
levels of human skin. Most of the available evidence suggests
that nuclear DNA of the epidermis is the main and probably the
initial target for UV radiation (Tyrrell, 1978). First of all
pyrimidine dimers are produced in high yield after exposure of
the skin to UV (Pathak et al., 1972) but other photoproducts
are also formed. As for model lesions in DNA, the pyrimidine
dimers, three principal molecular mechanisms for dealing with
them - the excision and the postreplication repair, and the
photoreactivation - have now been well documented in human
cells as well (Hönigsmann et al., 1980).

The observations of Cleaver (1968) on xeroderma
pigmentosum (XP), a rare genetic skin disorder, were the first
indications of defective DNA repair in association with human
disease. Since that time the defect has been implicated in UV
carcinogenesis, mutagenesis and ageing (Brash and Hart, 1978).
Wide-ranging investigations have explored its genetic
heterogeneity and its relation to clinical features
(complementation groups in XP, ataxia telangiectasia, Cockayne
syndrome, etc.) It was also Cleaver (1970) who supposed that
defective repair is one of the reasons of photosensitivity.
Several findings have called attention to the immunological
aspects of the repair processes (Slor et al., 1976).

Autoradiographic and biochemical techniques have shown
that besides the DNA repair processes the semiconservative DNA

as well as RNA and protein synthesis and mitosis are inhibited within the first hour post irradiation. Recovery of DNA synthesis is followed by an accelerated activity eventually leading to epidermal hyperplasia (Epstein et al., 1969).

A wealth of information on these problems in XP and, to a lesser extent in other light sensitive disorders, has accumulated in the literature. Owing to the subtle details detected in the molecular and cellular reactions to UV radiation, correlations of UV light sensitivity, types of repair deficiencies and other abnormal reactions to UV light as well as predisposition to malignancies are rather complicated. Moreover, the findings in the literature are occasionally conflicting. Table 1 attempts to summarize the most important data available in this field.

One group of skin disorders displays intense light sensitivity and a defective repair mostly accompanied by a predisposition to malignancies. Of them all, XP has been most extensively studied clinically and experimentally. Its complementation groups connected with various types and severity of the clinical symptoms represent different levels of the defect in excision repair and post-replication repair, respectively (Fischer et al., 1982). In actinic keratosis, an epidermal cancer in situ both the peripheral lymphocytes (Lambert et al., 1976) and the fibroblasts (Sbano et al., 1978) display a reduced DNA repair synthesis which may be one of the important factors in the etiology of the disease. The published data concerning excision repair in lupus erythematosus precipitated by sunlight are conflicting. Cleaver (1970) demonstrated normal repair DNA synthesis in cultured fibroblasts of patients whereas Altmann et al. (1971) found an altered repair capacity in the bone marrow cells and the peripheral lymphocytes, parallel to an increased semiconservative DNA synthesis. The latter is considered to be in closely connected to the production of anti-DNA antibodies. In addition, Beighlie and Teplitz (1975) demonstrated evidence of enzymatic defects in the DNA repair in this disease. In trichothiodystrophy, an autosomal recessive disorder, characterized by brittle hair with reduced sulphur content, ichthyosis, peculiar face and mental and physical retardation, occasionally photosensitivity, Lehmann and Arlett (1988) found a variety of different responses in the detailed molecular and cellular study of the UV light effects (survival, excision repair, etc.) The heterogeneous molecular defect detected poses a number of questions about the relationship between the defective DNA repair and the clinical symptoms of XP and this disorder.

In the other group of light sensitive skin diseases the excision repair is normal but the cells show other abnormalities in their reaction to UV radiation. The photosensitivity in Cockayne syndrome characterized by a progeroid appearance, striking dwarfism and other physical and neurological abnormalities correlates with an increased sensitivity of fibroblasts to UV light when colony-forming ability is measured. Although the cells have normal levels of repair synthesis they do have an abnormally slow recovery of overall semiconservative DNA synthesis (Timme and Moses, 1988). Cells from Bloom syndrome with telangiectatic erythema, characteristic butterfly lesions of the face and

Table 1 Correlation of UV light sensitivity, predisposition to malignancies and repair deficiencies.

Disorder	Light sensitivity	Malignancies	Defective repair or other abnormal cellular reaction to UV rad.	Repair defect to X-ray ionizing rad. etc.
Xeroderma pigmentosum	+	+	+	-
Actinic keratosis	+	+	+	-
Lupus erythematosus	+	+	+	-
Trichothio-dystrophy	+	?	±	-
Morbus Darier	+	-	+	-
Cockayne syndrome	+	+	+	-
Bloom syndr.	+	+	+	-
Rothmund-Thomson syndr.	+	+	+	-
Melanoma mal.	+	+	-	-
Basal cell nevus syndr.	+	+	-	-
Psoriasis	+	-	-	-
Rosacea	+	-	-	-
Epidermodysplasia verruciformis	-	+	±	-
Dyskeratosis congenita	-	+	+	-
Ataxia telangiectasia	-	+	-	+
Progeria	-	-	-	+

photosensitivity display a temporal perturbation of DNA repair processes. Lymphoid cell lines from the patients have one-half to one-quarter the normal levels of DNA ligase I (Willis and Lindahl, 1987) and an increased frequency of sister chromatid exchanges. In Rothmund-Thomson syndrome with an autosomal recessive mode of inheritance and high incidence of tumours UV excision repair proved to be normal (Cleaver, 1970)

but the reconstitution of supercoiled DNA after UV radiation
was delayed (Altmann, 1980).

In a series of heterogeneous skin diseases in a certain
connection with sunlight the study on repair deficiencies has
yielded negative results independently of whether they are
predisposed to malignancies (basal cell nevus syndrome,
melanoma malignum - Friedberg et al., 1979) or not (psoriasis,
rosacea - Nunzi et al., 1981).

Some cutaneous disorders fail to show light sensitivity
but are characterized by a defective excision repair
(epidermodyplasia verruciformis - Hammar et al., 1976) or by
other abnormal cellular responses to UV light (dyskeratosis
congenita - Cleaver, 1970). The next group of such dermatoses
displays a normal excision repair though an abnormal one to
ionizing or X-ray radiation (ataxia telangiectasia, progeria,
etc - Friedberg et al., 1979). To make the list complete,
there are disorders with defective repair processes and
predisposition to malignancies without any signs of cutaneous
involvement (Fanconi anemia, chronic lymphatic leukemia, etc.
Friedberg et al. 1979). Thus it is apparent from a review of
literature that increasing attention is being directed toward
the possible roles that defective or altered DNA repair may
play in a number of human diseases.

Turning to the photodermatoses it has been briefly
mentioned that the correlation between the absence of DNA
repair and symptoms of extreme UV sensitivity in XP does
suggest that DNA repair is important in this respect (Cleaver,
1970). It has also been shown that DNA radiated with UV light
is a potent immunogen (Tan et al., 1969). Its antigen
determinant are supposedly thymine dimers the most important
DNA photoproducts. Davis (1977) studying these antibodies in
the sera of patients showed that they were detectable in
systemic lupus erythematosus (SLE) but not in the discoid form
(DLE) and in a variety of non-defined photodermatoses. In
addition, Slor et al. (1976) could find such antibodies in XP
but not in XP variant and in SLE. They presumed that the
photoproducts in the DNA of XP patients persist for a
relatively long period because of their defective DNA repair
system. If these cells die they will lyse and their UV-
damaged DNA will be introduced into the immune system,
triggering the production of anti UV-DNA antibodies. The fact
that antibodies can be formed to UV altered DNA leads one to
consider this mechanism in the photosensitivity diseases of
man which have not yet been satisfactorily elucidated (Tan et
al., 1969).

As natural insolation is of great intensity and rather
rich in UV radiation in the eastern part of Hungary, a lot of
light dermatoses occur in the patients of the Dermatological
Department in Debrecen. Considering the published data we
have been studying some of their pathogenetic problems, among
others the development of light sensitivity, with a special
view to the connection with UV induced DNA repair processes.
The problem seemed to be of special interest in polymorphic
light eruption (PLE), a clinical entity characterized by
pleomorphic eruptions first of all on sun-exposed areas, which
is considered to be a manifestation of photoallergy, mostly of

delayed type. However, no endogenous or exogenous photosensitizer or antigen responsible for initiating the photosensitive reaction could be detected so far. Furthermore, there were no data on whether DNA photoproducts brought about by UV light could function as antigens. Neither have the UV-induced repair processes been studied in respect to whether their possible defect could explain the intense light sensitivity characteristic of the disease.

Summing up our study and its most important results collected over almost 20 years, we can say the following. Investigations were performed both in the skin and the peripheral lymphocytes of PLE patients and, as a comparison, in other light sensitive dermatoses: cutaneous porphyrias provoked phototoxically by a hereditary or acquired enzymatic disturbance in the porphyrin metabolism, then photocontact dermatitis, further on DLE of autoimmune pathogenesis, and finally XP as a model disease for defective repair.

As for the skin the intensity of the UV induced excision repair DNA synthesis and the percentage of the cells taking part in the process were normal both in PLE and photocontact dermatitis but somewhat lower in porphyric and DLE cases, and especially in XP. Changes in semiconservative DNA synthesis took place normally in the symptom-free skin of each group tested (a slight depression in the number of heavily labelled epidermal cells being in S phase 2 hours after UV radiation and a significant increase 48 hours later). In the epidermis of the active clinical eruptions of PLE and DLE, as well as in that of the repeated phototest site of PLE, the semiconservative DNA synthesis was significantly higher than in the unirradiated control sites, whereas in cutaneous porphyrias no difference could be measured (Horkay et al. 1978).

The study of the excision repair by autoradiographic and liquid scintillation techniques in the peripheral lymphocytes revealed that cells derived from PLE patients displayed in general a more or less reduced repair capacity (Fig. 1). The results could not be explained by UV induced lethal damage of the cells, and were independent of the clinical stage of the disease. No significant difference occurrred in cutaneous porphyrias and in DLE although in the latter the thymidine uptake of the cells in a few cases was somewhat lower than in the controls (Horkay et al., 1973; 1982). Later on, Jung et al. (1974) studying the excision repair of the lymphocytes in PLE also found a heterogeneity in the repair capacity of the patients.

As a conclusion we can say that changes after UV radiation both in the semiconservative and the repair DNA synthesis of the epidermal cells of PLE patients occur in a normal way. Thus the possibility that the original cause of photosensitivity is a defective repair in the skin can be excluded. Presumably the photo-dimers do not exist permanently in the DNA molecules of the epidermal cells so they cannot serve as a basis for initiating an immune response. But the so-called minor photoproducts or other metabolites quite independent of DNA may come up as antigens developed in the epidermis.

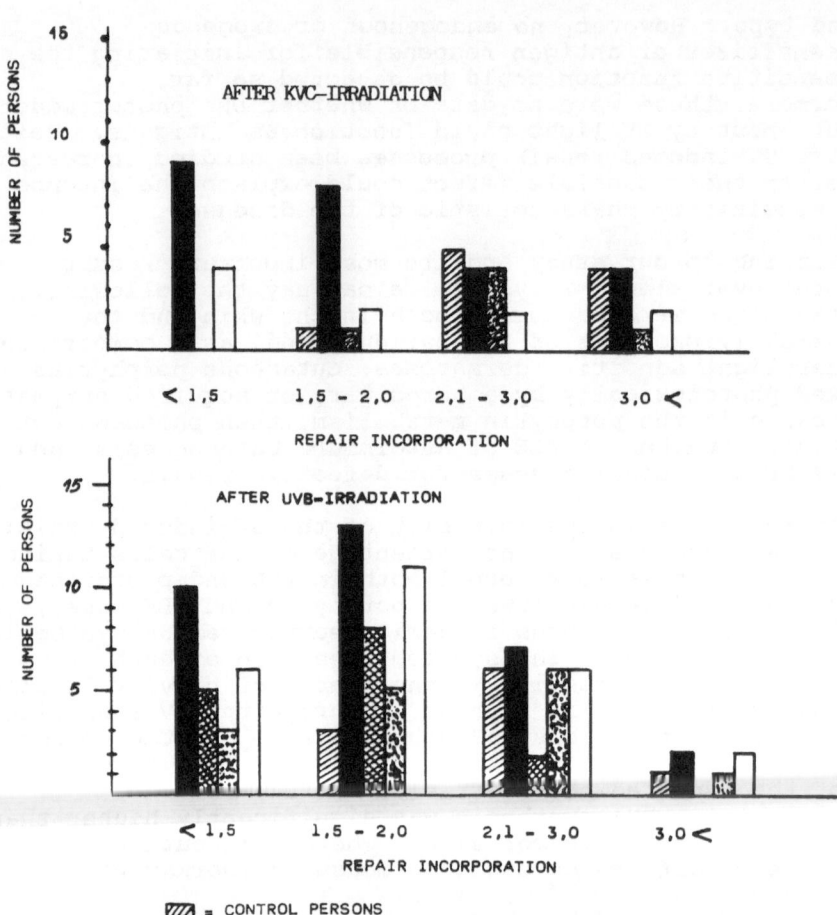

Fig. 1   Distribution of the repair incorporation in
peripheral lymphocytes 2 hours after UV-
irradiation.

As for the lymphocytes the next step of our study was to
find a possible explanation for the altered repair capacity of
these cells derived from PLE patients. On the basis of the
literature we decided to study the poly(ADP)ribose (PAR)
synthesis and some trace elements supposedly connected with
DNA synthesis. As for PAR, this unique homopolymeric chain in
the chromatin takes part not only in differentiation and
transformation processes of the cells but also in the
replicative and repair DNA synthesis (Klocker et al., 1983;
Altmann, 1983). An abrupt increase both in DNA repair and PAR
synthesis occurs in normal lymphocytes in response to
treatment with DNA damaging agents, among others UV radiation,
whereas cells from XP fail to show an increase in PAR
synthesis in response to UV light (Berger et al., 1980). As
for the trace elements, zinc is an important component of DNA
polymerase beta, one of the key enzymes of excision repair,
while copper ions especially bound to the chromatin are potent
inhibitors of the same enzyme (Assadian et al., 1982).

In our study basic and UV induced PAR synthesis in the lymphocytes as determined by Altmann's method (1983) using tritiated NAD. Contents of the trace elements in the chromatin of the cells were measured by neutron activation analysis. Mean values of the basic PAR synthesis of each group of patients did not differ significantly from those of the controls. The UV induced DNA damage of the cells was followed by a more or less increased PAR synthesis. The mean in PLE was significantly lower than in the controls. An even lower UV induced PAR synthesis was found in XP cells. On the other hand, in cutaneous porphyrias the stimulation of PAR synthesis after UV radiation was similar to the controls (Fig. 2). As for the individual values of PLE patients a considerable heterogeneity could be observed in the induction factor (i.e. difference between the UV light induced and basic PAR synthesis). As for the contents of trace elements studied in the chromatin of the lymphocytes, the mean zinc levels both in PLE and porphyrias did not differ from those of the controls. At the same time chromatin of XP patients displayed a very low ion concentration. The mean copper content was significantly higher than that of the controls not only in XP but in PLE as well, whereas in porphyrias no difference could be measured (Horkay et al., 1985).

How can we comment on these findings? Based on published data the results suggest that presumably a close connection exists between the excision repair and PAR synthesis in light-dermatoses as well. In cutaneous porphyrias a normal UV induced PAR synthesis occurs in the peripheral lymphocytes displaying normal repair synthesis whereas the cells derived from PLE are characterized by a reduced repair capacity and PAR synthesis after UV radiation. It can be presumed that in PLE the endonuclease incision, if there is any, near pyrimidine dimers occurs with some delay which does not result in any activation of PAR synthesis. Another possibility to explain our findings is an inhibition of PAR polymerase by the high copper content in the chromatin of the cells which may contribute to the reduced UV light induced excision repair and PAR synthesis. Such a relationship between the copper content, the enzyme activities and the excision repair processes has been observed in other diseases (Altmann et al.,

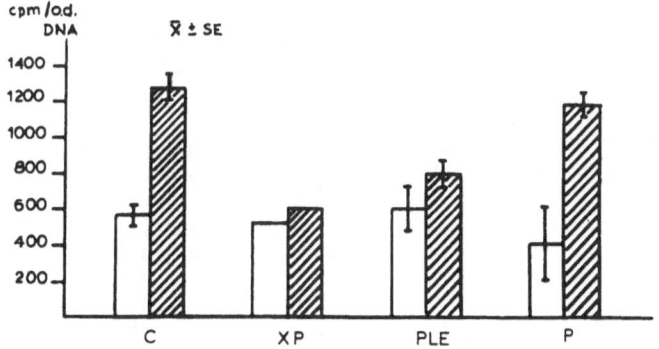

Fig. 2   Basic (blank) and UVC induced (shaded) PAR-sythesis in photodermatoses.

1983). The heterogeneity of the individual values of UV light
induced excision repair and PAR synthesis in PLE as well as
the occasional familial incidence of the disease e.g. among
American Indians (Birt and Davis, 1975) and also in our
patients suggest that PLE might not be one entity and might
have a genetic background.

The results presented suggest important correlations and
call attention to several interesting problems. But it can
also be seen that there are a number of unsolved questions.
Anyway it seems that the relationship of DNA damage and its
repair to human diseases may be wider, more complicated than
currently suspected. As a conclusion let us cite Frederick
Urbach (1976): "The biologically significant knowledge about
the interaction between photons and molecules may lead to a
better understanding of the mechanisms of DNA repair. Having
acquired a better understanding of these photochemical events
we can determine the mechanisms underlying some of the
photodermatoses."

REFERENCES

Altmann, H., 1983, Poly-(ADP-Ribose-)Synthese und
    Regulationsstörungen bei Erkrankungen, Wien. Klin.
    Wschr., 95:861.
Altmann, H. and Eberl, R., 1971, Untersuchungen über die DNS-
    Reparationskapazität von Knochenmarkzellen bei Lupus
    erythematodes, Wiss. Ztschr. Friedrich-Schiller-Univ.
    Jena, Math.-Nat. R., 20:477.
Altmann, H., 1980, DNA repair in diseases, in: "DNA repair and
    late effects", Altmann, H., Riklis, E., Slor, M. eds.,
    NRCN Israel.
Altmann, H., Teherani, D.K., Keck, M., Turanitz, K., Söltz-
    Szöts, S. and Syre, B., 1983, Untersuchungen zu
    somatisch-genetischen Veränderungen in Lymphozyten von
    Patienten mit Verrucae vulgares und Condylomata
    acuminata, Hautarzt Suppl. VI., 34:166.
Assadian, M.A., Teherani, D.K., Altmann, H. and Binder, W.,
    1982, Bestimmung von Zn, Cu und Mn im Chromatin aus
    bovinen Milch- und Blutlymphozyten, OEFZS Seibersdorf.
Beighlie, D.J. and Teplitz, R.L., 1975, Repair of damaged DNA
    in systemic lupus erythematodes, J. Rheumatol., 2:149.
Berger, N.A., Sikorski, G.W., Petzold, S.J. and Kurohara,
    K.K., 1980, Defective poly/adenosine diphosphoribose/
    synthesis in xeroderma pigmentosum, Biochemistry, 19:289.
Birt, A.R. and Davis, R.A., 1975, Hereditary polymorphic light
    eruption of American Indians, Internat. J. Derm., 14:105.
Brash, D.E. and Hart, R.W., 1978, DNA damage and repair in
    vivo, J. Envir. Path. Tox., 2:79.
Cleaver, J.E., 1968, Defective repair replication of DNA in
    xeroderma pigmentosum, Nature, 218:652.
Cleaver, J.E., 1970, DNA damage and repair in light-sensitive
    human skin disease, J. Invest. Dermatol., 54:181.
Davis, P., 1977, Antibodies to UV DNA and photosensitivity,
    Br. J. Derm., 97:197.
Epstein, W.L., Fukuyama, K. and Epstein, J.H., 1969, Early
    effects of ultraviolet light on DNA synthesis in human
    skin in vivo, Arch. Dermatol., 100:84.
Fischer, E., Thielmann, H.W., Neundörfer, B., Rentsch, F.J.,
    Edler, L. and Jung, E.G., 1982, Xeroderma pigmentosum

patients from Germany: clinical symptoms and DNA repair characteristics, <u>Arch. Dermatol. Res.</u>, 274:229.

Friedberg, E.C., Ehmann, U.K. and Williams, J. I., 1979, Human diseases associated with defective DNA repair, "Advances in Radiation Biology", Vol. 8. Academic Press, Inc.

Hammar, H., Hammar, L., Lambert, B. and Ringborg, U., 1976, A case report including EM and DNA repair investigations in a dermatosis associated with multiple skin cancers, epidermodysplasia verruciformis, <u>Acta. Med. Scand.</u>, 200:441.

Horkay, I., Tamasi, P. and Csongor, J., 1973, UV-light induced DNA damage and repair in lymphocytes in photodermatoses, <u>Acta. Dermatovener.</u> 53:105.

Horkay, I., Varga, L., Tamasi, P., and Gundy, S., 1978, Repair of DNA damage in light sensitive human skin diseases, <u>Arch. Derm. Res.</u>, 263:307.

Horkay, I., Varga, L. and Krajczar, J., 1982, UV-light induced DNA damage and repair in photodermatoses, <u>in</u>: "DNA-Reparatur und Chromatin", Altmann, H., Klein, G. eds., österreichisches Forschungszentrum Seibersdorf G.m.b.H.

Horkay, I., Teherani, D.K., Altmann, H. and Kosa, A., 1985, Determination of Cu, Zn and Mn by neutron activation analysis in the chromatin of lymphocytes from patients with skin diseases, <u>J. Radioanal. Nucl. Chem., Letters</u>, 93:81.

Hönigsmann, H., Jaenicke, K., Brenner, W., Rauschmeier, W., Gschnait, F. and Parrish, J.A., 1980, Kinetics of thymine dimer repair in normal human skin after single and combined doses of UV-A, UV-B and PUVA, <u>J. Invest. Dermatol.</u>, 74:458.

Jung, E.G. und Bohnert, E., 1974, Chronisch polymorphe Licht-dermatose, <u>Dermatologica.</u>, 148:209.

Klocker, H., Auer, B., Hirsch-Kauffmann, M., Altmann, H., Burtscher, H.J. and Schweiger, M., 1983, DNA repair dependent NAD$^+$ metabolism is impaired in cells from patients with Fanconi's anemia, <u>The EMBO Journal</u>, 2:303.

Lambert, B., Ringborg, U. and Swanbeck, G., 1976, Ultraviolet-induced DNA repair synthesis in lymphocytes from patients with actinic keratosis, <u>J. Invest. Derm.</u>, 67:594.

Lehmann, A.R. and Arlett, C.F., 1988, Trichothiodystrophy - a UV-sensitive disorder, <u>in</u>: DNA-Repair, Chromosome Alterations and Chromatin Structure under Environmental Pollutions, H. Altmann and G. Zasukhina, eds., österreichisches Forschungszentrum Seibersdorf G.m.b.H.

Nunzi, E., Rebora, A., Cornelis, J.J. and Cormane, R.H., 1981, UV-induced pyrimidine dimers and rosacea, <u>Br. J. Derm.</u>, 104:711.

Pathak, M.A., Krämer, D.M. and Güngerich, U., 1972, Formation of thymine dimers in mammalian skin by ultraviolet radiation <u>in vivo</u>, <u>Photochem. Photobiol.</u>, 15:177.

Sbano, E., Andreassi, L., Fimiani, M., Valentino, A. and Baiocchi, R., 1978, DNA-repair after UV-radiation in skin fibroblasts from patients with actinic keratosis, <u>Arch. Derm. Res.</u>, 262:55.

Slor, H., Nivy, S., Cleaver, J.E. and Friedberg, E.C., 1976, Anti-ultraviolet-irradiated DNA antibodies in Xeroderma pigmentosum patients, <u>in</u>: "DNA-repair and late effects", Altmann, H. ed., Rötzer-Druck, Eisenstadt.

Tan, E.M. and Stoughton, R.B., 1969, UV-light induced damage to deoxyribonucleic acid in human skin, <u>J. Invest. Derm.</u>, 52:537.

Timme, T.L. and Moses, R.E., 1988, Review: Diseases with DNA damage-processing defects, <u>Am. J. Med. Sci.</u>, 295:40.

Tyrrell, R.M., 1978, Molecular aspects of the interaction of near UV radiation with living matter, <u>in</u>: "Proceedings of the International Symposium on Current Topics in Radiobiology and Photobiology", R.M. Tyrrell, ed., Academia Brasileira de Ciencias, Rio de Janeiro.

Urbach, F., Forbes, P.D., Davies, R.E. and Berger, D., 1976, Cutaneous photobiology: past, present and future, <u>J. Invest. Derm.</u>, 67:209.

Willis, A.E. and Lindahl, T., 1987, DNA ligase I deficiency in Bloom's syndrome, <u>Nature</u>, 325:355.

COVALENTLY LINKED PORPHYRIN DIMERS AS MODEL SYSTEMS OF THE
PHOTOSYNTHETIC SPECIAL PAIR: SPECTROSCOPY, ENERGETICS AND
PHOTOCHEMISTRY

Edward Zenkevich, Alexander Shulga, Andrei Chernook,
Georgii Gurinovich and Eugenii Sagun

Institutue of Physics, BSSR Academy of Sciences
220602 Minsk, USSR

INTRODUCTION

The last decade has seen growing interest in chemical
dimers of porphyrins or chlorophylls in which monomer
molecules are covalently linked through saturated fragments of
different structure or are attached to electron donors or
acceptors. Such objects find application in a whole number of
areas: photoenergetics, medicine, molecular electronics, etc.
The photosynthetic aspect of the investigation of such systems
is due to the fact that the covalently linked porphyrin or
chlorphyll dimers differing in structure (Zenkevich et al.,
1985; Hunt et al., 1984), as well as diads (Wasielewski and
Niemczyk, 1984) and triads (Gust and Moore, 1987) with charge
transfer, including tetrapyrrole macrocycles, are convenient
systems in which one can model and comprehensively analyze the
excitation energy transfer and electron transport in natural
complexes of chlorophyll in vivo.

This paper presents the principal results of experimental
and theoretical studies of the processes of singlet and
triplet state deactivation at 77-293 K in covalently linked
porphyrin dimers of two types: 1) ethane-bisporphyrins with a
bond through mesopositions and, 2) dimers of
cyclopentanporphyrins with a covalent bond of monomers through
isocycles. These compounds are synthesized in our laboratory
and identified by NMR and mass-spectroscopy.

INTERMOLECULAR INTERACTIONS IN ETHANE-BISPORPHYRINS

Using NMR ^{1}H and Draiding structural models, we have
shown that in this type of dimer conformational mobility of
microcycles relative to the bond between them with variation
of the distance R between centres from 5.5 A$^\circ$ for fully
eclipsed to 10.4 A$^\circ$ for fully staggered conformations is
realised (see Fig. 1). At 293 K, in solution, there may also
exist an intermediate skew conformation with R = 6.6-7.2 A$^\circ$.

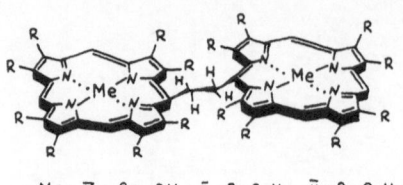

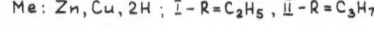

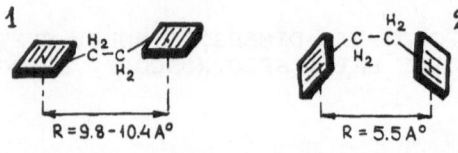

Fig. 1   Structure of ethane-bisporphyrins and
schematic representation of fully staggered
(1) and fully eclipsed (2) conformations.

It follows from the temperature dependence of the dimers NMR
^{1}H spectra that the most advantageous structure from the point
of view of energy is the fully staggered conformation
(Zenkevich et al., 1989a), therefore, it must dominate in
rigid solutions at 77 K.

The absorption bands of dimers in the visible range (Q-
bands), as well as the fluorescence and phosphorescence
spectra, are bathochromically shifted by 8-13 nm (200-500
cm^{-1}) relative to the monomer spectra, and in the Soret band
(B-bands) there is splitting by $\Delta \nu$ = 950-1200 cm^{-1} which is
most pronounced for Me-derivatives at 77 K (Gurinovich et al.,
1984).   Based on the calculations using the Fulton and
Gouterman theory (1964) for possible conformations of ethane-
bisporphyrins taking into account the different orientations
of NH-tautomers in the dimers of free bases and double
degeneration of electronic states in metal complexes, it is
shown that the vibronic band in symmetric dimers can be
treated as the intermediate case, i.e., it is caused by the
excitonic interaction of the dipole moments of intense B-
transitions of two molecules and by the intermolecular
electron-vibrational interaction ($\epsilon \approx \lambda$) and determines the
values of Soret-band splittings $240 \leq \Delta E_o \leq 830$ cm^{-1}.   The
bathochromic shift of electronic spectra of dimers relative to
the corresponding monomers can be associated with the exchange
interaction of π-conjugate macrocycles in any dimer
conformation.   The weakening of the exchange effects is
realised in the fully staggered conformation of ethane-
bisporphyrins at 77 K and appears as an increase in the
hypsochromic shift of Q-bands of dimers ($\Delta \nu$ = 9 nm) compared
to monomers ($\Delta \nu$ = 4 nm) as one goes from 293 K to 77 K
(Gurinovich et al., 1984).

Analysis of experimental data shows that in the complexes
under consideration the intermolecular interaction does not
lead to essential changes in quantum yields of fluorescence
and phosphorescence as well as in lifetimes of singlet and
triplet states as one passes from monomers to dimers (see
Table 1).   The main channel of S$_1$ state energy deactivation is
the intersystem crossing to the triplet state, and the ethane-
bisporphyrin fluorescence quenching by oxygen at 293 K is

Table 1 Parameters of singlet and triplet states for porphyrin monomers and corresponding dimers in EPIP mixture (degased solution). EPIP: Petroleum Ether: Diethyl Ether: Isopropanol 5:5:2.

Compound	T,K	B^a	τ_s,ns	τ_T,ms
OPP-mesoCH$_3$[b]	293	0.03	10.7	–
	77	0.04	15.5	7.6
(OPP)$_2$	293	0.025	8.2	–
	77	0.034	11.6	5.8
ZnOEP-meso CH$_3$	293	0.015	1.7	–
	77	0.020	2.5	31.4
(ZnOPP)$_2$	293	0.019	1.5	–
	77	0.030	2.5	33.9
CuOEP-meso CH$_3$	293	0.0001	0.0001	–
	77	0.0001	0.0001	0.16
(CuOPP)$_2$	293	0.0001	0.0001	–
	77	0.0001	0.0001	0.13
Cu(OPP)$_2$	293	0.0006	0.001	–
	77	0.0015	0.001	0.35
(Zn+Cu)(OPP)$_2$	293	0.00013	0.01	–
	77	0.0004	0.025	7.9
Ni(OEP)$_2$	293	0.0001	0.010	310^{-8}
OEP-cycle	293	0.08	15.2	–
	77	0.10	22.4	19.5
$3^1,5^1$-cyclodimer	293	0.09	17.2	–
	77	0.11	21.0	15.1
ZnOEP-cycle=CH$_2$	293	0.037	3.0	–
	77	0.045	3.50	44.5
$3^1,5^1$-Zn cyclodimer	293	0.038	2.8	–
	77	0.040	3.0	57.4

[a]B is the fluorescence quantum yield.
[b]OPP and OEP are octapropyl- and octaethylporphyrins.

realized with the rate constant $K_s = 1.5 \times 10^{10}$ lmol^{-1} s^{-1} as for monomers. The energy of dipole-dipole interactions of weak Q-transitions for free bases in ethane-bisporphyrins is 1.5 cm$^{-1} \leq V_{12} \leq 11$ cm^{-1}, i.e., the manifestation of exciton effects in the visible range is improbable. At the same time, calculations in Forster's approximation indicate that in symmetric dimers the probability of the inductive-resonance energy transfer for any conformation at all temperatures ($F_{ET}^{SS} = 1.5 \times 10^9 - 13.5 \times 10^{10}$ s^{-1}) turns out to be higher by at least one order of magnitude than the probability of excitation deactivation in any of the interacting molecules ($K_\Sigma = (0.9 - 1.2) \times 10^8$ s^{-1}). Consequently, the excitation in (OPP)$_2$ repeatedly migrates between different NH-tautomers of two interacting components of the dimer before deactivation of the S$_1$-states occurs. As a result, this appears as a "collapse" of fluorescence excitation polarization spectra of symmetrical dimers and a decrease in the limiting degree of fluorescence polarization for dimers compared to monomers.

The presence of Cu or Ni ions at the centre of one of the

components of hybrid ethane-bisporphyrins (the second component is the Zn-complex or a free base) manifests itself in a number of general laws (Zenkevich et al., 1989a). i) The fluorescence and phosphorescence of mixed dimers $Cu(OEP)_2$, $(Zn+Cu)(OPP)_2$, $Ni(OEP)_2$ is represented by luminescence bands of the components containing no Cu and Ni ions, and the singlet-singlet energy transfer from Cu- or Ni-complexes to the molecules of free bases in dimers is not observed. ii) For hybrid dimers $Cu(OEP)_2$, $Ni(OEP)_2$ and $(Zn+Cu)(OPP)_2$, fluorescence quenching greater by more than one order of magnitude compared to the symmetric dimers $(OPP)_2$ and $(ZnOPP)_2$ is characteristic (See Table 1). iii) As one passes from symmetric $(OPP)_2$ and $(ZnOPP)_2$ to hybrid dimers $Cu(OEP)_2$ and $(Zn+Cu)(OPP)_2$, the triplet state lifetime τ_T of the free base and Zn-complex decreases. According to picosecond flash-photolysis data in the mixed dimer $Ni(OEP)_2$ at 293 K the duration of excited states is determined by the values $\tau \approx 15\pm5$ ps. The main channels of the excitation deactivation in hybrid dimers may be presented as follows. The fluorescence quenching ($B_o/B = 38$) in $Cu(OEP)_2$ and the main portion of quenching ($B_o/B = 37$) in $(Zn+Cu)(OPP)_2$ results from the same cause: d-π exchange effects associated with the interaction of the unpaired d-electron of Cu ion in one dimer component with the π-conjugate system of the second component and leading to an increase in the probabilities of intersystem crossing in the second component of the dimer, i.e., to the reduction of τ_s and τ_t of porphyrin ligands. The absence of phosphorescence and induced T-T absorption of Cu-porphyrin in the hybrid dimers $Cu(OEP)_2$ and $(Zn+Cu)(OPP)_2$ is due to exchange-resonance triplet-triplet energy transfer to the component containing no Cu ion with the probability $F_{et}^{TT} \geq 2 \times 10^7$ s^{-1}. In the case of the mixed dimer $Ni(OEP_2)$ the interaction of components involving the proper S_1 and T_1 states of the porphyrin macrocycle does not lead to strong quenching ($B_o/B = 200$) of these states for the free base. We are apt to believe that the ultrafast deactivation of exciation in the free-base molecule of the mixed dimer $Ni(OEP)_2$ at 293 K is caused by the exchange-resonance energy transfer from this molecule to the lower-lying S_d-level of the Ni ion of the second component with the probability $F_{ET}^{SS} \approx 10^{11}$ s^{-1}.

THE PHOTOPHYSICS OF CYCLOPENTANPORPHYRIN DIMERS

The investigation of this type of dimer is interesting, since the isocycle as a structural element of chlorophyll and its natural analogs plays an important role in the formation of pigment aggregates. The composition of cyclodimers involves one molecule of the OEP-cycle and one molecule of the OEP-cycle=CH_2 which have different and frequency-shifted absorption and luminescence spectra (as well as their Zn- and Cu-complexes). In the $3^1,3^1$- and $3^1,5^1$-cyclodimers the covalent bond of monomers is realized at different positions of the isocycle in the OEP-cycle molecule and the distance between the centres of the macrocycles is R = 12.3 A^o (see Fig. 2). It is shown by NMR ^{1}H methods that the planes of porphyrin macrocycles in dimers do not overlap, and the analysis of data on linear dichroism and polarized fluorescence with the use of Draiding structural models has made it possible to substantiate the oscillator model of

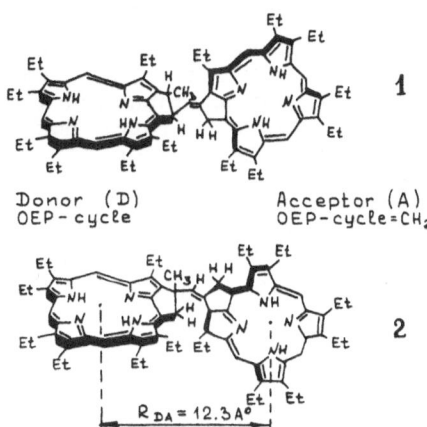

Fig. 2 Structures and geometric peculiarities of
cyclopentanporphyrin dimers: 1 - $3^1,5^1$-
cyclodimer, 2 - $3^1,3^1$-cyclodimer.

cyclopentaneporphyrins and calculate the angles between the
interacting oscillators in dimers (Zenkevich et al., 1989b).
Weak overlapping of π-electron macrocycles significantly
weakens the exchange interaction in dimers, which appears as a
red shift of electronic Q-bands by only 1-2 nm ($\Delta\nu$= 30-60
cm^{-1}) compared to the spectra of individual molecules. As
with the symmetric ethane-bisporphyrins, the intermolecular
interactions in cyclodimers do not practically change the
probabilities of intramolecular transitions in each dimer
component. The splitting of the Soret-bands in the absorption
spectra of cyclodimers ($\Delta E_q \approx$ 300-900 cm^{-1}) can also be
attributed to the exciton interactions of the dipole moments
of intensive B-transitions. At the same time, the dipole-
dipole interaction of weak Q-transitions ($\nu_{12} \leq$ 3 cm^{-1}) under
the conditions of frequency separation of the electronic
spectra of the components in cyclodimers shows up as
additional deactivation of the excited states of one of the
molecules of the complexes under consideration. The
resolvability in the dipole approximation of electronic
transitions, the absence of exciton effects in the region of
cyclopentaneporphyrin Q-bands, and the weak overlapping of
porphyrin macrocycles lead us to assume that in dimers the
predominant deactivation process of S-states of the donor, D,
(OEP-cycle or Zn-OEP-cycle) is the inductive-resonant energy
transfer to the acceptor, A (OEP-cycle = CH$_2$ or Zn-OEP-cycle =
CH$_2$). Comparison of experimental data and theoretical
calculations for Zn-complexes and free bases leads to the
following results.

In the case of Zn-cyclodimers, we have shown by the
method of selective laser spectroscopy at 4.2 K that in D and
A molecules the asymmetry caused by the isocyclic substitution
leads to the splitting of lower excited S-states into two
states and, therefore, the transfer processes can involve
differently polarized S_0-S_1 and S_0-S_2 states of interacting
molecules. The experimental quantum efficiency of D → A
transfer is Φ_{DA} = 0.99 at 293 K and Φ_{DA} = 0.95 at 77 K, and
the coincidence of the results on D fluorescence quenching and
A luminescence sensitization points to the absence of quantum

losses in transfer processes. Based on the calculations of integrals of D fluorescence and A absorption spectra overlapping ($J = 7.6 \times 10^{-16}$ to 3.6×10^{-14} cm^6mol^{-1}) and on the rigorous estimation of the orientation factors ($K^2 = 0.07-2.0$), taking into account the oscillator model of the $3^1,5^1$-cyclodimer for all possible combination of interacting levels, it has been found that at 293 K the determining role in transfer processes is played by the channels $S^D_2 \rightarrow S^A_1$ ($R_0 = 27$ A$^\circ$, $F^{SS}_{ET} = 4.1 \times 10^{10}$ s$^{-1} \gg K^D_\Sigma = 3.9 \times 10^8$ s^{-1}) and $S^D_1 \rightarrow S^A_1$ ($R_0 = 23$ A$^\circ$, $F^{SS}_{ET} = 1.1 \times 10^{10}$ s^{-1}), while at 77 K the interaction between D and A is only due to one process $S^D_1 \rightarrow S^A_1$ ($R_0 = 20$ A$^\circ$, $F^{SS}_{ET} = 6.6 \times 10^9$ s$^{-1} \gg K^D_\Sigma = 3.5 \times 10^8$ s^{-1}). The dynamics of S $\rightarrow$ S energy transfer in Zn-cyclodimers is very dependent on the temperature. However, at any T the energy transfer occurs after thermal relaxation of S_1 - and S_2 - states of D and A, i.e., it is not "hot". The good correlation of experimental data with theoretical calculations on D fluorescence quenching at 77 K $(B/B_0)_{exp} = (B/B_0)_{cal} = 0.05$ indicates that in this case the pair energy transfer is fully described by Forster's theory of inductive resonance within calculation and experimental error.

For free bases of cyclopentanporphyrins we have revealed the existence, at all temperatures, of an equilibrium mixture of two NH-tautomers caused by migration of internal protons and having different types of electronic absorption and fluorescence spectra (Zenkievich et al., 1984). In this case, the process of NH-tautomerism in excited S_1 and T_1-states does not practically compete with other deactivation processes of these states. As a result, at 77 K in the $3^1,5^1$-cyclodimer the best resonance conditions between the main tautomers of D and A ($J = 0.8 \times 10^{-14}$ cm^6mol^{-1}) lead to the fact that the S-S energy transfer is the regulating factor in the directional character of photoinduced NH-tautomerism in the A molecule at excitation to the absorption bands of D. The experimental value of the D $\rightarrow$ A transfer quantum efficiency at 77 K is $\Phi_{DA} = 0.97$, and the transfer probability is $F^{SS}_{ET} = 1.9 \times 10^9$ s^{-1}. However, calculations using Forster's theory for this dimer taking into account the inversion in the energy scale of the combining electronic states when going from metal complexes to free bases and rigorous estimation of orientation factors lead to lower values for the transfer probability $F^{SS}_{ET} = (0.1-5.7) \times 10^8$ s^{-1} (at $R_0^{theor} = 14.5$ A$^\circ$ for 77 K). The observed difference between experimental and theoretical transfer parameters can be associated with the manifestation of exchange processes caused by the weak overlapping of peripheral portions of π-conjugate systems. This is confirmed by the following experimental facts.

In Cu-cyclodimers, due to the fast intersystem crossing to the T-state ($r = 3 \times 10^{13}$ s^{-1}) the S-S transfer between D and A with the probability $F^{SS}_{ET} \leq 4 \times 10^{10}$ s^{-1} is not realized. However, for these systems at 77 K D phosphorescence quenching by ~ 10^3 times has been observed. This corresponds to the probability of exchange-resonance energy transfer involving the T-states $F^{TT}_{ET} > 7 \times 10^6$ s^{-1} which is much higher than the probability of D T-states deactivation ($K^D_T = 6.7 \times 10^3$ s^{-1}). In the case of Zn-cyclodimers the strong quenching of D luminescence at 77 K is due to two causes. As a result of the S-S inductive-resonance energy transfer competing with the intersystem crossing to the T-state of D ($F^{SS}_{ET} = 1.1 \times 10^{10}$ s^{-1}

>> r = 3.5 x 10^8 s^{-1}), only 5% of D molecules that have absorbed the light in the composition of the dimer go to the T_1-state. In turn, the deactivation of the remaining triplet molecules of D is realized due to the exchange-resonance T-T transfer to A with the quantum efficiency Φ_{ET} = 1.0. Thus, we have substantiated experimentally that in this geometry of dimers with a distance R = 12.3 A$^\circ$ between the centres of porphyrin macrocycles, the manifestation of exchange effects becomes possible which leads to an effective T-T transfer and to an increase in the deactivation of the singlet states of D due to the exchange-resonance S-S transfer to A under the conditions of weakening of dipole-dipole interactions.

The establishment of reliable correlations between the spectral-energetic characteristics of dimers and the mutual geometry of interacting systems makes it possible to carry out correct comparison of obtained results with theoretical calculations and, therefore, is the basis for detailed analysis of the laws and mechanisms of pigment-pigment interation in different chlorophyll complexes _in vivo_.

REFERENCES

Brookfield, R.L., Ellul, H., Harriman, A. and Porter, G., 1986, Luminescence of porphyrins and metalloporphyrins. Part II. Energy transfer in zinc-metal-free porphyrin dimers, J. Chem. Soc. Farad. Trans. 2, 82:219.
Fulton, R.L., and Gouterman, M., 1964, Vibronic coupling. II Spectra of dimers, J. Chem. Phys., 41:2280.
Gurinovich, G.P., Zenkevich, E.I., Shulga, A.M., Sagun, E.I. and Suisalu, A., 1984, Electronic excitation deactivation in chemical porphyrin dimers, Zh. Prikladn. Spektrosk., 41:446.
Gust, D., and Moore, T.A., 1987, Electron transfer in model systems for photosynthesis, in: "Supramolecular Photochemistry." V. Balzani, ed., D. Reidel Publsihing Company, Dordrech/Boston/Lancaster/Tokyo, NATO ASI Series, 241-267.
Hunt, J.E., Katz, J.J., Svirmickas A., and Hindman, J.C., 1984, Photoprocesses in Photosystem I model systems, J. Amer. Chem. Soc., 106:2242.
Wasielewski, M.R., and Niemczyk, M.P., 1984, Photoinduced electron transfer in meso[1]-triphenyltriptycenylporphyrin-quinones. Restricting donor-acceptor distances and orientations, J. Amer. Chem. Soc., 106:5043.
Zenkevich, E.I., Shulga, A.M., Chernook, A.V., and Gurinovich, G.P., 1984, Spectral peculiarities of NH-tautomerism in isocycle-containing porphyrins and their covalently linked dimers, Chem. Phys. Lett., 109:306.
Zenkevich, E.I., Shulga, A.M., Sagun, E.I., Gurinovich, G.P. and Chernook, A.V., 1985, Energetics and Photophysical processes in covalently linked porphyrin dimers, in: "Proc. 3rd Symp. Optical Spectroscopy," D. Fasler, K. H. Feller and B. Wilgelmi, eds., Teubner-Texter zur Physik, 4:297, Leipzig.
Zenkevich, E.I., Shulga, A.M., Chernook, A.V., Sagun, E.I., and Gurinovich, G.P., 1989a, Intermolecular interactions in ethane-bisporphyrin complexes with transition metal ions, Khim. Fiz., 8:842.

HOLE BURNING SPECTROSCOPY OF CHROMOPROTEINS

J. Friedrich

Univ. Mainz, Inst. Phys. Chem., Welder Weg 11
D6500 Mainz, Germany

ORDER AND DISORDER PHENOMENA IN OPTICAL SPECTRA OF BIOPOLYMERS

Since the first successful determination of the crystal
structure of a protein by Kendrew (1963), the level of
information on the structure and functioning of proteins has
increased immensely (Huber, 1988; Frauenfelder, 1984). An
important point towards an understanding of how a protein
functions seems to be the interrelation between order and
disorder. The highly regular X-ray scattering patterns
indicate a well defined structure of globular proteins. On
the other hand, there is a lot of experimental evidence, that
the structure of proteins is not strictly determined. In
fact, it seems that a protein can attain many conformational
states, very similar to glasses. Evidence comes from specific
heat experiments (Yang and Anderson, 1986), from ultrasound
and microwave saturation experiments (Singh et al., 1984), and
from optical experiments (Frauenfelder, 1986; Iben et al.,
1989; Kohler et al., 1988): The specific heat is linear in
temperature below 1K, like in glasses; there is saturable
ultrasound and microwave absorption, like in glasses; the
optical lines are inhomogeneously broadened and optically
labeled conformational states show thermally irreversible
features (Friedrich and Kohler, 1989), very much like glasses.

It seems that disorder is essential to a proper
functioning of proteins. The conformational freedom which a
protein has as a consequence of the rather ill defined
structure on a microscopic level, allows for rather irregular
motions (Karplus and McCammon, 1982). These irregular motions
support the functionally important modes. An artist's view of
how the energy of a protein along some fictive conformational
coordinate may look, is shown in Fig. 1. Structural
relaxation can occur between the various conformational states
represented by the local minima. In the simplest way these
relaxation processes can be modeled as relaxation processes in
a double well potential with barrier V, energy asymmetry Δ and
spatial extension d.

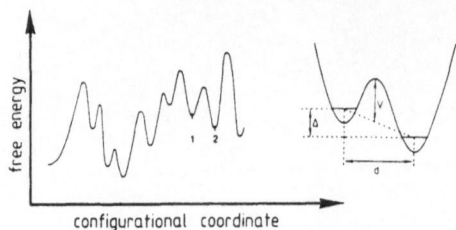

Fig. 1   Section through the configuration space of a
large protein along a fictive coordinate.
The simplest element which allows for conformational relaxation (e.g. from 1 to 2) is
a double well potential.

Structural disorder on a molecular scale hampers straight
forward high resolution optical spectroscopy.   Disorder
usually leads to large inhomogeneous line broadening.   For
most chromoproteins this broadening is on the order of several
hundred wavenumbers, thus obscuring all the details reflected
by a highly resolved molecular lineshape profile, namely the
homogeneous lineshape of the zero-phonon transition, the
homogeneous linewidth, the shape and intensity of the phonon
sideband.   There are, however, several techniques which allow
for extremely highly resolved optical spectroscopy despite
structural disorder (Friedrich and Haarer, 1984).   Among these
techniques, hole burning is most powerful.   In fact, most of
the high resolution spectroscopy of proteins has been done by
employing hole burning techniques.

HIGH RESOLUTION TECHNIQUES: SPECIAL FEATURES OF SPECTRAL HOLE
BURNING

The absorption profile of a chromophore embedded in a
perfect crystalline environment looks very much as depicted in
Fig. 2a (Friedrich and Haarer, 1984).   It consists of an
intense and narrow purely electronic line (the so called zero-
phonon line) and a rather broad phonon sideband.   Transitions
into the phonon sideband occur because the environment starts
vibrating as a consequence of the chromophore excitation.   The
relative intensity of the zero-phonon line is called the
Debye-Waller factor α.   It is a measure of the coupling
between the electrons of the chromophore and the vibrations
(the phonons) of the whole assembly consisting of chromophore
plus environment.   The electron-phonon coupling strength is
commonly expressed in terms of the Huang-Rhys factor S, which
itself is related to the Debye-Waller factor by

$$\alpha = \frac{\text{Intensity in zero-phonon line}}{\text{total intensity}} = \exp[-S] \qquad (1)$$

S<1 means that the electron-phonon coupling is weak; S>>1
means that the electron-phonon coupling is strong.   In the
latter case the spectrum is dominated by the broad phonon
sideband, whereas in the former case it is dominated by the
sharp zero-phonon line.   The most interesting quantity of the
zero-phonon line is its width.   If this width is truly
homogeneous, then, the lifetime T_1 and the pure dephasing time

T_2* can be determined from it, using the relation

$$\Delta W_h = \frac{1}{2T_1} + \frac{1}{T_2*} \qquad (2)$$

At very low temperatures T_2* gets rather long, hence ΔW_h is determined by the lifetime T_1 of the electronic state excited. As the temperature increases, T_2 decreases and eventually gets much shorter than T_1. In this case ΔW_h is solely determined by T_2*. It seems that the determination of T_1 and T_2* from a linewidth experiment is straightforward. It is, however, in no way trivial to ensure that a measured linewidth is really homogeneous. The reason is twofold: First, the resolution power of the experiment has to be on the order of 10^7, in order to measure optical lines with a width of several tens of MHz. Second, and this point is more severe, in systems characterized by structural disorder, like proteins, structural relaxation processes occur which tend to broaden the zero-phonon lines on time scales longer than T_2* (Friedrich and Kohler, 1989). This phenomenon is called spectral diffusion. Hence, to measure T_1 or T_2* by a frequency domain experiment one has to ensure that spectral diffusion does not play any role.

Fig. 2b shows the principle how spectral hole burning works. As discussed above, disorder may lead to a large dispersion of absorption frequencies. Suppose the chromophores are photoreactive. Then, shining narrow bandwidth laser light onto the sample, those molecules are bleached which happen to absorb at the laser frequency. They are transferred to the product state and, consequently, a hole will appear in the absorption because the number of molecular absorbers is diminished. The hole may show the sharp zero-phonon feature right at the laser frequency in case the electron-phonon coupling is not too large. However, there are also rather broad features to the left and to the right of the sharp hole, which reflect the phonon sidebands (Friedrich and Haarer, 1984).

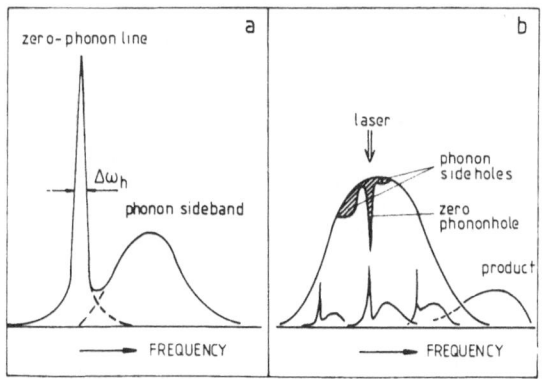

Fig. 2a) Line shape of a chromophore in a perfect
lattice. ΔW_h is the homogeneous linewidth.
b) Inhomogeneous line broadening and the
principle of hole burning.

What are the special features of hole burning?

Hole burning allows for a determination of the zero-phonon lineshape right at the exciting laser frequency. Fluorescence techniques, for example, do not have this capability unless very short excitation pulses and gated detection is used. Hence, hole burning can be used to measure fast relaxation times. For example, if the excited state has a lifetime of 1 psec and phase relaxation can be neglected (e.g. at low temperatures) its associated width is 5 cm^{-1}, which is easy to measure.

Apart from measuring fast relaxation times, spectral holes offer an even more interesting application. They can be used as extremely sensitive probes for the dynamics of the conformational substates of chromoproteins. As mentioned above, biopolymers are thought to be disordered (at least on microscopic scales), like glasses. If so, they are metastable and may relax from one conformational substate to another. These processes may be extremely slow, i.e. occur on time scales of minutes, hours, even weeks. They can be measured by measuring the change of the shape, the width and the area of the hole as a function of time or temperature or any other parameter which induces conformational transitions. Since the holes may be extremely sharp, the detection of such processes may be very sensitive.

APPLICATION OF HOLE BURNING IN THE SPECTROSCOPY OF PROTEINS

One of the first hole burning experiments on biopolymers was performed on antenna pigments of bacteriophotosynthesis (Friedrich et al., 1981). As a matter of fact most hole burning experiments have been carried out on photosynthetic pigments. Three major research fields of hole burning on proteins can be distinguished: The dynamics of the excited antenna states, the dynamics around the primary donor state of the reaction centre and the dynamics of the protein as a whole in its electronic ground state. Whereas the first two fields are concerned with the physics of photosynthesis, the latter deals with the basic features of the protein state of matter independent of the special pigment investigated. I will briefly review these three fields.

a) The Antennas

I will focus in this section on the phycobiliproteins which seem to be the best characterized antenna complexes. The most important feature of antennas is their capability of directional energy transfer. An optimized level structure within the pigments, which is provided to a high degree by the specific interaction between chromophore and apoprotein, is a prerequisite for spatially and energetically directional energy transfer.

The various pigments, namely phycoerythrin, phycocyanin and allophycocyanin are linked together by so called linker polypeptides to form highly organized supramolecular aggregates, the so called phycobilisomes. The absorbed energy is transported from the outer pigment phycoerythrin to phycocyanin and from there to the core pigment

348

allophycocyanin. In the phycobilisomes the pigments are
hexameric biliprotein complexes with a disk-like overall shape
of a rather high symmetry (Schirmer et al., 1985). It is not
a straight forward experiment to measure the energy transfer
times in a time domain experiment, mainly because the response
of the system is, due to its complexity, multiexponential. As
mentioned above, hole burning can provide information on the
relaxation times of states selected by the exciting laser. An
example of hole burned spectrum is shown in Fig. 3 (Kohler et
al., 1988): The lower trace in Fig. 3 represents the
absorption spectrum of phycobilisomes of the algae
Mastigocladus laminosus. The upper trace is the difference
spectrum (ΔOD) which remains after hole burning at 6370 Å into
the phycocyanin pigment. The sharp spike is the zero-phonon
line. ΔOD > 0 characterizes areas where product appears,
whereas ΔOD < 0 characterizes areas where educt is bleached.
This figure shows already that the phycobilisomes are an
efficient energy transfer system: As can be seen, most of the
reaction occurs in the allophycocyanin chromophore which
absorbs around 6550 Å and which is populated from phycocyanin
via energy transfer processes. Obviously, these processes do
not conserve energy correlation so that most of the
inhomogeneous band of the energy accepting pigment is
populated. The same conclusion was reached in a combined hole
burning - fluorescence line narrowing study on isolated
phycocyanin pigments (Kohler et al., 1988). Apart from the
photoreaction in the energy acceptors, there is also a
resonant reaction in the excited phycocyanin band as indicated
by the sharp zero-phonon line. As to the nature of the
reaction, not much is known. Since the product appears close
to the laser frequency, it was argued that the reaction might
by photophysical in nature, which means that the chromophore
is not directly affected, instead its immediate environment is
changed by the absorption of laser light. Since more educt is
burnt than product appears, it might be that, in addition to
the photophysical processes, some real photochemistry occurs.

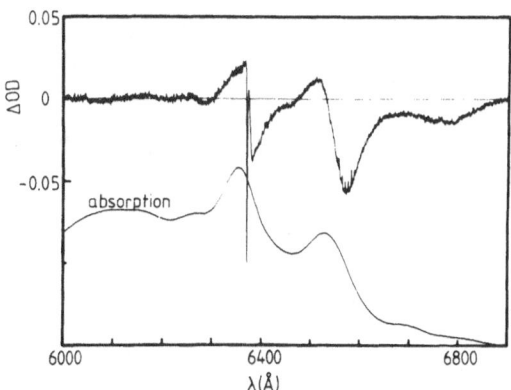

Fig. 3   Low temperature absorption (4K) and ΔOD
         spectrum of phycobilisomes of Mastigocladus
         laminosus. Laser excitation was performed
         at 6380 Å, as documented by the sharp zero-
         phonon hole (Kohler et al. 1988).

Fig. 4 shows holes in the phycocyanin pigment of intact
phycobilisomes as compared to the isolated phycocyanin
pigment, on an enlarged scale. The hole-width of the
phycobilisomes extrapolated to zero energy absorption is 0.34
cm^{-1}. The related relaxation time is 32 psec. If the width
were solely determined by energy transfer processes, we would
deduce, according to equ. 2, an energy transfer time on the
order of 16 ps. Since, at 4.2 K, the holewidth still depends
on temperature, this value of 16 ps should be considered as a
lower limit of the energy transfer time. Experiments show
that this value is independent of the spectral position of
hole burning (Kohler et al., 1988). The corresponding
holewidth of the isolated pigment (see Fig. 4) is
significantly narrower due to the fact that energy transfer is
blocked by the absence of a proper acceptor.

An interesting observation was very recently made by
Johnson and Small (1989). In the antenna complex of the
bacterium _Prostecochloris aestuari_ they found evidence for an
excitionic-like coupling between chromophores. Fig. 5 shows
that the resonant hole at the laser frequency (arrow) is
accompanied by a broad hole at somewhat higher energies with a
width on the order of 50 cm^{-1}. The two holes are associated
with two excitonically coupled chromophore states. From the
width of the upper level a related lifetime on the order of
250 fs is deduced.

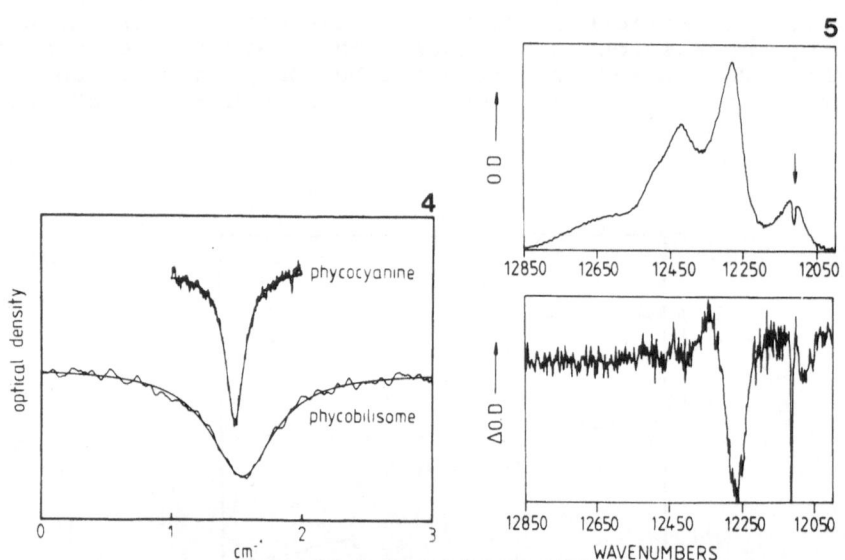

Fig. 4   Zero-phonon holes of isolated phycocyanin
         trimers and phycobilisomes (Kohler et al.,
         1988).

Fig. 5   Absorption and ΔOD-spectra of the antenna
         complex of _Prostecochloris aestuari_
         (Johnson et al., 1989).

b)   Reaction Centres

In recent years there has been a lot of activity in hole burning spectroscopy of reaction centres. This activity was triggered by a paper by Boxer et al. in 1986. Meanwhile several of the problems addressed have been clarified, mainly by work performed by Small's group (for example see Tang et al., 1989). Hole burning was applied to the primary donor states of Rhodobacter sphaeroides (P870), Rhodopseudomonas viridis (P960), photosystem I (P700) and photosystem II (P680).

The hole burning spectra of the primary donor states are in striking contrast to those of the antennas (Fig. 6): They are dominated by a broad feature on the order of several 100 cm^{-1}. When a sharp structure is observed, it is usually of rather low intensity. Whether this sharp feature is observed or not depends for some systems strongly on the burn wavelength.

Several possible solutions were suggested to account for these specific features in the hole burned spectra of the primary donor states: 1) The broad holes are due to ultrafast relaxation processes on the order of 100 fs, the nature of which is not yet clear. This interpretation was supported by time resolved experiments by Meech et al. (1985). Won and Friesner (1988), suggested that a strong coupling between the primary donor state to close lying charge transfer states may lead to an ultra fast relaxation which, in turn, may lead to broad structureless holes. Also, recent hole burning experiments by Shuvalov et al. (1988) seem to be in line with this interpretation. These authors report that narrow holes (≤ 1 cm^{-1}) only appear in case the electron transfer chain is

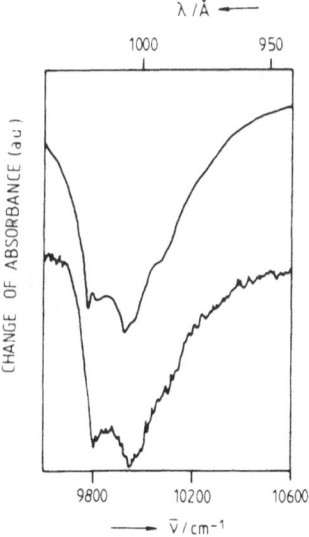

Fig. 6   Hole-burned spectrum of P960* of Rhodopseudomonas viridis. Burning was performed
around 9800 cm^{-1}. The upper trace is a
simulated spectrum (Tang et al., 1989).

blocked. Otherwise the hole-width is on the order of 100 cm^{-1}. This large width is attributed to relaxation into a charge separated precursor state in which, however, only the primary donor and the bacteriochlorophyll molecules are involved. 2) Since the sharp zero-phonon features have meanwhile been observed for several intact reaction centres (P870, P960, P680, see, for instance Tang et al., 1989) another view emerges: The broad features can be well accounted for by assuming (for P870 and P960) a moderately strong electron-vibration coupling (S-factor between 1 and 2). The vibration is on the order of 150 cm^{-1} and is characteristic for the special pair ("pair marker"). The progression of the pair marker mode including the zero-zero transition couples moderately strongly to low frequency phonons (protein modes), thus accounting for the low intensity zero-phonon hole. The widths of the zero-phonon holes are on the order of 10 cm^{-1} which reflects a lifetime of the zero point level of the primary donor state on the order of 1 psec in good agreement with time domain experiments. The 1 psec lifetime is associated with the formation of the charge separated state. Vibrational thermalization occurs on a much faster time scale prior to charge separation.

Contrary to the reaction centres of bacterial photosynthesis, no marker mode could be identified in P680, the reaction centre of photosystem II (Jankowiak et al., 1989). This suggests that the structure of the special pair (if there is any) may be different. Note, that the crystal structure of photosystem II has not yet been determined. The reaction centre of photosystem I lacks the sharp zero-phonon feature which suggests that the electron-phonon coupling is much larger than in P870 and P960 as well as in photosystem II. An S-factor on the order of 5 accounts for the experimental facts (Gillie et al., 1989). The relevant parameters (mode frequencies, S-factors, linewidths) are obtained with reasonable accuracy from simulation of the hole burned spectra (Hayes et al., 1988)

c)   The protein state of matter

The molecular weight of proteins may cover a range from several 10 kDa to several thousand kDa. The phycobilisomes of Mastigocladus laminosus, for example, have a molecular weight on the order of 5000 kDa. Though, for spectroscopic purpose, these proteins are dissolved in a glassy matrix, it seems that the chromophores do not significantly interact with the glassy solvent since they seem to be well shielded by the huge biopolymer. Hence, the chromophores probe the proteinaceous environment rather than the host glass. A spectral hole burnt into the absorption of a specific chromophore can be considered as a characteristic label for a specific protein state. As the temperature is changed, or as time goes on, this state undergoes conformational relaxation which is reflected in a broadening and a recovery of the burnt-in hole. Hence, relaxation processes of the whole protein in its groundstate can be investigated with a high level of precision just by observing the changes of a burnt-in hole as a function of a proper parameter. In this application mode the hole burning technique is rather unique because there is almost no other spectroscopic technique with the capability of measuring groundstate relaxation processes of structurally disordered materials (for a review, see Friedrich and Kohler, 1989).

An example (Kohler and Friedrich, 1989) is shown in Fig. 7. The insert shows part of the absorption of the core pigment, namely allophycocyanin, of intact phycobilisomes together with a burnt-in hole at 6573 Å (see also Fig. 3). This hole serves as a label for the starting state of the protein. The actual experiment is a temperature cycling experiment, which is performed in the following way: The hole is always measured at the starting termperature where it was burnt. In the case considered here, this temperature is 4.2 K. The protein is, however, exposed to termperature cycles, where the so-called cycling temperature (abscissa in Fig. 7) is steadily increased in steps of, say 1 K, or so. When the temperature is increased the protein relaxes into a manifold of new conformational states (see Fig. 1) and eventually gets trapped in one of them when the temperature is lowered again. The change in the microenvironment results in slight changes in the interaction between protein and chromophore which in turn leads to a broadening of the burnt-in hole. I think that an experiment of this type is the most direct demonstration that a protein indeed undergoes relaxation between conformational substates, even at temperatures as low as a few K. The change of the hole width as a function of cycling temperature can be modeled within the frame of spectral diffusion theories modified in a proper way to account for thermal irreversibility (Friedrich and Kohler, 1989; Kohler and Friedrich, 1989, 1988). The data can be accounted for by assuming that the conformational relaxation processes occur by tunneling at temperatures below 35 K and are activated above. Tunneling processes lead to a linear increase of the hole width whereas activated processes lead to an increase proporional to $T^{3/2}$. These temperature laws essentially result from a distribution of conformational barrier heights and associatd energies of the conformational states as a consequence of local structural disorder. In this respect, proteins are very similar to glasses (Friedrich and Kohler, 1989), a result also confirmed by other authors (Frauenfelder, 1984; Yang and Anderson, 1986; Singh et al., 1984; Iben et al., 1989).

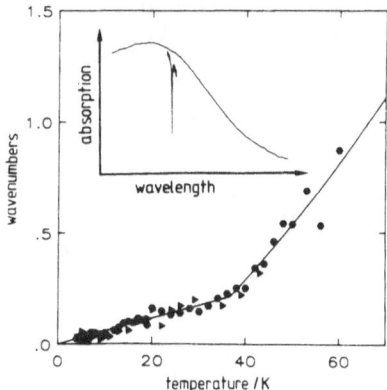

Fig. 7 Thermally irreversible broadening of an
optical hole (insert) as measured by
a temperature cycling experiment. Sample:
phycobilisomes of <u>Mastigocladus laminosus</u>
(Kohler and Friedrich, 1989).

SUMMARY

In this chapter I have reviewed the activities in the field of hole burning spectroscopy on chromoproteins. I restricted the frame of this survey to "persistent" and transient spectral holes based on photochemical or photophysical reactions. I stress, however, that the term "hole burning" is applied to a much wider field of spectroscopic phenomena, very often to any case of optical inhomogeneity which is subject to frequency selective experiments (Campbell et al., 1987). Also, though I confined myself to proteins, the hole burning technique is by no means restricted to this class of biopolymers. Recently, it was demonstrated that holes can be burnt into daunomycin-DNA intercalation complexes (Floser and Haarer, 1988). Summarising, I think that the technique of spectral hole burning with its rich variety of applications in relaxation spectroscopy on extremely short as well as on extremely long time scales is very powerful in elucidating special features of biopolymers and complicated assemblies like the photosynthetic apparatus.

ACKNOWLEDGEMENT

I gratefully acknowledge the support by the Deutsche Forschungsgemeinschaft (SFB 262 and Fr 456/12-1). I would also like to thank Prof. G. J. Small for sending his recent papers on hole burning in reaction centres to me.

REFERENCES

Boxer, S.G., Lockhart, D.J. and Middendorf, T.R., 1986, Photochemical hole burning in photosynthetic reaction centres, Chem. Phys. Lett., 123:476.

Campbell, B., Chance, M.R and Friedman, J.M., 1987, Linkage of functional and structural heterogeneity in proteins: Dynamic hole burning in carboxymyoglobin, Science, 238:373.

Floser, G. and Haarer, D., 1988, The photochemistry of daunomycin in solution and intercalated into DNA studied by photochemical hole burning, Chem. Phys. Lett., 147:288.

Frauenfelder, H., 1984, From atoms to biomolecules, Helvetica Physica Acta, 57:165.

Friedrich, J., Scheer, H., Zickendraht-Wendelstadt, B. and Haarer, D., 1981, High resolution optical studies on C-phycocyanine via photochemical hole burning, J. Am. Chem. Soc., 103:1030.

Friedrich, J., Scheer, H., Zickendraht-Wendelstadt, B. and Haarer, D., 1981, Photochemical hole burning: A means to observe high resolution optical structures in phycoerythrin, J. Chem. Phys., 74:2260.

Friedrich, J. and Haarer, D., 1984, Photochemisches Lochbrennen und optische Relaxationsspektroskopie in Polymeren und Glasern, Ang. Chemie, 96:96, Int. Ed. Engl. 23:113.

Friedrich, J. and Kohler, W., 1989, Relaxation in glasses and proteins in "Dynamical processes in condensed molecular systems", ed. by J. Klafter, J. Jortner, A. Blumenn, World Scientific Publ. C., Singapore.

Gillie, J.K., Lyle, P.A., Small, G.J. and Goldbeck, J.H., 1989, Spectral hole burning of the primary electron donor state of photosystem I., in press.

Hayes, J.M., Gillie, J.K., Tang, D. and Small, G.J., 1988, Theory for spectral hole burning of the primary electron donor state of photosynthetic reaction centres, Biochimica et Biophysica Acta, 932:287.

Huber, R., 1988, Beweglichkeit und Starrheit in Proteinen und Protein-Pigment-Komplexen, Ang. Chemie., 100:79.

Iben, E.T., Braunstein, D., Doster, W., Frauenfelder, H., Hong, M.K., Johnson, J.B., Luck, S., Ormos, P., Schulte, A., Steinbach, P.J., Xie, A.H. and Young, R.D., 1989, Glassy behaviour of a protein, Phys. Rev. Lett., 62:1916.

Jankowiak, R., Tang, D., Small, G.J. and Seibert, M., 1989, Transient and persistent hole burning of the reaction centre of photosystem II, J. Phys. Chem., 93:1649.

Johnson, S.G. and Small, G.J., 1989, Spectral hole burning of a strongly exciton-coupled bacteriochlorophylla antenna complex, Chem. Phys. Lett., 155:371.

Karplus, M. and McCammon, J.A., 1982, The dynamics of proteins, Scientific American, 254(4):42.

Kendrew, I.C., 1963, Myoglobin and the structure of proteins, Science, 139:1259.

Kohler, W., Friedrich, J., Fischer, R and Scheer, H., 1988, Site-selective spectroscopy and level ordering in C-phycocyanine, Chem. Phys. Lett., 143:169.

Kohler, W., Friedrich, J., Fischer, R. and Scheer, H., 1988, High resolution frequency selective photochemistry of phycobilisomes at cryogenic temperatures, J. Chem. Phys., 89:871.

Kohler, W., Friedrich, J., Fischer, R. and Scheer, H., 1988, Low temperature spectroscopy of cyanobacterial antenna pigments, in: "Photosynthetic Light Harvesting Systems, Organization and Function", ed. by H. Scheer and S. Schneider, W. de Gruyter-Verlag, Berlin, NY.

Kohler, W. and Friedrich, J., 1988, Irreversible features of thermal line broadening in glasses as probed by persistent optical holes, Europhys. Lett., 7:517.

Kohler, W. and Friedrich, J., 1989, Probing of conformational relaxation processes of proteins by frequency labeling of optical states, J. Chem. Phys., 90:1270.

Meech, S.R., Hoff, A.J. and Wiersma, D.A., 1985, Evidence for a very early intermediate in bacterial photosynthesis. A photon-echo and hole burning study of the primary donor band in Rhodopseudomonas Sphaeroides, Chem. Phys. Lett., 121:287.

Schirmer, T., Bode, W., Huber, R., Sidler, W. and Zuber, H., 1985, X-Ray crystallographic structure of the light harvesting biliprotein C-phycocyanin from the thermophilic cyanobacterium Mastigocladus Laminisus and its resemblance to globin structures, J. Mol. Biol., 184:257.

Shuvalov, V.A., Klevanik, A.V., Ganago, A.O., Shkuropatov, A.Ya. and Gubanov, V.S., 1988, Burning of narrow spectral holes at 1.7K in the absorbance band of the primary electron donor of Rhodopseudomonas Viridis reaction centres with blocked electron transfer, FEBS Lett., 237:57.

Singh, G.P., Schink, H.J., Lohneisen, H.V., Parak, F. and
    Hunklinger, S., 1984, Excitation in metmyoglobin crystals
    at low temperatures, Z. Phys. B - Condensed Matter,
    55:23.
Tang, D., Johnson, S.G., Jankowiak, R., Small, G.J. and Tiede,
    D.M., 1989, Structure and marker mode of the primary
    donor electron state absorption of photosynthetic
    bacteria: Hole burned spectra, paper presented at the
    22nd Jerusalem Symposium.
Won, Y. and Friesner, R.A., 1988, Theoretical studies of
    photochemical hole burning in photosynthetic bacterial
    reaction centres, J. Phys. Chem., 92:2214.
Yang, I.-S. and Anderson, A.C., 1986, Specific heat of melanin
    at temperatures below 3 K, Phys. Rev., B34:2942.

THE JAHN-TELLER EFFECT IN PORPHYRINS AND HEMOPROTEINS

Isaac B. Bersuker and Solomon S. Stavrov

Institute of Chemistry, Academy of Sciences, MoSSR
Kishinev 277028, USSR

INTRODUCTION

The Jahn-Teller Effect (JTE) is the consequence of vibronic interactions and is a phenomenon of basic importance when interpreting dynamics, ligand binding characteristics and photoprocesses in hemoproteins (Bersuker, 1984; Bersuker and Polinger, 1989; Englman, 1972). The main statement here is that the instability of the high-symmetry configuration of any molecular (or crystal) system is due to (and only to) and controlled by either the degeneracy of the electronic ground state in the case of proper JTE, or the vibronic mixing of the ground and low-lying excited state in the case of the pseudo Jahn-Teller effect (PJTE). Furthermore, there are some simple relations between the parameters of the system which allow one to predict the possible instabilities and their dependence on composition of the local centre and on its environment. In particular, in the most widespread case of the PJTE the high-symmetry configuration becomes unstable with respect to symmetrized nuclear displacements if

$$F^2/\delta \geq K_0 \tag{1}$$

where δ is the energy gap between the ground state 1 and excited state 2, mixing under displacement q, $F=<1|(dH/dq)_0|2>$ is the vibronic coupling constant (H is the Hamiltonian) and $K_0=<1|(d^2H/dq^2)_0|2>$ is the force constant in the q direction in the absence of vibronic coupling.

When more than one excited state is involved in the vibronic mixing, the condition (1) changes to

$$\sum_i F_i^2/\delta_i \geq K_0 \tag{1a}$$

where i numerates excited states.

The condition (1) or (1a) for the configuration

instability emerges from the fact that as a result of the vibronic mixing of the ground state with the excited ones, the force constant of the former reduces to

$$K = K_0 - \sum_i F_i / \delta_i \qquad (2)$$

and becomes negative when (1a) is valid.

The instabilities arising due to the JTE and PJTE, being their primary consequences, initiate a series of new effects and regularities in the observable properties of polyatomic systems important to many areas of physics and chemistry of molecules and crystals including spectroscopy, stereochemistry, crystal chemistry and physics, phase transitions, chemical activation and reactivity, electron-conformational effects in biological systems (Bersuker, 1984; Bersuker and Polinger, 1989; Englman, 1972). Concerning the latter, the condition of instability (1) is of special significance, since these systems possess many low-frequency vibrations (soft modes) providing a small K_0 value for appropriate nuclear displacement q. In combination with non-filled (open shell) close-in-energy d states of transition metal active centres (small δ) the condition of instability (1) becomes most favourable. Having soft modes, the local distortion resulting from the vibronic coupling moves easily along the system producing conformational changes. If the instability occurs as a result of the appearance (disappearance, excitation) of an electron or a ligand, this effect is manifest as an electron conformational transition.

In this communication some of the vibronic effects inherent in metalloporphyrins and hemoproteins are discussed, especially those related to the chromophore-matrix interaction. Some of the results have a general meaning and can be applied to other biological systems with similar conditions. For a more detailed presentation of the background and applications of this approach the reader is referred to some other publications (Bersuker, 1984; Bersuker and Polinger, 1989; Englman, 1972; Bersuker and Stavrov, 1988).

INTERACTION BETWEEN METAL POSITION, SPIN STATE AND LIGAND COORDINATION

Consider first the instability of planar metalloporphyrins, Me(P), without extra ligands having D_{4h} symmetry with respect to the out-of-plane displacement of the metal atom which belongs to the A_{2u} representation of the D_{4h} point group. The selection rules for the matrix element $F_{ij} = \langle i/(dH/dq)_0/j \rangle$ restrict the possible states j which may contribute to the change of the i state force constant K_0 according to Eq. (2) to those that are formed by a one-electron transfer between the molecular orbitals (MO) shown in Fig. 1. The highest occupied MOs (HOMOs) in this figure corresponds to the low-spin states [in $Co^{II}(P)$, $Ni^{II}(P)$, $Cu^{II}(P)$, $Zn^{II}(P)$] or to those of intermediate spin values [in $Mn^{II}(P)$, $Fe^{II}(P)$]. The high-spin states can be obtained for $Mn^{II}(P)$, $Fe^{II}(P)$, and $Co^{II}(P)$ by means of an $e_g(d_\pi) \rightarrow b_{1g}(d_{x^2-y^2})$ excitation, and for $Ni^{II}(P)$ by an $a_{1g}(d_{z^2}) \rightarrow b_{1g}(d_{x^2-y^2})$

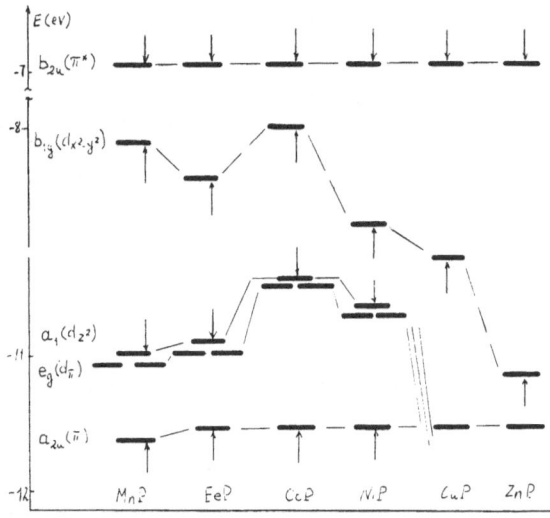

Fig. 1 Highest occupied and lowest unoccupied MOs
which contribute to the PJT mixing.

excitation. Thus all the states under consideration can be
divided into four classes: (1) partly occupied $a_{1g}(d_z2)$ and
unoccupied $b_{1g}(d_x2_{-y}2)$ MOs [the ground state of $Mn^{1g}(P)$,
$Fe^I(P)$, $Co^{II}(P)$ and maybe $Fe^{II}(P)$]; (2) occupied $a_{1g}(d_z2)$ and
unoccupied $b_{1g}(d_x2_{-y}2)$ MOs [ground state of $Ni^{II}(P)$, $Co^I(P)$,
$Fe^0(P)$, and possibly $Fe^{II}(P)$]; (3) occupied $a_{1g}(d_z2)$ and partly
occupied $b_{1g}(d_x2_{-y}2)$ MOs [ground state of $Cu^{II}(P)$ and $Zn^{II}(P)$];
(4) partly occupied $a_{1g}(d_z2)$ and $b_{1g}(d_x2_{-y}2)$ MOs [excited high-
spin states of $Mn^{II}(P)$, $Fe^{II}(P)$, $Fe^I(P)$, $Fe^0(P)$, $Co^{II}(P)$,
$Co^I(P)$, and $Ni^{II}(P)$].

The adiabatic potential surfaces for the electronic
states of these classes have been investigated earlier
(Bersuker and Stavrov, 1988). It was shown that the main
contribution to the softening of the out-of-plane displacement
of the metal atom in all these cases is given by the interaction with the upper state formed by one-electron $a_{2u}(\pi)$ >
$a_{1g}(d_z2)$ excitation. Therefore in the cases when the $a_{1g}(d_z2)$
MO is occupied, an essential softening and hence instability
of the in-plane position of the metal atom is not expected.
This determines the planar configuration of the
metalloporphyrins in the states of classes (2) and (3), in
particular, the case of $Zn^{II}(P)$ where due to the fully
occupied $a_{1g}(d_z2)$ MO the metal remains in the plane of the
porphyrin ring, even distorting it by displacing the pyrrole
nitrogen atoms.

In case of class (1) the possible stable configuration is
determined by condition (1): the triplet state of $Mn^{II}(P)$
(Fig. 1) is unstable, whereas the states of $Co^{II}(P)$ and
$Fe^{II}(P)$ just soften, most essentially in the last case. As
seen from Fig. 1, for the states from the class (4) there are
two low-lying upper states which admix in by the A_{2u}
displacements: one of them is the same as in the previous case
obtained by the $a_{2u}(\pi)$ >$a_{1g}(d_z2)$ excitation, the other is

formed by the $b_{1g}(d_{x^2-y^2}) > b_{2u}(\pi*)$ excitation. The second contribution enhances the softening of the in-plane position of the metal atom making it unstable. However these states become ground states of the system only if the John-Teller stabilization energy is larger than the excitation energy. Just this condition is realized in the high-spin $Mn^{II}(P)$.

In the $Me(P)(L_1)(L_2)$ system with axial ligands L_1 and L_2 the influence of the latter may be essential. It can be estimated by means of perturbation theory (Bersuker and Stavrov, 1988), that in the second order gives the following expression for the out-of-plane displacement of the metal atom:

$$q_o = (\sum_{i,j} p_i f_{ij}(A_{ij} + B_{ij})/\delta_{ji})/(K_0 - \sum_{i,j} p_i f_{ij}^2/\delta_{ji}) \qquad (3)$$

where F_{ij} is the orbital vibronic constant (Bersuker, 1984), the matrix element of the operator of vibronic interactions on the i and j MO of the $Me(P)$; p_i is the occupation number of the i-th MO; δ_{ij} is the energy gap between the appropriate MOs which differs from that of the free $Me(P)$ by the totally symmetric contribution of the interaction between $Me(P)$, L_1 and L_2; A_{ij} and B_{ij} are the matrix elements of the A_{2u} component of the crystal field of ligands L_1 and L_2, respectively.

It follows from Eq. (3) that $q_o \neq 0$ even in the case when condition (1a) is not fulfilled. This means that under certain conditions, the influence of ligands L_1 and L_2 may lead to the out-of-place position of the metal atom even when without these ligands the in-plane position is stable, provided that it is soft enough.

As seen from Eq. (3), the main contribution to the q_o value comes from the A_{ij} and B_{ij} terms that are due to the mixing of the most diffuse MO (in the axial direction) of $Me(P)$, namely $a_{2u}(\pi)$ with $a_{1g}(d_z 2)$ and $a_{1g}(d_z 2)$ with $a_{2u}(4p_z)$, by the crystal field of L_1 and L_2. On the other hand, q_o is the larger, the softer is the force constant of the A_{2u} displacement. These two conditions are most favorable in the cases of single occupation of the $a_{1g}(d_z 2)$ and $b_{1g}(d_{x^2-y^2})$ MO, and therefore the largest q_o is expected just in high spin states (provided all the other conditions remain the same). This conclusion is confirmed by the experimental data on high-spin complexes with one extra ligand $Mn^{II}(P)(L)$ and $Fe^{II}(P)(L)$ ($B_{ij}=0$). Note that in the case of high-spin $Fe(P)(L)$ q_o is roughly independent on the oxidation state, since the change $Fe^{II} \rightarrow Fe^{III}$ does not change the occupation numbers of the MO under consideration.

Eg. (3) allows one to derive some conclusions about the dependence of the out-of-plane displacement q_o on the nature of the metal atom Mn, Fe, and Co, ceteris paribus. The interchange of the atoms alters mostly the values A_{ij} (f_{ij} and δ_{ji} being much less influenced) which depend mainly on the axial diffuseness of the $a_{1g}(d_z 2)$ and $a_{2u}(4p_z)$ orbitals of the metal atom. Since these latter decrease in the order Mn>Fe>Co, the q_o values are also expected to be reduced in the same way, provided the ligand L remains the same. This conclusion is also confirmed by the experimental data: in the case of L=NO q_0=0.34 A, 0.21 A, and 0.09 A and for L=imidazole q_0=0.52 A, 0.42 A, and 0.13 A for Mn, Fe, and Co complexes, respectively.

The coordination of the second ligand reduces the q_0 value for two reasons. First, the crystal fields of the two ligands have opposite directions, and therefore $|A_{ij}+B_{ij}| < |A_{ij}|$ (in particular, when the two ligands are equal $A_{ij}+B_{ij}=0$). Second, the presence of the second ligand increases the δ_{ji} values which also reduces the q_0 value. This effect is especially important in the case of extra coordination of strong field ligands such as CO, O_2, CS etc.

In general, the vibronic approach allows one to explain a wide spectrum of experiments on the change of spin and geometry of metalloporphyrins by coordination of ligands of different nature, including the changes of geometry of the heme in hemoglobin, peroxidase and cytochrome P-450 by coordination of CO, O_2, and other ligands, as well as the absence of such changes by the change of the oxidation state of the iron atom in cytochrome c, etc.

GEOMETRY OF COORDINATION AND PHOTOLYSIS OF DIATOMICS

It is known that the coordination of diatomics, such as CO, NO, O_2, to metalloporphyrins and hemoproteins may have different geometries among which the linear (perpendicular to the porphyrin ring) Me(P)(L)[1] and bent Me(P)(L)[b] seem to be the most important. Consider first the stability of the linear case with respect to its possible distortion toward the bent configuration.

The Me(P)(L)[1] complex has C_{4v} symmetry and the bending of the ligand L is a E type symmetrized displacement of this point group. Taking this into account and the HOMOs and LUMOs of the complexes shown in Fig. 2, one can see that the contribution to the softening of the E bending under consideration is due to the mixing of the $e(d_\pi+\pi^*_L)$ and $e(d_\pi-\pi^*_L)$ MOs with $a_1(d_z2)$ one. If this softening is large enough so that the condition (1a) is fulfilled, then the linear coordination is unstable and the bent position is expected.

In the case of the CO ligand the energy gap between the $e(d_\pi+\pi^*_L)$ and $a_1(d_z2)$ MO is too large (Fig. 2) and the linear coordination is expected to be stable. Its instability could be expected if the $e(d_\pi-\pi^*_L)$ is populated. But this is not the case in all known real complexes, and this explains in a natural way the observed linear coordination in these systems (Bersuker and Stavrov, 1988).

In the case of Me(P)(O_2)[1] the mixing of $e(d_\pi+\pi^*_L)$ with the $a_1(d_z2)$ MOs may already be enough for its instability, but the population of the $e(d_\pi-\pi^*_L)$ MO makes this effect much stronger. As can be seen from Fig. 2, this MO becomes populated when the number of 3d electrons exceeds four. Therefore for complexes with $3d^n$ electrons with n>4 (Mn[II], Fe[III], Fe[II], and Co[II]) the coordination of O_2 has to be strongly bent.

In Me(P)(NO)[1], the two energy gaps $2\delta_1$ and $2\delta_2$ are almost

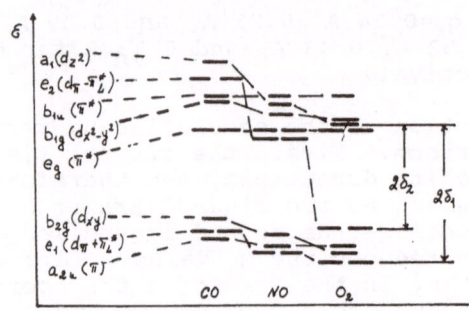

Fig. 2   Schematic MO energy diagram for Me(P)(L)[1]
(L=CO, NO, O_2)

the same as in Me(P)(CO)[1] but the appropriate MO populations
are different. In the $3d^n$ complexes with the NO ligand, the
$e(d_\pi-\pi*_L)$ MO becomes populated if n>5 resulting in a bent
coordination, and it remains unpopulated for n≤5 that leads to
undistorted linear coordination. This conclusion is confirmed
by X-ray investigations of the complexes of Fe^{III}(P), Mn^{II}(P)
(n=5), Fe^{II}(P) (n=6), and Co^{II}(P) (n=7) with NO as an extra
ligand (Bersuker and Stavrov, 1988).

     In general it is seen that essential distortion of the
linear coordination is expected when the $e(d_\pi-\pi*_L)$ MO is
populated. We intend to show now that just the population of
this MO is also responsible for the characteristic features of
photodecay of complexes Me(P)(L). To do this, let us consider
the possible photodecay along the E displacement described
above, that bends the linear coordination. We assume that the
photodecay takes place directly from the excited state in
which the system occurs after light absorption, as it follows
from experimental and theoretical investigations (Bersuker and
Stavrov, 1988). These states are formed by a one-electron
excitation from a nondegenerate MO of the ground state to the
e type MO having essential $e(d_\pi-\pi*_L)$ nature. The population
of this MO, as stated above, results in significant
instability and distortion of the linear coordination. The
adiabatic potential curve E_1 for this case is illustrated in
Fig. 3, the E_2 term determining the decay of the system being
also shown. Owing to the repulsion of these two terms which
(judging on the high quantum yield of the photodecay of
Me(P)(L)[1]) is strong enough, the shape of the adiabatic
surface of the state populated by light absorption coincides
with that of E shown in Fig. 3.

     Fig. 3 shows that if the system after photoexcitation
falls near to the point of linear coordination (ρ=0) the
photodecay takes place immediately since the point ρ=0 is
higher than the maximum of the activation barrier of decay (in
the state populated by light). Another situation emerges if
the system after light absorption occurs near the bottom of
the well (the minimum) of the adiabatic potential of the
excited state. In this case the efficiency of the photodecay
is determined by the competition between the two possibilities
of the system to either overcome the activation barrier of
photodecay, its rate being dependent on the thermal population

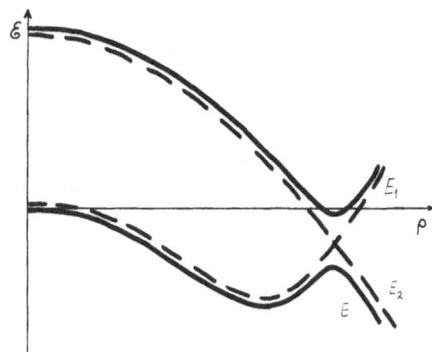

Fig. 3   Potential energy curves of excited and decay
terms with (solid lines) and without (broken
lines) interaction between them.

of excited vibrational states in the well, or to relax from
the excited electronic state to a lower-in-energy and
nonactive (with respect to photodecay) state. The high
relaxation rates ($\approx 10^{-12}$ - 10^{-13} sec) result in low quantum
yields of photolysis in this case.

The system Me(P)(L)[1], having in the ground state a
minimum for the adiabatic potential at ρ=0, after a Franck-
Condon light absorption process falls into the same point ρ=0
and hence undergoes a direct activationless photodecay. On
the contrary, the system Me(P)(L)[b] with a minimum displaced
from the point ρ=0, after the light absorption occurs near the
minimum of the excited state, results in an activation process
with low quantum yield of photodecay. These conclusions
coincide with the well known experimental data (Bersuker and
Stavrov, 1988; Hoshino and Kogure, 1989). It follows also
from the mechanism of photolysis suggested here that in the
cases of metalloporphyrins with diatomics the latter emerge
from the process of photodecay in an excited rotational state
appropriate to the E displacement. This prediction can be
verified experimentally by picosecond infrared or resonance
Raman spectroscopy.

VIBRONIC EFFECTS IN EXCITED ELECTRONIC STATES

In the symmetry point group D_{4h} of Me(P) the excited B
and Q states that provide the characteristic B and Q bands of
light aborption belong to the E_u representation. It was shown
(Canters and van der Waals, 1978) that in the Q state of
Mg(Por) and Zn(Por) (Por=porphine) the vibronic interaction
with the b_{1g} displacements contributes essentially to a series
of observable magnitudes. Below we consider the influence of
low-symmetry external fields - the crystal field of a
Shpol'ski matrix or of the protein globule, porphyrin ring
substitutions, metal substitution by protons in free bases, -
in combination with the b_{1g} displacements, on the splitting of
zero-phonon lines and vibrational satellites of the B and Q
bands of optical absorption. The investigation is carried out
in the framework of the Gouterman four-orbital model which
relates the origin of the absorption lines under consideration

to transitions between the ground state and two excited B and Q states formed by the exchange interaction of two one-electron transitions, $a_{2u} \rightarrow e_g(\pi*)$ and $a_{1u} \rightarrow e_g(\pi*)$. The shape of the adiabatic potential of the four components of these two E_u states, B and Q, were found under the assumption that they do not interact strongly (Bersuer and Stavrov, 1988). It was shown that the vibronic interaction makes the force constant different

$$K_Q = K_0(1-p); \quad K_B = K_0(1+p) \tag{4}$$

where p is a magnitude, characterizing the strength of the PJT mixing of the B and Q states: $p = F^2/(K_0\delta)$ $(p \geq 0)$, and the ratio of the splittings ΔE of each of the B and Q states due to the vibronic interaction is

$$\mu = |\Delta E_Q| / |\Delta E_B| = (1+p)/(1-p) \geq 1 \tag{5}$$

In other words the splitting of the Q state by b_{1g} external field due to the vibronic coupling with b_{1g} displacements $|\Delta E_Q|$ is always larger than or equal to the similar splitting of the B state, $|\Delta E_B|$. The case of p=0 coincides with the Gouterman four-orbital model which completely ignores vibronic interactions.

The p value can be determined from experimental data on the Q state (Bersuker and Stavrov, 1988). For Zn(Por) p=0.4, while for Cd(TBP) (TBP=tetrabenzporphine) p≈0, and hence μ=3.33 and μ=1 for these two systems respectively. The experimental values are respectively, μ=2.1 and μ=0.9, also confirming the theoretical predictions. This approach allows one to explain the origin of the μ value in $H_2(P)$, since the substitution Me H_2 can be described by a field of b_{1g} symmetry. It follows that the splittings of the Q and B bands are expected to be rather different in $H_2(Por)$ and almost the same in $H_2(TBP)$. This conclusion is also in good agreement with the experimental data.

It can be shown that the field of b_{2g} symmetry, distinguished from that of b_{1g} symmetry, is reduced by the vibronic interaction with the b_{1g} displacements (Bersuker and Stavrov, 1988), the vibronic reduction being much larger for the Q state, than for the B one. Therefore, the larger splitting of the B band (or its lines), as compared with the Q one, testifies to the presence of external fields of b_{2g} symmetry as the reason for the observed splittings.

The consideration of simultaneous influence of b_{1g} and b_{2g} fields allows one to make conclusions about the origin of the line positions in the absorption spectra and to consider the invert problem. In particular, the characteristic position of the zero-phonon lines of the Q band in $H_2(mesoP)$ in different matrices and in proteins obtained (Fidy et al., 1989) allows to conclude that the contribution of the b_{1g} and b_{2g} components of the external fields to the observed line splittings are approximately equal. In other words the interactions of all the porphyrin substituents with the matrix contribute to the characteristic features of the optical absorption spectrum. ·

REFERENCES

Bersuker, I.B., 1984, "The Jahn-Teller Effect and Vibronic
    Interactions in Modern Chemistry", Plenum, New York.

Bersuker, I.B. and Stavrov, S.S., 1988, Structure and
    properties of metalloporphyrins and hemoproteins: The
    vibronic approach, <u>Coord. Chem. Rev.</u>, 88:1.

Bersuker, I.B. and Polinger, V.Z., 1989, "Vibronic
    Interactions in Molecules and Crystals", Springer-Verlag,
    Berlin.

Canters, G.W. and van der Waals, J.H., 1978, High-Resolution
    Zeeman Spectroscopy of Metalloporphyrins, <u>in</u>: "The
    Porphyrins", Vol. III, D. Dolphin, ed., Academic, New
    York.

Englman, R., 1972, "The Jahn-Teller Effect in Molecules and
    Crystals", Wiley, New York.

Fidy, J., Paul, K.-G. and Vanderkooi, J.M., 1989, The
    mechanism of phototautomerization in mesoporphyrin
    horseradish peroxidase. Studies by fluorescence line-
    narrowing spectroscopy, <u>J. Phys. Chem.</u>, 93:2253.

Hoshino, M. and Kogure, M., 1989, Photochemistry of nitrosyl
    porphyrins in the temperature range 180-300 k and the
    effects of pyridine on photodenitrosylation of
    nitrosyliron tetraphenylporphyrin, <u>J. Phys. Chem.</u>,
    93:5478.

FLUORESCENCE SITE SELECTION SPECTROSCOPY ON HEME PROTEINS

Judit Fidy and Jane M. Vanderkooi

Department of Biochemistry & Biophysics, School of
Medicine, University of Pennsylvania, Philadelphia
PA 19104, USA, and
Inst. of Biophysics, Semmelweis
Medical University, H-1088 Budapest, Puskin u.9
Hungary

INTRODUCTION

Site selection spectroscopic techniques allow one to
obtain high resolution optical absorption or emission spectra
of molecules (atomic groups) embedded in specific matrices.
For organic systems, the fluorescence site selection
spectroscopic technique (or fluorescence line narrowing: FLN)
was first applied by Personov et al. (1972) and the absorption
spectroscopic technique, spectral hole burning, by Kharlamov
et al. (1974) and by Gorokhovskii et al. (1974). The
application of the method, however, for molecules embedded in
biological matrices of native structure (e.g. proteins) has
been fairly limited (Friedrich et al., 1981a, b; Avarmaa et
al., 1984; Renge et al., 1984, 1988; Gillie et al., 1987;
Boxer et al., 1987). In our laboratory, fluorescent
substitution for iron porphyrins has been used for heme
proteins to produce molecules of native conformation
appropriate for the application of luminescence spectroscopy.
Our studies on myoglobin, hemoglobin, cytochrome c and
horseradish peroxidase demonstrated that high resolution
emission spectra with site selection characteristics can be
obtained for small monomeric heme proteins (Angiolillo et al.,
1982; Horie et al., 1985; Vanderkooi et al., 1985). In this
paper, further evidence for the applicability of the FLN
technique for heme proteins is presented and applications of
the technique for studying substrate binding effects are
demonstrated.

MATERIALS AND METHODS

The FLN technique requires the use of laser excitation at
cryogenic temperatures and the determination of high
resolution emission spectra. As excitation source, a

continuous wave tunable Argon ion - rhodamine dye laser system
(Coherent, Palo Alto, CA) was used, and the emission spectra
were measured at a resolution of 1 cm^{-1} via a J-Y Ramanor HG2
double holographic monochromator equipped with a computerized
data collection and evaluation system (Instruments SA, Edison,
NJ). Low temperature was achieved by a liquid helium dewar
(Air Products, Allentown, PA); the samples were brought to
cryogenic termperatures within several seconds to avoid
structural relaxation of the protein during cooling.
Substituted versions of cytochrome c (cyt c) and horseradish
peroxidase (HRP) were synthesised as described previously
(Vanderkooi et al., 1985). The fluorescent porphyrin used for
substitution was mesoporphyrin IX (MP) shown in Figure 1. The
samples were prepared in 50 mM ammonium acetate at a
concentration of 1-50 μM, and 50% glycerol was added to assure
transparency at low temperature. Aromatic substrates for HRP
were from Sigma Chemical Co. except for naphthohydroxamic acid
which was synthesised in Dr. G. R. Schoenbaum's Laboratory
(Children's Hospital, Memphis TN), and kindly donated for our
purposes.

RESULTS AND DISCUSSION

Evidence for site selection spectroscopy

      a) Highly resolved relaxed emission spectra. The FLN
spectrum of Zn-cyt c measured by vibronic excitation in S_1 is
shown in Figure 2A and that measured by 0← 0 excitation in
Figure 2B. Arrows in the figure indicate the position of
excitation energies. Upon vibronic exciation (Figure 2A), the
highest energy emission lines originate from $S_1^{\,0}$ levels after
relaxation and lead to $S_0^{\,0}$ (0← 0 emissions); these are the
lines around 17000 cm^{-1}. Emissions from $S_1^{\,0}$ to higher $S_0^{\,v}$
vibrational levels in the ground state are represented with
lower resolution. It is important to note that above the
range of the emission lines around 17000 cm^{-1}, no comparable
emission intensity was detected, indicating the fact of
relaxed emissions (the small intensities above 17000 cm^{-1} were
proved to originate from denatured cyt c molecules
contributing a background to the spectra). If the excitation
leads to $S_1^{\,0}$, only the transitions to the $S_0^{\,v}$ levels are
represented but, with increased resolution (Figure 2B).

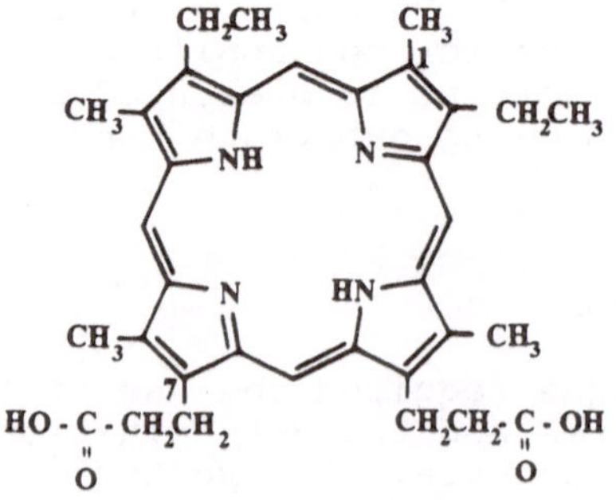

Fig. 1 Mesoporphyrin IX

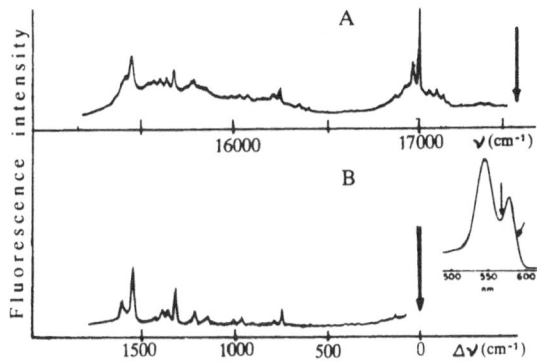

Fig. 2　FLN spectra of Zn-cyt c at 5.2 K, arrows
indicate the excitation wavelengths.　2A:
excitation energy is 17600 cm^{-1}; 2B:
excitation energy is 17000 cm^{-1}.　On the
abscissa the difference between excitation
and emission energy is indicated.　Inset.
absorption spectra at room temperature,
arrows indicate the excitation energies.

b) Inhomogeneous distribution of the chromophore can be
determined. Use the sharp $0 \leftarrow 0$ emission line intensities
measured by vibronic excitation in FLN was suggested by
Fuenfschilling et al. (1981, 1982) for determination of the
inhomogeneous distribution of the chromophore in the matrix.
If these line intensities are related to the n(E)
inhomogeneous distribution function (IDF) of the chromophore
transition energies (E), then the $0 \leftarrow 0$ intensity originating
from the excitation of the same vibronic level in function of
the excitation energy will follow the functional form of IDF
as

$$I = K\epsilon n(E)f$$

where K is a constant, I is the line intensity of a sharp
$0 \leftarrow 0$, ϵ is the transition probability from ground state to the
chosen vibrational level (without phonon interaction), f is
the emission probability from S_1^0 to S_0^0 (without phonon
interaction), n is the IDF and E is the $0 \leftarrow 0$ transition
energy, identical with the difference of excitation energy and
chosen vibrational energy in S_1.　The measurement should lead
to the same functional form independent of the S_1^v level
chosen for the measurement.　If the $0 \leftarrow 0$ line intensities
satisfy this requirement, site selection is proved, and IDF
describes the true inhomogeneously broadened $Q_x(0,0)$ line
shape at low temperatures.

The change in $0 \leftarrow 0$ line intensities is shown in Figure 3A
for ZnMP-HRP upon changing the excitation energy.　The lines
seen in the spectral range of $0 \leftarrow 0$ emissions originate from
the excitation of the $v = 727$ cm^{-1}, $v = 771$ cm^{-1} and $v = 694$
cm^{-1} vibrational level in S_1.　The measurement for the
determination of n(E) for ZnMP-HRP was performed by varying
the excitation energy in steps of 10 cm^{-1}; the data for all
three lines lead to the same function as plotted in Figure 3B.
The data points then were fitted by a Gaussian distribution

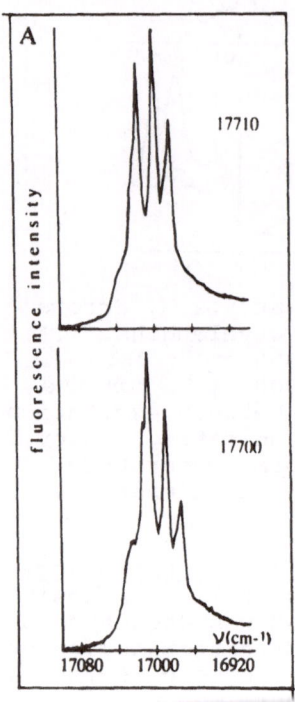

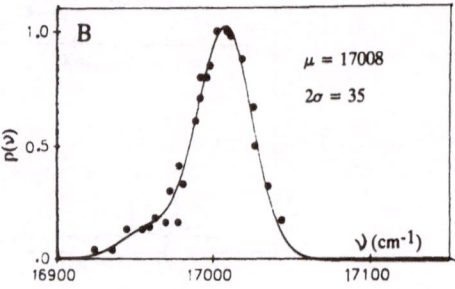

Fig. 3 Determination of IDF for ZnMP-HRP. A: FLN
spectra in the range of 0←0 energies
measured at 7.8 K by the same flux but
different excitation photon energies as
indicated. B: normalized data for IDF and
fitted Gaussian plotted as a function of the
0←0 transition energy. Fitting parameters
are given in cm^{-1}.

Table 1 Parameters μ and 2σ of Gaussian fit to the
data points for an inhomogeneous distribu-
tion function expressed in cm^{-1} for
different HRP complexes. In the case of
energy splitting, values of μ and 2σ for
both functions are presented.

Sample	μ (cm^{-1})		2σ (cm^{-1})	
MP-HRP	16100	16000	52	52
ZnMP-HRP	17008		35	
MP-HRP/BHA	16168		56	
MP-HRP/NHA	16175		43	
MP-HRP/HQ	16215	16175	28	32
MP-HRP/β-RE	16245		60	
ZnMP-HRP/NHA	17240		60	

function as $(\exp-(E-\mu)^2/2\sigma)$, and the IDF was characterized by the mean $0\leftarrow0$ energy μ and width of distribution 2σ as shown in the Figure and in Table 1. The parameter values refer only to the main component of ZnMP-HRP; the other population, being present at a level of 10% and yet not identified, was neglected.

Characterization of MP-HRP

The IDF for free base MP-HRP (Fidy et al., 1989a) is shown in Figure 4A, the parameters can be found in Table 1.

Based on our previous studies (Fidy et al. 1989a, Zollfrank et al. 1991, Fidy et al. 1991b), we believe that the four spectral bands correspond to different tautomeric forms of MP, stabilized by the protein. Form 1 is the stable form at room temperature, at cryogenic temperatures, it transforms under the exciting power of 40 mW into a small population of form 2 and a large population of form 3 within a few seconds. The sample being then mostly of type 3, transforms into form 4 within a time scale of 20 min. at 5 K. Around 30 K, form 4 can be reverted into form 3 in dark. A proper excitation wavenumber at this temperature may lead to the excitation of both populations, as is shown in Fig. 5A; where form 3 is represented by an emission line at 16100 cm^{-1}, while the population of form 4 contributes to the FLN spectrum of MP-HRP with the line at 16000 cm^{-1}. The time course of transformation of form 3 into form 4 at lower temperatures is shown in in Fig. 5B when irradiation by a power of 40 mW.

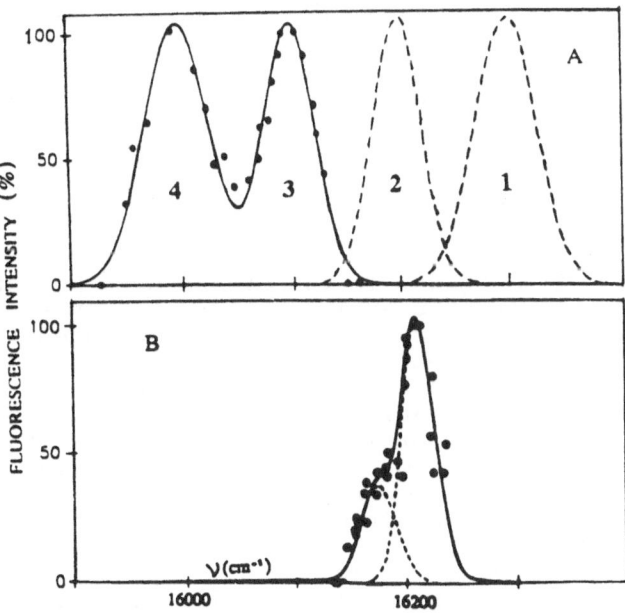

Fig. 4   IDF for MP-HRP.   A: free protein B: 25 mM HQ is added.   Exciting power - 40mW.

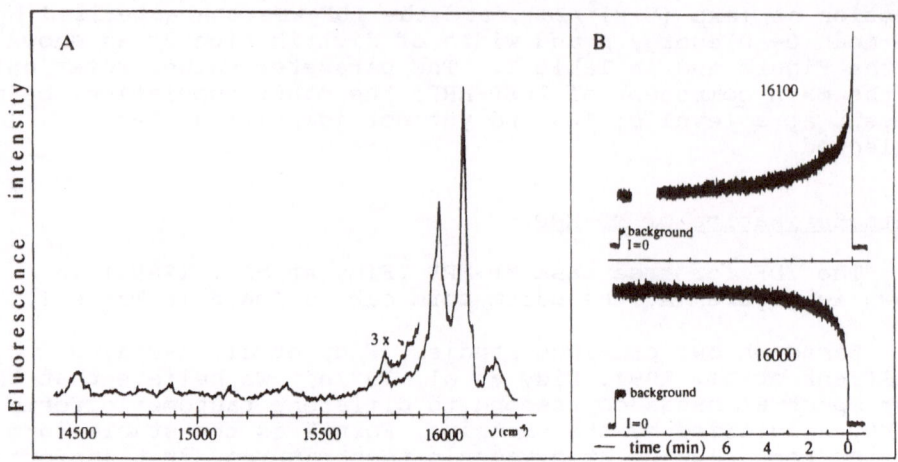

Fig. 5   A: FLN spectra of MP-HRP measured by
excitation at 17310 cm^{-1} at 30 K. B: line
intensity changes as a function of time at
16000 cm^{-1} and 16100 cm^{-1} as indicated
when irradiated at 17310 cm^{-1} at 5K.

The mean transition energies for the spectral components
and values for the inhomogeneous band width are given in Table
1.   The comparison of the width values for MP and Zn-MP in HRP
shows that the presence of a metal capable of achieving 5
coordination reduces the structural variability of the heme
pocket.   In case of ZnMP-HRP, fluorescence lifetime
measurements (Fidy et al. 1991a) indicated that within the
single maximum IDF in Figure 3, two populations are present,
not resolved in energy.   For Zn porphyrins, it is expected
that vibronic interactions may lead to deformational
resolution of the energy degeneracy of the S_1 excited state
(Hoffman & Ratner, 1978).   However, the IDF of this molecule
shows that the asymmetry of the heme pocket potential in HRP
does not lead to energy speration for the in-plane deformed
forms.

Substrate binding effects

In Figure 4B, a change in the MP-HRP heme pocket
structure upon binding the aromatic hydrogen donor
hydroquinone (HQ) is indicated through the changed IDF of 0< 0
transition energy (Fidy et al., 1989b).   We have studied the
effect of a series of aromatic substrates, benzohydroxamic
acid (BHA), naphthohydroxamic acid (NHA), isomeric resorcyclic
acids (e.g. β-RE) and resorcinol (RE).   All aromatic
substrates strongly reduced or abolished the energy separation
for the tautomeric forms of MP in HRP.   Changes in the solvent
shift is hard to evaluate without identifying the structure of
the tautomeric conformations 1 - 4 for the free MP-HRP.   We
suggest that the position of form 2 be compared to that of the
band for MP-HRP/HQ.   Then, a slight shift towards higher
energies is seen.   The effect is also shown by the ZnMP-
HRP/NHA complex, which is an indication that the effect is not
the result of the nature of MP, but is related to substrate

binding (Fidy et al. 1991a). The substrate, also reduced the motional freedom of the protein structure in the pocket as shown by the reduction in width of the IDF. The narrow distribution functions in the case of HQ as shown in Figure 4B allows an estimation of the energy separation as 40 cm^{-1}, which was not possible to resolve in the other substrate-bound cases. The results show that the substrate changes the electric potential within the heme pocket significantly, either by shielding polar groups or inducing a conformational change which allows new polar groups to be expressed near the heme.

Neither conventional spectroscopic studies nor reported resonance Raman and NMR studies (discussed in Fidy et al., 1989b) reveal differences between the effects of binding of the aromatic molecules studied. The NMR results (Sakurada et al., 1986; Thanabal et al., 1988) indicating a steric disturbance of the porphyrin ring by the substrate around the 8-methyl and 7-propionate group of the heme were confirmed by our spectral data which reveal steric disturbances through the phonon coupling of vibronic transitions (Fidy et al., 1989b).

ACKNOWLEDGEMENTS

The authors thank Dr. G. Holtom from the Regional Laser and Biotechnology Laboratories, Department of Chemistry, University of Pennsylvania for valuable discussions.

REFERENCES

Angiolillo, P.J., Leigh, Jr. J.S. and Vanderkooi, J.M., 1982, Resolved Fluorescence Emission Spectra of Iron-Free Cytochrome c, Photochem. Photobiol., 36:133.

Avarmaa, R., Renge, I. and Mauring, K., 1984, Sharp-line Structure in the Fluorescence and Excitation Spectra of Greening Etiolated Leaves, FEBS Letts., 167:186.

Boxer, S.G., Gottfried, D.S., Lockhart, D.J. and Middendorf, Th., R., 1987, Nonphotochemical Hole Burning in a Protein Matrix: Chlorophyllide in Apomyoglobin, J. Chem. Phys., 86:2439.

Fidy, J., Koloczek, H., Paul, K.-G. and Vanderkooi, J.M., 1987, The pH Dependence of Phototautomerism in Horseradish Peroxidase Monitored by Fluorescence Site Selection Sepctroscopy, Chem. Phys. Letts., 142:562.

Fidy, J., Paul, K.-G. and Vanderkooi, J.M., 1989a, The Mechanism of Phototautomerization in Mesoprphyrin Horseradish Peroxidase. Studies by Fluorescence Line-Narrowing Spectroscopy, J. Phys. Chem., 93:2253.

Fidy, J., Paul, K.-G. and Vanderkooi, J.M., 1989b, Differences in the Binding of Aromatic Substrates to Horseradish Peroxidase Revealed by Fluorescence Line Narrowing, Biochemistry, 28:7531.

Fidy, J., Holtom, G.R., Paul, K.-G. and Vanderkooi, J.M., 1991, The binding of naphthohydroxamic acid to horseradish peroxidase montiored by Zn-mesoporphyrin fluorescence line narrowing, J. Phys. Chem., in May.

Friedrich, J., Scheer, H., Zickendracht-Wendelstadt, B. and Haarer, D., 1981a, Photochemical Hole Burning: A Means to Observe High Resolution Optical Structures in

Phycoerythrin, <u>J. Chem. Phys.</u>, 74:2260.

Friedrich, J., Scheer, H., Zickendracht-Wendelstadt, B. and Haarer, D., 1981b, High Resolution Optical Studies on C-Phycocyanin via Photochemical Hole Burning <u>J. Amer. Chem. Soc.</u>, 103:1030.

Fuenfschilling, J. and Zschokke-Graener, I., 1982, The Determination of the Distribution of Site Energies of Chlorophyll b in an Organic Glass, <u>Chem. Phys. Lett.</u>, 91:122.

Fuenfschilling, J. and Zschokke-Graener, I. and Williams, D.F., 1981, The Determination of the Site-energy Distribution of Organic Molecules Dissolved in Glassy Matrices, <u>J. Chem. Phys.</u>, 75:3670.

Gillie, J.K., Hayes, J.M., Small, G.J. and Golbeck, J.H., 1987, Hole Burning Spectroscopy of a Core Antenna Comlex, <u>J. Phys. Chem.</u>, 91:5524.

Gorokhovskii, A.A., Kaarli, R.K. and Rebane, L.A., 1974, Burnout of the Dip in the Contour of a Purely Electron Line in Spolskii Systems, <u>JETP Lett.</u>, 20:216.

Hoffman, B.M. and Ratner, M.A., 1978, Jahn-Teller Effects in Metalloporphyrins and Other Four-Fold Symmetric Systems, <u>Molec. Phys.</u>, 35:901.

Horie, T., Paul, K.-G. and Vanderkooi, J.M., 1985, Study of the Active Site of Horseradish Peroxidase Isoenzymes A and C by Luminescence, <u>Biochemistry</u>, 24:7935.

Kharlamov, B.M., Personov, R.I. and Bykovskaya, L.A., 1974, Stable Gap in Absorption Spectra of Solid Solutions of Organic Molecules by Laser Irradiation, <u>Opt. Commun.</u>, 12:191.

Personov, R.I., Al'shits, E.I. and Bykovskaya, L.A., 1972, Effect of Fine Structure Appearance in Laser Excited Fluorescence Spectra of Organic Compounds in Solid Solutions, <u>Opt. Commun.</u> 6:169.

Renge, I., Mauring, K. and Avarmaa, R., 1984, High Resolution Optical Spectra in vivo. Photoactive Protochlorophyllide in Etiolated Leaves at 5 K, <u>Biochem. Biophys. Acta.</u>, 766:501.

Renge, I., Mauring, K and Vladkova, R., 1988, Zero Phonon Transitions of Chlorophyll a in Mature Plant Leaves Revealed by Spectral Hole Burning Method at 5 K, <u>Biochem. Biophys. Acta.</u>, 935:333 (1988).

Romanovskii, Yu. V., Bykovskaya, L.A. and Personov, R.I., 1981, Fine-Structure Luminescence of Etio- Copro- and Mesoporphyrins on Selective Laser Excitation, <u>Biophysics</u> 26:631.

Sakurada, J., Takahashi, S. and Hosoya, T., 1986, Proton Magnetic Resonance Studies on the Iodide Binding by Horseradish Peroxidase, <u>J. Biol. Chem.</u>, 262:4007.

Thanabal, V., La Mar, G.N. and de Ropp, J.S., 1988, A Nuclear Overhauser Effect Study of the Heme Crevice in the Resting State and Compound I of Horseradish Peroxidase; Evidence for Cation Radical Delocalization to the Proximal Histidine, <u>Biochemistry</u> 27:5400.

Vanderkooi, J.M., Moy, V.T., Maniara, G., Koloczek, H. and Paul, K.-G., 1985, Site-Selected Fluorescence Spectra of Porphyrin Derivatives of Heme Proteins, <u>Biochemistry</u> 24:7931.

Zollfrank, J., Friedrich, J., Vanderkooi, J.M. and Fidy, J., 1991, Conformational relaxation of a low temperature protein as probed by photochemical hole burning: horseradish peroxidase, <u>Biophys. J.</u>, in February.

CORRELATION BETWEEN PROTEIN STRUCTURE AND ROOM TEMPERATURE
PHOSPHORESCENCE OF TRYPTOPHANS

Jane M. Vanderkooi[1], Sandor Papp[2], Jeffrey W.
Berger[1], Wayne W. Wright[1] and Walter Englander[1]

[1]Department of Biochemistry & Biophysics, School
of Medicine, University of Pennsylvania
Philadelphia, PA 19104
[2]University Medical School of Debrecen, Dept. of
Biophysics, H4012 Debrecen 12, Hungary

INTRODUCTION

All excited-state molecules are more reactive than their
ground-state counterparts, and therefore every molecule in the
excited-state can act as a new chemical species, with
properties distinct from those of the ground state molecule.
Whether the excited species undergoes chemical reactions
depends upon the excited-state lifetime, as well as the mutual
reactivity and accessibility of the excited-state molecule
with neighboring molecules. In this regard, in addition to
being considered as physical phenomena, fluorescence and
phosphorescence can be used as chemical relaxation techniques.

The sensitivity of an excited-state to the environment is
a function of the excited-state lifetime. Fluorescence, which
is a spin-allowed process, occurs on a time scale of
nanoseconds whereas spin-forbidden phosphorescence occurs on a
time scale of μseconds to seconds. This long lifetime is on
the time-scale of many molecular processes, including enzyme
turn-over, slow motions of protein domains and motions of
large complexes. Therefore, phosphorescence offers a
relatively unexploited time window for the study of many
interesting processes occurring in proteins.

The experiments described here are based upon the
following rationale: we form a reactive species in the protein
by exciting a particular residue with light and observe the
relaxation from the excited state by phosphorescence. The
decay of phosphorescence and interactions of the triplet state
with neighboring molecules are sensitive to the proteins's
structure, composition and dynamics, thus affording us a
method of studying these parameters.

EXPERIMENTAL

The paper by Calhoun et al. (1988) gives the sources of supplies and procedures for sample preparation. Oxygen was removed from the samples as described in the cited paper. Phosphorescence lifetimes were measured using the instrument previously described (Green et al., 1988). Protein concentrations were typically 1 to 2 mg/ml and the temperature was 22°.

Phosphorescence lifetimes were obtained from decay profiles by analysis for exponential decay. The experimental procedure selects for the longest-lived component of the decay, and the decay of phosphorescence could generally be well fit by a single-exponential function. Quenching was monitored by decrease in lifetime, and the quenching rate constant, k_{qe}, was computed from:

$$\tau_o/\tau = 1 + k_{qe} \tau_o [Q]$$

as described by Stern and Volmer (1919) but modified for lifetimes. Here τ_q is the lifetime in the absence of quencher and τ is the lifetime in the presence of the externally added quencher at concentration [Q].

Coordinates for alkaline phosphatase were obtained from Dr. H.W. Wyckoff (New Haven CT); other protein coordinates were obtained from the Protein Data Bank (Brookhaven National Laboratory, Upton L.I., N.Y). Three-dimensional representations of the structures were plotted using Biograph (BioDesign, Pasadena, CA).

RESULTS AND DISCUSSION

Emission from proteins

Extrinsic labels can be used as triplet-state probes of proteins. Nearly every molecule that fluoresces will also phosphoresce, and so there are many potential probe molecules. Yields of intersystem crossing are enhanced by incorporating heavy atoms such as iodide or bromine. The use of such probes has been reviewed (Vanderkooi and Berger, 1989). Although extrinsic or modified intrinsic labels have advantages in high quantum yields, they add an uncertainty to the experiment because the protein needs to be altered, and it is often difficult to estimate the effect of alteration on the protein structure. It is therefore advantageous to use the phosphorescence from intrinsic chromophores in the protein.

Anecdotal evidence for phosphorescence at room temperature from proteins has been cited for a long time. For instance, in Renaissance Italy an art form arose whereby egg white albumin was used as a basis of paint for pictures. The pictures were exposed to light and then brought into a darkened room. A long-lived emission, lasting several seconds, was observed (Harvey, 1957). It is likely that this is tryptophan phosphorescence emission, but unfortunately this art form is lost so we cannot be sure.

In spite of such early observations of long-lived emission from proteins under room temperature conditions, it was generally recognised only very recently that phosphorescence could be observed from proteins at room temperature. In 1974, Saviotti and Galley published phosphorescence excitation and emission spectra that convincingly demonstrated that some proteins exhibit phosphorescence at room temperature. In Figure 1, the emission spectrum of liver alcohol dehydrogenase (ADH) is shown. The emission is broader than observed at low temperature, but characteristic resolved peaks occurring at 405 and 445 nm convincingly prove that the emission is tryptophan phosphorescence.

Two reasons can be adduced as to why phosphorescence from proteins at room temperature was not observed earlier. One reason is that the yields are very low and the emission from ADH, for example, was only seen after a gating system was used to eliminate the fluorescence (Figure 1). The second reason is that oxygen quenches phosphorescence. The Stern-Volmer quenching constant for liver alcohol dehydrogenase is 3×10^8 $M^{-1} s^{-1}$ and τ_o, the lifetime in the absence of oxygen, is 300 ms. The oxygen level must be reduced to ~ 1 nM to achieve less than 10% reduction in lifetime. It is apparent that heroic efforts must be made to reduce the oxygen to this level, but with care it is technically possible to do so on a routine basis.

One question is whether proteins, in the absence of oxygen, will generally show long-lived phosphorescence at room temperature. We surveyed 40 proteins and found 30 of them to have luminescence with lifetimes greater than 1 ms (Vanderkooi et al., 1987). The lifetimes of tryptophans reported in the literature are given in Table 1. The conclusion is that most proteins that have tryptophans will phosphoresce with long lifetimes at room temperature.

In the absence of oxygen quenching, it is evident that the lifetimes of tryptophan phosphorescence have tremendous variability in proteins, ranging from microseconds for the fully exposed tryptophan to seconds, characteristic of tryptophan in frozen solution. This contrasts with variations in fluorescence lifetimes that typically range from about 2 to 6 nanoseconds.

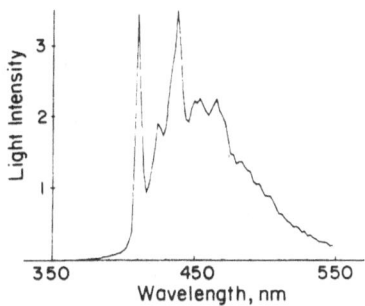

Fig. 1 Phosphorescence from liver ADH

What does the remarkable variability in the phosphorescence lifetime of tryptophan from protein to protein tells us about its environment?

Basically, we need to know what causes deactivation of the excited triplet state. We can separate the possible causes of deactivation (Figure 2) as follows:

1. Non-radiative deactivation, k_{nr}: These processes include the vibrational relaxation by which energy is lost to the medium as heat.

2. Deactivation by internal quenchers, k_{qi}: We include these as moieties that are fixed distances from the tryptophan. They can include heavy atoms such as metal centres or sulfurs, or they can be coloured centres such as heme.

3. Deactivation by external quenchers, k_{qe}: These are molecules or ions that are not part of the protein molecule. In this case the observed quenching is a function of the quencher concentration, as indicated in the figure. The value of k_{qe} is of primary interest because it reflects how the protein structure and dynamics influence a chemical reaction between a buried residue and a neighbouring molecule.

Table 1 Phosphorescence from tryptophans in proteins at ~ 22°

Protein	Lifetime (ms)
actin, rabbit muscle	15
albumin, bovine serum	0.9
albumin, human serum	0.9
albumin, chicken egg	15
alcohol dehydrogenase, liver	300
aldolase	45
alkaline phosphatase, Escherichia coli	1700
asparaginase	50
ATPase, sarcoplasmic reticulum	30
azurin	400
cellulase	200
edestin, hemp seed	500
glutamate dehydrogenase, bovine liver	1200
Escherichia coli	650
glucose oxidase	120
keratin	1400
lactic dehydrogenase	25
Beta-lactoglobulin	15
myosin, rabbit muscle	100
nuclease, micrococcal	400
parvalbumin (with calcium)	5
protease, Streptomyces griseus	100
Bacillus subtilis	10
ribonuclease T1	14
streptokinase	50
trypsin, bovine pancrease	1.4

Original references can be found in Papp and Vanderkooi, 1989.

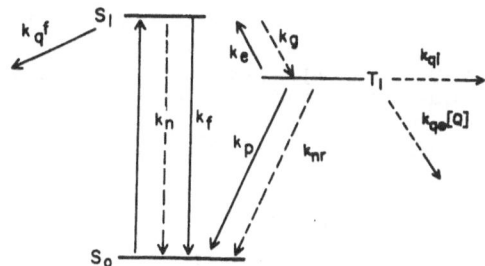

Fig. 2   Energy diagram. S_0 is the ground state, S_1
is the excited singlet state and T_1 is the
excited triplet state.

Phosphorescence and the location of tryptophans

By examining the x-ray structure of proteins, we can draw
conclusions about how structure is related to emission
properties. In general, we have found that tryptophans
located in α helixes or β-barrels show long lifetimes, whereas
tryptophans that are in more unstructured regions of the
protein show short lifetimes. As expected if the secondary
structure is determining the lifetime, denaturation results in
a decrease in phosphorescence lifetime. This was shown by
Strambini and Gonnelli (1986) for several proteins and by
Berger and Vanderkooi (1989) for lens crystallins. That rigid
environment should contribute to long lifetime of
phosphorescence is supported by the viscosity dependence of
indole luminescence (Strambini and Gonnelli, 1985). The
correlation between structure and phosphorescence is shown for
tryptophans in four proteins in Table 2 in which the location
of tryptophans with respect to surface and their
phosphorescence lifetimes are presented.

Interaction of tryptophan with external quencher molecules

In the absence of added quenchers, the phosphorescence
lifetime of the tryptophan moiety is essentially a function of
the rigidity of the site. However, many molecules added to
the solution quench the phosphorescence. This process is in
itself most interesting because it allows us to ask the
generalized question of how a molecule or ion in solution
interacts with a reactive species buried in the protein.

Three models have been proposed for fluorescence
quenching and we use them also to examine phosphorescence
quenching:

1.   Penetration: the quencher diffuses through the protein
      matrix to the tryptophan.

2.   Opening: the tryptophan is transiently exposed to the
      solvent, where it can interact with the quencher.

3.   Long range: the quencher interacts with tryptophan
      through space and direct contact is not necessary.

It is obvious that these models make different
experimental predictions. For instance, if penetration is

required, then the size, shape and charge of the quencher should drastically affect the quenching. If opening is required, then the quenching rate will depend upon the redox parameters of the quencher and the distance between the tryptophan and the surface of the protein. These models are discussed in more detail in a review (Papp and Vanderkooi, 1989).

In a recent paper (Calhoun et al., 1988), we reported studies of the effect of ten quenchers on the phosphorescence of nine proteins. Two classes of quenchers were found. For the diatomic molecules, O_2 and CO, the quenching constant did not vary a great deal from protein to protein. We consider that these compounds penetrate the protein. In contrast, for larger quenchers the quenching constant for seven proteins was remarkably the same for quenchers of different size, shape and charge, but varied drastically from protein to protein. Furthermore, in independent experiments we showed that the quenching constants did not depend upon the viscosity as predicted by Stokes law. In summary, external quenchers have widely-ranging quenching constants that depend upon the protein and not the details (i.e. size, charge) of the quencher.

These experiments seem to indicate that on the time scale of phosphorescence the protein does not undergo deformations that allow large molecules to penetrate. The data are also inconsistent with the idea that tryptophan needs to be exposed transiently to the solvent in order for it to interact with a quencher in solution.

We are left with the model that the excited-state indole ring can interact with a quencher through space. The quenching of excited states by radiationless transfer can occur either through a dipolar or electron exchange mechanism (Turro, 1978). The dipolar mechanism, which requires spectral overlap, cannot be operating, because the quenchers chosen for our study do not absorb in the region where the phosphorescence is occurring. For an exchange/transfer mechanism the rate decreases exponentially with distance (R) between donor emission and acceptor

$$k_q = k_o \exp[-(R)] \tag{1}$$

where k_q is the quenching constant at van der Waals contact.

When the quenching rate is slow, then the fast-diffusion limit will be obtained and reaction will depend upon the concentration, but not the diffusion of the quencher through the medium. Under these conditions, the second order quenching constant will be proportional to the first order transfer rate. It then follows that, if quenching is indeed due to long range transfer, the Stern-Volmer quenching constant will reflect the distance from the surface to the tryptophan.

This idea was tested for the four proteins given in Table 2. The Stern-Volmer quenching constants were determined for three quenchers: nitrite, methylvinylketone and cinnamamide. The results are shown in Figure 3. A good relationship between distance from the surface and quenchability by the externally added quencher was observed.

The observation that the logarithm of the quenching constant is inversely proportional to the distance to the surface, is strong evidence that long-range transfer is occurring by an exchange mechanism. Confirmation that quenchability reflects primarily the distance of the tryptophan to the surface requires examination of more proteins and quenchers. However, if electron transfer/exchange is indeed the mechanism of quenching, we note that the quenchability of internal tryptophans will be very sensitive to distance. For instance, a change of only 5A produced a four orders of magnitude difference in the quenchability (Figure 3). It then follows that small changes in conformation should have large effects on the quenchability.

Table 2 Room temperature phosphorescence of
tryptophans in proteins

Protein name	Tryptophan number	Location in structure	Distance from surface (A)	Lifetime (s)
RNase, T1	59	in β-sheet	2.2	0.014
ADH, liver	314	in β-sheet next to α-helices	5.3	0.3
	15	in random coil (emission not detected)		
Azurin	48	in β-sheet next to β-sheets	10.2	0.4
Alkaline phosphatase	268	in random coil (emission not detected)		
	220	in random coil (emission not detected)		
	109	in β-helix next to β-sheets	-15	1.2

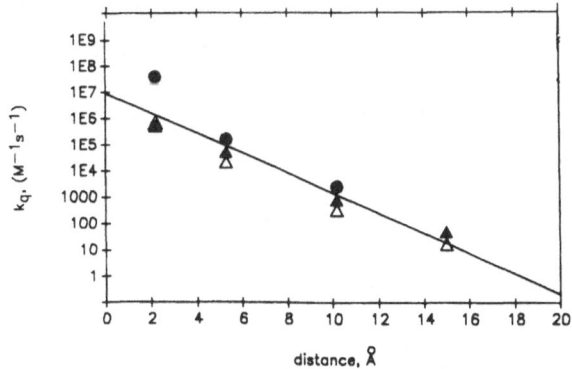

Fig. 3 Relationship between Stern-Volmer quenching constants and distance from the surface for the proteins described in Table 2. Circles: cinnamamide; closed triangles: nitrite; open triangles: methylvinylketone.

In summary, phosphorescence is a useful tool to study protein structure. Full details of how the environment of tryptophan affects its emission properties awaits further comparisons with known three dimensional structures, but it appears that tryptophans in rigid environments phosphoresce with long lifetimes. A fascinating finding of these studies is that there is such a wide variability in phosphorescence lifetimes for tryptophans in proteins at room termperature. It follows that, in proteins, tryptophans can be located in environments that resemble anything between essentially free solution to a rigid glass. Externally added quenchers affect the lifetime of intrinsic phosphorescence, and a major factor determining the rate constant for molecules with >4 atoms is the distance of the tryptophan from the surface. We conclude that phosphorescence of proteins at room temperature is especially useful as a chemical relaxation technique that can tell us about protein structure.

REFERENCES

Berger, J.W. and Vanderkooi, J.M., 1989, Characterization of lens crystallin tryptophan microenvironments by room temperature phosphorescence spectroscopy, Biochemistry, 28:5501.

Calhoun, D.B., Englander, S.W., Wright, W.W. and Vanderkooi, J.M., 1988, Quenching of room temperature phosphorescence by added small molecules, Biochemistry, 27:8466.

Green, T.J., Wilson, D.F., Vanderkooi, J.M. and DeFeo, S.P., 1988, Phosphorimeters for analysis of decay profiles and real time monitoring of exponential decay and oxygen concentrations, Anal. Biochem., 174:73.

Harvey, E.N., 1957, In: A History of Luminescence, (American Philosophical Society, Philadelphia).

Papp, S. and Vanderkooi, J.M., 1989, Tryptophan phosphorescence at room temperature as a tool to study protein structure and dynamics, Photochem. Photobiol., 49:775.

Saviotti, M.L. and Galley, W.C., 1974, Room temperature phosphorescence and the dynamic aspects of protein structure. Proc. Natl. Acad. Sci. U.S.A., 71:4154.

Stern, O. and Volmer, M., 1919, Uber die Ablingungzeit der Fluoreszenz, Phys. Z., 20:183.

Strambini, G.B. and Gonnelli, M., 1985, The indole nucleus triplet-state lifetime and its dependence on solvent microviscosity, Chem. Phys. Lett., 115:196.

Strambini, G.G. and Gonnelli, M., 1986, Effects of urea and guanidine hydrochloride on the activity and dynamical structure of quine liver alcohol dehyrogenase, Biochemistry, 25:2471.

Turro, N.J., 1978, In: Modern Molecular Photochemistry, The Benjamin/Cummings Publishing Co., Inc., Menlo Park CA.

Vanderkooi, J.M. and Berger, J.W., 1989, Excited triplet states used to study biological macromolecules at room temperature. Biochim. Biophys. Acta. 976:1.

Vanderkooi, J.M. Calhoun, D.B. and Englander, S.W., 1987, Of the prevalence of room temperature protein phosphorescence, Science, (Washington, DC), 236:568.

DYNAMIC BEHAVIOR OF CELL SURFACE RECEPTORS AS REVEALED BY LASER EXCITED FLUORESCENCE SPECTROSCOPY

László Mátyus, László Takács, László Pohubi, Margit Balázs, János Szöllösi, János Matkó, Lajos Trón* and Sándor Damjanovich

Department of Biophysics and Biomedical Cyclotron Laboratory*, University Medical School of Debrecen Debrecen Nagyerdei krt 98. H-4012 Hungary

INTRODUCTION

Dynamic interactions between proteins and lipids in the cell membrane govern those processes which are essential in the transduction of signals of the extracellular environment initiating or stimulating the specific responses of the cell as a whole living organism. Without the help of sophisticated photophysical techniques a great deal of these processes would have remained yet undiscovered. The cell machinery is highly dependent upon such dynamic interactions of the cell membrane constituents and those of the outside and the inside of the cell. Ions and nutrients are exchanged between the two compartments and the particular distributuon as well as stimulated redistribution of the cell surface elements are instrumental in the signal transfer.

The best known events are those which follow the mitogenic or antigen induced stimulation of lymphocytes which play a key role in the cellular as well as humoral immunity. The initial step in this process is the activation of phospholipase C enzyme that splits the phosphatidylinositol phosphates of the plasma membrane producing two interacting second messengers, namely inositol polyphosphates and diacylglycerol. The elevation of intracellular free calcium due to the action of inositol tris and possibly tetrakisphosphates together with diacylglycerol can activate the enzyme protein kinase C (Berridge, 1987). The activated enzyme then phosphorylates some key proteins which carry the message of the stimulation further towards the cell nucleus. Several other biochemical pathways with second messenger function are known also to be involved, like the cAMP or cGMP system.

Surprisingly enough, quite recently it has been demonstrated that a more physiologic way of activation of the same kind of lymphocytes (both T and B) follows a completely different, although uncharacterized pathway. The lymphokine (like interleukin 2) induced activation does not demand either phosphatidylinositols or calcium, and can also be initiated in lymphocyte mutants lacking protein kinase C. These findings accentuate the significance of any kind of difference between the early events in the mitogen or lymphokine induced activation (Tigges et al., 1989; Valge et al., 1988; Mills et al., 1988).

After this brief biological introduction we intend to describe here some of the optical methods suitable for studying the earliest events of signal transduction (for a more comprehensive treatment of the subject see Szöllösi et al., 1987b; Matkó et al., 1988; Trón et al., 1987).

FLUORESCENCE RESONANCE ENERGY TRANSFER (FRET) MEASUREMENTS

FRET is a useful technique for investigating intra- and intermolecular distance relationships in the range 1-10 nm (Stryer, 1978; Dale et al., 1981; Cantor and Schimmel, 1980; Lakowicz, 1983). According to the theory of Förster, a donor molecule in an excited state can transfer energy to an acceptor molecule by nonradiative energy transfer if donor and acceptor molecules meet several criteria: 1. the donor should have sufficiently high quantum efficiency; 2. the emission spectrum of the donor should overlap the excitation spectrum of the acceptor; 3. the distance between the donor and acceptor molecules should be in a given range depending on the lifetime and the spectral overlap integral.

Figure 1 shows a schematic diagram of absorption and emission spectra of a typical donor-acceptor pair. Such a pair for example can be fluorescein as donor and rhodamine as acceptor. The shaded area indicates the spectral overlap between the donor's emission spectrum and the acceptor's absorption spectrum. The energy transfer efficiency (E) can be expressed as follows.

$$E = r^6 / (r^6 + R^6)$$

where r is the actual distance between the donor and acceptor molecules, and R is a characteristic distance for a given donor acceptor pair, giving rise to an energy transfer efficiency of 50%. R depends on the spectral overlap integral between the donor emission and the acceptor absorption, the refractive index of the medium, the quantum yield of the donor and the relative orientation of donor emission and acceptor absorption dipoles. Energy transfer efficiency is a very sensitive parameter of small changes in donor-acceptor separation distance and consequently of amplitude changes in local fluctuations of appropriately labelled macromolecules.

This method was used by Somogyi et al. (1984) to study the fluctuations of the ribonuclease T_1-pyridoxamine-phosphate conjugate. They demonstrated that the averaged energy transfer efficiency normalized to the donor quantum yield reports on intramolecular oscillations. This normalized energy transfer efficiency is expected to be relatively insensitive to temperature in rigid structural regions.

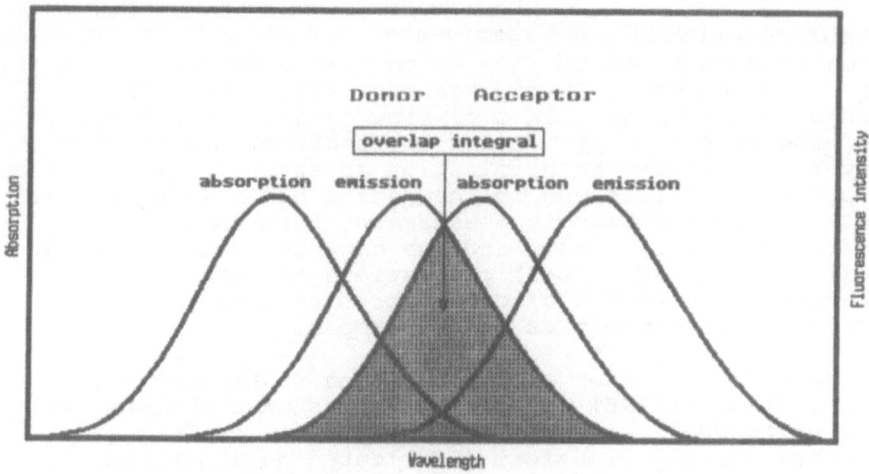

Fig. 1   Spectral requirements for fluorescence
resonance energy transfer.

However, temperature sensitivity increases with the
flexibility of the sampled region of the macromolecule.
O'Hara et al. (1988) obtained experimental data on calmodulin
and transferrin in accordance with the theoretical predictions
of Somogyi. This approach was extended to membrane proteins
as well. Matkó et al. (1989) used this technique to monitor
internal dynamics of the Ca^{2+}-ATPase in sarcoplasmic reticulum
membrane vesicles, using specific labelling of functional
sites on the protein. They found that some regions of the
protein exerted similar thermal fluctuations to small soluble
proteins, while other regions seemed to be extremely rigid.
The latter finding underlines the potential of this approach
in revealing spatial heterogeneity in the internal dynamics of
soluble and membrane bound proteins.

    The numerical value of E can be determined by measuring
the quenching of the donor fluorescence, the enhancement of
the acceptor fluorescence, the decrease of emission anisotropy
in the acceptor fluorescence, or the decrease of the lifetime
of the donor fluorescence. Of these possibilities the
quenching of the donor and/or the enhancement of the acceptor
fluorescence can be used in flow cytometric measurements. It
is difficult to directly relate transfer efficiencies measured
on cell surfaces to distance relationships, because among
other factors the localization of the fluorophores is
restricted to two dimensions and the relative orientations are
not known (Wolber and Hudson, 1979; Dewey and Hammes, 1980).
Despite these limitations the FRET method is useful for
differentiating random from nonrandom distributions of cell
surface molecules or for showing direct interactions between
different cell surface components (Chan et al., 1979). The
FRET method has been applied successfully to several
biological systems using steady-state fluorimeters or
microscopes (Uster and Pagano, 1986; reviewed by Szöllösi et
al., 1987b). Recently the FRET method has been adapted to
flow cytometers as well (Trón et al., 1984; Szöllösi et al.,
1984; Damjanovich et al., 1983; Damjanovich et al., 1988).

Without describing the experimental conditions, we summarize the major advantages of flow cytometric FRET determinations over other methods. Energy transfer efficiency can be calculated from mean values of flow cytometric histograms. This procedure is very similar to spectrofluorimetric methods; however, it allows the elimination of several error sources. The presence of free dye in the medium has a much smaller effect on flow cytometric measurement than on spectrofluorimetric ones, and the data collection can be specifically gated on cell populations of interest, eliminating the effect of cell debris or dead cells and revealing inhomogeneities in the sample.

Recently, another method has been elaborated for the determination of FRET efficiency values on a cell-by-cell basis using dual-laser flow cytometers called FCET (Flow cytometric energy transfer). Collected data are stored in list mode and FRET efficiency values are calculated for each cell.

A simplified version of the FCET method permits, in case of homologous ligands or of ligands binding to different epitopes of the same molecule, the determination of FRET efficiency values on a cell-by-cell basis employing flow cytometers equipped with single laser excitations (Szöllösi et al., 1987c). This modified flow cytometric energy transfer method enabled them to demonstrate the uniform density of Concanavalin A binding glycoproteins in the plasma membrane of normal and Gross virus leukemic mouse cells.

If the measurement is done on a flow cytometer with a high-speed computer, real time FRET efficiency histograms can be collected, and moreover, cell populations with different FRET efficiency can be sorted. The flow cytometric energy transfer method has certain limitations. Autofluorescence is often a problem; if it is bigger than 5-10% of the specific intensity, it causes unacceptable error in the FRET efficiency. For presently used donor-acceptor pairs, like fluorescein/rhodamine or fluorescein/Texas Red, $1-5 \times 10^5$ binding sites per cell are necessary to get detectable transfer efficiency.

Although many of the cell surface receptors of interest are expressed at much lower density, numerous problems can be solved by this method. Szöllösi et al. (1987a) proved the association between a 95-kDa peptide termed T27 and the 55 kDa Tac peptide on the surface of HuT 102B2 cells. Previous experiments (immunoprecipitation) were inconsistent, though suggested some physical interaction between the two proteins. Later Edidin et al. (1988) reached the same conclusion based on lateral mobility measurements of the same antigens. Their data suggest that the Tac and T27 peptide not only are in close vicinity but also interact physically. Both experiments were carried out on intact, live cells, excluding the possiblity of many artifacts.

Szöllösi et al. (1989) mapped the supramolecular structure of the MHC class I and class II antigens using the FCET method. They found physical interaction between class I and class II molecules.

MEMBRANE POTENTIAL MEASUREMENTS BY FLOW CYTOMETRY

An electric potential difference, the membrane potential, exists between the cell membrane interior and exterior. The origin of the membrane potential is related to the impermeability of a cell membrane to charged protein molecules, the action of the Na-K ATPase and the differential permeability of the membrane to different ions. Many cellular events (like cell activation) are accompanied by membrane potential changes, therefore monitoring membrane potential frequently offers an insight into these phenomena. Three main methods of membrane potential measurement can be distinguised among the most widely used ones: direct electrophysiological measurements (with electrodes), determination of the distribution of radioisotope-labelled charged molecules, and fluorescence measurements.

The membrane potential-sensitive fluorescent dyes can be divided into three groups: merocyanines, carbocyanines and oxonols (Waggoner, 1979; Tsien, 1989). Merocyanines change their spectroscopic properties with changing electric field; but these changes are too small to be detected in spectrofluorimeters or in flow cytometers. Their response time is extremely fast, so they are suitable dyes for following even action ptoentials, although the measurement requires a specially designed detection system.

Carbocyanines and oxonols belong to the so called translocating dyes; i.e. their distribtuion depends upon and therefore follows the transmembrane electrochemical gradient. Carbocyanines are widely applied for measuring potential difference in cells and in different organelles (Shapiro et al., 1979; Waggoner, 1976). They are positively charged lipophilic symmetric molecules, and their membrane permeability varies with the length of the side chain. The equilibrium distribution of the dye across the cell membrane depends upon the lipid/water partition coefficient of the dye (resulting in membrane potential-independent, background fluorescence) and the membrane potential of the cell (potential-sensitive part of the fluorescence).

The calibration of the fluorescence signal to the membrane potential is done by changing the membrane potential by using buffers with different potassium concentrations, and by using different ionophores. When the membrane potential is increased, i.e. the cells are hyperpolarized, they take up more dye, while depolarized cells release the dye. The direction of the change in the fluorescence intensity depends on the dye concentration; when relatively high dye concentrations are applied, the dye uptake leads to dye aggregation, resulting in self-quenching and a decrease in the fluorescence intensity. At lower concentrations (nM), practically no aggregation takes place, and the higher quantum efficiency in the apolar environment results in a higher fluorescence signal. The dye concentration should be individually titrated for every cell type, and for every experimental configuration, because carbocyanine dyes stick to surfaces, and the dye distribution is affected by mitochondria as well as by surface membrane potential. Mitochondrial potential is usually higher than the plasma membrane potential, and it also may respond to drugs changing the plasma membrane potential. Since the optical properties of dye can be altered by chemical modification of the structure, carbocyanines for flow cytometric measurements are also available. These dyes were successfully applied to follow early membrane potential changes caused by various drugs, chemotactic factors, ionphores, and other ligands of biological significance.

Cyclosporin A, a cyclic endecapeptide antibiotic used for immunosuppression in organ transplanation, causes dose-dependent depolarization of human peripheral lymphocytes (Damjanovich et al., 1987). Similar data were gained by Damjanovich et al. (1985;1986) and by Mátyus et al. (1986), with mouse spleen cells and thymocytes. These findings are the earliest detected effects of the drug, indicating an immediate change in the membrane structure and function. Though there are doubts regarding the interpretation of the data, carbocyanines seem to be suitable dyes for membrane potential determination, mainly for cells with no or few mitochondria.

Oxonols like carbocyanines, also have symmetrical structure, but their delocalized charge is negative, so their uptake differs from carbocyanines. Oxonols are excluded from hyperpolarized cells, whereas depolarized cells accumulate them. This property results in a great advantage compared to carbocyanines, because the highly polarized mitochondria do not affect the dye distribution. The spectrofluorimetric application of oxonols was somewhat restricted, because of the smaller intensity changes involved. Their flow cytometric utilization has recently started, after oxonol derivatives with spectral properties suitable for argon ion laser excitation were synthesized, so their widespread use is expected.

A recent interesting application of oxonol in flow was done by Pieri et al. (1989). They showed that bretylium tosylate - a sodium channel opener - resulted in an increase in membrane potential of depolarized human, rat and mouse T and B lymphocytes, which led to the novel conclusion that sodium channels are common in all kinds of lymphocytes and that they may play an important role in the regulation of lymphocyte activation.

Recently Gelfand et al. (1987) confirmed the above demonstrated depolarizing effect of cyclosporins using oxonol dye. This strongly supports that the two dye families, though they are quite different in nature, monitor the same changes.

Flow cytometric measurements are superior to spectrofluorimetric measurements here, as in general, because the free extracellular dye has less influence on the measured signal and inhomogeneities in the population can also be revealed. Because of the higher sensitivity, much lower dye concentrations are necessary for flow cytometry, thus decreasing the toxic effect of the dye. By gating it is possible to eliminate the effect of dead cells and cell debris, and using cell surface markers membrane potential changes can be correlated with defined subpopulations.

ACKNOWLEDGEMENT

This work was carried out as part of research programs sponsored by the Hungarian Academy of Sciences (Grant No. OTKA 112, OTKA 663 and OKKFT 1.1.4.2.).

REFERENCES

Berridge, M.J., 1987, Inositol trisphosphate and diacylglycerol: two interacting second messengers, Annu. Rev. Biochem., 56:159.

Cantor, C.R. and Schimmel, P.R., 1980, Biophysical Chemistry, pp. 448-454, W.H. Freeman and Company, San Francisco.

Chan, S.S., Arndt-Jovin, D.J. and Jovin, T.M., 1979, Proximity of Lectin Receptors on the Cell Surface Measured by Fluorescence Energy Transfer in a Flow System, J. Histochem. Cytochem., 27:56.

Dale, R.E., Novros, J., Roth, S., Edidin, M. and Brand, L., 1981, Application of Förster Long-Range Exciation Energy Transfer to the Determination of Distributions of Fluorescently-Labelled Concanavalin A-Receptor Complexes at the Surfaces of Yeast and of Normal and Malignant Fibroblasts, in: Fluorescent Probes, (Beddard, G.S. and West, A. eds.), pp. 159-189, Academic Press, London.

Damjanovich, S., Aszalós, A., Mulhern, S.A., Balázs, M. and Mátyus, L., 1986, Cytoplasmic membrane potential of mouse lymphocytes is decreased by cyclosporins, Molec. Immunol., 23:175.

Damjanovich, S., Aszalós, A., Mulhern, S.A. Marti, G., Balázs, M. and Mátyus, L., 1985, Cyclosporin A influences membrane potential of human and mouse lymphocytes. A critical comparison of steady-state fluorimetric and flow cytometric measurements, Biophys. J., 47:271a.

Damjanovich, S., Aszalós, A., Mulhern, S.A., Szöllösi, J., Trón, L. and Fulwyler, M.J., 1987, Cyclosporin depolarizes human lymphocytes: earliest observed effect on cell metabolism, Eur. J. Immunol., 17:763.

Damjanovich, S., Balázs, M., Szöllösi, J., Trón, L. and Somogyi, B., 1988, Protein dynamics and function, J. Mol. Catal., 47:155.

Damjanovich, S., Trón, L., Szöllösi, J., Zidovetzky, R., Vaz, W.L.C., Regateiro, F., Donna Arndt-Jovin, J., and Jovin, Th.M., 1983, Distribution and mobility of murine histocompatibility H-2K^k antigen in the cytoplasmic membrane, Proc. Natl. Acad. Sci. USA, 80:5985.

Dewey, T.G. and Hammes, G.G., 1980, Calculation of Fluorescence Resonance Energy Transfer on Surfaces, Biophys. J., 32:1023.

Edidin, M., Aszalós, A., Damjanovich, S. and Waldmann T.A., 1988, Lateral diffusion measurements give evidence for association of the Tac peptide of the IL-2 receptor with the T27 peptide in the plasma membrane of HuT-102-B2 T cells, J. Immunol., 141:1206.

Gelfand, E., Cheung, R.K. and Mills, G.B., 1987, The cyclosporins inhibit lymphocyte activation at more than one site, J. Immunol., 138:1115.

Lakowicz, J.R., 1983, Principles of Fluorescence Spectroscopy, pp. 303-339, Plenum Press, New York.

Matkó, J., Jóna, I. and Martonosi, A., 1989, Structural dynamics of the Ca^{2+}-ATPase of sarcoplasmic reticulum:

Temperature profiles of fluorescence polarization and intramolecular energy transfer, (manuscript in preparation).

Matkó, J., Szöllösi, J., Trón, L. and Damjanovich, S., 1988, Luminescence spectroscopic appraches in studying cell surface dynamics, Quart. Rev. Biophys., 21:479.

Mátyus, L., Balázs, M., Aszalós, A., Mulhern, S.A. and Damjanovich, S., 1986, Cyclosporin A depolarizes cytoplasmic membrane potential and interacts with Ca^{2+} ionophores, Biochim. Biophys. Acta, 886:353.

Mills, G.G., Girard, P., Grinstein, S., and Gelfand E.W., 1988, Interleukin-2 induces proliferation of T lymphocyte mutants lacking protein kinase C, Cell, 55:91.

O'Hara, P., Gorski, K.M., and Rosen, M.A., 1988, Energy transfer as a probe of protein dynamics in the proteins transferrin and calmodulin, Biophys. J., 53:1007.

Pieri, C., Recchioni, R., Moroni, F., Balkay, L., Márián, T., Trón, L. and Damjanovich, S., 1989, Ligand and voltage gated sodium channnels may regulate electrogenic pump activity in human, mouse and rat lymphocytes, Biochem. Biophys. Res. Com., 160:999.

Shapiro, H.M., Natale, P.J. and Kamentsky, 1979, Estimation of membrane potentials of individual lymphocytes by flow cytometry, Proc. Natl. Acad. Sci. USA, 76:5728.

Somogyi, B., Matkó, J., Hevessy, J., Welsh, G.R., and Damjanovich, S., 1984, Förster-type energy transfer as a probe for changes in local fluctuations of the protein matrix, Biochemistry, 23:3403.

Stryer, L. 1978, Fluorescence Energy Transfer as a Spectroscopic Ruler, Ann. Rev. Biochem., 47:819.

Szöllösi, J., Damjanovich, S., Balázs, M., Nagy, P., Trón, L., Fulwyler, M.J. and Brodsky, F.M., 1989, Physical association between MHC Class I and Class II molecules detected on the cell surface by flow cytometric energy transfer, J. Immunol., 143:208.

Szöllösi, J., Damjanovich, S., Goldman, C.K., Fulwyler, M.J., Aszalós, A.A., Goldstein, G., Rao, P., Talle, M.A. and Waldmann, T.A., 1987a, Flow cytometric resonance energy transfer measurements support the association of a 95-kDa peptide termed T27 with the 55-kDa Tac peptide, Proc. Natl. Acad. Sci. USA, 52:7246.

Szöllösi, J., Damjanovich, S., Mulhern, S.A. and Trón, L., 1987B, Fluorescence energy transfer and membrane potential measurements monitor dynamic properties of cell membranes: A critical review, Prog. Biophys. Molec. Biol., 49:65.

Szöllösi, J., Mátyus, L., Trón, L., Balázs, M., Ember, I., Fulwyler, M.J., and Damjanovich, S., 1987c, Flow cytometric measurements of fluorescence energy transfer using single laser excitation, Cytometry, 8:120.

Szöllösi, J., Trón, L., Damjanovich, S., Helliwell, S.H., Donna Arndt-Jovin, And Jovin Th.M., 1984, Fluorescence energy transfer measurements on cell surfaces: A critical comparison of steady-state fluorimetric and flow cytometric methods, Cytometry, 5:210.

Tigges, M.A., Casey, L.S., Koshland, M.E., 1989, Mechanism of interleukin-2 signaling: Mediation of different outcomes by a single receptor and transduction pathway, Science, 243:781.

Trón, L., Szöllösi, J., and Damjanovich, S. Helliwell, S.H., Arndt-Jovin D.J., and Jovin, T.M., 1984, Flow cytometric measurement of fluorescence resonance energy transfer on cell surfaces, <u>Biophys. J.</u>, 45:939.

Trón, L., Szöllösi, J., and Damjanovich, S., 1987, Proximity measurements of cell surface proteins by fluorescence energy transfer, <u>Immunol. Lett.</u>, 16:1.

Tsien, R.Y., 1989, Fluorescent probes of cell signaling, <u>Ann. Rev. Neurosci.</u>, 12:227.

Uster, P.S. and Pagano, R.E., 1986, Resonance energy transfer microscopy: Observations of membrane-bound fluorescent probes in model membranes and in living cells, <u>J. Cell. Biol.</u>, 103:1221.

Valge, V.E., Wong, J.G.P., Datlof, B.M., Sinskey, A.J., and Rao, A., 1988, Protein kinase C is required for responses to T cell receptor ligands but not to interleukin-2 in T cells, <u>Cell</u>, 55:101.

Waggoner, A., 1976, Optical Probes of Membrane Potential, <u>J. Membr. Biol.</u>, 27:317.

Waggoner, A.S., 1979, Dye Indicators of Membrane Potential, <u>Ann. Rev. Biophys. Bioeng.</u>, 8:47.

Wolber, P.K. and Hudson, B.S., 1979, An Analytic Solution to the Förster Energy Transfer Problem in Two Dimensions, <u>Biophys. J.</u>, 28:197.

RESOLVED Q_X AND Q_Y ELECTRONIC TRANSITIONS IN CHLOROPHYLL a

M. Fragata

Université du Québec à Trois-Rivières, Centre de
recherche en photobiophysique, Trois-Rivières
Québec, G9A 5H7, Canada

INTRODUCTION

The present trend in photosynthesis research of green
plants and algae is the determination of the structure-
function relationships in the thylakoid membranes of the
chloroplast (see, e.g. Trebst 1989). One of the major
unsolved questions concerns the role of molecular orientations
and electronic symmetry of chlorophyll (Chl) and analogs in
energy and electron transfer. In this connection, a recent
linear dichroism study of Chl a (Fragata et al., 1988)
permitted the identification of the electronic transitions
responsible for the spectral activity in the visible region
which are thought to be relevant in the primary photophysical
phenomena. Although the data agree well with results
published previously (cf. Table 2), close scrutiny of the
literature reveals some discrepancies which deserve to be
analysed. The study reported here is directed toward this
end.

DICHROIC CHARACTERISTICS OF CHL a

The spectra of chlorphylls and analogs can be described
in terms of a four orbital model by analogy to the spectra of
the more symmetrical porphyrins and chlorins (Gouterman 1959,
1961) where $\pi \rightarrow \pi*$ transitions are of $E\mu$ symmetry, meaning
that they consist of a degenerate pair of states which follow
from two equivalent dipole transitions in the X and Y
directions in the plane of the molecule (cf. Fig. 1). In the
red side of the spectra the excited states as well as the
observed transitions were labelled Q_x and Q_y, respectively
(Gouterman 1959).

In Fig. 1 the molecular framework of Chl a is represented
with two systems of geometrical (or reference) axes. That is,
(i) X_1, Y_1, where X_1 passes through the C_7 atom, taking into
account that the orientation of the pigments is due to the

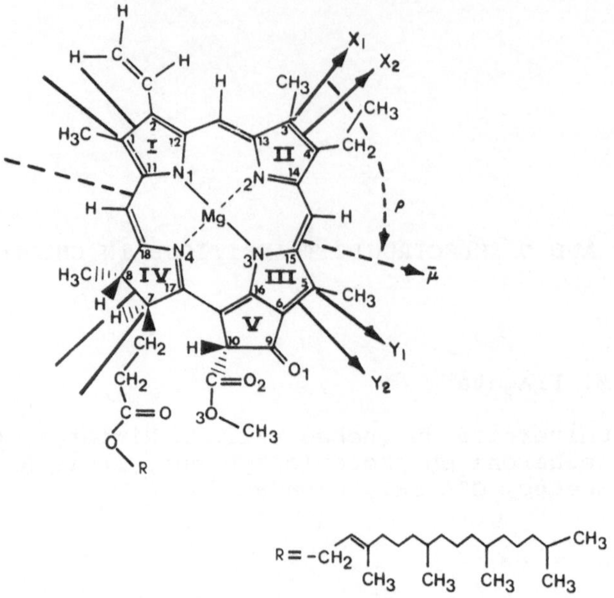

Fig. 1 Chemical structure of chlorphyll $\underline{a}$ with X_1, Y_1 or X_2, Y_2 as the geometrical (or reference) axes of the molecular framework. μ is a transition dipole moment given clockwise at an angle ρ from X_1 or X_2.

alignment of the phytyl chain in the orientation matrix, and (ii) X_2, Y_2 are the axes of electronic symmetry of the π-electron system. Though the two representations are widely used in the literature, the electronic symmetry system of axes is most probably a better choice. This is an interesting point that should be addressed in future studies. Fig. 1 also displays a transition dipole moment μ oriented at an angle ρ given clockwise from the X_1 or X_2-axis of the molecular frame. The reason for this is that the orthogonality of Q_x and Q_y in chlorophylls and analogs is not substantiated by a number of fluorescence polarization and linear dichroism studies (see, e.g. van Gurp 1988; Gouterman and Stryer 1962; Fragata et al., 1988). The lack of orthogonality may result from overlap of upper vibronic components of Q_y, or from loss of C_{2v} symmetry by the chlorin ring (Hofrichter and Eaton 1976).

Fig. 2 presents the dichroic characteristics of Chl $\underline{a}$. The molecules were aligned in a lamellar phase of glycerylmonooctanoate /H_2O (70/30, w/w) binary mixture (Fragata et al., 1988). The spectra measured are (i) the linear dichroism $LD = A_{\parallel} - A_{\perp}$, where $A_{\parallel}$ and $A_{\perp}$ are the absorbances of light parallel and perpendicular, respectively, to the plane of polarization of the incident radiation, (ii) the isotropic absorbance, A_{iso}, and (iii) the reduced linear dichroism $LD^r = LD/A_{iso}$ (see details in Fragata et al., 1988). Finally, the LD^r spectrum of Chl $\underline{a}$ spectrum can be used to decompose A_{iso} into X- and Y-polarized spectra according to a procedure described by Matsuoka and Nordén (1982) for transitions polarized in the plane of a planar chromophore. These data have been reported before (cf. Fig. 3 of Fragata et

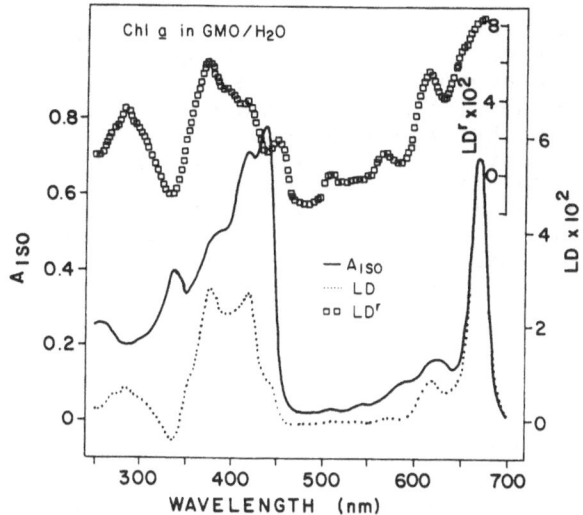

Fig. 2   Isotropic (A_{ISO}), linear dichroism (LD) and
reduced linear dichrosim (LD^r) spectra of
chlorophyll <u>a</u> oriented in a lamellar phase
glycerylmonooctanoate / H_2O (see text).

al., 1988): in summary, the most significant spectral features
in the region above 480 nm are (i) a very intense Y-polarized
band with a maximum at about 670 nm, and (ii) a weaker X-
polarized band at the same wavelength which accounts for
approximately 14% of the total intensity. From this it is
possible to give a new interpretation of the Chl <u>a</u> spectrum in
the red region which excludes a description in terms of two
transitions of mutually perpendicular polarizations as has
been usually done. This is discussed in the next section.

RESOLVED ELECTRONIC TRANSITIONS

Deconvolution of the X- and Y-polarized spectra of Chl <u>a</u>
was carried out as described by Kurucsev (1978). In brief, it
is assumed that (i) the observed absorbance, $A(\bar{\nu})$, at any
wave-number   may be due to the absorbances, $A_i(\bar{\nu})$, of several
overlapping transitions, i.e., $A(\bar{\nu}) = \sum_i A_i(\bar{\nu})$ and (ii) each
transition may be approximated by a progression of a single
vibrational mode. The vibronic progression is described as a
displaced harmonic oscillator with Gaussian bands of constant
bandwidth leading to the five-parameter equation.

$$A_i(\bar{\nu}) = \sum_m \frac{X_i^m}{m!} I_{00i}\left(1 + \frac{mV_i}{\bar{\nu}_{00i}}\right)$$
$$\exp\left[-\frac{4\ln 2}{b_i^2}(\bar{\nu} - \bar{\nu}_{00i} - mV_i)^2\right]$$

where I_{00i} is the intensity, $\bar{\nu}_{00i}$ the position of the (0,0)
band of the i-th transition, V_i the separation between the
harmonic bands, X_i the ratio of (1,0) to (0,0) band
intensities and b_i the Gaussian bandwidth. In the curve-
fitting procedure the summation in the above equation was
truncated after eight terms.

The X- and Y-polarized spectra of Chl $\underline{a}$ could be fitted to 7 and 4 vibronic progressions, respectively (see detailed discussion in Fragata et al., 1988). The optimum separation between the bands, V_i, was found to be 1294 cm^{-1} for all progressions in the region of 690-490 nm and 1301 cm^{-1} in the region below about 460 nm. Table 1 gives a summary of the data: it includes the positions, intensities relative to progression 1 and the polarization directions in the XY plane of the transitions. The angle, ρ, between the transition moments in the molecular frame (Fig 1) is given as $\cot^2 \rho = (I_{ooi})_x / (I_{ooi})_y$, where $(I_{ooi})_x$ and $(I_{ooi})_y$ are the intensities of the (0,0) band of the i-th X and Y transitions, respectively. The following discussion will focus on the electronic transitions which are particularly relevant to the photophysical phenomena bearing a connotation to the primary processes of photosynthesis, i.e., the Q_x and Q_y transitions in the low frequency region of the spectrum.

A first conclusion of this study is that Chl $\underline{a}$ has transitions polarized at a ρ angle of 70° (C1). Although this is not in agreement with previous interpretations where Q_x and Q_y transitions are assumed to be perpendicularly polarized to each other (e.g. Seely 1977), it is noteworthy that the 70°

Table 1   Energies and Polarisations of Resolved Electronic Transitions in Chlorphyll $\underline{a}$.

Label	Energy (cm^{-1})	Band	Relative Intensity	Polarization (degrees)
C1	14925	(0,0)	100	
	16207	(1,0)	11	70°
	17513	(2,0)	1	
C2	15408	(0,0)	14	
	16722	(1,0)	5	Y
	17986	(2,0)	1	
C3	15748	(0,0)	9	X
	17036	(1,0)	9	
C4	17953	(0,0)	2	X
C5	22936	(0,0)	73	X
C6	23310	(0,0)	39	50°
C7	25707	(0,0)	17	X
C8	26525	(0,0)	28	Y
C9	28902	(0,0)	28	X

[a]Angles are given clockwise from X_1-axis of the molecular framework such as represented in Fig. 1.

Table 2 Resolved electronic transitions in
Chlorophyll a.

Q_y-polarized transitions

System[a]	Label		Band	Wavelength	Ref.
DH			(0,0)	673	b
LCM			(0,0)	671	c
LCM			(0,0)	670	d
GMO/H$_2$O	C1	70°	(0,0)	670	e
CO			(0,0)	667	f
THF			(0,0)	664	g
MH			(0,0)	662	b
DH			(1,0)	620	b
CO			(1,0)	619	f
GMO/H$_2$O	C1	70°	(1,0)	617	e
THF			(1,0)	615	g
MH			(1,0)	611	b
THF			(2,0)	593	g
CO			(2,0)	585	f
GMO/H$_2$O	C1	70°	(2,0)	571	e
GMO/H$_2$O	C2		(0,0)	649	e
GMO/H$_2$O	C2		(1,0)	598	e
GMO/H$_2$O	C2		(2,0)	556	e

Q_x-polarized transitions

System	Label	Band	Wavelength	Ref.
DH		(0,0)	640	b
GMO/H$_2$O	C3	(0,0)	635	e
THF		(0,0)	626	g
MH		(0,0)	624	b
CO		(0,0)	619	f
LCM		(0,0)	582	c
LCM		(0,0)	580	d
DH		(1,0)	595	b
GMO/H$_2$O	C1	(1,0)	587	e
CO		(1,0)	585	f
MH		(1,0)	578	b
THF		(1,0)	540	g

[a] CO, castor oil. DH, dihydrate. GMO/H$_2$O, glycerylmonooctan-
oate/H$_2$O. LCM, liquid crystal mixture. MH, monohydrate. THF,
tetrahydrofuran. [b] Belkov and Losev (1978). [c] Frackowiak et
al., (1987). [d] Bauman and Wrobel (1980). [e] Fragata et al.,
(1988). [f] van Metter (1978). [g] Hyninen and Sievers (1981).

orientation calculated by Fragata et al., (1988) is supported by the earlier work of Bauman and Wrobel (1980) and the more recent studies of van Gurp (1988) and van Gurp et al., (1989).

Another major conclusion is that the absorption spectrum of Chl $\underline{a}$ above 480 nm cannot be described in terms of just two progressions as has been attempted to date (see, e.g., Weiss 1972; Petke et al., 1979). It is noted, however, that progressions 1 and 2 may belong to the same electronic transition, i.e., Q_y. Another interesting point is the previous belief (e.g. Fragata et al., 1988; van Metter 1978) that a band near 635 nm should be assigned to the Q_x (0,0) transition. This has now been confirmed by Belanger et al., private communication) in a study of fluorescence and absorption of Chl $\underline{a}$ in various solvents at 295 and 77 K. A matter arising is that this band has an energy of about 1300 cm^{-1} below that of the next lowest energy X-polarized transition, i.e., near 590 nm (Table 1) which has been traditionally assigned to Q_x (0,0) (e.g., Petke et al., 1979; Bauman and Wrabel 1980; Frackowiak et al., 1987) but now clearly identified as Q_x (1,0) (Fragata et al., 1988). Table 2 summarizes and clarifies further the Q_y and Q_x data which were discussed above.

ACKNOWLEDGEMENTS

Several grants from the N.S.E.R.C. Canada, the Fonds F.C.A.R. du Quebec, and the Université du Québec à Trois-Rivières are gratefully acknowledged.

REFERENCES

Bauman, D. and Wrobel, D., 1980, Dichroism and polarized fluorescence of chlorophyll a, chlorophyll c and bacteriochlorophyll $\underline{a}$ dissolved in liquid crystals, Biophys. Chem., 12:83.

Belkov, M.V. and Losev, A.P., 1978, On the location of the electronic transitions of chlorphyll $\underline{a}$ and protochlorophyll $\underline{a}$ depending on the degree of solvate state, Spect. Lett., 11:653.

Frackowiak, D., Bauman, D., Manikowski, H., Browett, W.R., and Stillman, M.J., 1987, Circular dichroism and magnetic circular dichroism spectra of chlrophylls $\underline{a}$ and $\underline{b}$ in nematic liquid crystals. II. Magnetic circular dichroism spectra, Biophys. Chem., 28:101.

Fragata, M., Nordén, B. and Kurucsev., 1988, Linear dichroism (250-700 nm) of chlorophyll $\underline{a}$ and pheophytin $\underline{a}$ oriented in a lamellar phase of glycerylmonooctanoate/H_2O. Characterization of electronic transitions, Photochem. Photobiol., 47:133.

Gouterman, M. 1959, Study of the effects of substitution on the absorption spectra of porphin, J. Chem. Phys., 30:1139.

Gouterman, M., 1961, Spectra of porphyrins, J. Mol. Spectroscopy, 6:138.

Gouterman, M., and Stryer, L., 1962, Fluorescence polarization of some porphyrins, J. Chem. Phys., 37:2260.

van Gurp, M., 1988, "Molecular orientation of natural dyes in ordered systems studied with polarized light", Ph.D. Thesis, University of Utrecht.

van Gurp, M., van der Heide, U., Verhagen, J., Piters, T., van Ginkel, G. and Levine, Y.K., 1989, Spectroscopic and orientational properties of chlorophyll a and chlorophyll b in lipid membranes, Photochem. Photobiol., 49:663.

Hofrichter, J., and Eaton, W.A., 1976, Linear dichroism of biological chromophores, Annu. Rev. Biophys. Bioeng., 5:511.

Hynninen, P. and Sievers, G., 1981, Conformations of chlorphylls a and a' and their magnesium free derivatives as revealed by circular dichroism and proton magnetic resonance, Z. Naturforsch., 36b:1000.

Kurucsev, T., 1978, Vibronic analysis of the visible absorption and fluorescence spectra of the fluorescein dianion, J. Chem. Educ., 55:128.

Matsuoka, Y., and Nordén, B., 1982, Linear dichroism study of 9-substituted acridines in stretched poly (vinyl alcohol) film Chem. Phys. Lett., 85:302.

van Metter, R.L., 1978, "A study of the optical properties of chlorophyll in solution and in a protein complex", Ph.D. Thesis, University of Rochester.

Petke, J.D., Maggiora, G.M., Shipman, L. and Christoffersen, R.E., 1979, Stereoelectronic properties of photosynthetic and related systems - V. Ab initio configuration interaction calculations on the ground and lower excited singlet and triplet states of ethylchlorophyllide a and ethyl pheophorbide a, Photochem. Photobiol., 30:203.

Seely, G.R., 1977, Chrophyll in model systems: clues to the role of chlorphyll in photosynthesis, in: Primary Processes of Photosynthesis, J. Barber, ed., Elsevier, Amsterdam.

Trebst, A., 1989, The topology of the reaction center polypeptides of photosystem II, Physiol. Plant., 76:A14.

Weiss, C., Jr., 1972, The pi electron structure and absorption spectra of chlorphylls in solution, J. Mol. Spectroscopy, 44:37.

LASER-INDUCED TIME-RESOLVED FLUORESCENCE STUDIES OF BIOLOGICAL SYSTEMS

Roberta Ramponi, Rinaldo Cubeddu and Carlo A. Sacchi*

Centro di Elettronica Quantistica e Strumentazione Elettronica CNR, Istituto di Fisica, Politecnico Milano, Italy

INTRODUCTION

The measurement of fluorescence and of its temporal characteristics is a powerful technique in the investigation of biological systems. Both endogenous (Cubeddu et al., 1990; Docchio, 1989) and exogenous (Bottiroli et al., 1979; Yamashita et al., 1984; Larsen and Johansson, 1988; Malatesta and Andreoni, 1988; Docchio, 1989) fluorophores can be considered. In the first case, the fluorescence of biomolecules themselves is studied, to obtain direct indications of their nature, behaviour and possible alterations. In the second case, fluorescent probes that bind specifically to the biological substrate to be investigated are used and their localization or the modifications induced in their fluorescence parameters by the substrate are studied. In the past, the fluorescence analysis of biological systems was essentially based on continuous wave (cw), i.e. steady state, techniques. However, these suffer from some intrinsic limitations both in selectivity and in sensitivity. Indeed, in complex systems, the spectra of different fluorophores often overlap, making their discrimination through their cw excitation and/or emission characteristics difficult to achieve. Moreover, the contribution of a single fluorescent component to the overall emission spectrum depends on both the relative abundance of its molecules in the excited state and its fluorescence quantum yield. Thus, quantitative information on the relative abundance of the different chromophores is difficult to obtain and components with a very low fluorescence efficiency can hardly be detected.

Time-resolved fluorescence analysis overcomes the main limitations of cw studies. In fact, fluorophores with

---

*Professor Carlo A. Sacchi passed away shortly after concluding this work.

overlapping spectra usually differ in their fluorescence decay time, and can thus be identified through the temporal behaviour of their emission. Moreover, in the fluorescence decay waveform of a complex system, the relative peak amplitudes of the different decay components give a direct indication of the relative abundances of the corresponding excited chromophores, independently of their fluorescence quantum yield, thus quantitating their presence and allowing the detection of poorly fluorescent molecules.

State-of-the-art systems for time-resolved fluorescence measurements developed in the last decade mainly utilize short-pulse lasers as the excitation sources. Indeed, several lasers can generate pulses much shorter than those obtainable with conventional flashlamps, thus permitting the measurement of short fluorescence decay times without resort to deconvolution procedures, provided that the time resolution of the detection apparatus is sufficient. Moreover, since the laser beam is usually diffraction limited, it can be focussed to dimensions comparable to those of single cells or cellular structures, thus making it possible to excite microstructures selectively. The intrinsic monochromaticity of the laser beam also allows selective excitation of the different chromophores when molecules with non-overlapping excitation spectra are simultaneously present. Finally, in the case of tunable lasers, such as the dye-lasers, the emission efficiency of different components in the sample can be optimized by tuning the laser wavelength to the excitation peaks of the various fluorophores.

Several systems for time-resolved fluorescence analysis both in solution and at a microscopic level have been developed in recent years. The fluorescence detection can be based on different techniques, depending on the time-resolution needed, the repetition rate of the laser used as the excitation source, and the intensity of the signal to be measured (Thaer and Sernetz, 1973). In particular, some systems are based on light flux techniques (Bottiroli et al., 1979; Andreoni and Cubeddu, 1984; Docchio et al., 1984a), some on single-photon timing techniques (Kinoshita et al., 1981; Cova et al., 1983; Andreoni and Cubeddu, 1984; Rodgers and Firey, 1985; Schneckenburger, 1985), others on the phase-shift method (Haar and Hauser, 1978; Lakowicz et al., 1984), or on the use of a streak-camera as the detector (Yamashita et al., 1984). Some instruments allow acquisition of emission spectra together with that of the fluorescence decay waveform (Ramponi and Rodgers, 1987; Cubeddu et al., 1988; Yamashita et al., 1988).

If many fluorescence photons are detectable per excitation pulse, it is possible to use a fast photomultiplier (PMT) and to digitize its output to reproduce directly the emission decay waveform by using sampling techniques. The digitized waveforms are then summed up and averaged over many repetitions of the fluorescence pulse. The laser repetition rate required to keep the measurement time acceptable is not high (50-100 Hz) and, since the time-resolution is limited by the FWHM of the PMT (~400 ps for microchannel-plate photomultipliers (MCP PMT), ≥ 1 ns for typical commercial fast photomultipliers with high amplification), the requirements on the laser pulse duration are not particularly severe. Thus,

N_2-pumped dye-lasers (Cubeddu et al., 1980) can be used as the excitation source, with considerable cost reduction as compared to systems using mode-locked lasers.

When fluorescence signals are very low and/or a better time-resolution is needed, single-photon timing techniques are more appropriate. As the light detector, single-photon avalanche photodiodes (SPAD), or MCP or other fast PMT's can be used. The time-resolution of the detection apparatus is given by the time jitter of the detector itself (~70 ps for SPAD, <50 ps for MCP PMT's). The light-source repetition rate required is high, thus, to match with the good time-resolution necessary to study the fluorescence of biological samples, mode-locked lasers, which provide ps pulses at up to a few GHz repetition rates, are the optimal excitation sources.

In the phase-shift method, the excitation light is sinusoidally modulated. Thus, the fluorescence emission is also sinusoidally modulated at the same frequency, but shifted in phase and with a decreased degree of modulation. The phase shift, at a given modulation frequency, depends on the fluorescence decay characteritics of the sample and its measurement allows determination of the decay times of the sample itself. The phase shift approach is in principle equivalent to the single-photon-timing method, since frequency and time domain are related by a Fourier transform. However, in practice, in the presence of more than two exponential decays, the experimental data obtained by phase shift measurements are usually more difficult to analyze: phase and modulation determinations over a range of frequencies are generally required (Lakowicz et al., 1984).

Finally, fluorescene detection using a streak-camera provides the best time-resolution. However, its overall time-scale does not allow the contemporaneous measurements of decay time constants in the subnanosecond <u>and</u> 10 ns range. Thus, it is often inadvisable to use this method in the case of complex systems with a significant spread of the decay times to be measured. Moreover, its cost is much higher than that of the other light detectors considered above.

In this work we describe two systems for time-resolved fluorescence measurements developed in our laboratory, one based on light flux techniques (Docchio et al., 1984a), the other on time-correlated single-photon counting (Cubeddu et al., 1988). In particular, the latter was developed to acquire directly both fluorescence decay waveforms and time-integrated and time-gated emission spectra.

The performances of the two instruments are discussed and some measurements performed on cultured cells treated with a derivative of hematoporphyrin (Hp) are shown as an example.

AUTOMATIC PULSED-LASER MICROFLUOROMETER

In connection with time-resolved fluorescence microscopy studies (Bottiroli et al., 1979; Docchio et al., 1982), a pulsed-laser microfluorometer has been developed in our laboratory. In its most recent version (Docchio et al., 1984a), it makes use of a highly versatile and easy to operate

microprocessor acquisition and control unit with a high degree
of data pre-eleaboration capability. A block-diagram of the
system, which is based on the light flux technique, is shown
in Figure. 1.   The excitation source is a pulsed nitrogen
laser-pumped dye-laser, producing pulses of about 100 ps at a
repetition rate up to about 100 Hz (Cubeddu et al., 1980).
This rather inexpensive source provides tunable excitation
over the whole spectrum from near UV to near IR, thus being
well suited for fluorescence studies of molecules of
biological interest. The laser beam, after collimation, is
sent, via a beam splitter, to the microscope (Leitz Compact)
and to a reference photodiode, which provides a signal
proportional to the excitation pulse energy for normalization
purposes. A light chopper in the laser beam path alternately
transmits and stops the pulses, so as to allow on-line
background subtraction from the fluorescence curve. The
synchronisation between the laser pulses and the chopper is
derived from the chopper, whose status is registered
electronically and used to drive the nitrogen laser. The beam
enters the side-window of the microscope, and is focused on
the sample by the objective. The spot size in the object
plane depends on the objective focal length, the minimum being
of the order of the laser wavelength (0.3-0.5 μm).

The fluorescence signal from the sample is collected
throuh the microscope optics, passes through barrier or
interference filters that cut off laser scattered light and
select the spectral region of observation, and is detected by
a fast PMT (Hamamatsu 1564U-01 or Philips XP 1210). The
electric signal from the PMT is analyzed by a dual-time-scale,
microprocessor-controlled averager, developed in our
laboratory. The averager consists of: (i) a control and

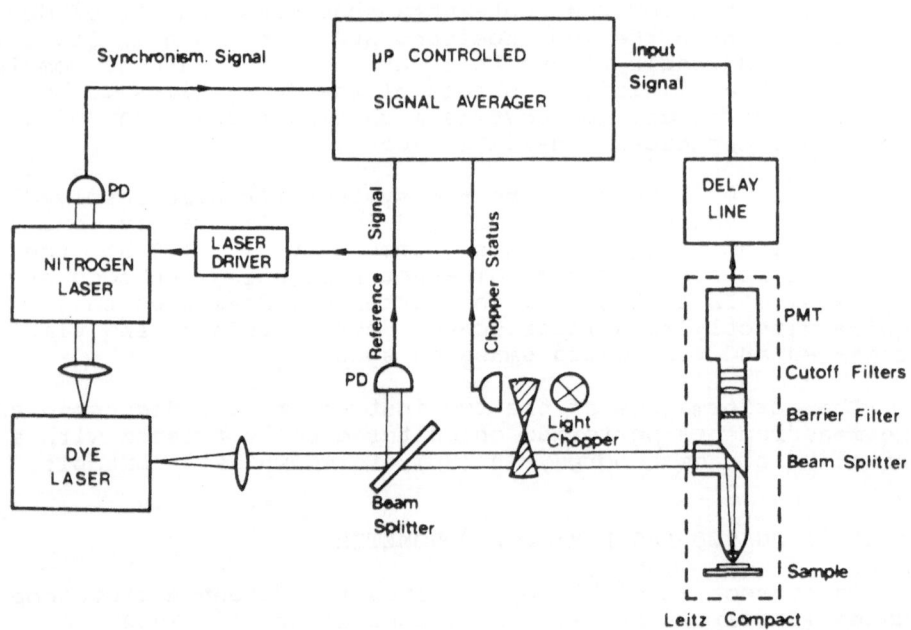

Fig. 1   Block diagram of the pulsed laser micro-
          fluorometer.

averaging unit based on the 8-bit Z-80 microprocessor and containing both standard Zilog boards and self-designed dedicated boards, (ii) a system console, (iii) a sampling oscilloscope as the fast acquisition unit, (iv) a monitoring oscilloscope, and (v) a local mass memory.

The sampling oscilloscope creates a "slowed-down" replica of the fluorescence waveform by sampling successive repetitions of the signal, each having a different time delay with respect to the trigger pulse provided by the synchronisation photodiode. The signal from the sampling oscilloscope is digitized by the microprocessor control unit with a resolution of 10 bits and stored in the system memory. At each point of the waveform, the instrument performs background subtraction and normalization of the digitized value to the energy of the excitation pulse. An averaging over 100-200 repeated scans of the waveform is typically required, due to the weakness of the fluorescence normally emitted by the biological samples and to the presence of stray light, fluctuations, and noise. An on-line display is provided by means of an oscilloscope driven in the x-y mode by the control unit. At the end of the measurement, the digitized and averaged waveform is stored in the local mass memory or transferred to an off-line computer for data analysis and graphics.

It is worth noting that, in this system, the sampling time with respect to the trigger pulse is controlled by the microprocessor itself, with consequent advantages in terms of system versatility. In fact, this allows a dual- or multiple-scale operation. Thus, it is possible to increase the density of the sampled points selectively in the waveform region characterized by a fast transient. In this way, the precision in the determination of fast decay times is significantly improved, with a minimum increase in the overall number of excitation pulses. Moreover, the points of the waveform can be sampled in a random sequence, thus overcoming the systematic errors that would be normally introduced in decay time measurements by photodecomposition of susceptible substances. The system described performs the required random sampling operations by means of a software-implemented, pseudo-random number-generator (Docchio et al., 1984b).

To perform time-resolved spectral analysis in the case of complex molecular systems, it is possible to repeat the fluorescence decay waveform measurements with the same number of sampled points and of averages at different wavelengths, and then reconstruct the spectrum by deconvolution of the decay curves detected at the various wavelengths. However, this procedure is time-consuming and, moreover, the fluorescence of the biological samples is often too low to permit light-flux measurements at a single wavelength with a bandwidth sufficiently narrow for reasonable spectral analysis.

In conclusion, one point should be stressed. Despite the drawbacks typical of light flux techniques as compared with those based on single-photon timing in terms of noise, electromagnetic interference, background intensity, dependence of the waveform on the PMT response, etc., the instrument described is adequate for all those applications where the

requirements on time-resolution are not too strict. On the
other hand, it makes use of an excitation source with the
maximum tunability together with subnanosecond pulses, is
rather compact, easy to operate and has a reasonable cost,
characteristics which make it suitable for many laboratories.

TIME-RESOLVED SPECTROFLUOROMETER WITH PROGRAMMABLE GATING

The availability of picosecond lasers with high
repetition rate and of photomultipliers with low transit-time
jitter (<50 ps) allows the performance of time-resolved
fluorescence measurements with time-correlated single-photon
counting techniques. An instrument based on these techniques
(see Figure 2: in this first version the acquisition unit is
actually a multichannel analyzer and the personal computer is
connected off-line) has been developed in our laboratory to
perform fluorescence measurements with ps time-resolution
(Andreoni and Cubeddu, 1984). The excitation source is a
mode-locked laser. Depending on the wavelength and on the
pulse duration required, either an ion laser (argon or
krypton), with a pulse duration of ~120 ps, or a
synchronously-pumped dye-laser, with pulses of ~5 ps, are
used. The repetition rate of 75-80 MHz is reduced to 750-800
kHz through an acousto-optic pulse-picker on the output beam
of the ion laser or through a cavity-dumper within the dye-
laser cavity. Different confirgurations can be used for
excitation and collection of the fluorescence depending on the
sample to be examined. Samples in solution and cell
suspensions are placed in a 1 cm path quartz cuvette, and the

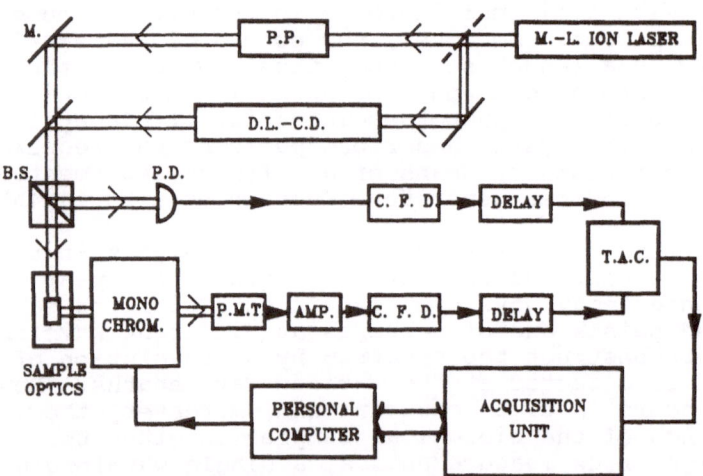

Fig. 2   Block diagram of the system for time-gated
         fluorescence spectroscopy (M-L, mode-locked;
         PP, pulse picker, M, Mirror: DL-CD dye-laser
         and cavity dumper; BS, beam splitter; PD,
         photodiode; CFD, constant fraction
         discriminator; TAC, time-to-amplitude
         converter; PMT, photomultiplier; AMP, pre-
         amplifier and amplifier.

fluorescence is observed at 90° to the excitation beam direction. In the case of microscopic samples (e.g. single cells or other microscopical preparations) the system can be coupled to a microscope. In both configurations, the fluorescence emitted by the sample is measured through a monochromator, to select the observation wavelength, by a MCP PMT (Hamamatsu 1564U-01) in photon-counting mode. If the spectrum-integrated decay waveform has to be measured, a cut-off filter is inserted instead of the monochromator to eliminate scattered laser light. The experimental data are obtained by the time-correlated single-photon counting technique (Cova et al., 1983). The single-photon emission probability following an excitation pulse is well known to reproduce the intensity-vs-time profile of the decay waveform (O'Connor and Phillips, 1984). The excitation pulse (detected by a fast PIN photodiode (PD)) is used to trigger the measurement, whereas the stop pulse is provided by the first fluorescence photon detected by the MCP PMT. To avoid distortions, the laser intensity has to be adjusted to produce a counting rate ≤1% of the excitation flash rate. To increase the possible data acquisition rate, the instrument is usually operated in an inverse configuration, i.e. the fluorescence photons provide the start pulses, and the excitation pulses the corresponding stop, allowing enough time for the time-to-amplitude converter (TAC) to reset between start pulses. The signals from the PD and from the PMT are suitably delayed and shaped by constant-fraction-discriminators (CFD), and then sent to a home-made TAC that generates an output voltage proportional to the time-interval between excitation and fluorescence emission. The signal from the TAC drives a multichannel analyser (MCA) operated in the multi-scaler mode as pulse-height analyser. Thus, at the end of the measurement, the histogram contained in the MCA corresponds to the probability distribution versus time of the fluorescence photons, i.e. to the emission decay waveform. The experimental data are then sent to a computer for numerical evaluation and graphical presentation. The decay time constants are evaluated by fitting the fluorescence waveforms with one or more exponential decays, using a non-linear least squares procedure. The quality of the fitting is estimated on the basis of the weighted residuals and their autocorrelation function. The time resolution of the detection apparatus is limited by the PMT jitter and is ~45 ps, so that the overall time resolution of the system ranges from ~45 ps to ~150 ps depending on the laser source used. In many practical situations, this allows analysis of the measurements without the need of deconvolution.

In a more recent version, the instrument has been modified by the addition of an acquisition unit capable of also performing time-resolved spectroscopy (Cubeddu et al., 1988). Due to the high sensitivity of the single-photon counting method, decay waveforms of even poorly fluorescent samples can be observed at single emission wavelengths within a rather narrow bandwidth. To evaluate the emission spectra of single fluorophores in the case of complex samples, two methods can be followed (Meech et al., 1981): (i) indirect reconstruction of the spectra of deconvoluted decay waveforms measured at single emission wavelengths, and (ii) direct acquisition of the spectra within suitable time-gates with respect to the excitation pulses. The modified instrument developed in our laboratory acquires the fluorescence decay waveform <u>together</u> with the time-gated spectra. A block-diagram of the apparatus is shown in Figure 2. Apart from the acquisition unit, the set-up is the same as described above. This unit consists of the same MCA as used for the emission decay waveform measurements, of a three-input-channel scaler and of a two-channel computer-controlled pulse-selecting-board (PSB) developed in our laboratory. One of the three channels of the scaler receives the signals directly from the TAC, the two remaining channels from the PSB that selects only the TAC pulses having amplitudes within two digitally determined boundaries set by the operator at the beginning of the measurement. A portable HP-9807 computer, programmed in BASIC, provides full automatic control of the instrument. In particular, it drives the monochromator stepping motor, allows the setting of the measurement parameters (spectral region, spectral resolution, gate width and delay from the excitation, measurement-time per point, etc.), operates the PSB, synchronizes the scaler, acquires and stores the experimental data, and is used for their numerical evaluation and graphical presentation. The system described is thus able to perform for each sample the simultaneous acquisition of (i) the spectrum-integrated decay waveform obtained as the sum, at all the wavelengths considered, of the fluorescence counts as a function of time; (ii) the time-integrated emission spectrum obtained by storing all the fluorescence counts as a function of the wavelength, and (iii) two time-gated spectra obtained by counting, again as a function of the wavelength, only the fluorescence pulses falling within preselected time intervals. When required, single-wavelength waveforms can be measured by operating the MCA at the selected wavelength for a suitable time.

TIME-RESOLVED FLUORESCENCE OF CELLS TREATED WITH A DERIVATIVE OF HEMATOPORPHYRIN

The photophysical properties of hematoporphyrin (Hp) and of some of its derivatives containing different percentages of the aggregated tumor localizing fraction (TLF) have been investigated in solution, in model systems, and in several biological samples (cell cultures and microscopic preparations) using both the systems described above. The main results are reviewed in Ramponi et al. (1990).

Single-photon-counting measurements allow resolution of the fluorescence of the most aggregated fraction of Hp derivative which is too short to be detected with light flux techniques. As an example of a real application, some measurements performed on cell suspensions containing an Hp derivative, using the second instrument, are described. Figure 3 shows the time-integrated spectrum and two time-gated spectra of L1210 cells, both untreated and after 30 min of treatment with 5 μg/ml DHE (the purified form of HpD containing 99.5% TLF). The laser intensity was adjusted to have a counting rate of ~1% at the fluorescence peak wavelength in both cases. Thus, only relative comparisons can be made among the fluorescence intensities measured in the two samples. The fluorescence decay waveform of the treated cells could be fitted by three exponential components (0.67 ns, 4.77 ns, and 14.82 ns, with relative peak amplitudes of 39.1%,

UNTREATED   DHE 5 µg/ml

                             30 min

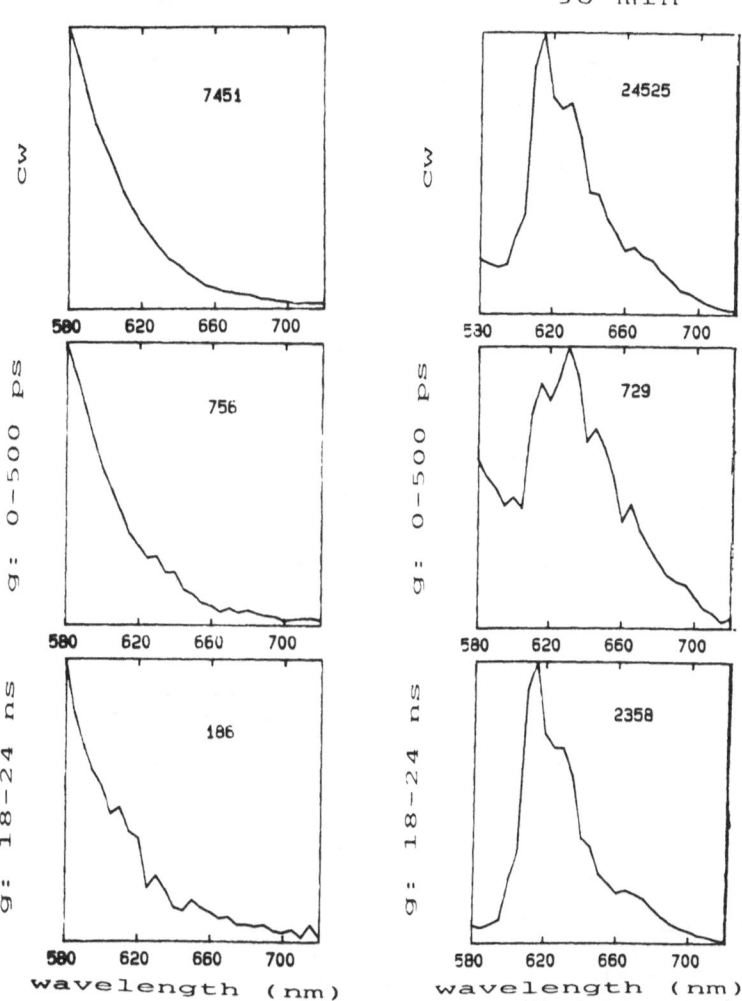

Fig. 3 Time-integrated (cw) and time-gated spectra
        (gate limits 0-500 ps and 18-24 ns) of L1210
        cells incubated in drug-free culture medium
        (left) and incubated in culture medium
        containing DHE (right) for excitation at
        364 nm. The numbers inside the frames of
        the single spectra correspond to the counts
        at the peak. The laser intensity was
        adjusted to have a counting rate of ~1% in
        both cases.

28.3%, and 32.6%, respectively). Thus, the two gates were
chosen with 18 ns delay from the peak (6 ns width) to
discriminate the long-decaying component, and with no delay
(500 ps width) to detect the spectral contribution of all the
fluorescent components almost independently of their
fluorescence efficiency. The same gating parameters were used
for untreated cells. Actually, the fluorescence decay
waveform of the untreated cells is much more rapid, the

longest decay time constant being around 6 ns with a relative peak amplitude of 23.7%. As a result, the ratio of peak values of the 18 ns delayed gated spectra of treated over untreated cells is about 4 times higher than that of the time-integrated spectra of treated over untreated cells. This suggests a possible application of time-gated emission spectroscopy to increase the sensitivity and selectivity of fluorescence-based tumor diagnosis employing HpD. Indeed, the contribution of the tissue autofluorescence in a time-delayed spectrum should be negligible, whereas in the cw measurements it may sometimes prevent a correct interpretation.

ACKNOWLEDGEMENTS

The authors are deeply indebted to Dr. G. Bottiroli and to Prof. F. Docchio for their valuable contribution to the development of the instruments described in this paper.

The financial support of the Italian C.N.R. Special Projects ("Tecnologie biomediche" and "Oncologia") and of CARIPLO (Cassa di Risparmio delle Provincie Lombarde, Milan, Italy), through the donation "Progetto Laser" to the Politecnico of Milan, are gratefully acknowledged.

REFERENCES

Andreoni, A. and Cubeddu, R., 1984, Time-resolved fluorescence measurements on porphyrins and furocumarins, Med. Biol. Envir., 12:421.

Bottiroli, G., Prenna, G., Andreoni, A., Sacchi, C.A. and Svelto, O., 1979, Fluorescence of complexes of quinacrine mustard with DNA. I. Influences of the DNA base composition on the decay time in bacteria, Photochem. Photobiol., 29:23.

Cova, S., Longoni, A., Andreoni, A. and Cubeddu, R., 1983, A semiconductor detector for measuring ultraweak fluorescence decays with 70 ps FWHM resolution, IEEE J. Quantum Electron., QE-19:630.

Cubeddu, R., De Silvestri, S. and Svelto, O., 1980, Subnanosecond amplified spontaneous emission pulses by a nitrogen-pumped dye laser, Opt. Commun., 34:460.

Cubeddu, R., Docchio, F., Liu, W.-Q., Ramponi, R. and Taroni, P., 1988, A system for time-resolved laser fluorescence spectroscopy with multiple picosecond gating, Rev. Sci. Instrum., 59:2254.

Cubeddu, R., Docchio, F., Ramponi, R. and Boulton, M., 1990, Time-resolved fluorescence spectroscopy of the retinal pigment epithelium: age related studies, IEEE J. Quantum Electr., in press.

Docchio, F., 1989, Ocular fluorometry: principles, fluorophores, instrumentation, and clinical applications, Lasers in Surg. and Med., 9:515.

Docchio, F., Ramponi, R., Sacchi, C.A., Bottiroli, G. and Freitas, I., 1982, Time-resolved fluorescence microscopy of hematoporphyrin-derivative in cells, Lasers in Surg. and Med., 2:21.

Docchio, F., Ramponi, R., Sacchi, C.A., Bottiroli, G. and Freitas, I., 1984a, An automatic pulsed laser microfluorometer with high spatial and temporal resolution, J. Microsc., 134:151.

Docchio, F., Ramponi, R. and Zaraga, F., 1984b, Subnanosecond waveform analysis in the presence of fading by means of a random sampling technique, Rev. Sci. Instrum., 55:365.

Haar, H.P. and Hauser, M., 1978, Phase fluorometer for measurement of picosecond processes, Rev. Sci. Instrum., 49:632.

Kinoshita, S., Ohta, H. and Kushida, T., 1981, Subnanosecond fluorescence-lifetime measuring system using photon counting method with mode-locked laser excitation, Rev. Sci. Instrum., 52:572.

Lakowicz, J.R., Lacko, G., Cherek, H., Gratton, E. and Limkeman, M., 1984, Analysis of fluorescence decay kinetics from variable-frequency phase shift and modulation data, Biophys. J., 46:463.

Larsen, M. and Johansson, L.B.A., 1988, Time-resolved fluorescence properties of fluorescein and fluorescein glucuronide, Exp. Eye Res., 48:477.

Malatesta, V. and Andreoni, A., 1988, Dynamics of anthracyclines/DNA interaction: a laser time-resolved fluorescence study, Photochem. Photobiol., 48:409.

Meech, S.R., O'Connor, D.V., Roberts, A.J. and Phillips, D., 1981, On the construction of nanosecond time-resolved emission spectra, Photochem. Photobiol., 33:159.

O'Connor, D.V. and Phillips, D., 1984, "Time-correlated single photon counting," Academic Press, London.

Ramponi, R. and Rodgers, M.A.J., 1987, An instrument for simultaneous acquisition of fluorescence spectra and fluorescence lifetimes from single cells, Photochem. Photobiol., 45:161.

Ramponi, R., Sacchi, C.A. and Cubeddu, R., 1990, Present status of research on hematoporphyrin derivatives and their phtophysical properties, in: "Laser applications in medicine and biology (vol. 5), "M.L. Wolbarsht, ed., Plenum Press, New York, in press.

Rodgers, M.A.J. and Firey, P.A., 1985, Instrumentation for fluorescence microscopy with picosecond time resolution, Photochem. Photobiol., 42:613.

Schneckenburger, H., 1985, Time-resolved microfluorescence in biomedical diagnosis, Opt. Eng., 24:1042.

Thaer, A. and Sernetz, M., eds., 1973, "Fluorescence techniques in cell biology," Springer, Berlin.

Yamashita, M., Nomura, M., Kobayashi, S., Sato, T. and Aizawa, K., 1984, Picosecond time-resolved fluorescence spectroscopy of hematoporphyrin derivative, IEEE J. Quantum Electron., QE-20:1363.

Yamashita, M., Tomono, T., Kobayashi, S., Torizuka, K., Aizawa, K. and Sato, T., 1988, Picosecond fluorescence spectroscopy on incorporation processes of hematoporphyrin derivative into malignant tumor cells in vitro, Photochem. Photobiol., 47:189.

LASER-INDUCED MULTI-STEP PROCESSES IN CHROMOPHORES OF BIOMEDICAL INTEREST

Alessandra Andreoni

Centro Endocrinologia ed Oncologia Sperimentale
C.N.R., and Department of Biology and Cellular and
Pathology, University of Naples, Via Sergio
Pansini, 5-80131 Napoli

INTRODUCTION

Following the successful use of high-power lasers to
perform selective photochemistry via multi-step excitation
(Letokhov, 1978), the possibility of non-linear photobiology
has been attracting the attention of a number of researchers
at different laboratories. The first proposals of multi-
quantum excitation to promote interactions with selected
biomolecules paralelled some established schemes of selective
laser photochemistry and relied on the preparation of the
selected biomolecule in a specific vibronic state from which
the second excitation step was made to occur (Letokhov, 1976).
A difficuly with this approach arises from the fact that the
selectivity of the first step which, in the photochemistry of
atoms and small molecules, is warranted by a fine tuning of
narrow-band exciting light to a resonance of the selected
species is lost on a picosecond time scale in the case of
large biomolecules owing to the fact that their roto-
vibrational mode thermalization is very fast in the liquid
phase. However, the advent of picosecond laser sources which
provided high-intensity UV excitations stimulated the rapid
development of two-quantum UV photochemistry of nucleic acids
(Kryukov et al., 1979). The interest of studying this type of
photochemistry of nucleic acid components stems from the fact
that the sequential absorption of two photons (two-step
excitation, Fig. 1a) initiates chemical reactions similar to
those taking place under ionizing radiation. Furthermore, it
is worthwhile noting that it would be impossible to perform
single-quantum excitation of nucleic acids to the same high-
lying electronic states in aqueous environments because of the
strong absorption of water in the region of vacuum-UV
(Oraewsky and Nikogosyan, 1985).

Actually, the capability of high-intensity pulses to
relese, via multi-step interactions, the necessary excitation
energy to endogenous chromophores that absorb at wavelengths
too short to penetrate the overlaying tissues or media is a

unique feature of this type of excitation technique that may open new fields of application of non-linear photobiology in vivo. For instance, the modern techniques of eye surgery utilizing near IR or visible laser pulses have this rationale, basically. Light pulses at relatively long wavelengths are made to interact non-linearly with chromophores lying beyond natural cut-off filters such as the cornea or the lens which do not allow light of wavelengths shorter than about 295 or 400 nm, respectively, to penetrate the eye. In this connection, a further application might arise from the work of Bodaness and co-workers (Bodaness and King, 1985: Bodaness et al., 1986) in which tumor localization by two-step laser-induced fluorescence of hematoporphyrin derivative, HpD,is proposed as alternative to the tumor photolocalization techniques utilizing porphyrins and blue-light excitation (Profio, 1984). These authors showed that HpD excitation into the near-UV Soret band could be achieved in solution by using either a Q-swithed Nd:YAG laser emitting at 1064 nm or a Q-switched alexandrite laser providing pulses in the 730-770 nm spectral range (Fig. 1b). Finally, Andreoni et al. (1987) found that the efficacy of tumor photochemotherapy with porphyrins could be enchanced in vivo by adopting a non-linear photoactivation scheme utilizing red and near-IR pulses that penetrate tissues deeper than single photons with an energy equivalent to the energy sum of the two-pulse photons.

From a fundamental point of view, multi-step excitation provides not only the possibility of driving photoreactions via photons of relatively low energy - see, for instance, Kochevar et al. (1989) but, especially in conjunction with the use of photosensitizers which are specifically developed to activate reactions upon excitation in vivo with deeply penetrating light, can also initiate novel photoreactions from high-lying excited states that would not be populated in the linear absorption regime (Andreoni, 1980; Andreoni et al., 1980a: Karu et al., 1981). This aspect, which has been the key to the success of many applications of non-linear laser photochemistry, seems to be the most promising also in the field of multi-step laser photobiology. In principle, multi-step excitation could allow a broad spectrum of photosensitized reactions (Scaiano et al., 1988) to become feasible both in vitro and in vivo. Furthermore, the selectivity of these reactions for specific biological targets

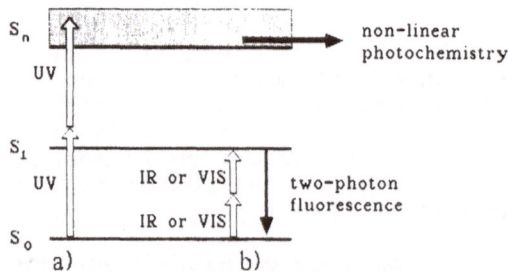

Fig. 1 Two-photon absorption transitions leading to a) photochemical reactions, b) fluorescent emission.

can be refined and improved, compared to that originating
from the specificity of the photosensitizer
affinity for selected biological components or cell
compartments, if suitable timings of the different steps of
the interaction are adopted. The examples reported in this
paper have been chosen to illustrate the two points of above.
It is important to mention that when a photosensitizer,
instead of a biomolecule, is used as the direct target of a
multi-step excitation, nanosecond rather than picosecond light
pulses are usually sufficienctly short to activate the desired
reaction efficiently, because it is much more common that the
lifetime of the excited state working as the intermediate
state of the multi-step excitation process falls in the
nanosecond range for photosensitizers than it is for
biomolecules, besides the fact that the latter have their
excited states at much higher energies above the ground state
(Andreoni, 1980).

NEW SENSITIZATION MECHANISMS INDUCED BY MULTI-STEP EXCITATION OF PHOTOSENSITIZERS

The primary steps in the activation of a photosensitizer
with continuous wave (cw) light are its excitation to the S_1
state and the conversion to the usually long-lived triplet
state T_1 in which photosensitized substrate oxygenations can
be activated (Foote, 1980). Actually, the relatively low
intensities typically achieved with cw sources as compared to
those available with pulsed lasers, together with the fact
that the S_1 state is usually much shorter lived than T_1, even
when the latter is efficiently quenched by the mechanisms
leading to substrate sensitization, makes any further
excitation of the photosensitizing molecule very unlikely
occur during the S_1 lifetime under cw pumping (Fig. 2). On
the contrary, multi-step excitation may be achieved by
irradiating with high intensity. At this point it is worth
mentioning that the use of short light pulses to obtain multi-
step excitation of high-lying singlet states (see pathway in
Fig. 2) is not only instrumental but, to some extent, it is
necessary to compete with the rate k_{ST} of intersystem-
crossing.

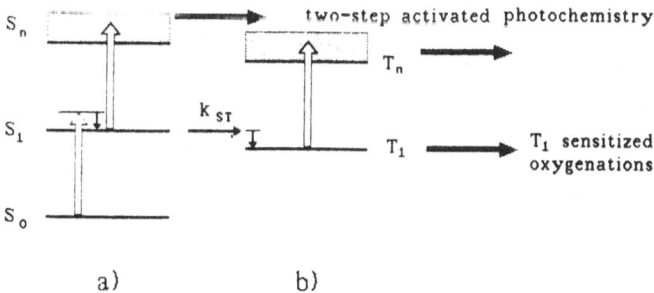

Fig. 2   Competition between two-step photoactivation
         a) and linear activation of T_1-sensitizer
         mediated photoprocesses.

The most straightforward way of assessing whether multi-step excitation to high-lying singlet states produces substrate photosensitization via pathways different from the T_1-mediated mechanisms may consist in studying the response of the substrate upon activation with different pulse-peak intensities quantitatively. As most experiments of multi-step activation are performed with laser beams, for which the propagation direction is well defined, it is more convenient to discuss this point in terms of substrate response to a given fluence (i.e., light dose per unit irradiated area) than relative to the light dose delivered. While it is obvious that the substrate response to a given fluence is, over a wide range of values, independent of intensity if the relevant mechanisms proceed via T_1 sensitizers, it must be expected to exhibit a strong dependence upon intensity when multi-step excitation of the photosensitizer plays a critical role in determining the effect observed on the substrate. In the latter case administering a fixed fluence with high rather than low intensity, for instance, will produce both a proportionally higher population in the intermediate S_1 state and enhance its promotion to the upper singlet state. Equivalently, the time rate at which the damage is produced would be linear with the intensity in the former case but more than linear in the latter. In principle, if the effects on the substrate were fully mediated by two-step excited sensitizers, the photoproducts should be formed at a rate proportional to the square of the intensity. Correspondingly, a constant fluence should give a damage proportional to the intensity. That the occurrence of multi-quantum absorption is the primary step of a number of photosensitized reactions was demonstrated by this method. Photodamage of an acridine dye - quinacrine mustard, QM - that required the absorption of two photons from a nitrogen laser pulse was reported by Andreoni et al. (1980b). Acridines are dyes used as specific stains of DNA both _in vitro_ and _in vivo_ that bind to this biomolecule by intercalating their chromophore moiety between adjacent DNA base pairs. While their absorption and fluorescence spectra are not perturbed by the binding to DNA as much as those of other dyes, their S_1 lifetimes are commonly much dependent upon the base-pair composition of the intercalation site. In fact, the presence of a Guanine residue adjacent to the chromophore may shorten the fluorescence lifetime, and correspondingly lower the fluorescence quantum yield, by a factor of up to 10, while the intercalation between two Adenine-Thymine pairs tends to increase both S_1 lifetime and fluorescence yield (Andreoni, 1980 and later in this paper). However, when dissolved in 0.2 M acetate buffer solution pH 4.6, QM exhibits a fluorescence lifetime of 3.9 ns and is photodamaged by irradiation with 5 ns pulses at 337.1 nm with an efficacy, per unit fluence, that increases linearly with the square of the pulse peak intensity up to a value of about 10 MW/cm^2. Andreoni et al. (1980b) also noted a tendency for the efficacy to increase only linearly with intensity above this value as a consquence of the saturation of the first step (excitation to S_1) of this non-linear process.

A similar effect was observed by the same group for the photochemotherapeutic drug HpD (Andreoni et al., 1982). The drug, which is known to self-associate in aqueous solution and to form aggregates with short-lived S_1 states, was dissolved with 0.1 M phosphate buffer, pH 7.4, to a relatively low concentration value (1.67 x 10^{-6} M) so as to be virtually monomeric and have an S_1 lifetime of about 15 ns. Irradiation with 10 ns pulses from a nitrogen laser produces irreversible photodamage of HpD with a rate increasing linearly with the square of the intensity up to about 5 MW/cm^2 and linearly with intensity at higher values (see Fig. 3). Interestingly, the photoproducts were found to be cytotoxic. The cytotoxicity was tested on an epithelial cell strain of rat thyroid origin. Irradiating the cell cultures which were exposed to 2 x 10^{-5} M HpD for 2 h prior to irradiation with 65 mW/cm^2 from either the nitrogen laser or from a cw argon-ion laser tuned at 333.6 nm produced cell kill, but the cytocidal efficacy of equal light fluences was higher with the pulsed than with the cw laser by as much as a factor of 1.5. Since differences in the HpD absorbance at the slightly different wavelengths of the two lasers cannot account for the size of the effect, the conclusion can be drawn that the pulsed-laser photons, which are absorbed by the S_1 sensitizer to perform the second step of the non-linear interaction, have a better chance of inducing cell kill than those of the cw laser which, being absorbed by ground-state HpD, can only increase the sensitizer population in the T_1 state and hence contribute to the linear mechanism of photodynamic action. Though the nature of the photosensitizing pathways initiated by HpD in the high-lying singlet states has not yet been elucidated, these pathways might be exploited to improve tumor phototherapy provided that these HpD states can be populated efficiently using nanosecond pulses at longer wavelengths that are endowed with deeper penetration in tissues compared to near-UV wavelengths. Unfortunately, all experimental evidences available today are against this perspective. In fact, not only has it been shown that tryptophan photodegradation sensitized by hematoporphyrin, HP, photoactivated by nanosecond pulses at 630 nm proceeds via a pure singlet oxygen mechanism sensitized by T_1 up to intensity values as high as 20 MW/cm^2 (Andreoni, 1987; Andreoni and Ghiglione, 1987), but also <u>in vivo</u>

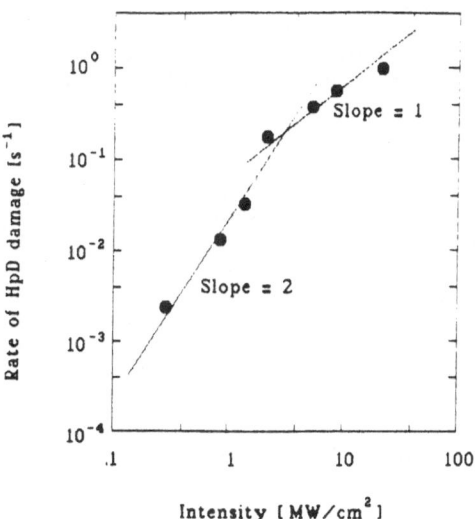

Fig. 3   Rate of HpD damage as a function of the
         pulse peak intensity (log-log plot).

experiments of tumor phototherapy with HpD excited by
nanosecond pulses at 631 nm point at excluding any improvement
in tumor regression based on these non-linear mechanisms
(Andreoni et al., 1987).

An intensity dependent phototoxic response of human
bladder carcinoma cells to treatment with deoxycycline, DOTC,
i.e. the most potent in vitro photosensitizer among
tetrayclines, and exposure to 351 nm pulses from a XeF excimer
laser has been reported by Shea et al. (1988). The cell
culture dishes were irradiated with peak intensities of 18.5,
125 and 1250 MW/cm^2 following 30 min. incubation with 20.7 x
10^{-6} M DOTC and the cell sensitivity to damage (Mitochondrial
vacuolization) was quantified in terms of fluence threshold.
A threshold of 3 J/cm^2 was determined for the two lowest peak-
intensity values and 6 J/cm^2 for the highest one. The authors
related the cause of this phenomenon to the generation of
superoxide anion by electron transfer from the excited DOTC
which is capable of rapid catalytic quenching of singlet
oxygen, the major species responsible for DOTC-photosensitized
mitochondrial damage under cw irradiation (Shea et al., 1986).
On the one hand, the shortness of the DOTC S_1 lifetime, about
0.4 ns (Hasan, private communication) compared to the duration
of the pulse used to photoactivate DOTC, about 20 ns, makes it
questionable that only high-lying singlet states are populated
during the light pulse of the highest intensity (see below),
on the other hand it makes it unlikely that ground state
depletion is a significant factor in the observed decrease in
mitochrondrial damage at increasing intensity. The same
group, however, repeated the experiment using 24 ps pulses at
355 nm obtained from a frequency-tripled mode-locked Nd:YAG
laser with an improved technique for measuring the
mitochondrial damage (Hefetz et al., 1989). The fluence value
causing mitochondrial damage in 50% of cell cultures was found
to increase, again by a factor of about 2, when the peak
intensity was increased from 130 to 760 MW/cm^2 and, in the
latter case, became higher than the (constant) value obtained
for cw irradiation with an argon-ion laser at 333.6 nm and
intensity values below 1 W/cm^2. The results of these
experiments will be discussed in the next section because
their interpretation involves the relations between pulse
duration and sensitizer excited-state kinetics that will be
illustrated later in this paper.

A more direct way of showing the occurrence of multi-step
photoactivation processes than by studying the dependence of
equal-fluence efficacy upon intensity could be easily devised
for sensitizers exhibiting wavelength dependent
photobiological effects. In fact, comparing the sensitization
obtained by directly exciting into an upper state, presumably
with UV photons, with that obtained by exciting into S_1 with a
high-intensity light of longer wavelength, one should be able
to assess the occurrence of a multi-step activation of the
photosensitizer as well as evaluate its relevance in terms of
photobiological effects. It is surprising that this approach
has been attempted only in the case of rose bengal, RB
(Fluhler et al., 1989) because, even disregarding the case of
furocoumarins, wavelength dependent photochemistry has been
demonstrated for a number of drugs, some of which have
photosensitizing capabilities (see, for instance, Motten et
al.,(1985)) for the phenothiazine drug, chlorpromazine; Li and

418

Chignell (1987) for the anthracycline drugs, adriamycin and daunomycin). The xanthene dye RB, which has been widely used in studies of photooxidation and photodynamic action, has been shown recently to cause higher phototoxicity upon absorption of photons at 308 nm than at 514 nm (Fluhler et al., 1989; Allen and Kochevar, 1989). Fluhler et al. (1989) investigated the possibility of activating the mechanisms sensitized upon 308 nm excitation via non-linear excitation using photons of lower energy. In particular, these authors studied the intensity dependence of RB photosensitized inhibition of red blood cell acetylcholinesterase upon excitation with pulses of 40 ps duration at 532 nm. The pulses were generated by a frequency-doubled mode-locked Nd:YAG laser and focused onto samples of erythrocyte ghost suspensions (0.03 mg/ml membrane protein in 5×10^{-3} M PBS/10^{-3} M EDTA) added to 2×10^{-6} M RB. Delivering a total fluence of 104.2 J/cm^2 with different intensities, ranging from 10 MW/cm^2 to 12.2 GW/cm^2, resulted in an approximately constant (68%) enzyme photinhibition up to an intensity of 40 MW/cm^2 while, at higher intensities, the RB-mediated inhibition decreased until it reached less than 1 at 12.2 GW/cm^2. Parallel experiments in which either RB fluorescence or ground state absorption were montiored as a function of the incident peak intensity showed that ground state depletion occurred for RB in the 20 to 80 MW/cm^2 range and allowed the authors to conclude that the decrease in enzyme inhibition at high intensities was a consequence of ground state depletion. Thus, in spite of the high intensity of the pulse whose duration is, by the way, shorter than the S_1 lifetime (Rodgers, 1981), this state seems to have such a low absorption at 532 nm that it cannot be used as the intermediate state of a non-linear pumping process to reach the more reactive upper state.

SELECTIVITY OF MULTI-STEP PHOTOSENSITIZED REACTIONS

A necessary condition for the occurrence of multi-step photoactivation of a sensitizer into a high-lying singlet state is that the light pulse duration is shorter than the sensitizer S_1 - state lifetime, longer pulses would give rise to further absorption from the ground state and, possibly, from the T_1 state. One can take advantage, however, of the latter event to create a population in some high-lying triplet state of the photosensitizer and activate novel reactions with the substrate. Moreover, since binding to the biosubstrate often affects the S_1 lifetime to an extent rather specifically related to the biosubstrate, the duration of the pulse used for the interaction may be easily chosen so as to be shorter than the S_1 lifetime of the sensitizer bound to a site but longer than that of the same sensitizer bound to a different site. Thus non-linear interactions with pulses of suitable duration may produce selective photosensitization of biological targets.

It has already been mentioned that acridine dyes have shorter S_1 lifetimes when they bind DNA adjacent to a Guanine base than when they are either free in solution or bound to DNA between two Adenine-Thymine base pairs. Andreoni et al. (1980b) irradiated QM (10^{-6} M in 0.2 M acetate buffer solution, pH 4.6) bound to the synthetic polynucleotides poly dA-poly dT of poly dG-poly dC (10^{-4} M nucleotide phosphorus) with 5 ns pulses from a nitrogen laser, that is with pulses of duration in between the S_1 lifetime values for QM bound to the two polynucleotides and at a wavelength, 337.1 nm, equally absorbed by QM ground state molecules in both cases. Irradiation with about 10 MW/cm^2 peak intensity produced QM photodegradation both in poly dA-poly dT and in poly dG-poly dC but at a rate which was 4.73 times faster in the former case compared to the latter one. This demonstrates that the two-step photodamage of QM occurring in the high-lying singlet state populated by the pulse when the dye is free (see previous section) still predominates over any high-lying triplet state photochemistry whose activation would be favoured by the fast intersystem-crossing that QM experiences when it binds poly dG-poly dC.

One can interpret the results obtained by Hefetz et al. (1989), and mentioned in the previous section on the sensitivity of mitochrondria of human bladder carcinoma cells to treatment with DOTC and near-UV light generated by either cw (argon-ion laser at 333.6 nm) or nanosecond (XeF excimer laser at 351 nm) or picosecond (Frequency-tripled mode-locked Nd:YAG at 355 nm) lasers in the same light. These sources allow DOTC-photosensitized mitochondrial damage to be studied upon activating the drug with intensity values covering an extremely broad range as well as via a variety of pumping schemes. In fact, while cw light can only activate photodynamic action by T_1 DOTC energy transfer (Fig. 4a), the 24 ps pulses from the Nd:YAG laser may produce phototoxicity via non-linear excitation of high-lying singlet states (Fig. 4b) and the XeF laser pulses, having 00 ns duration, i.e. longer than the DOTC S_1 lifetime (see above), may pump high-lying triplet states of the drug (Fig. 4c). Recently this group of researchers (Shea et al., 1989) made accurate determinations of the fluence values, D_{50}, causing mitochrondrial injury in 50 cell cultures upon irradiation with 10 to 1000 mW/cm^2 intensity with the cw laser or with peak intensities between 18.5 and 1250 kW/cm^2 for the 20 ns pulses or between 5.5 and 760 MW/cm^2 for the 24 ps pulses. Their results demonstrate than an efficient non-linear pumping of high-lying singlet states of DOTC is obtained with intensities above 1 MW/cm^2 (24 ps pulses) and that the consequent photoxicity is higher than that of photodynamic action (cw light). Moreover, whatever the intensity of the 20 ns pulses, higher D_{50} fluence values are measured compared to irradiation with both cw light and picosecond pulses. The difference relative to cw irradiation may indicate that the 20 ns pulses, even at the lowest intensity adopted, remove DOTC molecules from the T_1 state, thus causing a decrease in the population available to sensitize the linear mechanism, and promote them to some high-lying triplet state which is less efficient than T_1 at sensitizing mitochondrial injury. Furthermore, with the 20 ns pulses, a steep increase in D_{50} (from 2 to 5.2 J/cm^2) is observed on going from .125 to 1.25 MW/cm^2 intensity, which might reflect the onset of the DOTC promotion to further triplet states of even lower sensitizing efficacies. Finally, the fact that, with the picosecond pulses a tendency of D_{50} values to increase develops only at intensities much higher than with the nanosecond pulses (between 130 and 760 MW/cm^2) is in keeping with the observation of lower D_{50} values for picosecond compared to nanosecond pulse irradiation, which might imply a faster

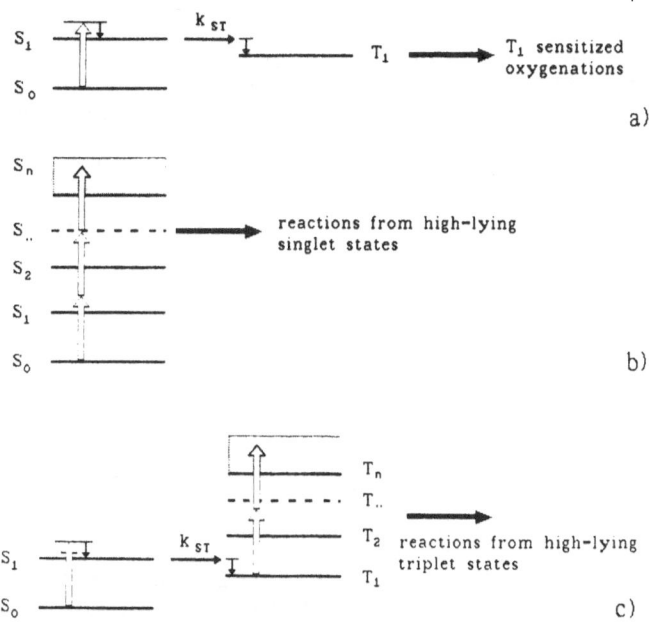

Fig. 4   Photoactivation pathways of DOTC under
irradiation with a) cw light, b) ps pulses
and c) nonosecond pulses.

depletion of high-lying singlet than triplet state by the
photochemical reactions responsible for the biological effect.

Schemes of more general validity than the previous ones,
in which selectivity is achieved by using pulses of durations
shorter or longer than the S_1 lifetime, have been developed to
photoactivate selectively either high-lying singlet or triplet
states of photosensitizers to perform non-linear photobiology.
Such schemes utilize two pulses, both of durations shorter
than the S_1 lifetime at different wavelengths; one, at a
wavelength within the ground state absorption of the
photosensitizer, is utilized to pump it to the S_1 state while
the other is utilized for the second step and, when it is
possible, is chosen at a wavelength not absorbed by the ground
state sensitizer. To promote the S_1 sensitizer to a high-
lying singlet state the second pulse is sent simultaneously to
the first one, whereas, to populate high-lying triplet states,
it is conveniently delayed with respect to the first pulse to
let the molecules cross over to T_1. The wavelength of the
second pulse is suitably chosen to optimize the efficiency of
the desired activation process.

One scheme of non-linear activation of this type was
adopted for an acridine dye, proflavine (PF), which, at
variance with QM, does not undergo photodegradation upon
intense pumping with pulses at the wavelength (430 nm) of its
ground state absorption peak (Andreoni et al., 1980c).
However, PF could be efficiently damaged by combined
irradiation with both 430 nm pulses (duration 250 ps)
generated by a nitgrogen-pumped dye laser and the pulses of
the pump nitrogen laser itself (500 ps duration) at a
wavelength, 337.1 nm, where ground state PF does not absorb.

Using the experimental set-up described by Andreoni et al. (1980c), the two beams could be made spatially coincident in the sample and the nitrogen laser pulses were made to arrive at the sample with a delay adjustable from 0 to 8 ns with respect to the dye laser pulses. The damaging efficacy of this two-pulse excitation of PF free in solution was found to decrease at increasing delays, thus showing that, when PF molecules absorb the nitrogen laser pulse during their lifetime in the S_1 state (4.3 ns), they reach a more reactive (singlet) state compared to when they are promoted to an upper (triplet) state by the pulse arriving when they have already crossed over to T_1. For PF bound to poly dA-poly dT, a very similar decrease in the efficacy of PF damage at increasing delays was observed in agreement with the fact that this acridine does not experience substantial lengthening of its S_1 lifetime upon binding between Adenine-Thymine base pairs compared to when it is free in solution. On the contrary, the S_1 state is strongly quenched by the binding to poly dG-poly dC. In this case, our experiments of two-step photodamage showed a more dramatic decrease, compared to the previous case, in the efficacy of the process, by a factor of 5, even for delays as short as 1 ns. This reflects that T_1 molecules are rapidly quenched by some mechanisms that put them out of resonance for the occurrence of the second step. It is likely that PF decays back to S_0, from where the photons at 337.1 nm are not absorbed, via direct electron transfer to Guanine which is the most readily oxidizable DNA base.

It has already been mentioned that multi-step activation of porphyrins excited in the red absorption band utilized for tumor phototherapy failed to work both *in vitro* and *in vivo* (Andreoni and Ghiglione, 1987; Andreoni et a.., 1987). However, in an attempt to find photosensitization pathways not requiring the mediation of oxygen, the reactivity of porphyrins excited to either singlet states higher in energy than S_1 or triplet states higher than T_1 has been investigated using the powerful pulses of excimer-pumped dye lasers (Andreoni and Ghiglione, 1987; Andreoni et al., 1987; Andreoni, 1987). In these experiments, the output beam of XeCl excimer laser was split into two beams to pump the oscillators of two dye lasers generating 5 ns pulses at different wavelengths. Here, we will discuss only experiments in which the first step consisted of an exciation in the red. Andreoni and Ghiglione (1987) performed experiments in which pulses at 700 nm were shed onto air-equilibrated samples of 2.1×10^{-5} M HP/10^{-4} M tryptophan, Trp, in a 30% methanol - 70 phosphate buffer mixture either simultaneously or delayed (0 to 24 ns delay) with respect to pulses at 630 nm. The transient absorption of the pulse at 700 nm was measured and found to be maximal when the delay was 16.4 ns. Since the S_1 lifetime of HP was determined to be 14.5 ns in this environment (Andreoni, 1987), the result indicates that T_1 HP absorbs more at 700 nm when compared to S_1 HP. Accordingly, it was found that, while the rate of Trp photodegradation was only slightly increased when the 700 nm beam, which is not absorbed by ground state HP, was superimposed on that at 630 nm and the pulse at the two wavelengths were coincident in time, a sizeable increase was observed if the pulses at 700 nm were made to interact with the sample 16.4 ns later than those at 630 nm. Thus reactions initiated by HP pumped to high-lying triplet states seem to lead more effectively to Trp photodegradation compared to photooxidation mediated by singlet oxygen. It is worth noting that experiments performed in vivo on a murine sarcoma tumor model (Andreoni et al., 1987) utilizing Photofrin, i.e. the commercial HpD drug, and pulses at these wavelengths showed a higher phototherapeutic efficacy than that of cw irradiation with the same light dose.

The photosensitizing pathways initiated by highly excited HP were investigated by Andreoni (1987) on the same HP/Trp system upon activation with 630 and 481 pulses generated by an excimer-pumped laser set-up similar to the previous one because stronger non-linearities were detected by performing the second step with 481 nm instead of 630 nm pulses. At variance with the previous case, however, ground state HP absorbs at 481 nm and irradiation with a single beam at this wavelength produced Trp degradation as well as irradiation with the beam at 630 nm alone. The relative rates of Trp photodamage in the two cases were proportional to the corresponding HP ground state absorption cross-sections. Furthermore, transient absorption measurements showed stronger absorption of S_1 than T_1 HP at 481 nm. Correspondingly, sending the pulses at 481 nm and those at 630 nm simultaneously resulted in a faster Trp degradation than sending the pulses at 481 nm with a delay. In both cases, however, the effects on Trp were more than additive and, also, higher than those obtained by irradiating with both beams but with no time correlation between the pulses at the two wavelengths. Activation of HP to high-lying triplet states, achieved with 16.4 ns delay, though less potent at sensitizing Trp damage compared to activation to high-lying singlet states, initiated mechanisms less sensitive to the presence of oxygen. Conditions of low oxygen concentrations at which the effects of irradiating with simulateneous pulses were completely suppressed permitted high amounts of Trp damage to be produced when the pulses at 481 nm were delayed by 16.4 ns with respect to those at 630 nm. The fact that bubbling the sample solutions with an electron scavanger such as N_2O suppressed the photochemistry sensitized by high-lying triplet HP is a strong indication that two-step excitation with the delayed pulses activates HP photoionization and direct substrate oxidation.

ACKNOWLEDGEMENTS

The continuing encouragements of Prof. O. Svelto, director of the laser laboratory at the Engineering Faculty of the Politecnico of Milano, where the author worked for years with the most valuable cooperation of the colleagues, as well as the warmest courtesy of friends such as Drs. I. E. Kochevar, T. Hasan and Y. Hefetz from the Wellman Laboratories at the Massachusetts General Hospital, Harvard Medical School, Boston, are greatly acknowledged. In particular, the author thanks Tayyaba Hasan and Yaron Hefetz for the useful discussions about their most recent results and for letting her include some of their still unpublished data in the present work.

REFERENCES

Allen, M.T. and Kochevar, I.E., 1989, Is rose bengal
    photosensitization of red blood cell acetylcholinesterase
    wavelength dependent?, Abstract WPM-E9, 17th Annual
    Meeting of the Am. Soc. for Photobiol. Boston, MA, 2-6
    July 1989, Photochem. Photobiol., 49:76S.

Andreoni, A., 1980, Lasers in microfluorometry and selective
    photobiology, in: "Lasers in Biology and Medicine", F.
    Hillenkamp, R. Pratesi and C.A. Sacchi, eds., Plenum
    Publishing Corp., New York.

Andreoni, A., Cubeddu, R., De Silvestri, S., and Svelto, O.,
    1980a, Laser selective photobiology: dye-biomolecule
    complexes, in: "Lasers in Photomedicine and
    Photobiology", R. Pratesi and C.A. Sacchi, eds.,
    Springer-Verlag, Berlin, Heidelberg, New York.

Andreoni, A., Cubeddu, R., De Silvestri, S. and Laporta, P.,
    1980b, Two-step laser-induced photodamage of quinacrine
    mustard, Chem. Phys. Letters, 72:448.

Andreoni, A., Cubeddu, R., De Silvestri, S., Laporta, P. and
    Svelto, O., 1980c, Time-delayed two-step selective
    photodamage of dye-biomolecule complexes, Phys. Rev.
    Letters, 45:431.

Andreoni, A., Cubeddu, R., De Silvestri, S., Laporta, P. and
    Svelto, O., 1982, Two-step laser activation of
    hematoporphyrin derivative, Chem. Phys. Letters, 88:37.

Andreoni, A., 1987, Two-step photoactivation of excimer-pumped
    dye-laser pulses, J. Photochem. Photobiol., B: Biol.,
    1:181.

Andreoni, A. and Ghiglione, A.R., 1987, Photoactivation of
    hematoporphyrin via non-linear excitation with ns tunable
    laser pulses, in: "1986 European Conference on Optics,
    Optical Systems and Applications," Vol. 701, S. Sottini
    and S. Trigari, eds., SPIE, Bellingham.

Andreoni, A., Cubeddu, R., Svelto, O., Canti, G.F. and Ricci,
    L., 1987, In vivo photochemotherapy with hematoporphyrin
    derivative via two-step exciation, in: "1986 European
    Conference on Optics, Optical Systems and Applications,"
    Vol. 701, S. Sottini and S. Trigari, eds., SPIE,
    Bellingham.

Bodaness, R.S. and King, D.S., 1985, The two-photon induced
    fluorescence of the tumor localizing photosensitizer
    hematoporphyrin derivative via 1064 nm photons from a 20
    ns Q-switched Nd:YAG laser, Biochem. Biophys Res.
    Commun., 126:346.

Bodaness, R.S., Heller, D.F., Krasinski, J. and King, D.S.,
    1986, The two-photon laser-induced fluorescence of the
    tumor localizing photosensitizer hematoporphyrin
    derivative. Resonance-enhanced 750 nm two-photon
    excitation into the near-UV Soret band, J. Biol. Chem.,
    261:12098.

Fluher, E.N., Hurley, J.K. and Kochevar, I.E., 1989, Laser
    intensity and wavelength dependence of RB photosensitized
    inhibition of red blood cell acetyicholinesterase,
    Biochim. Biophys. Acta, 990:269.

Foote, C., 1980, Mechanistic characterization of
    photosensitized reactions, in: "Photosensitization.
    Molecular, Cellular and Medical Aspects," NATO-ASI Series
    H: Cell Biology, Vol. 15, G. Moreno, R.H. Pottier, and
    T.G. Truscott, eds., Springer-Verlag, Berlin.

Hefetz, Y., Shea, C.R., Wimberly, J., Gillies, R. and Hasan, T., 1989, Intensity dependence of doxycycline phototoxicity to human bladder carcinoma cells in vitro, Abstract WPM-C1, 17th Annual Meeting of The Am. Soc. for Photobiol., Boston, MA, 2-6 July 1989, Photochem. Photobiol., 49:65S.

Karu, T.I., Dryukov, P.G., Letokhov, V.S., Matveetz, Yu.A. and Semchishen, V.A., 1981, High quantum yield of photochemical reactions of protoporphyrin IX induced by powerful ultrashort laser pulses, Appl. Phys., 24:245.

Kochevar, I.E., Buckley, L.A., Hefetz, Y., Hillenkamp, F. and Deutsch, T., 1989, Cyclobutylpyrimidine dimers and strand breaks formed in DNA using 532 picosecond laser radiation, Abstract MPM-C11, 17th Annual Meering of the Am. Soc. for Photobiol., Boston, MA, 2-6 July 1989, Photochem. Photobiol., 49:19S.

Kryukov, P.G., Letokhov, V.S., Nikogosyan, D.N., Borodavkin, A.V., Budowsky, E.J. and Simukova, N.A., 1979, Multiquantum photoreactions of nucleic acid components in aqueous solution by powerful ultraviolet picosecond radiation, Chem. Phys. Letters, 61:375.

Letokhov, V.S., 1976, Future applications of selective laser photophysics and photochemistry, in: "Tunable Lasers and Applications," A. Mooradian, T. Jaeger and P. Stokseth, eds., Springer-Verlag, Berlin, Heidelberg, New York.

Letokhov, V.S., 1978, Laser selective photophysics and photochemistry, in: "Progress in Optics," Vol. XVI, E. Wolf, ed., North-Holland, Amsterdam, New York, Oxford.

Li, A.S.W. and Chignell, C.F., 1987, Spectroscopic studies of cutaneous photosensitizing agents. X: A spin-trapping and direct electron spin resonance study of the photochemical pathways of daunomycin and adriamycin, Photochem. Photobiol., 45:565.

Motten, A.G., Buettner, G.R. and Chignall, C.F., 1985, Spectroscopic studies of cutaneous photosensitizing agents. VIII. A spin-trapping study of light induced free radicals from chlorpromazine and promazine, Photochem. Photobiol., 42:9.

Oraewsky, A.A. and Nikogosyan, D.N., 1985, Picosecond two-quantum UV photochemistry of thymine in aqueous solution, Chem. Phys., 100:429.

Profio, A.E., 1984, Laser excited fluorescence of hematoporphyrin derivative for diagnosis of cancer, IEEE J. Quantum Electron., QE-20:1502.

Rodgers, M.A.J., 1981, Picosecond fluorescence studies of rose bengal in aqueous micellar dispersions, Chem. Phys. Letters, 78:509.

Scaiano, J.C., Johnston, L.J., McGimpsey, W.G. and Wier, D., 1988, Photochemistry of organic reaction intermediates: novel reaction paths induced by two-photon laser excitation, Acc. Chem. Res., 21:22.

Shea, C.R., Wimberly, J. and Hasan, T., 1986, Mitochondrial phototoxicity sensitized by doxycycline in cultured human carcinoma cells, J. Invest. Dermatol., 87:338.

Shea, C.R., Long, F.H., Deutsch, T.F., Hillenkamp, F. and Hasan, T., 1988, Doxycycline-sensitized phototoxicity following excimer-laser irradiation: effects of irradiance, Lasers in the Life Sciences, 2:29.

Shea, C.R., Hefetz, Y., Gillies, R., Wimberley, J., Dalickas, G. and Hasan, T., 1989, manuscript in preparation.

CLED-PUMPED SOLID-STATE LASERS (MICROLASERS): POTENTIAL APPLICATIONS TO MEDICINE

R. Pratesi

Istituto di Elettronica Quantistica del CNR
Dipartimento di Fisica dell'Universita', Firenze
Italy

INTRODUCTION

The most interesting and important developments in the field of laser sources during the last few years are represented by the new generation of high power semiconductor diode lasers and by the new class of all-solid-state lasers, in which diode lasers have replaced the traditional flashlamp pumping. The principles of operation of semiconductor diode lasers (or coherent light emitting diodes, CLEDs) and their potential applications to medicine have already been reviewed (Pratesi, 1984 and 1986; Brancato and Pratesi, 1987), together with the initial developments of diode-laser-pumped solid-state lasers. In this paper an updated review of CLED-pumped solid-state lasers and of their use in medicine will be presented.

CLED-PUMPED SOLID-STATE LASERS (MICROLASERS)

CLED-pumped solid-state lasers (or microlasers after Amoco Laser Co.) (Chumbley, 1989) represent a new class of miniaturised, all-solid-state devices based on the emerging technology of diode-laser pumping of a solid-state crystal. This type of laser incorporates a high-power CLED in place of a flashlamp to excite the laser crystal.

The advantages of microlasers range from improved performance that has been unattainable with many earlier lasers, to greater ease of manufacturing the products. The all-solid-state structure of these lasers allows the improvement of the performances typical of solid-state lasers (SSLs), namely: excellent beam profile, high stability, single frequency operation, low noise.

Merits of CLED Pumping

The replacement of flashlamps with CLEDs permits

important improvements of SSL performances. First of all, the CLED represents the most efficient device to convert electrical power into optical power (efficiencies in excess of 60% have been reported) (Sakamoto, et al., 1989): therefore, the overall efficiency of conversion of the electrical power to the output power of the SSL can be greatly increased. Moreover, the emission spectrum of the CLED is very narrow compared to the spectra of incoherent optical sources, and CLED pumping can, then, be very efficient. In Fig. 1 the absorption spectrum of Nd:YAG crystal and the emission spectra of flashlamp and CLED are compared.

Since the CLED output wavelength can be finely tuned to the absorption peak of the laser medium to excite a pump level that lies only slightly above the upper laser level, the density of waste heat generated is several times lower than that produced by flashlamps. This drastically reduces the amount of external cooling required and permits the design of lasers with diffraction-limited output beams of extremely high stability.

Because of their directional emission, the efficiency of transport to, and depostion of CLED pump radiation in the solid-state gain element is several times higher than can be achieved with flashlamps that emit broadband radiation into 4π steradians.

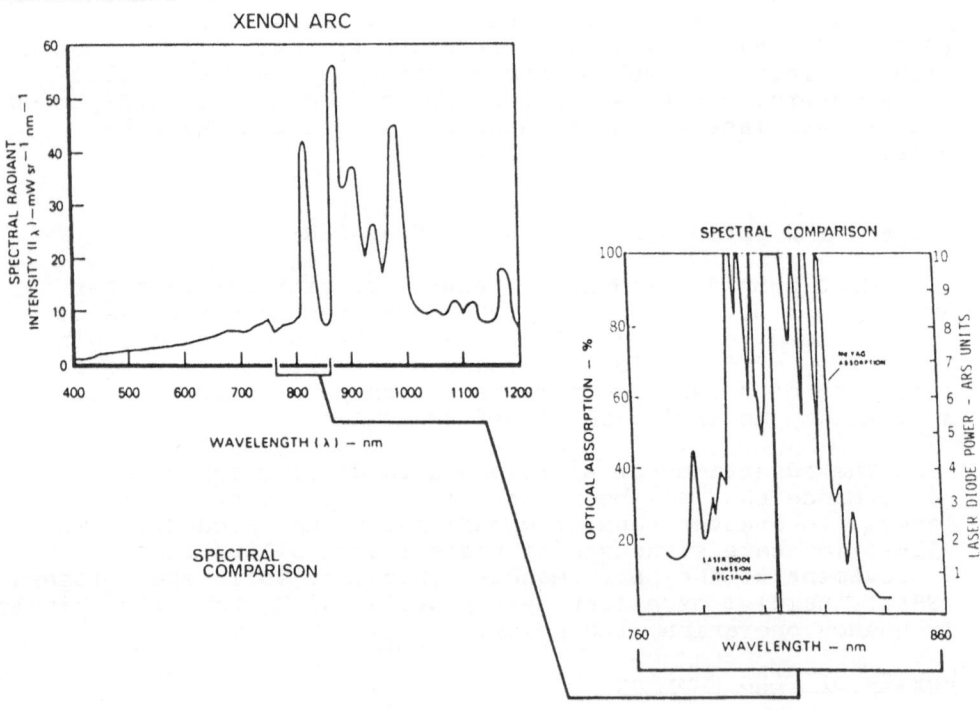

Fig. 1   Spectral comparison of Nd:YAG absorption
          flashlamp and CLED emission (adapted from
          Begley et al., 1986).

Finally, another important merit of CLED pumping is represented by the possibility of remote pumping. CLED pump sources, in fact, can also be efficiently coupled into a SSL gain element via multimode fibers, keeping the pump and power conditioning equipment remote from the laser head and work area. This permits improved laser stability and higher output power operation with still very compact laser heads.

CLED-Pumping Geometry

The key parameter for high "wall-plug" efficiency of microlasers is represented by the matching between the laser mode and the gain (i.e. pump) distribution. In general, CLED pumping takes two forms: end pumping and surface pumping, while the two most popular configurations of the solid-state medium are: cylindrical rod, and geometric slab.

End-Pumping of Rod Microlaser

End-pumping of laser rods has produced the highest efficiency from any configuration. The optical mode from the pump CLED can be matched to the lasing mode of the active solid-state material, thus maximising the coupling efficiency between pump and laser. The output power from this type of laser is, in general, limited by the power obtainable from the single CLED emitter. Although high power CLED devices are available, the output radiation pattern cannot be efficiently coupled to the active solid-state material. Figure 2 shows a typical schematic diagram of an end-pumped microlaser. The maximum overall conversion efficiency reported is 11% with output power of 0.43 W in the fundamental spatial mode (Berger et al., 1987). The highest (unpublished) wallplug efficiency is 15.7% at a power level of 0.5 W using neodymium-doped yttrium vanadate as the laser gain medium (Krupke, 1988); optical-to-optical efficiency as high as 50% was achieved, approaching the theoretical limit of 78% given by the pump/output photon energy defect.

The output power regime with end-pumping ranges from the milliwatt level to less than 10 W.

Side-Pumped Zig-Zag Slab Microlaser

Slab lasers are pumped at a geometrical surface as illustrated in Figure 3. The geometry of the slab results in the laser mode propogation through the material by total internal reflection. An advantage of this configuration is an increased effective length compared with the physical length of the active material and scalability to high output power. Wall-plug efficiencies are generally lower than those obtained with end-pumping, while higher average output power levels are achieved.

Surface pumping of slab lasers is associated with the use of two dimensional (2D) CLED arrays. Scaling to high power regimes is obtained by increasing slab sizes and pumping diode number. Present-day duty factors are limited by the rate at which waste heat is removed from the densely packed arrays. The development of low-cost fiber bundles will allow the delivery of higher power densities from less-densely packed diode arrays. Microchannel cooling structures

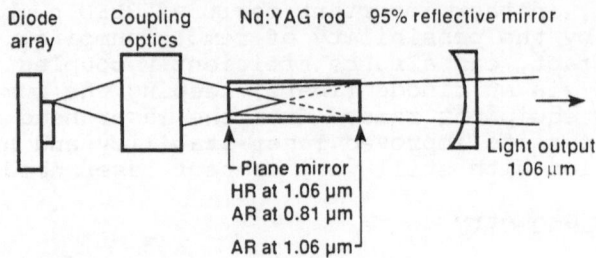

Fig. 2   Schematic diagram of end-pumped microlaser.
         (Berger et al., 1987).

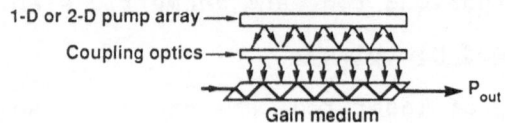

Fig. 3   Schematic drawing of transverse-pumped slab
         laser (by courtesy of W. F. Krupke).

integrated into densely packed diode arrays are also being
developed.

The power levels attainable with this pumping geometry
range from about 10 W to a few kilowatts

Multi-CLED Pumping

The high cost of assembling two-dimensional CLED arrays
and the commercial availability of 1-W cw CLED sources with
26% efficiency (such as the model SDL 2460-C) suggest the
construction of high power SSLs with a large number of 1-W
CLEDs coupled to optical fibers to route the power into a rod
or slab geometry SSL.   The CLEDs can be mounted remotely from
the SSL to facilitate electrical driver design and cooling.
Figures 4 and 5 show two recent schemes introduced for cw
single-spatial mode operation.   In Fig. 4, seven CLED arrays
coupled to a fiber bundle have been used to end-pump a Nd:YAG
rod, obtaining 0.66 W output with 35% and 4.4% optical-to-
optical and overall efficiencies, respectively (Berger et al.,
1988).   A further increase of the pumping power can be
achieved by coupling a multi-array CLED bar to a fiber bundle
with fibers linearly distributed at one end and circularly
distributed at the other end (Streifer et al., 1988).

Figure 5 illustrates a recent approach for transversely
pumping a slab microlaser via optical fibres (Watanabe et al.,
1989).

HIGH-PERFORMANCE MICROLASERS

Several promising, novel designs for microlasers with
both cw high-output power and diffraction limited beam quality
have recently been introduced by Lightwave Electronics,
Spectra Physics and NEC.

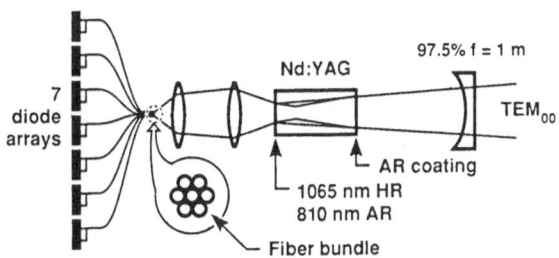

Fig. 4   Schematic representation of 7 CLED array
          pumped microlaser (Berger et al., 1988).

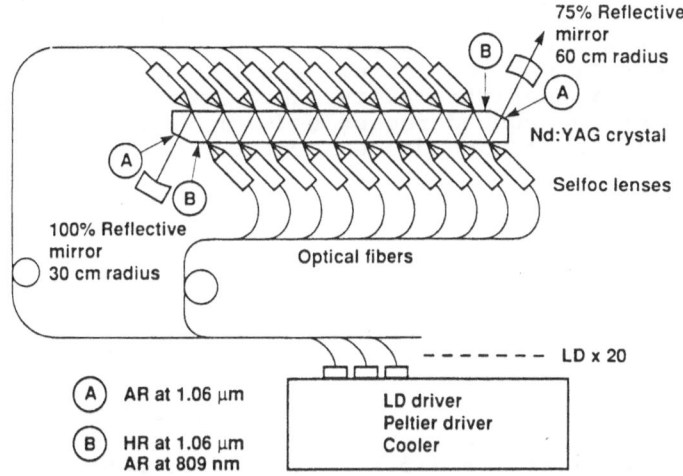

Fig. 5   Single-spatial-mode slab microlaser, side-
          pumped by fibre coupled CLEDs (Watanabe et
          al., 1989).

Monolithic, Non-Planar Ring Microlaser

A miniature (12 x 9 x 3mm) monolithic Nd:YAG ring
resonator with non-planar ring light path (Fig. 6) has been
successfully operated. This device combines the
characteristics of a CLED into a monolithic ring structure to
achieve single axial mode, unidirectional oscillation, with
extremely narrow bandwidth (Fan and Byer, 1988). In this
microlaser, selection of one of the two counter propagating
waves takes place at the curved, partially transmitting face,
and the need for discrete intracavity elements, necessary in
conventional linear or standing wave lasers, is avoided.

Recent progress has led to output powers up to 53 mW cw
in both fundamental spatial and axial modes, with a
measurement limited bandwidth of less than 3 kHz over 100 ms
measurement time (Byer, 1989).

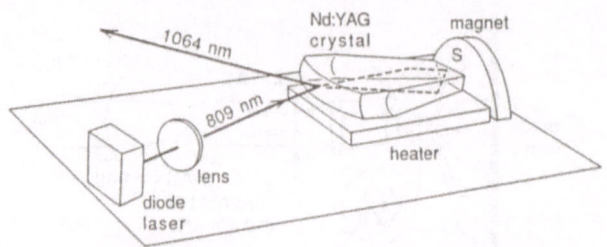

Fig. 6   Schematic of solid-state, unidirectional,
         nonplanar ring microlaser (Fan and Byer,
         1988).

Side-Pumped, Tightly Folded Resonator Microlaser

A 10 W CLED-bar has been used to side-pump a small (5 x 5
x 20 mm) specially coated Nd:YAG block (Fig. 7).   The pumping
light from the diode bar is coupled into the Nd:YAG block via
a small piece of optical fibre.   The two cavity mirrors are
aligned so that the intracavity laser beam bounces back and
forth through the pumped region between opposing sides with
vertices aligned at each diode location.

The output power is 3.8W when pumped with 10.9 W from the
diode bar.   The output beam is a diffraction limited Gaussian
single-spatial mode.   The conversion efficiency is 35%, a high
value for a side-pumped geometry, comparable to that typically
achieved with end pumping (Baer and Head, 1989).

A conceptually similar side-pumped design is illustrated
in Fig. 5.   In this laser the output from twenty 1-W CLEDs is
used to pump a 59 x 8 x 3 mm 0.8% Nd:YAG crystal.   With an
input power to the crystal of 11.3 W, the single spatial mode
ouput power achieved is 3.5 W, with 32% of optical-to-optical
efficiency (Baer and Head, 1989).

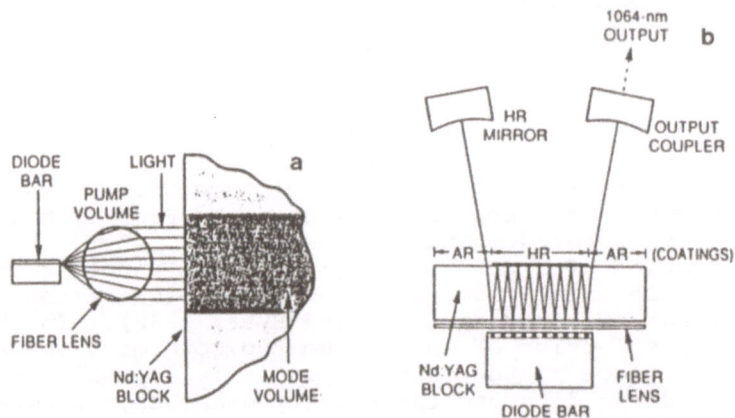

Fig. 7   Side-pumped, tightly-folded resonator
         microlaser: a) fibre lens micro-collimator;
         b) schematic of the laser cavity (Baer and
         Head, 1989).

SHORT PULSE OPERATION AND NONLINEAR FREQUENCY CONVERSION

As is known, SSLs can be operated to generate i) high-peak power, short duration pulses with the technique of Q-switching; ii) ultrashort pulses by modelocking a multimode cavity; iii) new frequencies by nonlinear harmonic generation and frequency mixing.

The miniaturization process involved in the construction of microlasers has led to corresponding research to obtain small size, possibly monolithic components for actively pulsing and frequncy shifting microlasers.

Q-Switched Microlasers

For giant pulse operation a shutter must be inserted into the laser cavity, but, even so, the overall length of the microlaser can be maintained below 50 mm. The shutter can be either an acousto-optical switch or an electro-optical switch. The electro-optical switch is particularly appealing due to its very small size that permits a very short (25mm) cavity. In particular, Spectra Physics has developed a special Q-switched microlaser for a laser processing system devoted to the fast growing market of memory chip production lines. Another promising device has been recently developed by Lightwave Electronics: the microlaser is very compact (40mm cavity length) and is able to generate very intense pulses (0.25 MW) with a high energy content (>1 mJ). As will be discussed later, a further increase in the output energy of this microlaser will make it suitable for replacing standard Q-switched Nd:YAG lasers in eye micro-surgery.

Second Harmonic Generation

In early experiments, second harmonic generation (SHG) of microlaser output was achieved by placing the nonlinear crystal within the laser resonator to take advantage of the high circulating power in the laser. SHG that produced milliwatts of green output using either KTP or LiNbO as the nonlinear crystal was demonstrated. More recently, 180 mW power output at 532 nm in a nearly diffraction limited beam has been reported by intracavity SHG in a microlaser with the pumping geometry shown in Fig. 4 (Berger et al., 1988).

Recent advances in single-frequency laser operation and in nonlinear materials have also led to greatly improved conversion efficiencies using external resonant SHG. In a recent experiment the combined advantages of a single-frequency nonplanar microlaser with the stability of a monolithic resonator fabricated onto a lithium niobate nonlinear crystal have led to efficient external resonant harmonic generation (Byer 1989, Kozlovsky et al., 1988). Single frequency, narrow bandwidth green power as high as 30 mW cw has been generated, with harmonic conversion efficiency of 56%, the highest value for cw operation reported so far. This result opens the possibility of stabilising the optical frequency of the nonplanar microlaser by locking onto a hyperfine component of the iodine absorption spectrum by menas of sub-Doppler saturation spectroscopy (Byer, 1988).

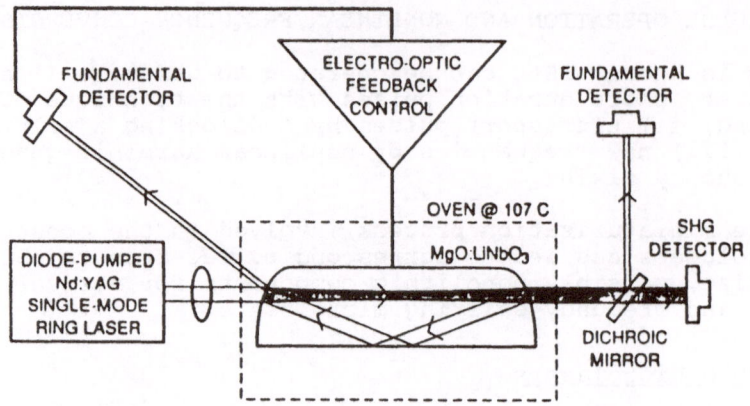

Fig. 8   Schematic of external resonant second
         harmonic generation (Byer 1988).

RARE EARTH ION MICROLASERS

The number of possible laser transitions under CLED
pumping is limited by the number of rare earth (+3) ions with
absorption at CLED output wavelengths. With present high
power CLEDs emitting in the 750-860 nm spectral region the
rare earth ions are limited to Nd, Pm, Dy, Er, and Tm (Fan and
Byer, 1988). The Nd ion is an excellent dopant for
microlasers: it exhibits strong absorption in the emission
bands of GaAs, GaAlAs, GaAsP LEDs and CLEDs. Nd:YAG
microlasers have been operated at room temperature at 1.06,
1.32 and 0.95 μm.

Room temperature laser action at 2.1 and 2.3 μm in CLED-
pumped Tm,Ho:YAG and Tm:YAG, respectively, has also been
reported. Still longer laser wavelength, 2.8 μm, has been
generated by pumping an Er:YAG crystal with 797 nm CLEDs.
These microlasers are of interest in medicine due to the
differences in absorption coefficients of water at these
wavelengths with respect to Nd:YAG and CO_2 lasers. New
transitions in rare earth ion doped crystals are also
currently being investigated (Fan and Byer, 1988).

MEDICAL APPLICATIONS

Following the initial prediction (Pratesi, 1984), the
application of CLEDs to medicine is expanding rapidly. The
introduction of CLED sources in ophthalmology has been
welcomed as the beginning of a new era for retinal
photocoagulation (Raven et al., 1989). With the already
existing power levels, endoscopic photocoagulative techniques
can be tried, as well as CLED photochemotherapy with long
wavelength photosensitizers. CLED-thermia and tissue welding
may also be investigated.

Microlasers are also expected to be of considerable
interest in medicine due to their improved performances with
respect to standard flashlamp lasers. Nd:YAG microlasers with
cw multiwatt output powers are becoming commercially

available: eye microsurgery will offer, once again, the most promising applications. A further increase in output power will permit the replacement of traditional endoscopic laser photocoagulators in gastroenterology and bronchology with more compact and reliable microlaser photocoagulators. Unlike CLEDs, microlasers can generate high peak power and high energy pulses: commercial Q-switched Nd:YAG microlasers have already the right numbers to test their feasibility as new sources for "cold" laser eye microsurgery.

Future developments of rare earth ion microlasers emitting in the 1-3 μm spectral range will permit new photosurgical and photoablative techniques (Pratesi, in preparation).

ACKNOWLEDGEMENTS

The author is extremely indebted to Drs. W. F. Krupke (Lawrence Livermore Laboratories), D. R. Scifres (Spectra Diode, Inc), A. Mooradian (Lincoln Laboratory), and to Amoco Laser and Lightwave Electronics for the supply of updated technical material.

Work supported by the Special CNR-Project "Tecnologie Elettro-ottiche".

REFERENCES

Berger, J., Welch, D.F., Scifres, D.R., Streifer, W. and Cross, P.S., 1987, High power, high efficient Nd:YAG laser end pumped by a laser diode array, Appl. Phys. Lett., 51:1212.

Berger, J., Welch, D.F., Streifer, W., Scifres, D.R., Smith, J.J., Hoffman, H.J., Preisley, D. and Radecki, D., 1988, Seven laser diode end-pumped Nd:YAG laser via a fibre bundle, Opt. Lett., 13:306.

Begley, D.L., Krebs, D.J. and Ross, M., 1986, Diode pumped Nd:YAG lasers and their unique features, in: "Lasers in Medicine", Proc. of the SPIE Vol. 712:42.

Brancato, R. and Pratesi, R., 1987, Applications of diode lasers in ophthalmology, Lasers Ophthalmol., 1:119.

Baer, T.M. and head, D.F., 1989, High efficiency diode bar pumped solid state laser using a tightly folded resonator, Digest of Technical Papers, CLEO'89, paper FJ5.

Byer, R.L., 1988, Diode laser-pumped solid-state lasers, Science 239:742.

Byer, R.L., 1989, Nonlinear frequency conversion enhances diode-pumped lasers, Laser Focus World p. 77.

Chumbley, P.E., 1989, Microlasers offer new reliability for R & D, Res. Dev., p.72.

Fan, T.S. and Byer, R.L., 1988, Diode laser-pumped solid-state lasers, IEEE J. Quant. Electr., 24:895.

Kozlovsky, W.J., Nabors, C.D. and Byer, R.L., 1988, Efficient second harmonic generation of diode-laser-pumped cw Nd:YAG laser using monolithic MgO:LiNbO$_3$ external resonant cavities, IEEE J. Quant. Electr., 24:913.

Krupke, W.F., 1988, prospects for diode-laser-pumped solid-state lasers, Las. & Optron., p. 79.

Pratesi, R., 1984, Diode lasers in photomedicine, <u>IEEE J. Quant. Electr.</u>, 20:1433.

Pratesi, R., 1987, Semiconductor (diode) lasers: basic principls and potential applications in the biomedical field, <u>in</u>: "Proceedings of the 1st Congress of the European Society for Photobiology", Grenoble Sept. 7-12, <u>Photobiochem. Photobiophys.</u> Suppl.57.

Pratesi, R., Advances in miniature lasers. Initial and potential applications to Medicine and Biology, in preparation.

Raven, A., Lee, R.M. and Keeler, C.R., 1989, Diode lasers: a new era in retinal photocoagulation. I. Basic principles and instrumentation, Second International Congress on Laser Technology in Ophthalmology, Lugano, May, 24-27.

Sakamoto, M., Welch, D.F., Endriz, J.G., Scifres, D.R. and Streifer, W., 1989, 76 W cw monolithic laser diode arrays, <u>Appl. Phys. Lett.</u>, 54:2299.

Streifer, W., Scifres, D.R., Harnagel, G.L., Welch, D.F., Berger, J. and Sakamoto, M., 1988, Advances in diode laser pumps, <u>IEEE J. Quant. Electr.</u>, 24:883.

Watanabe, S., Kudo, S., Yamane, T. and Washio, K., 1989, Efficient and high power Nd:YAG laser multiple-facet end pumped by laser diodes, <u>Digest of Technical Papers</u>, CLEO'89, paper PD8-1.

PHOTOSENSITIZED LUMINESCENCE OF SINGLET MOLECULAR OXYGEN: MECHANISMS AND APPLICATION FOR PHOTOBIOLOGY AND PHOTOMEDICINE

Alexander A. Krasnovsky Jr.

Department of Biology, Moscow State University
Moscow, 119899, USSR

INTRODUCTION

Excited oxygen molecules in the singlet $^1\Delta_g$ state ($^1O_2$) are known to determine photosensitized oxygenation of various biologically important compounds. Therefore photosensitized $^1O_2$ formation may lead to photodynamic killing of normal and tumor cells and to a wide variety of other photobiological effects (Foote, 1976; Krinsky, 1984; Straight and Spikes, 1985). To understand the real role of singlet oxygen in photobiological processes one needs to develop reliable methods of $^1O_2$ detection in model systems and directly in living cells and tissues. At present, two methods are the most widely used: the method of chemical traps (Foote and Wexler, 1964; Foote, 1976) and direct measurement of the infrared $^1O_2$ luminescence accompanying $^1\Delta_g - ^3\Sigma_g^-$ transition (Krasnovsky, 1976, 1979).

It is noteworthy that the first measurements of this luminescence were performed in the gas phase: in twilight airglow (Jones and Harrison, 1958) and under discharge in helium-oxygen mixture (Noxon, 1961). Snelling (1968) observed this emission under photosensitized $^1O_2$ generation in the gaseous mixture of oxygen and benzene vapors. In 1976 this author managed to detect this luminescence under photosensitized $^1O_2$ production in air-saturated solutions of photosensitizers (Krasnovsky, 1976, 1979a). It made possible the application of $^1O_2$ luminescence measurements for the analysis of photosensitized processes in solutions and biological systems. The present paper is a short review of the contemporary studies in this area.

TECHNIQUE FOR LUMINESCENCE MEASUREMENTS

Initial measurements of $^1O_2$ luminescence in solutions were performed using home-made set-ups with mechanical phosphoroscopes, cooled S-1 photmultipliers and electronics for registering the photocurrent intensity (Krasnovsky 1976, 1977 1979a). These set-ups allowed us to measure excitation and emission spectra and lifetimes of the $^1O_2$ luminescence. However, the relatively low time resolution of phosphoroscopes limited their use to a small number of solvents where the $^1O_2$ lifetimes (τ_Δ) exceeded 0.5 ms. Using this equipment the $^1O_2$ luminescence was observed in solutions of photosensitizers in CCl_4, CS_2 and freons (Krasnovsky, 1979a).

In the equipment of the next generation the phosphoroscopes were eliminated and measurements were carried out in the steady-state mode similar to a fluorimeter. S-1 photomultipliers or semiconductor photodetectors were employed for the luminescence registration. Using this equipment $^1O_2$ luminescence was detected in aqueous solutions of photosensitizers (Krasnovsky, 1979b; Khan and Kasha, 1979). Further development of the steady-state technique with the use of modulated laser beams and lock-in amplifiers made it possible to observe $^1O_2$ luminescence directly in porphyrin loaded mouse tumors (Parker, 1987).

The first time-resolved measurements under pulsed laser excitation were reported by Salokhiddinov et al. (1979). They detected luminescence by S-1 photomultipliers through an interference light filter. Kinetic curves were recorded after a single exciting flash using storage oscilloscopes. Then, similar time-resolved set-ups with photodiode detectors were reported (Salokhiddinov et al. 1980; Ogilby and Foote 1982; Hurst et al., 1982; Parker and Stanborn, 1982; Rodgers and Snowden, 1982). We built a laser set-up of different type (Egorov and Krasnovsky, 1983). It comprised nitrogen lasers, S-1 photomultipliers, a photon-counting electronic system and a multichannel analyzer for averaging signals over many laser flashes. The sensitivity of this set-up was so high that we could place a monochromator between the photomultiplier and the sample and use excitation laser flashes of very low intensity (about 0.02-0.04 mJ). The detailed description of this set-up was published recently by Krasnovsky et al. (1988).

Time resolution of all the above instruments was about 0.5-2 μs. Recently new equipment with improved time-resolution was constructed (Kai-Kong and Ogilby, 1987; Egorov et al., 1988, 1989). In our spectrometer the principle of time correlated photon counting was employed, and 10 ns resolution was achieved. In addition to the above microsecond set-up, the new spectrometer comprises a time-to-amplitude converter and a more powerful copper-vapor laser with a high frequency of flash repetition (Egorov et al., 1988, 1989). With this set-up we have measured the rise and decay kinetics of the $^1O_2$ luminescence in H_2O at different oxygen pressures and in aqueous suspensions of living cells loaded with porphyrin.

SPECTRAL PARAMETERS OF THE $^1O_2$ LUMINESCENCE

Measurements have shown that in CCl_4 the main maximum of the photosensitized $^1O_2$ luminescence lies at 1272±2 nm. The half bandwidth is about 17 nm (Krasnovsky, 1981). Obviously, this maximum belongs to the $^1\Delta_g(0)-^3\Sigma_g^-(0)$ transition. The maximum of the 0-1 transition was detected at 1590 nm. Its intensity was lower by a factor of 50 (Salokhiddinov et al., 1980; Khan and Kasha, 1980). With the use of a phosphoroscope, additional maxima at the shorter wavelengths were observed (Krasnovsky, 1979a). Their intensity was proportional to the square of the intensity of excitation. Previously we proposed that this emission was photochemiluminescence accompanying photodestruction of photosensitizers (Krasnovsky, 1979a). Recently we came to the conclusion that it is caused by dimols $(^1O_2)_2$ (Krasnovsky and Neverov, 1988). Excitation spectra of both long and short wavelength emissions correspond to absorption spectra of sensitizers.

LIFETIME AND QUANTUM YIELD OF $^1O_2$ LUMINESCENCE IN SOLUTIONS

The first measurements of the lifetime of $^1O_2$ luminescence were performed using set-ups with mechanical phosporoscopes. It was found that τ_Δ was about 30 ms in CCl_4 (Krasnovsky 1977, 1979a). This value was higher by the factor of 40 than that obtained earlier with the use of $^1O_2$ chemical traps (Merkel and Kearns, 1972). Recently Shmidt et al. (1989) showed that τ_Δ reaches 87 ms after careful purification of this solvent.

The reported quantum yields (Φ_r) of the $^1O_2$ luminescence in CCl_4 are summarized in Table 1. The values of the $^1O_2$ radiative lifetime (τ_r) presented in this table were calculated by the formula:

$$\tau_r = \tau_\Delta \Phi_\Delta / \Phi_r \tag{1}$$

where Φ_Δ is the quantum yield of $^1O_2$ generation by photo-sensitizers. It is seen from the Table that τ_r is much higher than τ_Δ. However, τ_r is close to that of $^1O_2$ in collision complexes with solvent molecules ($\tau_r \approx 5$ s) (Long and Kearns, 1973) and differs by three orders of magnitude from τ_r of free oxygen molecules in the gas phase at the low pressures ($\tau_r = 3850$ s) (Badger et al., 1965). Analysis of these data leads to the assumption that in solution all $^1O_2$ molecules are in collision complexes with solvent molecules and every collision strongly increases $^1O_2$ radiative deactivation. Some of collisions cause nonradiative $^1O_2$ quenching, and the efficiency of this quenching determines the τ_Δ value. Thus, solvents play the role of $^1O_2$ quenchers and amplifiers of $^1O_2$ light emission.

Table 1. Values of Φ_r, τ_r and τ_Δ in CCl_4.

	Krasnovsky, 1981	Losev et al., 1988	Shmidt et al., 1989
Φ_r	$(6.3^{\pm}2.7) \times 10^{-3}$	$(22^{\pm}5.5) \times 10^{-3}$	$(9.0^{\pm}3.6) \times 10^{-3}$
τ_r, s	$4^{\pm}2$	$1.25^{\pm}0.3$	$1.0^{\pm}0.4$
τ_Δ, ms	$26^{\pm}3$	28	$87^{\pm}3$

It was of interest to study the relation between τ_Δ and τ_r in different solvents. First estimation did not reveal reliable differences between τ_r values in the solvents in which τ_Δ values varied by 4 orders of magnitude (Krasnovsky, 1981). Recently several groups reported that τ_r is solvent dependent (Krasnovsky et al., 1987, 1988; Darmanyan, 1987; Gorman et al., 1987; Scurlock and Ogilby, 1987; Losev et al., 1988). Our data are summarised in Table 2. They were obtained in solutions of tetraphenylporphin on the basis of the assumption that the Φ_Δ value was constant in all solvents investigated. τ_Δ values were obtained using the set-up with pulsed laser excitation.

Table 2 shows that τ_r is weakly dependent on solvents as compared with τ_Δ. This supports the conclusion that τ_Δ values are caused by different efficiencies of $^1O_2$ quenching by solvents. To test this conclusion, we carried out a systematic study of quenching the $^1O_2$ luminescence by addition of different solvents to photosensitizer solutions in CCl_4 (Krasnovsky, 1980, 1981, 1982). It was found that quenching obeys the Stern-Volmer equation, and τ values in neat solvents are well approximated by the formula:

$$\tau_\Delta = 1/(k_q \times C_{max}) \tag{2}$$

where k_q is the rate constant of bimolecular quenching of singlet oxygen by solvent molecules and C_{max} is the molar concentration of neat solvents. Experimental data obtained are illustrated by Table 3. One can conclude that formula (2) yields satisfactory results for organic solvents. Later this conclusion was confirmed by Hurst and Schuster (1983).

When solutions contain additional $^1O_2$ quenchers this equation can be transformed into:

$$\tau_\Delta = 1/\Sigma(k_q \times C_q) \tag{3}$$

where k_q and C_q are rate constants and concentrations of quenchers. This equation makes it possible to estimate τ_Δ values for multicomponent solutions without recourse to measurement. If one assumes that the organelles of living cells are multicomponent solutions one can estimate the lifetime of singlet oxygen in living structures (Krasnovsky, 1986, 1988).

Table 2. Values of τ_r and τ_Δ in different solvents (Krasnovsky, 1981; Krasnovsky et al, 1987, 1988).

Solvents	τ_Δ, μs	τ_r, arb. units
CCl_4	26000	1
$CHCl_3$	250	1
Acetone	51	1.2
Ethanol	15	1.4
Benzene	31	0.5
D_2O	67	2.9

Table 3 Comparison of τ_Δ values with those calculated by the formula (2). (k_q values were measured in CCl_4, Krasnovsky, 1980, 1982).

Solvents	C_{max}, M	k_q, $M^{-1}s^{-1}$	$1/(k_q \times C_{max})$, μs	τ_Δ, μs
Chloroform	12.4	380	210	250
Acetone	13.7	1500	49	51
Methanol	24.7	4300	9.5	10.5
H_2O	55.4	2600	7.0	3.09

PHOTOSENSITIZED $^1O_2$ LUMINESCENCE IN WATER

The first measurements of the photosensitized $^1O_2$ luminescence in aqueous solutions were made under stationary excitation (Krasnovsky, 1979b; Khan and Kasha, 1979). Then, time-resolved measurements followed which were performed using a microsecond detection technique (Ogilby and Foote, 1982; Hurst et al., 1982; Rodgers and Snowden, 1982; Egorov and Krasnovsky, 1983; Salokhiddinov et al., 1983; Parker and Stanboro, 1984). This technique provided a reliable detection of the $^1O_2$ decay kinetics in D_2O. Measurements in H_2O were strongly restricted by the insufficient time resolution and sensitivity of this equipment. Recently, using the set-up with nanosecond time resolution, we managed to measure with high precision the rise and decay kinetics of $^1O_2$ luminescence in H_2O at different oxygen pressures (Egorov et al., 1988, 1989). These measurements allowed us to determine an accurate value of τ_Δ equal to $3.09 \pm 0.06 \mu s$.

One can see from Table 3 that this value does not coincide with that obtained from the formula (2). Experiments have shown, however, that it is a consequence of the change of k_q for H_2O in different environments. When we measured k_q by addition of H_2O to D_2O the k_q value of 4300 $m^{-1}s^{-1}$ was obtained. This leads to $\tau_\Delta \approx 4$ μs which is closer to reality (Krasnovsky, 1982, 1986, 1988).

It is noteworthy that τ_r in water is higher and, hence luminescence intensity is lower than in organic solvents (Table 2).

PHOTOSENSITIZED $^1O_2$ LUMINESCENCE IN LIVING SYSTEMS

Parker (1987) reported the first experimental observation of photosensitized $^1O_2$ luminescence in living systems. A very weak signal was obtained under continuous illumination of a porphyrin loaded tumor directly on the skin of a mouse. The luminescence lifetime was estimated to be few milliseconds. The time resolved measurement with flash laser excitation was reported in our paper (Egorov et al., 1988, 1989). The signal was detected in aqueous suspensions of yeast cells containing tetraphenylporphin. The luminescence lifetime of $3\mu s$ coinciding with the $^1O_2$ lifetime in water was obtained. The data suggest that this luminescence belongs to $^1O_2$ molecules located outside of cells. A similar result was obtained with

a model system: suspensions of erythrocyte ghosts in heavy
water (Firley and Rodgers, 1988). The luminescence lifetime
in this system was close to that of $^1O_2$ in heavy water.
Analysis of these data show that the $^1O_2$ luminescence
intensity in living membranes is much lower than in solutions.

SOME APPLICATIONS FOR PHOTOBIOLOGY AND PHOTOMEDICINE

Studies of $^1O_2$ photogeneration

     Measurements of $^1O_2$ luminescence made it possible to
observe photogeneration of singlet oxygen by many
photosensitizers important for photobiology and photomedicine.
Strong $^1O_2$ generators were shown to be monomeric molecules of
chlorophyll and pheophytin (Krasnovsky 1977, 1979a; Djagarov
et al., 1987), bacteriochlorophylls and bacteriopheophytins
(Krasnovsky, 1977, 1979a; Krasnovsky et al., 1985; Borland et
al. 1987), protochlorophyll and protochlorophyllide
(Krasnovsky, 1979a; Egorov et al., 1988), pheophorbide a
(Krasnovsky et al., 1988), retinals (Krasnovsky and Kagan,
1979), flavins (Krasnovsky, 1979b, 1982, 1988; Chacon, 1988),
metal-free porphyrins and chlorins and their complexes with
magnesium, tin, aluminum, zinc and palladium (Krasnovsky,
1979a; Venediktov and Krasnovsky, 1982; Lambert et al., 1986;
Keene et al., 1986; Keir et al., 1987; Krasnovsky et al.,
1987, 1988; Djagarov et al., 1987; Egorov et al., 1988; Brault
et al., 1988; Kochubeev et al., 1989), porphycene (Redmont et
al., 1989), phthalocyanines (Keir et al., 1987; Valgula et
al., 1988; Lagorio et al., 1989), naphthalocyanines (Firley et
al., 1988), cercosporin and 4-thiouridine (Foote and
Dobrowolsky, 1984). Moderate $^1O_2$ generators were shown to be
furocoumarins (Krasnovsky et al., 1983, 1986; Beaumont et al.,
1985; Knox et al., 1986, 1988), β-caroboline and kinurenine
derivatives (Egorov et al., 1987) and tetracycline and some
other antibiotics (Hasan and Khan, 1986; Li et al., 1987;
Egorov et al., 1987). Photsensitizing activities of all these
substances strongly depended on their aggregation state.
Aggregated molecules were usually less active than monomeric.

     The most efficient mechanism of $^1O_2$ generation was shown
to be energy transfer to oxygen from triplet molecules of the
photosensitizers. Quantum yield of this process in air
saturated solutions is usually close to the quantum yield of
formation of the photosensitizer triplet state.

Studies of $^1O_2$ quenching

     Measurement of $^1O_2$ luminescence is a convenient tool for
studying rate constants (k_q) for quenching singlet oxygen by
biomolecules. Application of this tool made it possible to
increase the precision of k_q determination and to measure k_q
for many compounds which had not been studied previously using
chemical traps (Foote, 1976, 1979). It was observed that
strong physical $^1O_2$ quenchers are chlorophylls and pheophytins
(Krasnovsky, 1977, 1979a; Krasnovsky et. al., 1982; Tanelian
an Wolf, 1988), bacteriochlorophylls and bacteriopheophytins
(Krasnovsky, 1977, 1979a; Krasnovsky et al., 1982, 1985;
Borland, 1987) and other natural and synthetic porphyrins
(Krasnovsky, 1979a; Venediktov and Krasnovsky, 1979;
Krasnovsky et al., 1982). This means that these pigments

combine high photosensitizing activity with the ability to protect membranes against damage by singlet oxygen. K_q values were obtained for saturated and unsaturated fatty acids, lipids and cholesterol (Krasnovsky and Kagan, 1979; Krasnovsky et al., 1983; Chacon et al., 1988, Vever-Bizet et al., 1989), aminoacids, thiols, proteins and nucleotides (Reddi et al., 1985; Rougee and Bensasson, 1986; Egorov and Krasnovsky, 1986; Krasnovsky, 1986; Rougee et al., 1988), vitamin C (Byteva et al., 1979; Chou and Khan, 1983; Rougee and Bensasson, 1986; Egorov and Krasnovsky, 1986), carotenoids, retinals and vitamin A (Krasnovsky and Kagan, 1979; Krasnovsky and Paramonova, 1983; Borland, 1989), saccharides, melanines and ommochromes, NAD and NADH (Egorov and Krasnovsky, 1986; Egorov et al., 1987). Tables 4 and 5 represent some of the k_q values obtained in our laboratory.

Table 4   Rate constants of general (k_q) and chemical (k_{ox}) quenching of singlet oxygen by lipophylic substances in CCl_4.

Quenchers	k_q,$M^{-1}s^{-1}$	k_{ox}/k_q	Quenchers	k_q	k_{ox}/k_q
β-carotene	7×10^9	5×10^{-4}	Bacterio-chlorophyll A	10^9	3×10^{-2}
Zeaxantin	7×10^9	4×10^{-4}	Bacterio-pheophytin A	1.2×10^7	2×10^{-3}
Chlorophyll A	7×10^8	3×10^{-3}	Bacterio-chlorphyll B	1.5×10^9	10^{-1}
Chlorophyll B	3×10^8	2×10^{-3}	Bacterio-pheophytin A	2×10^8	1.5×10^{-2}
Pheophytin A	2×10^7	1.4×10^{-3}	Phosphatidyl-choline	10^5	0.5

(Krasnovsky and Paramonova, 1983; Krasnovsky, 1979a; Krasnovsky et al., 1982, 1983, 1985).

Table 5   Rate constants of $^1O_2$ quenching by water-soluble substances in D_2O (Krasnovsky, 1986; Egorov et al., 1986, 1987).

Quenchers	k_q,$M^{-1}s^{-1}$	Quenchers	k_q,$M^{-1}s^{-1}$
Tryptophan	5×10^7	Saccharose	2.5×10^4
Human serum albumin	1.7×10^8	CTP	$<2\times10^4$
Sodium ascorbate	4×10^7	ATP	4×10^4
DOPA-melanine	1.2×10^8	NAD	3×10^4
Ommochrome	3×10^6	NADH	2×10^7

Estimation of τ_Δ in living structures

Introduction of the obtained k_q values into formula (3) allowed us to estimate the τ_Δ values in different structures of living cells. As a result of these calculations we obtained values of τ_Δ = 300-500 ns in nonphotosynthetic structures and τ_Δ = 5 ns in photosynthetic apparatus (Krasnovsky, 1986, 1988). The heterogeneity of living cells makes such calculations only approximate. At present, however, there are no other methods for τ_Δ estimation in living cells. It is not clear if time-resolved luminescence measurements will answer this question. The lifetime of the triplet state of photosensitizers is known to vary from $6\mu s$ up to 20 ms in aerated cells loaded with phorphyrins or phthalocyanines (Firely and Rodgers, 1988; Firely et al., 1988; Takemura et al., 1989). According to our estimation, τ_Δ is much lower, hence, decay kinetics of the $^1O_2$ luminescence in vivo should be determined by the decay of the photosensitizer triplet state rather than by $^1O_2$ deactivation. The $^1O_2$ luminescence lifetime obtained in a mouse tumor is consistent with this proposal (Parker, 1987).

$^1O_2$ in the photosynthetic apparatus

Studies of $^1O_2$ in chloroplasts were presented by Foote (1976) and subsequent in papers by other authors (Knox and Dodge, 1985). Recently this author reconsidered this problem using the results of $^1O_2$ luminescence measurements (Krasnovsky, 1986). Quantum yield of $^1O_2$ formation was estimated to be 0.1-0.03% in normal photosynthetic apparatus. The main $^1O_2$ generators are likely triplet molecules of "young" chlorophyll spatially separated from carotenoids. The main $^1O_2$ quenchers are carotenoids and chlorophylls which protect lipids and proteins against photodamage and decrease τ_Δ down to 5 ns. It is highly probable that photosensitized $^1O_2$ formation underlies photodestruction of the photosynthetic apparatus under intensive sun light. In agreement with this, we observed that there is a quantitative correlation between the efficiency of chloroplast photodestruction and the stationary concentration of the triplet chlorophyll in the photosynthetic apparatus of normal and mutant plants (Egorov et al., 1985). Similar effects are likely to be valid for photosynthetic bacteria.

$^1O_2$ and vision

$^1O_2$ is assumed to destroy retinal cells under strong illumination and may participate in light-induced cataractogenesis (Delmelle, 1979; Emanuel and Ostrovsky, 1981; Andley, 1987; Kagan 1988). These assumptions are supported by the efficient photosensitization of the $^1O_2$ luminescence by retinals and kinurenine derivatives in solutions (Krasnovsky and Kagan, 1979; Egorov et al., 1987). Application of formula (3) made possible the estimation that τ_Δ = 300-500 ns in outer segments of retinal rods, the main quenchers being amino acids of rodopsin. According to this calculation, tocopherol protects lipids and does not protect proteins of rod membranes (Krasnovsky, 1988). An ability of melanine to quench $^1O_2$ might be important for protection of eyes against photodamage (Table 5).

444

$^1O_2$ and photomedicine

Photosensitized $^1O_2$ formation is supposed to be the main initial stage of photodynamic tumor therapy and some photodynamic diseases (Foote, 1976; Straight and Spikes, 1985; Dougherty, 1987). It might also be important in the side phototoxicity of some drugs (Li et al., 1987). Therefore, measurements of $^1O_2$ luminescence is now widely used as a method to study photosensitizing activities of medically important compounds and mechanisms of their action. These studies are presented in the section dealing with $^1O_2$ generation. Obviously, measurements of luminescence quenching might be a promising tool for revealing biologically important protectors against damage by singlet oxygen.

Other applications

In addition, one can apply oxygen luminescence to a wide variety of photophysiological problems: action of fungi toxins, plant phytoallexins and photodynamic herbicides, blue light reception, photoaxis and phototropism, and for many other photoprocesses which take place under the combined action of light and oxygen (Foote, 1976, Knox and Dodge, 1985; Senger, 1982; Schuchard and Nultsch, 1984).

REFERENCES

Andley, U.P., 1987, Photodamage of eye, Photochem. Photobiol., 46:1057.
Badger, R.M., Wright, A.C. and Whitlock, R.F., 1965, Absolute Intensities of the discrete and continuous absorption bands of oxygen gas at 1.26 and 1.065μm and radiative lifetime of the $^1\Delta_g$ state of oxygen, J. Chem. Phys., 43:4345.
Beaumont, P.C., Rodgers, M.A.J., Parsons, B.J. and Philips, G.O., 1985, Singlet oxygen production by some furocoumarin derivatives in the presence of DNA: Time-resolved luminescence measurements, Photochem. Photobiol., 42:605.
Borland, C.F., Cogdell, R.J., Land, E.J. and Truscott, T.G., 1989, Bacteriochlorophyll A triplet state and its interaction with bacterial carotenoids and oxygen, J. Photochem. Photobiol., B: 3:237.
Borland, C.F., McGarvey, D.J., Truscott, T.G., Cogdell, R.G. and Land, E.J., 1987, Photophysical studies of bacteriochlorophyll A and bacteriophenophytin A - singlet oxygen generation, J. Photochem. Photobiol., B: 1:93.
Brault, D., Vever-Bizet, C., Rougee, M. and Bensasson, R., 1988, Photophysical properties of a chlorin a potent sensitizer for photochemotherapy, Photochem. Photobiol., 47:151.
Chacon, J.N., McLearie, J. and Sinclair, R.S., 1988, Singlet oxygen yields and radical contributions in the dye-sensitized photooxidation in methanol of esters of polyunsaturated fatty acids (oleic, linolenic and arachidonic), Photochem. Photobiol., 47:647.
Chou, P-T., and Khan, A.U., 1983, L-ascorbic acid quenching of singlet delta molecular oxygen in aqueous media: generalized antioxidant property of vitamin C, Biochem. Biophys. Res. Comm., 115:932.

Darmanyan, A.P., 1987, Effect of solvents on the quantum yield of singlet oxygen generation by anthracene and on the radiative lifetime of oxygen $^1\Delta_g$-state, <u>Khimich. Fizika (USSR)</u>, 6:1192.

Delmelle, M., 1979, Possible implication of photooxidation reactions in retinal photodamage, <u>Photochem. Photobiol.</u>, 29:713.

Djagarov, B.M., Gurinovich, G.P., Novichenkov, V.E., Salokhiddinov, K.I., Shulga, A.M. and Ganja, B.A., 1987, Photosensitized formation of singlet oxygen and quantum yield of intersystem crossing in molecules of porphyrins and metal porphyrins, <u>Khimich. Fizika (USSR)</u>, 6:1069.

Djagarov, B.M., Sagun, E.I., Ganja, V.A., and Gurinovich, G.P., 1987, Mechanisms of quenching the triplet state of chlorophyll and related compounds by molecular oxygen, <u>Khimich. Fizika (USSR)</u>, 6:919.

Dougherty, T.J., 1987, Photosensitizers: therapy and detection of malignant tumors, <u>Photochem. Photobiol.</u>, 45:879.

Egorov, S.Yu., Babizhayev, M.A., Krasnovsky, A.A., jr. and Shvedova, A.A., 1987, Photosensitized generation of singlet molecular oxygen by the endogenous substances of the lens crystallines, <u>Biofizika (USSR)</u>, 32:184.

Egorov, S.Yu., Dontsov, A.E., Krasnovsky, A.A., jr and Ostrovsky, M.A., 1987, Quenching of singlet oxygen by screening pigments: melanins and ommochromes, <u>Biofizika (USSR)</u>, 32:685.

Egorov, S.Yu., Kamalov, V.F., Kozoteev, N.I., Krasnovsky, A.A., jr., Toleutaev, B.N. and Zinukov, S.V., 1989, Rise and decay kinetics of photosensitized singlet oxygen luminescence in water. Measurements with nanosecond time-correlated photon counting, <u>Chem. Phys. Lett.</u>, 163-421.

Egorov, S.Yu., and Krasnovsky, A.A., jr., 1983, Photosensitized luminescence of singlet oxygen under pulsed laser excitation. Decay kinetics in aqueous solutions, <u>Biofizika (USSR)</u>, 28:497.

Egorov, S.Yu., and Krasnovsky, A.A., jr., 1986, Quenching of singlet molecular oxygen by components of the media used for isolation of chloroplasts and testing their photosynthetic activity, <u>Fiziol rast. (USSR)</u>, 33:10.

Egorov, S.Yu., Krasnovsky, A.A., jr. and Kulakovskaia, L.I., 1985, Investigation of the mechanism of chloroplast photodestruction: participation of the chlorophyll triplet state, <u>Fiziol. rast. (USSR)</u>, 32:668.

Egorov, S.Yu., Krasnovsky, A.A., jr., Safronova, I.A., Bystrova, M.J. and Krasnovsky, A.A., 1988, Photogeneration of singlet molecular oxygen by pigments-precursors of chlorophyll, <u>Dokl. AN SSSR</u>, 299:1266.

Egorov, S.Yu., Zinukov, S.V., Kamalov, V.F., Koroteev, N.I., Krasnovsky, A.A. jr. and Toleutaev, B.N., 1988, Kinetic measurements of photosensitized singlet molecular oxygen luminescence with nanosecond resoluton, <u>Opt. Spektrosc. (USSR)</u>, 65:899.

Emanuel, N.M. and Ostrovsky, M.A., 1981, Antioxidants in photobiology, <u>Vestnik AN SSSR</u>, 9:66.

Firley, P.A., Ford, W.E., Sounik, J.R., Kenney, M.E., and Rodgers, M.A.J., 1988, Silicon Naphthalocyanine triplet state and oxygen: a reversible energy transfer reaction, <u>J. Am. Chem. Soc.</u>, 110:7626.

Firley, P.A., and Rodgers, M.A.J., 1988, Photochemical

properties of erythrocite ghosts containing porphyrin, <u>Photochem. Photobiol.</u>, 47:615.

Firley, P.A., Jones, T.W., Jori, G. and Rodgers M.A.J., 1988, Photoexcitation of zinc phthalocyanine in mouse myeloma cells: the observation of triplet states but not of singlet oxygen, <u>Photochem. Photobiol.</u>, 48:357.

Foote, C.S., 1976, Photosensitized oxygenation and singlet oxygen, <u>in</u>: "Free radicals in biology", W.A. Prior ed., Academic Press, New York.

Foote, C.S., 1979, Quenching of singlet oxygen, <u>in</u>: "Singlet oxygen", H.H. Wasserman and R.W. Murray eds., Academic Press, New York.

Foote, C.S. and Dobrowolski, D.C., 1984, Singlet oxygen production from photodynamic sensitizers, <u>in</u>: "Oxygen radicals in chemistry and biology", W. Bors, M. Saran, D. Tait eds., Walter de Gruyter, Berlin.

Foote, C.S. and Wexler, S., 1964, Olefin oxidations with excited singlet molecular oxygen, <u>J. Amer. Chem. Soc.</u>, 86:3879.

Gorman, A.A., Hamblett, J., Lambert, C., Prescott, A.L., Rodgers M.A.J., Spence, H.M., 1987, Aromatic ketone-naphtalene systems as absolute standards for the triplet sensitized formation of singlet oxygen, O_2 $(^1\Delta_g)$, in organic and aqueous media: A time-resolved luminescence study, <u>J. Amer. Chem. Soc.</u>, 109:3091.

Hasan, T., Khan, A.U., 1986, Phototoxicity of tetracyclines: photosensitized emission of singlet delta oxygen, <u>Proc. Natl. Acad. Sci., USA</u>, 83:4604.

Hurst, J.R., MacDonald, J.D. and Schuster C.B., 1982, Lifetime of singlet oxygen in solution directly determined by laser spectroscopy, <u>J. Amer. Chem. Soc.</u>, 104:2065.

Hurst, J.R. and Schuster, G.B., 1983, Nonradiative relaxation of singlet oxygen in solution, <u>J. Am. Chem. Soc.</u>, 105:5756.

Jones, A.V. and Harrison A.W., 1958, $^1\Delta_g$-$^3\Sigma_g$-O_2 infrared emission band in the twilight airglow spectrum, <u>J. Atmosph. Terr. Phys.</u>, 13:45.

Kagan, V.E., 1988, Lipid peroxidation in biological membranes, CRC Press, Boca Raton, Florida, USA.

Kai-Kong, Ju. and Ogilby, P., 1987, A time-resolved study of singlet molecular oxygen $(^1\Delta_g O_2)$ in a solution-phase photosensitized reactions. A new experimental technique to examine the dynamics of quenching by oxygen, <u>J. Phys. Chem.</u>, 91:1611.

Keene, J.R., Kessel, D., Land, E.J., Redmont, R.W., and Truscott, T.G., 1986, Direct detection of singlet oxygen sensitized by hematoporphyrin and related compounds, <u>Photochem. Photbiol.</u>, 43:117.

Keir, W.E., Land, E.J., McLenan, A.H., McGarvey, D.J. and Truscott, T.G., 1987, Pulsed radiation studies of photodynamic sensitizers: the nature of DHE, <u>Photochem. Photobiol.</u>, 46:587.

Khan, A.U., Kasha, M., 1979, Direct spectroscopic observation of singlet oxygen emission at 1268 nm excited by sensitizing dyes of biological interest in liquid solution, <u>Proc. Natl. Acad. Sci.</u>, 76:6047.

Khan, A.U., 1980, Direct spectroscopic observation at 1,27 μm and 1.58 μm emission of singlet $(^1\Delta_g)$ molecular oxygen in chemically generated and dye-photosensitized liquid solutions at room temperature, <u>Chem. Phys. Lett.</u>, 72:112.

Knox, C.N., Land, E.J. and Truscott, T.G., 1986, Singlet oxygen generation by furocoumarin triplet states-I. Linear furocoumarins (psoralens), Photochem. Photobiol., 43:359.

Knox, G.N., Land, T.J. and Truscott, T.G., 1988, Triplet state properties and triplet state-oxygen interactions of some linear and angular furocourmarins, J. Photochem. Photobiol., B: 1:315.

Knox, P. and Dodge, A.D., 1985, Singlet oxygen in plants, Photochemistry, 24:889.

Kochubeev, G.A., Frolov, A.A. and Gurinovich, G.P., 1989, Chlorin e_6. Spectral and energy parmeters and generation of singlet molecular oxygen in home homogeneous and heterogeneous media, Khimich. Fizika (USSR), 8:1184.

Krasnovsky, A.A., Jr., 1976, Photosensitized luminescence of singlet oxygen in solution, Biofizika (USSR), 21:748.

Krasnovsky, A.A., jr., 1977, Photoluminescence of singlet oxygen in solutions of chlorophylls and pheophytins, Biofizika (USSR), 22:927.

Krasnovsky, A.A. jr., 1979a, Photoluminescence of singlet oxygen in pigment solutions, Photochem. Photobiol., 29:29.

Krasnovsky, A.A., jr., 1979b, Photosensitized luminescence of singlet oxygen in aqueous solutions, Biofizika (USSR), 24:747.

Krasnovsky, A.A., jr., 1981, Quantum yield of photosensitized luminescence and radiative lifetime of singlet ($^1\Delta_g$) molecular oxygen in solutions, Chem. Phys. Lett., 81:433.

Krasnovsky, A.A., jr., 1982, Luminescence under photosensitized formation of singlet oxygen in solutions, in: "Excited molecules. Kinetics of transformations", A.A. Krasnovsky ed. "Nauka", Leningrad, p. 32.

Krasnovsky, A.A., jr., 1986, Singlet Oxygen in photosynthesizing organisms. J. All Union Mendeleev Chemical. Soc. (USSR), 31:562.

Krasnovsky, A.A., jr., 1988, Mechanism of formation and role of singlet oxygen in photobiological processes, in: "Molecular mechanisms of biological action of optical radiation. A.B. Rubin ed., Nauka, Moscow, p.23.

Krasnovsky, A.A., jr., Egorov, S.Yu., Nasarova, O.V., Yartsev, E.I. and Ponomarev, G.V., 1987, Photogeneration of singlet molecular oxygen by water-soluble porphyrins, Biofizika (USSR), 32:982.

Krasnovsky, A.A., jr., Egorov, S.Yu., Nasarova, O.V., Yartsev, E.I. and Ponomarev, G.V., 1988, Photosensitized formation of singlet molecular oxygen in solutions of water-soluble porphyrins. Direct luminescence measurements, Studia biophysica, 124:123.

Krasnovsky, A.A., jr. and Kagan, V.E., 1979, Photosensitzation and quenching of singlet oxygen by pigments and lipids of the retina, FEBS Lett, 108:152.

Krasnovsky, A.A., jr., Kagan, V.E. and Minin, A.A., 1983, Quenching of singlet oxygen luminescence by fatty acids and lipids, FEBS Lett., 155:233.

Krasnovsky, A.A., jr. and Neverov, K.V., Photosensitized dimol luminescence of singlet molecular oxygen in solutions, Biofizika (USSR), 23:884.

Krasnovsky, A.A., jr., Neverov, K.V., Egorov, S.Yu. and Roeder, B., 1988, Photophysical parameters of pheophorbide A: phosphorescence and generation of singlet oxygen, Opt. spectrosc. (USSR), 64:790.

Krasnovsky, A.A., jr. and Paramonova, L.I., 1983, Interaction of singlet oxygen with carotenoids: rate constants of physical and chemical quenching, <u>Biofizika (USSR)</u>, 28:725.

Krasnovsky, A.A., jr., Sukhorukov, V.L. and Potapenko, A.Ya., 1983, Photogeneration of singlet oxygen by psoralens, <u>Bull. exp. biol. med. (USSR)</u>, 9:59.

Krasnovsky, A.A., jr., Venedictov, E.A. and Chernenko, O.V., 1982, Quenching of singlet oxygen by chlorophylls and porphyrins, <u>Biofizika (USSR)</u>, 27:966.

Krasnovsky, A.A., jr., Vychegzanina, I.V., Drozdova, N.N. and Krasnovsky, A.A., 1985, Generation and quenching of singlet oxygen by bacteriochlophylls and bacteriopheophytins A and B. <u>Dokl. AN SSSR</u>, 283:474.

Krinsky, N.I., 1984, Biology and photobiology of singlet oxygen, <u>in</u>: "Oxygen radicals in chemistry and biology", W. Bors, M.Sarah, D. Tait eds., Walter de Gruyter Co., Berlin.

Lagorio, M.G., Dicello, L.E. and San-Roman, E.A. and Braslavsky, S.E., 1989, Quantum yield of singlet oxygen sensitization by copper (II) tetracarboxyphthalocyanine, <u>J. Photochem. Photobiol.</u>, B: 3:615.

Lambert, C.R., Reddi, E., Spikes, J.D., Rodgers, M.A.J. and Jori, G., 1986, The effect of porphyrin structure and aggregation state on photosensitized processes in aqueous ad miccelar media, <u>Photochem. Photobiol.</u>, 44:595.

Li, A.S.W., Chignell, C.F., and Hall, R.D., 1987, Cutaneous phototoxicity of tetracycline antibiotics: generation of free radicals and singlet oxygen during photolysis as measured by spin-trapping and the phosphorescence of singlet molecular oxygen, <u>Photochem. Photobiol.</u>, 46:379.

Long, C., Kearns, D.R., 1973, Selection rules for the intermolecular enhancement of spin forbidden transitions in molecular oxygen, <u>J. Chem. Phys.</u>, 59, 5729.

Losev, A.P., Byteva, I.M., Gurinovich, G.P., 1988, Singlet oxygen luminescence quantum yeilds in organic solvents and water, <u>Chem. Phys. Lett.</u>, 143:127.

Merkel, P.B. and Kearns, D.R., 1972, Radiationless decay of singlet molecular oxygen in solution. An experimental and theoretical study of electronic-to-vibrational energy transfer, <u>J. Amer. Chem. Soc.</u>, 94:7244.

Noxon, J.F., 1961, Observation of the $(b^1\Sigma_g - a^1\Delta_g)$ transition in $^1O_2$, <u>Can. J. Phys.</u>, 39:1110.

Ogilby, P.R. and Foote, C.S., 1982, Chemistry of singlet oxygen. 36. Singlet molecular oxygen $(^1\Delta_g)$ luminescence in solution following pulsed laser excitation, <u>J. Amer. Chem. Soc.</u>, 104:2069.

Parker, J.G., 1987, Optical monitoring of singlet oxygen generation during photodynamic treatment of tumors, <u>IEEE Circuits and Dev. Mag.</u>, 3:10.

Parker, J.C. and Stanboro, W.D., 1982, Optical determination of the collisional lifetime of singlet molecular oxygen $[^1O_2(^1\Delta_g)]$ in acetone and deuterated acetone, <u>J. Am. Chem. Soc.</u>, 104:2067.

Parker, J.G., and Stanboro, W.D., 1984, Optical determination of the rates of formation and of O_2 $(^1\Delta_g)$ in H_2O, D_2O and other solvents, <u>J. Photochem.</u> 25:545.

Reddi, E., Lambert, C.R., Jori, J. and Rodgers, M.A.J., 1987, Photokinetic and photophysical measurements of the sensitized photooxidation of the tryptophyl residue in N-

acethyl tryptophanaamide and human serum albumin,
Photochem. Photobiol., 45:345.

Redmont, R.W., Valduga, G., Nonell, S., Braslavsky, S.E. and
Schaffner, K., 1989, The Photophysical properties of
porphycene incorporated in small unilamellar lipid
vesicles, _J. Photochem. Photobiol._, B: 3:193.

Rodgers, M.A.J. and Snowden, P.T., 1982, Lifetime of O_2 ($^1\Delta_g$)
in liquid water as determined by time-resolved infrared
luminescence measurements, _J. Amer. Chem. Soc._, 104:5541.

Rougee, M. and Bensasson, R.V., 1986, Determination des
constants de vitesse de deactivation de l'oxygen singulet
($^1\Delta_g$) en presence de biomolecules, _C. R. Acad. Sci._,
Serie II, 302:1223.

Rougee, M., Bensasson, R.V., Land, E.J. and Parient, R., 1988,
Deactivation of singlet oxygen by thiols and related
compounds, possible protectors against skin
photosensitivity, _Photochem. Photobiol._, 47:485.

Salokhiddinov, K.I., Byteva, I.M. and Dzagezov, B.M., 1979,
Lifetime of singlet oxygen luminescence in solutions
under pulsed laser excitation, _Opt. spectrosc. (USSR)_,
47:881.

Salokhiddinov, K.I., Dzagarov, B.M., Byteva, I.M. and
Gurinovich, G.P., 1980, Photosensitized luminescence of
singlet oxygen solutions at 1588 nm, _Chem. Phys. Lett._,
76:85.

Salokhiddinov, K.I., Dzagarov, B.M., Yegorova, C.D., 1983,
Direct measurement of the lifetime of molecular oxygen in
singlet ($^1\Delta_g$) state generated in water by porphyrin
sensitizers, _Opt. spectrosc. (USSR)_, 55:71.

Schmidt, R., 1989, Determination of the phosphorescence
quantum yield of singlet molecular oxygen ($^1\Delta_g$) in five
different solvents, _J. Phys. Chem._, 93:4507.

Scurlock, R.D. and Ogilby P.R., 1987, Effect of solvent on the
rate constant for the radiative deactivation of singlet
molecular oxygen ($^1\Delta_g$ O_2), _J. Phys. Chem._, 91:4599.

Schuhardt, H. and Nultch, W., 1984, Possible role of singlet
molecular oxygen in the control of the photoactic
reaction sign of _Anabaena variabilis_, _J. Photchem._,
25:317.

Senger, H., 1982, The effect of blue light on plants and
microorganisms, _Photochem. Photobiol._, 35:911.

Snelling, D.R., 1968, Production of singlet oxygen in the
benzene oxygen photochemical system, _Chem. Phys. Lett_.
2:346.

Straight, C.R. and Spikes, J.D, 1985, Photosensitized
oxidation of biomolecules, _in_ "Singlet O_2", A.A. Frimer
ed., CRC Press, Boca Ratoh, Florida.

Takemura, T., Ohta, A., Nakajima, S. and Sakata, I., 1989,
Critical importance of the triplet lifetime of
photosensitizer in photodynamic therapy of tumors,
Photochem. Photobiol., 50:339.

Tanelian, C. and Wolf, C., 1988, Mechanism of physical
quenching of singlet molecular oxygen by chlorophylls and
related compounds of biological interest, _Photochem._
Photobiol., 48:3.

Valduga, G., Nonell, S., Reddi,E., Jori, G. and Braslavsky,
S.E., 1988, The production of singlet molecular oxygen by
Zn(II) phthalocyanine in ethanol and in unilamellar
vesicles. Chemical quenching and phosphorescence studies,
Photochem. Photobiol., 48:1.

Venedictov, E.A. and Krasnovsky, A.A., jr., 1979, Mechanisms
    of quenching of singlet molecular oxygen by porphyrins
    and their metal complexes, <u>Izv. Vuzov. Khimia i Khim.
    Technologia (USSR)</u>, 22:396.
Venekiktov, E.A. and Krasnovsky, A.A., jr., 1982, Efficiency
    of generation of singlet molecular oxygen by porphyrins,
    <u>Zh. Prik. Spektr. (USSR)</u>, 36:152.
Vever-Bizet, C., Dellinger, M., Brault, D., Rougee, M. and
    Bensasson, R.W., 1989, Singlet molecular oxygen quenching
    by saturated and unsaturated fatty acids and by
    cholesterol, <u>Photochem. Photobiol.</u>, 50:321.

STRUCTURE OF NUCLEIC ACID PHOTOPRODUCTS

Jean Cadet[1], Anthony Shaw[1], Lucienne Voituriez[1],
Paul Vigny[2], and Lou-sing Kan[3]

[1]Laboratoires de Chimie, DRF, Centre d'Etudes
Nucléaires de Grenoble, 85X, F-38041 Grenoble
Çedex, France
[2]LPCB, Institut Curie, 75231 Paris Cedex 05
France
[3]Division of Biophysics, The Johns Hopkins
University, Baltimore, MD 21205, USA

INTRODUCTION

Much effort has been devoted during the three last
decades to the characterization of the main DNA photoproducts
arising from exposure to UVC radiation and to the mechanism of
their formation (for early comprehensive reviews see: Wang
1976). It was clearly established that the nucleobases, more
precisely the pyrimidine moities, rather than the relatively
inert phosphodiester-osidic backbone, are the critical targets
for the far-UV induced photoreactions. One of the major early
findings has been the discovery of the [2 + 2]
cyclodimerisation of thymine upon exposure to 254 nm light in
frozen aqueous solutions (Beukers and Berends, 1960). This
provided the impetus for the search for cyclobutathymine in
cellular DNA and for the determination of the biological role
of this major photolesion including numerous DNA repair
studies. Significant progress was recently made in a better
understanding of various aspects of the photochemistry of
nucleic acids as emphasised in the present survey. This has
been facilitated by the availability of powerful molecular
biology methods (DNA sequencing techniques, overproduction of
DNA repair enzymes by gene cloning) and by the advent of
resolutive spectrometric techniques (pico- and femtosecond-
resolved techniques, high field mono- and two-dimensional NMR,
soft-ionisation mass spectrometries ...) as well as by the
improvement of the chemical synthesis methods of
oligonucleotides.

PYRIMIDINE PHOTOPRODUCTS

Further significant insights were recently gained into
the mechanism of formation and the chemical, conformational

and/or the photochemical features of the four main classes of pyrimidine photoproducts.

Cyclobutadipyrimidines

One of the major achievements in the chemistry of pyrimidine photoproducts was the sequence-specific incorporation of the cis-syn cyclobutadithymine Thy<>Thy into oligodeoxyribonucleotides using phosphoramidite-based solid-phase DNA synthesis technology (Taylor, et al., 1987). This allowed the synthesis of the bacteriophage containing a site-specific cis-syn Thy<>Thy (Taylor and O'Day, 1989) which appears to be a suitable tool for DNA repair and mutagenic studies. It should be noted that the related cis-syn diastereomers of d(GAT<>GG) (Cadet et al., 1987), d(T<>TT) and d(TTT<>T) (Rycyna et al., 1988) were prepared by acetone photosensitization and subsequently purified by reversed-phase high-performance liquid chromatography in the ion-suppression mode. Far-UV irradiation was also used for generating the cis-syn d(CGCT<>TGCG) (Kemmink et al., 1987). It should be noted that relevant structural information was inferred from fast atom bombardment (Ulrich et al., 1989), ^{252}Cf plasma desorption (Viari et al., 1989) and mass spectrometry analyses of the cis-syn and trans-syn diastereomers of d(T[p]T photodimers as well as for the (6-4) and related Dewar isomer photoadducts (vide infra).

It is interesting to note that the formation of non-adjacent cyclobutadipyrimidines was observed in the single-stranded alternating copolymers poly [d(G-T)] (Nguyen and Minton, 1988) and double-stranded structures including poly[d(G-G-A)]poly[d(T-C)] (Evans and Morgan, 1982) and poly[d(C-T)] (Brown et al., 1985).

Various experimental and theoretical attempts were made at determining the role of pyrimidine cyclodimerisation in terms of the conformational changes within dinucleoside monophosphates and longer oligonucleotides. The presence of a cis-syn cyclobutadithymine as inferred from the ^{1}H NMR study of d(GCGT<>TGCG), d(CGCAACGC) in aqueous solutions is likely to induce major distortion within the duplex only in its close vicinity (Kemmink et al., 1987). This is in agreement with the results of quantitative electron microscopic and polyacrylamide gel electrophoresis analyses (Husain et al., 1988) as well as of molecular mechanic studies (Rao et al., 1984) which show a lack of severe topological unwinding of the modified duplex oligonucleotides. However, a recent energy minimization conformational investigation of a duplex dodecanucleotide containing a cis-syn Thy<>Thy led to different conclusions, suggesting the occurrence of long range effects (Pearlman et al., 1985). Further studies which may take advantage of the recent X-ray crystallographic determination of the cyanoethyl ester of the cis-syn d(TpT) internal photodimer (Hruska, et al., 1986) are required to resolve these somewhat conflicting results.

A comprehensive mechanism for the E. coli photolyase-mediated photoreversal of cyclobutadipyrimidines has recently emerged (for a review see: Sancar, 1987). The 40,000 Da enzyme was found to be more efficient for photosplitting Thy<>Thy and heterodimers than cyclobutadicytosine (Myles et

454

al., 1987). Detailed photophysical investigations showed that the photoreduction of the flavin apoenzyme led to a 12-15 fold increase in the yield of Pyr<>Pyr monomerisation within the 350-450 nm range (Sancar et al., 1987).

(6-4) Pyrimidine-pyrimidone adducts

The apparent high mutagenic potential of (6-4) pyrimidine-pyrimidone photoadducts, which however remains to be confirmed (for a review see: Mitchell and Nairn, 1989), is likely to explain the recent upsurge of interest in this important class of pyrimidine photoproducts. Interestingly, the chemical structure of the (6-4) internal bipyrimidine adduct of thymidylyl-(3'-5')-thymidine (Rycyna and Alderfer, 1985), thymidylyl-(3'-5')-2'-deoxycytidine (Taylor et al., 1991), thymidylyl-(3'-5')-2'-deoxyuridine and 2'-deoxycytidylyl-(3'-5')-thymidine (Douki et al., 1991) was unambiguously assigned on the basis of extensive spectroscopic measurements including high field NMR and mass spectrometry analyses. A 5R, 6S stereochemistry was established for the chiral carbons of the pyrimidine moiety of the (6-4) bipyrimidine d(TpT) adduct (Rycyna and Alderfer, 1985).

It is noteworthy that the (6-4) thymine biadduct was found to be generated within far-UV irradiated DNA following mild hydrolysis and subsequent specific fluorescent detection of the released 5-hydroxy-6,4'(5'-methylpyrimidin-2'-one)-5,6-dihydrothymine (Voituriez et al., 1988). The alkali lability of the (6-4) pyrimidine-pyrimidone photoadducts has been used to induce phosphodiester bond cleavage in defined sequence oligonucleotides which are likely to occur on the 3'-side of the photoproducts (Lippke et al., 1981; Bourre et al., 1987). In a subsequent step the determination of the location of the photolesions achieved using the Maxam and Gilbert gel electrophoresis sequencing method. Piperidine-induced DNA cleavages were found to be predominantly located at cytosine sites 3' to thymine and to a lesser extent to cytosine in far-UV irradiated 342 base pair α sequence (Lippke et al., 1981) and SV40 fragments (Bourre et al., 1987). However, a more quantitative analysis would require further determination of the alkaline lability of the (6-4) photoadducts and of the corresponding Dewar isomers.

A major photochemical feature of the (6-4) pyrimidine-pyrimidone adducts which may be of biological significance is their quantitative conversion into a Dewar valence isomer upon exposure to UV-B radiation as shown for d(TpT) (Taylor and Cohrs, 1987), d(CpT), (Voituriez, et al., 1988) and d(TpU) (Voituriez and Cadet, 1989). It is noteworthy that the conformational properties of the Dewar isomer of d(TpT) as obtained from 500 MHz (Taylor et al., 1988a) and 611 MHz (Voituriez et al., 1988) ^{1}H NMR analysis, as well as from molecular modelling calculation (Taylor et al., 1988a), are almost similar to those of the (6-4) pyrimidine-pyrimidone precursor (Rycyna and Alderfer, 1985; Kan et al., 1988; Taylor et al., 1988b). However, it is likely that the isomerisation of the pyrimidone ring led to a more compact structure.

Spore photoproducts

Interesting information was inferred from the ^{1}H NMR

analysis of the 5R* and 5S* diastereomers of 5,6-dihydro-(α-thymidylyl)thymidine obtained by UVC irradiation of $[C^2H_3]$ thymidine in frozen aqueous solution (Shaw et al., 1988). A concerted mechanism which involves a stereospecific hydrogen transfer from one methyl group to the carbon 6 of the second thymine moiety is likely to explain the formation of the "spore photoproducts" under these conditions of irradiation.

Pyrimidine "photohydrates"

E. coli endonuclease III was found to incise far UV irradiated DNA at cytosine sites which are not flanked by pyrimidine (Weiss and Duker, 1986; Helland et al., 1986; Gallagher et al., 1989). Human and yeast redoxyendonucleases are also able to cleave DNAs at cytosine sequences following exposure to UVC radiation (Doetsch et al., 1986; Gossett et al., 1988). 6-Hydroxy-5,6-dihydrocytosine was recently found to be the photoproduct enzymatically removed by the DNA glycosylase activity of both E. coli and mammalian endonucleases. (Boorstein et al., 1989).

PURINE PHOTOPRODUCTS

Purine components have long been considered to be relatively resistant to exposure to far-UV light. However, it was inferred from the results of recent gel sequencing experiments that different repair enzymes are able to generate significant incisions at purine sites in UVC irradiated DNA fragments. E. coli and calf thymus endonuclease (Helland et al., 1986) as well as human (Doetsch et al., 1988) and yeast (Gossett, et al., 1988) redoxyendonucleases were found to cleave at some, but not all, guanine loci. DNA nicks were shown to occur at both adenine and guanine sites when M. luteus crude extracts (Duker and Gallagher, 1986) and endonuclease V, the gene product of the bacteriophage T4-infected E. coli recognising Pyr<>Pyr, were used as the enzymes (Gallagher and Duker, 1986, 1989). Further experiments are required to characterize these unknown photoadducts.

The UvrA, UvrB, and UvrC proteins which constitute the three subunits of the E. coli ABC excinuclease were obtained by gene cloning. This repair enzyme, which is able to excise a wide variety of lesions including Pyr<>Pyr and (6-4) pyrimidine-pyrimidone, was found to remove the dimeric photoproducts by hydrolysing the 8th phosphodiester bond 5' and the 4th or 5th phosphodiester bond 3' to the damaged site (for a review see, Sancar, 1987). A different mechanism which is likely to involve initial enzymic splitting of the phosphodiester bond of Pyr<>Pyr was proposed for human excision repair (Weinfeld, et al., 1986).

Photoionization processes

Detailed flash photolysis studies have recently shown that photoionization, which probably involves singlet excited states, is the predominant far-UV mediated-reaction of both adenine (Arce 1987) and guanine (Arce and Rivera, 1989) components in aqueous solutions. This received indirect confirmation from the characterization of the major

decomposition photoproduct of 2'-deoxyguanosine in aerated
aqueous solution (Berger and Cadet, unpublished results).
This modified compound which has a lactone structure is
identical to those generated by the type I photodynamic effect
leading to the initial formation of a purinyl radical cation
(Cadet, et al., 1986).

Photoadducts

The main photoaddition reaction occurring between two
adjacent adenine moities in dinucleosidemonophosphate and
poly(dA) was recently elucidated. The formation of the so-
called "Pörschke photoproduct" is likely to involve an
azetidine intermediate as the result of covalent binding of
the N(7) and C(8) atoms of the 5'-adenine moiety to the C(6)
and C(5) carbons of the adenine at the 3'-side (Kumar, et al.,
1987). The dimeric adenine photoadduct, isolated as 4,6-
diamino-5-guanidinopyrimidine (DGPY) upon acidic hydrolysis,
was found also to be produced in far-UV irradiated DNA,
although in a relatively lower yield than in poly(dA) (Sharma
and Davies, 1989). The quantum yield for the formation of the
adenine biadduct in single- and double-stranded DNA was
estimated to be 6 x 10^{-5} and 9 x 10^{-6} respectively. It should
be noted that adenine is not able to significantly photoreact
with an adjacent guanine nucleobase as shown from model
studies (Sharma and Davies, 1989).

PYRIMIDINE-PURINE PHOTOADDUCTS

Evidence for the far-UV induced covalent binding of a
thymine base to an adjacent adenine or inosine moiety was
recently obtained using dinucleosidemonophosphates and
alternating DNA copolymers as substrates (Kumar and Davies,
1987). The 5,6-thymine and the 1,6-adenine ethylenic bonds
are likely to be involved in this photoreaction which was
found to occur in both single- and double-stranded DNA.

PYRIMIDINE-AMINO ACID PHOTOADDUCTS

Early aspects of the main far-UV induced photoaddition
reactions of amino acids to DNA components including the
photoexchange reaction between thymidine at N(1) position and
various amino acids as well as the photobinding of arginine to
thymidine, have been surveyed in detail in two recent reviews
(Saito and Matsuura, 1985; Cadet, et al., 1986). Careful
reinvestigation of the photoreactions of thymine with L-
cysteine allowed a clarification of the major inconsistencies
in the previous structure assignments of some adducts (Shetlar
and Hom, 1987). In particular 5-S-L-cysteinylmethyl-5,6-
dihydrouracil is generated but it exhibits different
spectroscopic properties from those previously reported for a
putative identical structure (Varghese, 1973). In addition,
the mixture of the two cis and trans diastereomers of 5-S-L-
cysteinyl-5,6-dihydrothymine was resolved (Shetlar and Hom,
1987). Another interesting study dealt with the
identification of the main photoaddition products arising from
the 306 nm irradiation of 5-bromouracil with tryptophan,
tyrosine or histidine in aqueous solutions (Dietz and Koch,
1987). Regiospecific photocoupling which involves electron

transfer from the aromatic amino acids was found to occur
between uracil at the 5-position and the C(2) carbon of the
indole and phenol rings as well as the 5-position of the
imidazole ring of the respective amino acids (Dietz and Koch,
1987)

TWO QUANTUM PHOTO-IONISATION OF PURINE AND PYRIMIDINE COMPONENTS

A non-linear photochemistry of nucleic acid components
was found to be associated with the use of high power nano-
and picosecond UV laser sources (for a recent review see,
Nikogosyan, 1989). Decomposition of DNA constituents may
arise from the reaction of hydroxyl radicals as the result of
photoionisation and/or purine and pyrimidine photosensitized
dissociation of water molecules. Photodegradation of the
nucleobases is also likely to occur through two-photon
excitation to high-lying singlet or triplet states and
subsequent photoionisation. The formation of a guanine
radical cation was suggested to explain the observed DNA
strand cleavages in plasmid DNA upon exposure to 248 nm UV
light from a KrF excimer laser (Croke, et al., 1988).

DETERMINATION OF BASE PHOTOLESIONS WITHIN CELLULAR DNA

Various methods (gas chromatography - mass spectrometry,
HPLC associated with various spectroscopic detection including
amperometry, fluorescence, mass spectrometry), and biochemical
methods (immunoassays, endonucleases, ^{32}P post-labelling) are
now available for monitoring DNA photoleasions within
irradiated cells (for reviews see, Cadet and Vigny, 1988,
1990). Emphasis should be placed on the development of
chemical and/or biochemical assays which allow the measurement
of specific photolesions within non radiolabelled cellular DNA
and in single cells (Mori, et al., 1989).

REFERENCES

Arce, R., 1987, Characterization of the transient species in
the 266 nm-laser photolysis of adenine and its
derivatives, Photochem. Photobiol., 45:713-722.
Arce, R. and Rivera, J., 1989, Intermediates produced from the
room temperature 266 nm laser photolysis of guanines. J.
Photochem. Photobiol., A: Chemistry, 49, 219-237.
Beukers, R. and Berends, W., 1960, Isolation and
identification of the irradiation product of thymine.
Biochem. Biophys. Acta., 41:550-551.
Boorstein, R.J., Hilbert, T.P., Cadet, J., Cunningham, R.P.
and Teebor, G.W., 1989, UV-induced pyrimidine hydrates in
DNA are repaired by bacterial and mammalian DNA
glycosylase activities, Biochemistry, 28:6164-6170.
Bourre, F., Renault, G. and Sarasin, A., 1987, Sequence effect
on alkali-sensitive sites in UV-irradiated DNA, Nucleic
Acids Res., 15:8861-8875.
Brown, D.M., Gray, D.M. and Patrick, M.H., 1985, Photochemical
demonstrations of stacked C-C base pairs in a novel DNA
secondary structure, Biochemistry, 24:1676-1683.
Cadet, J. and Vigny, P., 1988, Biochemical and chemical assays

for monitoring the formation of DNA photolesions, J. Photochem. Photobiol. B, Biology, 2:282-286.

Cadet, J. and Vigny P., 1990, Photochemistry of Nucleic Acids, in: "Bioorganic Photochemistry", H. Morrison, ed., J. Wiley & Sons, New York, vol. 1, pp. 1-272.

Cadet, J., Berger, M., Decarroz, C., Wagner, J.R., van Lier, J.E., Ginot, Y.M. and Vigny, P., 1986, Photosensitized reactions of nucleic acids, Biochimie, 68:813-834.

Cadet, J., Vignes, M., Voituriez, L. and Kan, L.-S., 1987, Formation of cyclobutadipyrimidine photoproducts within di- and hexadeoxynucleotides by acetone photosensitization. 2nd Congress of the European Society for Photobiology. Padova, Italy, Book of Abstracts, p. 71.

Croke, D.T., Blau, W., Ohuigin, C., Kelly, J.M., and McConnell, D.J., 1988, Photolysis of phosphodiester bonds in plasmid DNA by high intensity UV laser irradiation. Photochem. Photobiol., 47:527-536.

Dietz, T.M. and Koch, T.D., 1987, Photochemical coupling of 5-bromouracil to tryptophan, tyrosine and histidine, peptide-like derivatives in aqueous fluid solution, Photochem. Photobiol., 46:972-978.

Doetsch, P.W., Helland, D.E. and Haseltine, W.A., 1986, Mechanism of action of a mammalian DNA repair endonuclease, Biochemistry, 25:2212-2220.

Doetsch, P.W., Helland, D.E. and Lee, K., 1988, Wavelength dependence for human redoxy-endonuclease-mediated DNA cleavage at sites of UV-induced photoproducts, Radiat. Res., 113:543-549.

Douki, T., Voituriez, L. and Cadet, J., 1991, Characterization of the (6-4) photoproduct of 2'-deoxycytidylyl-(3'-5')-thymidine and of its Dewar valence isomer, Photochem. Photobiol., 53:293-297.

Duker, N.J. and Gallagher, P.E., 1986, Recent advances in molecular pathology. Detection of DNA damage in human cells and tissues using sequencing techniques, Exp. Mol. Path., 44:117-131.

Evans, D.H. and Morgan, A.R., 1982, Extrahelical bases in duplex DNA, J. Mol. Biol., 160:117-122.

Franklin, W.A., Doetsch, P.W. and Haseltine, W.A., 1985, Structural determination of the ultraviolet light-induced tymine-cytosine pyrimidine-pyrimidone (6-4) photoproduct, Nucleic Acids Res., 13:5317-5325.

Gallagher, P.E. and Duker, N.J., 1986, Detection of purine photoproducts in a defined sequence of human DNA, Mol. Cell. Biol., 6:707-709.

Gallagher, P.E. and Duker, N.J., 1989, Formation of purine photoproducts in a defined human DNA sequence, Photochem. Photobiol., 49:599-605.

Gallagher, P.E., Weiss, R.B., Brent, T.P. and Duker, N.J., 1989, Wavelength dependence of DNA incision by a human ultraviolet endonuclease, Photochem. Photobiol., 49:363-367.

Gossett, J., Lee, K., Cunningham, R.P. and Doetsch, P.W., 1988, Yeast redoxyendonuclease, a DNA repair enzyme similar to Escherichia coli endonuclease III, Biochemistry, 27:2629-2634.

Helland, D.G., Doetsch, P.W. and Haseltine, W.A., 1986, Substrate specificity of a mammalian DNA repair endonuclease that recognises oxidative base damage, Mol. Cell. Biol., 6:1983-1990.

Husain, I., Griffith, J. and Sancar, A., 1988, Thymine dimers bend DNA, Proc. Natl. Acad. Sci. USA, 85:2558-2562.

Hruska, F.E., Voituriez, L., Grand, A. and Cadet, J., 1986, Molecular structure of the cis-syn photodimer of d(TpT) (cyanoethylester), Biopolymers, 25:1299-1417.

Kan, L.S., Voituriez, L. and Cadet, J., 1988, Nuclear magnetic resonance studies of cis-syn, trans-syn and 6-4-photodimers of thymidylyl (3'-5')thymidine monophosphate and cis-syn photodimers of thymidylyl (3'-5')thymidine cyanoethyl phosphotriester, Biochemistry, 27:5796-5803.

Kemmink, J., Boelens, R., Koning, T.M.G., Kaptein, R., van der Marel, G.A. and van Boom, J.H., 1987, Conformational changes in the oligonucleotide duplex d(GCGTTGCG). d(CGCAACGC) induced by formation of a cis-syn thymine dimers, Eur. J. Biochem., 162:37-43.

Kumar, S. and Davies, R.J.H., 1987, The photoreactivity of pyrimidine-purine sequences in some deoxydinucleoside monophosphate and alternating DNA copolymers, Photochem. Photobiol., 45:471-579.

Kumar, S., Sharma, N.D., Davies, R.J.H., Phillipson, D.W. and McCloskey, J.A., 1987, The isolation and characterisation of a new type of dimeric adenine photoproduct in UV-irradiated deoxyadenylates, Nucleic Acids Res., 15, 1199-1216.

Lippke, J.A., Gordon, L.K., Brash, D.E. and Haseltine, W.A., 1981, Distribution of UV light-induced damage in a defined sequence of human DNA: Detection of alkaline-sensitive lesions at pyrimidine nucleoside-cytidine sequences, Proc. Natl. Acad. Sci. USA, 78:3300-3392.

Mitchell, D.L. and Nairn, R.S., 1989, The biology of the (6-4) photoproduct, Photochem. Photobiol., 49:805-819.

Mori, T., Wani, A.A., D'Ambrosio, S.M., Chang, C-C. and Trosko, J.E., 1989, In situ pyrimidine dimer determination by laser cytometry, Photochem. Photobiol., 49:523-526.

Myles, G.M., Van Houten, B. and Sancar, A., 1987, Utilization of DNA photolyase, pyrimidine dimer endonucleases, and alkali hydrolysis in the analysis of aberrant ABC excinuclease incisions adjacent to UV-induced DNA photoproducts, Nucleic Acids Res., 15:1227-1243.

Nikogosyan, D.N., 1989, Two-quantum UV photochemistry of nucleic acids Int. J. Radiat. Biol., (in press).

Nguyen, H.T. and Minton, K.W., 1988, Ultraviolet-induced dimerization of non-adjacent pyrimidines, J. Mol. Biol., 200:681-693.

Pearlman, D.A., Holbrook, S.R., Pirkle, D.H. and Kim, S.-H., 1985, Molecular models for DNA damaged by photoreaction, Science, 227:1304-1308.

Rao, S.N., Keepers, J.W. and Kollman, P., 1984, The structure of d(CGCGAAT[]TCGCG). d(CGCGAATTCGCG): the incorporation of a thymine photodimer into a β-DNA helix, Nucleic Acids Res., 12:4789-4807.

Rycyna, R.E. and Alderfer, J.A., 1985, UV irradiation of nucleic acids: formation, purification and solution conformation analysis of the "the 6-4 lesion" of dTpdT, Nucleic Acids Res., 13:5949-5963.

Rycyna, R.E., Wallace, J.C., Sharma, M. and Alderfer, J.L., 1988, Ultraviolet irradiation of nucleic acids. Purification and solution conformational analyses of oligothymidylates containing cis-syn photodimers, Biochemistry, 27:3152-3163.

Sancar, A., 1987, DNA repair *in vitro*, *in*: "From Photophysics to Photobiology", A. Favre, R. Tyrrell and J. Cadet, eds., Elsevier, Amsterdam, pp. 301-315.

Sancar, G.B., Jorns, M.S., Payne, G., Fluke, D.J., Rupert, C.S. and Sancar, A., 1987, Action spectrum of *E. coli* photolyase. Photolysis of the ES complex and the absolute action spectrum, *J. Biol. Chem.*, 262:492-498.

Saito, I. and Matsuura, T., 1985, Chemical aspects of UV-induced cross-linking of proteins to nucleic acids. Photoreactions with lysine and tryptophan, *Acc. Chem. Res.*, 18:134-141.

Sharma, N.D. and Davies, R.J.H., 1989, Extent of formation of a dimeric adenine photoproduct in polynucleotides and DNA, *J. Photochem. Photobiol, B: Biology*, 3:247-258.

Shaw, A., Voituriez, L. and Cadet, J., 1988, Thymidine spore photoproducts. Isolation and mechanism of formation. 10th International Congress of Photobiology, Jerusalem, Israel, Book of Abstracts, p.54.

Shetlar, M.D. and Hom, K., 1987, Mixed products of thymine and cysteine produced by direct and acetone-sensitized photoreactions, *Photochem. Photobiol.*, 45:703-712.

Taylor, J.-S. and Cohrs, M.P., 1987, DNA, light and Dewar pyrimidinones: The structure and biological significance of TpT3, *J. Am. Chem. Soc.*, 109:2834-2835.

Taylor, J.-S. and O'Day, C.L., 1989, Synthesis of a bacteriophage containing a site-specific *cis-syn* thymine dimer, *J. Am. Chem. Soc.*, 401-402.

Taylor, J.-S., Brockie, I.R. and O'Day, C.L., 1987, A building block for the sequence-specific introduction of *cis-syn* thymine dimers into oligonucleotides. Solid-phase synthesis of TpT[c,s]pTpT, *J. Am. Chem. Soc.*, 109:6735-6742.

Taylor, J.-S., Garrett, D.S. and Cohrs, M.P., 1988a, Solution-state structure of the Dewar pyrimidinone photoproduct of thymidylyl-(3'-5')-thymidine, *Biochemistry*, 27:7206-7215.

Taylor, J.-S., Garrett, D.S. and Wang, M.J., 1988b, Models for the solution state structure of the (6-4) photoproduct of thymidylyl-(3'-5')-thymidine derived via a distance- and angle-constrained conformation search procedure, *Biopolymers*, 27:1571-1593.

Taylor, J.-S., Lu, H.-F. and Kotyk, J.J., 1990, Quantitative conversion of the (6-4) photoproduct of TpdC to its Dewar valence isomer upon exposure to simulated sunlight, *Photochem. Photobiol.*, 51, 161-167.

Ulrich, J., Cadet, J., Becchi, M. and Fraisse, D., 1989, FAB - Mass spectrometry of dipyrimidine photodimers, *Adv. Mass Spectrom.*, 11A:494-495.

Varghese, A.J., 1973, Properties of photoaddition products of thymine and cysteine, *Biochemistry*, 12:2725-2730.

Viari, A., Ballini, J.P., Vigny, P., Voituriez, L. and Cadet, J., 1989, Plasma desorption mass spectrometric study of UV-induced lesions within DNA model compounds, *Biomed. Environ. Mass Spectrom.*, 18:547-552.

Voituriez, L. and Cadet, J., 1989, Formation of the (6-4) pyrimidine-pyrimidone photoadduct of thymidylyl (3'-5') 2'-deoxyuridine and its photoconversion into a Dewar valence isomer. Third Congress of the European Society for Photobiology, Budapest, Hungary, Book of Abstracts, p. 240.

Voituriez, L., Voisin, C., Kan, L.-S. and Cadet, J., 1988, (6-4) Bipyrimidine dinucleoside-monophosphate photoproducts:

characterization and photochemical properties, 10th
International Conference on Photobiology, Jerusalem, Book
of Abstracts, p.83.

Wang, S.Y., ed., 1976, in: "Photochemistry and Photobiology of
Nucleic Acids", Academic Press, New York, Vol. I & II.

Weinfeld, M., Gentner, N.E., Johnson, L.D. and Paterson, M.C.,
1986, Photoreversal-dependent release of thymidine and
thymidinemonophosphate from pyrimidine dimer-containing
DNA excision fragments isolated from ultraviolet-damaged
human fibroblasts, Biochemistry, 25:2656-2664.

Weiss, R.B. and Duker, N.J., 1986, Photoalkylated DNA and
ultraviolet-irradiated DNA are incised at cytosines by
endonuclease III, Nucleic Acids Res., 14:6621-6631.

MAMMALIAN DNA REPAIR MUTANTS: RECENT PROGRESS IN THE GENETIC
AND MOLECULAR BIOLOGY OF THE PROCESSING OF DNA LESIONS INDUCED
BY UV AND BY DNA CROSS-LINKING AGENTS

Ethel Moustacchi

Institut Curie - Biologie, URA 1292 du CNRS, 26 Rue
d'Ulm, F-75231 Paris cedex 05 (France)

INTRODUCTION

A variety of enzymatic mechanisms exist for repairing or
tolerating lesions in cellular DNA, thereby promoting cell
survival. Most of the models of these processes are derived
from analysis of bacterial and yeast mutants as opposed to
wild type cells. More recently, the analysis of some
potential type of hereditary DNA repair defects has been
extended to rodent and human cells. Such studies on mammalian
cells led to two important notions:

a) A direct correlation between unrepaired DNA damage
and carcinogenesis in humans was observed following the
discovery that the cancer-prone hereditary disease, xeroderma
pigmentosum (XP), involved a defect in the incision of
pyrimidine dimers produced by 254 nm ultraviolet radiation
(UV) (for review Cleaver, 1983). Since then, other hereditary
diseases with predisposition to cancer have been shown to be
altered in their processing of DNA lesions induced by physical
and/or chemical agents. Such alterations can be accompanied
by either hypermutability at specific loci as in the case of
XP (Maher and McCormick, 1976) or by hypomutability as in the
case of ataxia telangectasia (AT) (Arlett, 1980) or of
Fanconi's anemia (FA) (Papadopoulo et al., 1990a). In such
diseases, however, a chromosomal instability, as revealed by
high frequencies of spontaneous and induced chromosomal
aberrations, has been reported. In connection with this
general aspect, it is becoming of great interest to determine
whether unrepaired or faulty repaired DNA lesions can lead to
activation of protooncongenes implicated in subsequent
tumorigenesis.

b) Various degrees of importance have been reported for
the different types of DNA lesions and the efficiency of their
repair by the excision-resynthesis process at particular
genomic sites, according to the location of these sites
(replications fork, nuclear matrix attachment), the chromatin

structure, the level of gene expression (active against inactive transcription sites), etc. As an example, it can be mentioned that similarly to XP, Cockayne's syndrome (CS) cells are hypersensitive to the lethal and mutagenic action of UV light. However, they demonstrate a specific deficiency in the ability to incise damage in actively transcribed regions of DNA (Mayne et al., 1988), the overall repair of UV lesions being normal. In CS no particular predisposition to cancer is observed.

A BRIEF OUTLINE OF THE DNA REPAIR PATHWAYS (for details see Friedberg, 1985)

Excision Repair

This involves incision by endonucleas(s) on both sides of a DNA adduct, filling in of the gap by polymerase(s) and rejoining with ligase(s). UV light, numerous chemical carcinogens, DNA cross-linking alkylating (mitomycin C, cis-platinum, bifunctional nitrogen mustards) or non alkylating (bifunctional psoralens plus 365 nm radiation) agents lead to lesions recognised by the excision repair system. It is accepted that the unscheduled DNA synthesis or UDS, as determined by ^{3}H-thymidine incorporation in non-S phase DNA following treatment, measures the gap filling step. Glycosylases recognise specific abnormal bases, such as alkylated adenine and guanine residues which are excised to leave apurinic sites (methylating agents for instance induce such altered bases). These are further processed by endonucleases followed by the further steps of the excision repair process.

It is generally believed that this pathway is error-free. When it is blocked as in rad3 type mutants of Saccharomyces cerevisiae or in XP human cells, the frequency of UV-induced mutants is increased.

Other error-free repair mechanisms

UV-induced pyrimidine dimers of the cyclo-butane type are directly reversed to pyrimidines by the combined action of visible light and the photolyase. On the other hand, a specific protein which directly dealkylates the O^6 position of guanine residues in DNA has been characterized in both bacteria and eukaryotes.

Recombinabtion repair

This involves strand exchange between either chromosomes or chromatids at the replication fork (also called in that case post-replication repair). It is implicated in double-strand break repair, i.e. in damage induced by X-rays and by X-mimetic drugs such as bleomycin. Following incision, this process also plays a role in DNA interstrand cross-link repair at least in bacteria and yeast. When this system is blocked as in rec A bacteria or in the rad52 type mutants of S. cerevisiae, recombination cannot be induced.

mispaired bases and preferentially repairs GT in favour of GC. When this system is blocked as in mut H, L etc. mutants of E. coli, a spontaneous mutator activity is detected and a hypersensitivity to cis-platinum and methylating agents is found.

In general, most cellular responses to DNA damage are constitutive while some specific responses appear to be induced by the replication fork blockage in the presence of lesions. The mechanism for the inducible SOS response is well understood in E. coli. Evidence for inducible responses in mammalian cells have been reported, although the real mechanism involved is still debated. It is clear that, in eucaryotic cells, such responses cannot be equated to the SOS processes (Rossman and Klein, 1985; Sarasin, 1985).

THE RODENT CELL LINES MUTANTS

A large variety of rodent cell lines which are defective in DNA repair have been isolated and are being used to identify the specific DNA repair step that is altered. Following mutagenization of the parent cellular population, mutants specifically sensitive to UV radiation damage or to ionizing radiation and/or to specific chemical agents have been selected. The cell lines most commonly isolated derive from Chinese hamster (CHO, V-79) or from murine (mouse lymphoma) lines. From the mutant isolation frequencies, it has been suggested that 100 to 200 genes are likely to be involved in the processing of DNA lesions in mammalian cells. This is not surprising since, in the yeast S. cerevisiae, a simpler unicellular eucaryote, around 100 genes specifically involved in processing of DNA lesions have already been identified.

Genetic classification

Using the somatic hybridization technique, Chinese hamster mutants, initially selected by their hypersensitivity to UV radiation, have been classified into 8 complementation groups (Thompson et al., 1987; Zdzienicka et al., 1988). Their UV sensitivity ranges from a factor of 2 to a factor of 6 when compared to normal cell lines. Most of these mutants demonstrate a cross-sensitivity to mitomycin C ranging from a factor of 2 to 90 (for instance, the mutants UV20 or UV41 belonging to complementation groups 1 and 4, respectively). Similarly, the CHO mutants selected for their hypersensitivity to ionizing radiations have been classified into 9 genetic complementation groups. Other mutants selected on the basis of their sensitivity to antitumoral drugs such as mitomycin C, bleomycin or adriamycin are available. Their cross-sensitivity profiles to other agents provide insights on the nature of their DNA repair defect. For mitomycin C hypersensitivity, 4 complementation groups have been identified in CHO cell lines (for review see Hickson and Harris, 1988) whereas only 2 complementation groups have been characterized in mouse lymphoma mutants (Hama-Inaba et al., 1983, 1988).

DNA repair defect

Mutants from groups 1 to 5 and from group 7 in Chinese hamster demonstrate a reduced capacity for UDS after treatment with UV radiation associated to a defect in the incision step of the excision repair of bulky DNA adducts. In this respect, these mutants are phenotypically similar to the classical XP cell lines. A mutant belonging to the complementation group 6, UV61, shows an almost normal rate of UDS after irradiation (Thompson et al., 1987). Interestingly, it was recently shown that this mutant normally removes the (6-4) photoproducts while cyclobutane type pyrimidine dimers are not. Since this mutant is more resistant to UV than mutants from the 1 to 5 genetic complementation groups, it is concluded that the (6-4) photoproducts are both cytotoxic and mutagenic in CHO cells and that the UDS observed early after UV-irradiation results from the repair of (6-4) photoproduct rather than cyclobutane dimers (Thompson et al., 1989). This study led also to the interesting suggestion that the (6-4) photoproducts may act to block cell division by arresting replication whereas the cyclobutane dimers may inhibit the expression of essential genes.

No characterization of the repair defect in the mutant of the 8th complementation group has yet been reported.

The cloning of human DNA repair genes by using rodent mutants

Human genomic DNA or a cDNA library in the suitable expression vector can be transfected into hamster or mouse mutant cells with relatively high efficiency. Restoration of normal cellular response to a cytotoxic agent (UV, mitomycin C, etc.) implies complementation of the repair defect in the host cell. This approach has led to the first successful cloning of a human DNA repair gene ERCCI (Rubin et al., 1985; Westerveld et al., 1984). Following transfection of human genomic DNA into the UV sensitive CHO mutant 43-3B (genetic complementation group 1), primary and secondary transformants with a normal survival to the cytotoxic effect of UV were isolated and the gene ERCCI involved in excision repair was characterized. The gene is 15 kb long (the corresponding protein includes 297 amino acids) with 10 exons, one of which can be alternatively spliced. The complete sequence of the ERCCI cDNA has been determined.

Since then, 5 other human genes involved in excision repair have been cloned according to the same methodology: ERCCI to ERCC5 correspond to the first five UV genetic complementation groups in hamster cells (groups 1 to 5). Two other human genes XRCCI and XRCC2 (for X-ray repair cross complementing) complement the EM9 and irs1 mutants. EM9, which is hypersensitive to X-ray, to methyl and ethyl methane sulfonate and chlorodeoxyuridine, has a greatly reduced efficiency in rejoining single strand breaks induced by ionizing radiation and chemicals. irs1 is two-fold hypersensitive to killing by ionizing radiation or by UV and it is 100 fold hypersensitive to mitomycin C. The precise nature of the defect is unknown.

Genes ERCC1 and ERCC2 are both located on the human chromosome 19. ERCC3, ERCC4 and ERCC5 are located on

chromosomes 2, 16 and 13, respectively. XRCC1 and XRCC2 are
located on human chromosomes 19 and 17 (Thompson et al., 1987;
Thompson, pers. commun.).

An extensive homology has been found between the deduced
amino acid sequence of the ERCC1 protein and that of the
cloned yeast repair gene RAD10 involved in excision repair.
34% identity in the region of highest homology, likely to
correspond to DNA-binding domains, was found. An additional
homology (31%) with the E. coli Uvr A protein was observed at
the carboxy-terminal end of ERCC1, beyond the region
homologous with RAD10 (Van Duin et al., 1986). More recently
it has been shown that the protein coding region of ERCC2
(Weber et al., 1988) has a 52% amino-acid identity with the
RAD3 gene in yeast (Weber, pers. commun.). The RAD3 gene is
essential in yeast and is involved in the incision step of the
excision repair process. It is therefore possible that ERRC2
is an essential gene.

Such similarities suggest conservation of function from
bacteria through yeast to man and point to the importance in
the cellular economy of the processes involved. None of the
human repair genes cloned at the moment complement any of the
human syndromes. Several of them are being currently tested
for their ability to correct XP cells (Van Duin et al., 1989).
Indeed all the processes necessary to achieve cloning of human
repair genes would be considerably facilitated by using mutant
rodent cells.

THE HUMAN HEREDITARY SYNDROMES WITH DEFECTS IN PROCESSING OF
DNA LESIONS

A number of excellent reviews deal with cancer-prone
hereditary syndromes with abnormalities in processing of DNA
lesions (Friedberg et al., 1979; Hanawalt and Sarasin, 1986).
We will focus here only on recent progress in relation to the
genetic and molecular biology of two of such diseases
xeroderma pigmentosum (XP) and Fanconi's anemia (FA).

Xeroderma pigmentosum

Cultured fibrolasts derived from XP patients display
hypersensitivity to the lethal, mutagenic and carcinogenic
effects of UV light (McCormick et al., 1986 and references
therein). These abnormal responses are associated in the nine
genetic complementation groups (designated A to I) with a
defect in the incision step of nucleotide excision repair
which acts to remove bulky DNA lesions (Kraemer et al., 1987).
The tenth complementation group, called variant, demonstrates
a deficiency in post-UV DNA processing (Lehmann et al., 1975)
and recent studies have suggested that it is also defective in
excision reapir (Wood et al., 1988). An understanding of the
roles these genes have in physiology awaits gene cloning.
Towards this goal different strategies have been pursued
including the introduction of tagged genetic information into
XP recipient cells either as DNA or chromosomes followed by
selection of marker(s) and increased resistance to UV
cytotoxic effects. The tendency of XP(A) lines to revert to
UV-resistance during the process of transfection and selection
has hampered for a while the cloning of gene(s) involved in

restoration of excision-repair (Royer-Pokora and Haseltine, 1984; Schulz et al., 1985). However, two recent reports have claimed restoration of UV-resistance in XP (complementation groups C and A, respectively) by either transfection of human cDNA clone library in the suitable vector (Teitz et al., 1987) or mouse genomic DNA (Tanaka et al., 1989). In this case, plasmid pSV2 gpt and mouse embryo DNA were co-transfected into an SV-40 transformed XP (group A) cell line. From about 1.6 x 10^5 pSV2 gpt-transformed colonies, two primary UV-resistant XP transfectants were isolated. An intermediate restoration of both UV-survival and of DNA repair ability were observed. This might be due to the heterologous nature (mouse DNA in human cells) of the transferred DNA. DNA from such primary transfectants was again transfected into XP cells and one secondary transfectant UV-resistant was isolated by screening 4.8 x 10^5 pSV2 gpt-transformed XP colonies. Such secondary transfectants retained fewer mouse repetitive sequences. A mouse gene that complements the defect of XP group A cells but not groups C, D, F and G was molecularly cloned. Northern blot analysis of poly(A)+RNA with a subfragment of cloned mouse DNA repair gene as the probe revealed that a 1 kilobase mRNA was transcribed in the donor mouse embryo, the XP secondary transfectants and in normal human cells (1 kb and 1.3 kb mRNA in that case). In the three XP group A cell lines examined, none of these mRNA was detected. In conclusion, the mouse homologue of the group A XP human gene has now been cloned. Interestingly, the size of mRNA that restored a normal level of UDS, following microinjection of poly(A)+RNA from normal cells into XP group A cells, was 1.2-1.4 kb in size (Hoeijmakers et al., 1988).

On the other hand, SV-40 transformed XP(A) have been converted to a UV-resistant phenotype by the introduction of either isolated human chromosomes via minicell fusion (Schultz et al., 1987) or chinese hamster (CHO) chromosomes via fusion with lethally X-irradiated CHO cells (Karentz et al., 1987) or with chinese hamster/human hybrid cytoplasts (Keijzer et al., 1987). Attempts to localize the genes, involved in complementation of the defect in XP using this last approach have not yet led to definitive answers. As mentioned above, specific human chromosomes were found to restore UV-resistance (or Xray) and DNA repair in each of the seven complementation groups of the repair-deficient hamster cells whereas all XP complementation groups that were tested appear to restore repair in all the CHO mutants (Thompson et al., 1988 and pers. commun.). Although sharing phenotypic similarities, these observations imply that none of the CHO mutants can be genetically equated to XP at the moment. On the other hand, normal human cells contain the genetic information necessary to supply the functions defective in the chinese hamster mutants.

Another approach examines the ability of cell extracts to nick irradiated plasmid DNA or to catalyse DNA synthesis on UV-irradiated plasmid DNA (Wood et al., 1988). In view of the long history of solution of the excision repair system in E. coli, including the recent unsuspected discovery concerning the functioning of the Uvr ABC complex (Orren and Sancar, 1989), it is not surprising that this promising approach with human cells extracts is still full of complications (see for discussion Kaufmann, 1988).

Fanconi's anemia

This inherited autosomal recessive disorder in man, is characterized by chromosomal instability, pancytopenia and bone marrow insufficiency which usually develops between 4 and 10 years of age and predisposition to malignancy, usually leukemia (for review see Schroeder-Kurth et al., 1989). Cells from individuals with FA have an increased sensitivity to agents that cause DNA interstrand cross-links (CL) such as mitomycin C (CL's between purines) or psoralens plus UVA (CL's between pyrimidines). It has been reported that FA cells in culture were only marginally susceptible to radiations or to monofunctional alkylating agents and it has been claimed that the primary defect in FA may be in the incision of DNA interstrand CL. However, this was not supported by the work of others. Genetic and phenotypic heterogeneity was recently demonstrated in FA. Indeed by somatic hybridization 2 complementation groups A and B were characterized (Duckworth-Rysiecki et al., 1985). Group B is indistinguisable from the normal type following treatment with a bifunctional psoralen (8-methoxypsoralen) plus UVA in terms of recovery of a normal rate of DNA semi-conservative synthesis (Moustacchi et al., 1987) and of incision of CL's as measured by alkaline elution (Papadopoulo et al., 1987) or by electron microscopic vizualisation of CL (Rousset et al., 1990).

In contrast to normal and group B cells, after the same treatment, FA cells from the genetic complementation group A do not demonstrate a recovery of a normal rate of DNA semi-conservative synthesis and are less efficient in the incision step. In terms of clonogenic cell survival, the group A cell lines are the most sensitive compared to normal ones and group B cells show an intermediate sensitivity when they are treated by bifunctional psoralens in combination with UVA. To our surprise in conditions in which solely monoadducts are induced, i.e. a treatment with 4,5',8-trimethylpsoralen (TMP) and 405 nm radiation, FA cells are more sensitive than normal cells to the cytotoxic effect of this agent. In contrast to the response to a cross-linking treatment, group B cells are more sensitive than group A cells which in turn are more sensitive than normal ones. Both FA cell lines, however, were found to be less able to repair TMP plus 405 nm induced cross-linkable monoadducts than normal cells (Averbeck et al., 1988). This may account for the hypersensitivity of such cell lines to the cytotoxic effect of TMP plus 405 nm.

In contrast to XP, the mutagenesis for 6-thioguanine resistance (6-TG^R), as measured by the limiting dilution technique in FA groups A and B lymphoblastoid cell line following treatment with bifunctional psoralens and either 365 nm (mixture of CL and monoadducts) or 405 nm (only monoadducts), was extremely reduced in FA relative to normal cells. Such hypomutability is true whether the frequencies of mutants are plotted against dose or survival level (Papadopoulo et al., 1990a). Moreover, mutagenesis studies in the Herpes virus thymidine kinase gene following the same DNA cross-linking treatment showed that the mutation rate in the progeny of psoralen-photodamaged viruses was unchanged after infection of FA cells, whereas it increased with dose in infected normal cells (Coppey et al., 1989). Interestingly, FA and ataxia telangectasia cells not only share the property

469

of being hypomutable but they also demonstrate aberrant
enhanced viral reactivation when infected with psoralen
photodamaged Herpes virus as in the case of FA (Coppey et al.,
1989) or with X or UV-damaged adenovirus as in the case of AT
(Bennett and Rainbow, 1988). Taken together with high
frequencies of chomosomal abnormalities in these two
syndromes, these observations suggest an alteration in
recombinational (or strand breaks) repair. In accordance with
this assumption, recent results from our group (Papadopoulo et
al., 1990b) indicate that the molecular nature of 6-TG^R
mutants differ in normal and FA cells. By Southern blot
analysis, it is shown that the major part of spontaneous and
psoralen photoinduced mutants at the HGPRT locus result from
large rearrangements in FA cells as opposed to point mutations
in normal cells. The FA cells may be then considered as
defective in an error-prone system which leads to modification
lower than 200 base pairs. Recombination repair would be
abortive and lead essentially to death of putative mutants
resulting in a low frequency of mutants at specific loci.

As a step to the cloning of the FA gene, complementation
of FA cells by either human genomic DNA (Diatloff-Zito et al.,
1986) or hamster DNA (Shaham et al., 1987) of the wild type
has been achieved. FA cells transfected with their own DNA in
the same condition did not demonstrate a corrected phenotype
(Diatloff-Zito et al., 1986). As in the case of XP, in some
instances the cells isolated after transfection of FA SV-40
transformed cell lines appear to be revertants rather than
true transfectants (Buchwald et al., 1987). The cells
demonstrated only partial (25 to 90% of wild type) resistance
to mitomycin C and did not contain the rodent DNA in the high
molecular weight DNA fraction.

The partial correction of mitomycin C sensitivity of FA
from group B cells by transfection of DNA from mouse lymphoma
L5178Y cells has been observed (Diatloff-Zito et al., 1990).
Southern blotting demonstrated the presence of DNA repetitive
sequences in FA primary transfectants and the response to the
cross-linking agent was stable for at least 8 months after the
transfection, while many of the mouse DNA sequences were lost
during that time. Secondary transfections still displayed
resistance to mitomycin C but to a lesser extent than the
primary ones. For the molecular cloning of the gene a genomic
library in lambda phage has been prepared from the primary
transfectants and four recombinant clones were found to
correct the FA group B defect (Diatloff-Zito, pers. commun.).
These clones are now being analysed. It is of interest to
note that microinjection of fractionated mRNA from Hela cells
into FA (group A) corrected their response to a psoralen
cross-linking challenge (Digweed and Sperling, 1989), namely
permanent inhibition of DNA synthesis. The mRNA responsible
for this effect contained around 650 bases which suggests that
the expected size of the coded polypeptide is in the order of
18 kDa. The estimated size of the DNA (0.6 kb) is compatible
with cloning in lambda phage. The existence of mouse mutants
which share with FA a number of phenotypic features (Hama-
Inaba et al., 1989) allows us to apply a strategy similar to
that used for the cloning of the ERCC genes. This possibility
is actually explored in our group. On the other hand, the
recent cloning of the yeast gene SNM1 (allelic to PSO2) (Haase
et al., 1989) opens new directions for the cloning of FA

genes. Indeed the snm1 (pso2-1) mutant displays most of the
cellular features characteristic of FA (Moustacchi et al.,
1983).

Finally, it should be mentioned that co-cultivation of FA
group A cells with either normal or mutant (MC$_{\pi}^{s}$ V) mouse cells
allows correction of the FA defect in terms of mitomycin C-
induced chromosomal aberrations (Rosselli and Moustacchi,
1990). This implies that a diffusible factor of relatively
low molecular weight produced by wild type cells allows
transient complementation.

CONCLUSION

In spite of the many pitfalls encountered, the molecular
cloning of the human ERCC genes and of XP genes has now been
achieved. Recent progress along the same line for the FA
gene(s) allows us to think that this goal will be reached also
for this other disease. However, much remains to be done in
order to identify the proteins involved and to understand
their regulation and precise mode of action in the processing
of DNA lesions and in the essential functions related to
cancer predisposition. The cloning taken together with the
developments of in vitro systems which permit the study of the
excision repair and the recombinogenic activities using human
cell extracts as well as the determination at the sequence
level of the nature of spontaneous and induced mutants in
normal and repair defective cells, opens new perspectives for
a better understanding of the family of cancer-prone
hereditary syndromes.

Note added in proofs: Since submission of this paper, several
new articles of interest especially in relation with the
cloning of XP genes have been published. These include Arrand
et al., 1989, Proc. Natl. Acad. Sci. USA, 86, 6997-7001; Teitz
et al., 1990, Gene, 87, 295-298 and Mutation Res., 1990, 236,
85-97.

ACKNOWLEDGEMENTS

The author is grateful to Drs. C. Diatloff-Zito, D.
Papadopoulo and C. Guillouf for communicating unpublished
results. This work was supported by grants from CEE (n° B10-
151F), ARC (No. 6381) and Ligue française contre le Cancer.

REFERENCES

Arlett, C.F., 1980, Mutagenesis in repair-deficient human cell
        strains, in: "Progress in Environmental Mutagenesis", M.
        Alacevic, Ed., Elsevier, North Holland Biochemical Press,
        Amsterdam.
Averbeck, D., Papadopoulo, D., and Moustacchi, E., 1988,
        Repair of 4,5',8-trimethylpsoralen plus light UVA damage
        in normal and Fanconi's anemia cell lines. Cancer Res.,
        48:2015.
Bennett, C.B., and Rainbow, A.J., 1988, Delayed expression of
        enhanced reactivation and decreased mutagenesis of UV-
        irradiated adenovirus in UV-irradiated ataxia

telangiectasia fibroblasts. <u>Mutagenesis,</u> 3:389.

Buchwald, M., Clarke, C., and Duckworth-Rysiecki, G., 1987, Studies of gene transfer and reversion to mitomicyn C resistance in Fanconi anemia cells. <u>Mutation Res.</u>, 184:153.

Cleaver, J.E., 1983, Xeroderma pigmentosum, <u>in</u>: "The Metabolic Basis of Inherited Disease", J. V. Stanbury, J. B. Wyngaarden, D.S. Frederickson, J. L. Goldstein and M.S. Brown, eds., 5th edn. McGraw Hill, New York.

Coppey, J., Sala-Trepat, M., and Lopez, B., 1989, Multiplicity reactivation and mutagenesis of trimethylpsoralen-damaged Herpes Virus in normal and Fanconi's anaemia cells. <u>Mutagenesis</u>, 4:67.

Diatloff-Zito, C., Papadopoulo, D., Averbeck, D., and Moustacchi, E., 1986, Abnormal response to DNA crosslinking agents of Fanconi anemia fibroblasts can be corrected by transfection with normal human DNA. <u>Proc. Natl. Acad. Sci. USA,</u> 83:7034.

Diatloff-Zito, C., Rosselli, F., Heddle, J. and Moustacchi, E., 1990, Partial complementation of the Fanconi anemia defect upon transfection by heterologous DNA. Phenotypic dissociation of chromosomal and cellular hypersensitivity to DNA cross-linking agents, <u>Human Genet.</u>, in press.

Digweed, M. and Sperling, K., 1989, Identification of a HeLa mRNA fraction which can correct the DNA-repair defect in Fanconi anaemia fibroblasts, <u>Mutation Res.</u>, 218:171.

Duckworth-Rysiecki, G., Hulten, M., Mann, J., and Taylor A.M.R., 1985, Identification of two complementation groups in Fanconi's anemia. <u>Somat. Cell. Mol. Genet.</u>, 11:35.

Friedberg, E.C., 1985, DNA repair. W.H. Freeman and Company, San Francisco.

Friedberg, E.C., Ehmann, U.K., and Williams, J.I., 1979, Human diseases associated with defective DNA repair. <u>Adv. Radiat. Biol.</u>, 8:85.

Haase, E., Riehl, D., Mack, M. and Brendel, M., 1989, Molecular cloning of <u>SNM1</u>, a yeast gene responsible for a specific step in the repair of cross-linked DNA, <u>Mol. Gen. Genet.</u>, 218:64.

Hama-Inaba, H., Hieda-Shiomi, N., Shiomi, T., and Sato, K., 1983, Isolation and characterization of mitomycin-C-sensitive mouse lymphoma cell mutants. <u>Mutation Res.</u>, 108:405.

Hama-Inaba, H., Sato, K., and Moustacchi, E., 1988, Survival and mutagenic responses of mitomycin C-sensitive mouse lymphoma cell mutants to other DNA cross-linking agents. <u>Mutation Res.</u>, 194:121.

Hanawalt, P.C., and Sarasin, A., 1986, Cancer-prone hereditary diseases with DNA processing abnormalities. <u>Trends in Genetics</u>, 2:124.

Hickson, I.D., and Harris, A.L., 1988, Mammalian DNA repair use of mutants hypersensitive to cytotoxic agents. <u>Trends in Genetics</u>, 4:101.

Hoeijmakers, J.H.J., Van Duin, M., Weeda, G., Van der Eb, A.J., Troelstra, C., Eker, A.P.M., Jaspers, N.G.J., Westerveld, A., and Bootsma, D., 1988, Analysis of mammalian excision repair: from mutants to genes and gene products, <u>in</u>: "Mechanisms and Consequences of DNA Damage Processing", E.C. Friedberg and P.C. Hanawalt, eds., UCLA symposium on Molecular and Cellular Biology, New Series, vol. 83, Alan R. Liss, Inc., New York.

Karentz, D., Mitchell, D., and Cleaver, J.E., 1987, Correction of excision repair in zeroderma pigmentosum by hamster chromosome fragments involves both classes of pyrimidine dimers. Som. Cell. Mol. Genet., 13:621.

Kaufmann, W.K., 1988, In vitro complementation of xeroderma pigmentosum. Mutagenesis, 3:373.

Keijzer, W., Stefanini, M., Bootsma, D., Verkerk, A., Geurts van Kessel, A.H.M., Jongkind, J.F., and Westerveld, A., 1987, Localization of a gene involved in complementation of the defect in xeroderma pigmentosum group A cells on human chromosome. Exp. Cell. Res., 169:490.

Kraemer, K.H., Lee, M.M., and Scotto, J., 1989, Xeroderma pigmentosum. Arch. Dermatol., 123:241.

Lehmann, A.R., Kirk-Bell, S., Arlett, C.F., Paterson, M.C., Lohman, P.H.M., De Weerd-Dastelein, E.A., and Bootsma, D., 1975, Xeroderma pigmentosum cells with normal levels of excision have a defect in DNA synthesis after UV-irradiation. Proc. Natl. Acad. Sci. USA, 72:219.

Maher, V.M., and McCormick, J.J., 1976, Effect of DNA repair on the cytoxicity and mutagenicity of UV irradiation and of chemical carcinogens in normal and xeroderma pigmentosum cells, in: "Biology of Radiation Carcinogenesis", J.M. Yuhas, R.W. Tennant and J.D. Regan, eds., Raven Press, New York.

Mayne, L.V. Mullenders, L.H.F., and Van Zeeland, A.A., 1988, Cockayne's syndrome: a UV-sensitive disorder with a defect in the repair of transcribing DNA but normal overall excision repair, in: "Mechanisms and Consequences of DNA Damage Processing", UCLA Symposium on molecular and cellular biology, New Series, vol. 83, Alan R. Liss, Inc., New York.

McCormick, J., Kately-Kohler, S., Watanabe, M., and Maher, V., 1986, Abnormal sensitivity of human fibroblasts from zeroderma pigmentosum variants to transformation to anchorage independence by ultraviolet radiation. Cancer Res., 46:489.

Moustacchi, E., Cassier, C., Chanet, R., Magana-Schwencke, N., Saeki, T., and Henriques, J.A.P., 1983, Biological role of photo-induced cross-links and monadducts in yeast DNA: Genetic control and steps involved in their repair, in: "Cellular Responses to DNA Damage", Alan R. Liss, Inc., New York.

Moustacchi, E., Papadopoulo, D., Diatloff-Zito, C., and Buchwald, M., 1987, Two complementation groups of Fanconi's anemia differ in their phenotypic response to a DNA-crosslinking treatment. Hum. Genet., 75:45.

Orren, D.K., and Sancar, A., 1989, The (A)BC excinuclease of Escherichia coli has only the UvrB and UvrC subunits in the incision complex. Proc. Natl. Acad. Sci. USA, 86:5237.

Papadopoulo, D., Averbeck, D., and Moustacchi, E., 1987, The fate of 8-methoxypsoralen-photoinduced DNA interstrand cross-links in Fanconi's anemia cell of defined genetic complementation group. Mutation Res., 184:271.

Papadopoulo, D., Porfirio, B. and Moustacchi, E., 1990a, Mutagenic response of Fanconi's anemia cells from a defined complementation group after treatment with photoactivated bifunctional psoralens, Cancer Res., 50:3289.

Papadopoulo, D., Guillouf, C., Mohrenweiser, H. and Moustacchi, E., 1990b, Hypomutability in Fanconi anemia cells is associated with increased deletion frequency at the <u>HPRT</u> lucus, <u>Proc. Natl. Acad. Sci. USA</u>, 87, in press.

Rosselli, F. and Moustacchi, E., 1990, Cocultivation of Fanconi anemia cells and of mouse lymphoma mutants leads to interspecies complementation of chromosomal hypersensitivity to DNA cross linking agents, <u>Human Genet.</u>, 84:517.

Rossmann, T.G., and Klein C.B., 1985, Mammalian SOS system: a case of misplaced analogies. <u>Cancer Invest.</u>, 3:175.

Rousset, S., Nocentini, S., Revet, B. and Moustacchi, E., 1990, Molecular analysis by electron microscopy of the removal of psoralen-photoinduced DNA cross-links in normal and Fanconi's anemia fibroblasts, <u>Cancer Res.</u>, 50:2443.

Royer-Pokora, B., and Haseltine, W.A., 1984, Isolation of UV-resistant revertants from a xeroderma pigmentosum complementation group A cell line. <u>Nature</u>, 311:390.

Rubin, J.S., Prideaux, V.R., Willard, H.F., Dulhanty, A.M. Whitmore, G.F., and Bernstein, A., 1985, Molecular cloning and chromosomal localization of DNA sequences associated with a human DNA repair gene. <u>Mol. Cell. Biol.</u>, 5:398.

Sarasin, A., 1985, SOS response in mammalian cells. <u>Cancer Invest.</u>, 3:163.

Schultz, R.A., Barbis, D.P., and Friedberg, E.C., 1985, Studies on gene transfer and reversion to UV resistance in xeroderma pigmentosum cells. <u>Som. Cell. Genet.</u>, 11:67.

Schultz, R.A., Saxon, P.J., Glover, T.W., and Friedberg, F.C, 1987, Microcell-mediated transfer of a single human chromosome complements xeroderma pigmentosum group A fibroblasts. <u>Proc. Natl. Acad. Sci. USA</u>, 84:4176.

Schroeder-Kurth, T.M., Auerbach, A.D., and Obe, G., (Eds), 1989, Fanconi anemia: Clinical cytogenetic and experimental aspects. Springer-Verlag, Berlin.

Shaham, M., Adler, B., Ganguly, S., and Changanti, R.S.K., 1987, Transfection of normal human and Chinese hamster DNA correcte diepoxybutane-induced chromosomal hypersensitivity of Fanconi anemia fibroblasts. <u>Proc. Natl. Acad. Sci. USA</u>, 84:5853.

Tanaka, K., Satokata, I., Ogita, Z., Uchida, T., and Okada, Y., 1989, Molecular cloning of a mouse DNA repair gene that complements the defect of group A xeroderma pigmentosum. <u>Proc. Natl. Acad. Sci. USA</u>, 86:5512.

Teitz, T., Naiman, T., Avissar, S.S., Bar, S., Okayama, H., and Canaani, D., 1987, Complementation of the UV-sensitive phenotype of a xeroderma pigmentosum human cell line by transfection with a cDNA clone library. <u>Proc. Nat. Acad. Sci. USA</u>, 84:8801.

Thompson, L.H., Salazar, E.P., Brookman, K.W., Collins, C.C. Stewart, S.A., Busch, D.B., and Weber, C.A., 1987, Recent progress with the DNA repair mutants of Chinese hamster ovary cells. <u>J. Cell. Sci.</u>, suppl 6:97.

Thompson, L.H., Shiomi, T., Salazar, E.P., and Stewart, S.A., 1988, An eighth complementation group of rodent cells sensitive to ultraviolet radiation. <u>Somat. Cell. Mol. Genet.</u>, 14:605.

Thompson, L.H., Mitchell, D.L., Regan, J.D., Bouffer, S.D., Stewart, S.A., Carrier, W.L., Nairn, R.S., and Johsnon, R.T., 1989, CHO mutant UV61 removes (6-4) photoproducts but not cyclobutane dimers. <u>Mutagenesis</u>, 4:140.

Van Duin, M., De Wit, J., Odijk, H., Westerveld, A., Yasul,
    A., Koken, M.H.M., Hoeijmakers, and Bootsma, D., 1986,
    Molecular characterization of the human excision repair
    gene ERCC-1 : cDNA cloning and amino acid homology with
    the yeast DNA repair gene RAD10.Cell, 44:913.
Van Duin, M., Vredeveldt, G., Mayne, L.V., Odijk, H.,
    Vermeulen, W., Klein, B., Weeda, G., Hoeijmakers, J.H.J.,
    Bootsma, D., and Westerveld, A., 1989, The cloned human
    DNA excision repair gene ERCC-1 fails to correct
    xeroderma pigmentosum complementation groups A through I.
    Mut. Res., 217:83.
Weber, C.A., Salazar, E.P., Stewart, S.A., and Thompson, L.H.,
    1988, Molecular cloning and biological characterization
    of a human gene, ERCC2, that corrects the nucleotide
    excision repair defect in CHO UV5 cells. Mol. Cell.
    Biol., 8:1137.
Westerveld, A., Hoeijmakers, J.H.J., Van Duin, M., De Wit, J.,
    Odijk, H., Pastink, A., and Bootsma, D., 1984, Molecular
    cloning of a human DNA repair gene. Nature, 310:425.
Wood, R.D., Robins, P., and Lindahl, T., 1988, Complementation
    of the xeroderma pigmentosum DNA repair defects in cell-
    free extracts. Cell, 53:97.
Zdzienicka, M.Z., van der Schans, G.P., and Simons, J.W.I.M.,
    1988, Identification of a new seventh complementation
    group of UV-sensitive mutants in Chinese hamster cells.
    Mutation Res., 194:165.

POLY ADP-RIBOSYLATION IN DNA DAMAGE PROCESSING

Felix R. Althaus, Pius Loetscher, Georg Mathis,
Hanspeter Naegeli, Phyllis Panzeter
and Claudio Realini

University of Zurich-Tierspital, Institute of
Pharmacology and Biochemistry, Winterthurerstrasse
260, CH-8057 Zurich, Switzerland

INTRODUCTION

Poly ADP-ribosylation is a posttranslational protein
modification with ubiquitous distribution among higher
eukaryotes. The biological function of this reaction is not
well understood, but numerous lines of evidence suggest that
it acts to modulate chromatin functions, particularly DNA
repair (for review see Althaus and Richter, 1987). The
eukaryotic poly ADP-ribosylation system is composed of three
enzymatic components: i) The DNA-dependent enzyme poly(ADP-
ribose)polymerase [EC2.4.2.30], which utilizes the respiratory
coenzyme NAD^+ as the substrate for the biosynthesis of a
homopolymer composed of ADP-ribosyl residues. We and others
(Leduc et al., 1986; Mathis and Althaus, 1987) have found this
enzyme to be closely associated with the nucleosomal core.
ii) The enzyme poly(ADP-ribose)glycohydrolase, which degrades
ADP-ribosyl polymers in an exoglycosidic reaction mode, and
iii) the enzyme ADP-ribosyl protein lyase, which removes the
protein-proximal ADP-ribosyl residue. The total ADP-ribose
processing capacity of this system in mammalian cells is quite
impressive, i.e. a total of 10 million ADP-ribosyl residues
per min per cell for poly(ADP-ribose)polymerase and likewise
for poly(ADP-ribose)glycohydrolase, and 72 million residues
for ADP-ribosyl protein lyase (for review see Althaus and
Richter, 1987). Recently, we have obtained evidence that
protein-bound ADP-ribosyl polymers may cause the release of
core DNA fragments from nucleosomal core particles (Mathis and
Althaus, 1987). In the present study, we have analyzed the
molecular properties of ADP-ribosyl polymers responsible for
this phenomenon. In addition, we have reconstituted an in
vitro poly ADP-ribosylation system with the goal of studying
the role of the enzyme poly(ADP-ribose)polymerase as a
possible modulator of DNA-protein interactions under defined
experimental conditions. This in vitro system has also
provided valuable insights into the mode of polymer
biosynthesis and the molecular factors which determine a given
pattern of posttranslational protein modification with these

polymers. Based on these results we propose a model for the biological function of poly ADP-ribosylation in chromatin. The key feature of this model is derived from the observation that the non-covalent association of DNA-binding proteins with ADP-ribosyl polymers may be part of a protein shuttle mechanism in DNA templates.

RESULTS AND DISCUSSION

Poly(ADP-Ribose)Polymerase prompts the release of core DNA from nucleosomal core particles in vitro

Isolated nucleosomal core particles with associated poly(ADP-ribose)polymerase were prepared from rat liver and incubated in the presence of $[^{32}P]-\beta$-NAD (for experimental details see Mathis and Althaus, 1987). Concomitant with the formation of protein-bound ADP-ribosyl polymers, we observed the release of 146 bp core DNA from its association with histones (Fig. 1). This release was completely blocked by benzamide, a NAD-competitive and DNA-noncompetitive inhibitor of poly(ADP-ribose)polymerase (McLick et al., 1987). Similarly, no release and no ADP-ribosyl polymer formation was observed in the absence of substrate or when β-NAD was substituted for with α-NAD, which is not a substrate for the polymerase (Fig. 1). These findings provided direct evidence for a mechanism by which posttranslational poly(ADP-ribose) modification may affect DNA-protein interactions at the nucleosomal level of chromatin organization. We next investigated the properties of poly(ADP-ribose) molecules which constitute the molecular signal for this release phenomenon. For this purpose, the polymers synthesized under the conditions shown in Fig. 1 were labelled with $[^{32}P]NAD^+$, detached from the proteins and purified by boronate affinity chromatography (Jacobson et al., 1984). The different polymer size classes were separated on sequencing gels and the relative frequency of each polymer size class was plotted using a 3-D computer graphics program (Fig. 2, for details see Naegeli et al., 1989). The results revealed a discontinuous polymer size pattern which was strictly maintained throughout the reaction. This pattern indicates that polymer elongation is completed at early reaction times and that subsequent polymer synthesis produces larger numbers of polymers rather than longer polymers. These results suggest that poly(ADP-ribose)polymerase modifies acceptor proteins in a processive mode. An obvious prediction from this concept is that the concentration of (ADP-ribose)$_n$-modified proteins should increase as larger numbers of otherwise identical polymers appear in the reaction. This prediction was tested in the experiment shown in Fig. 3. The results again fit the processivity model schematically outlined in Fig.4. Similar results were obtained with the purified enzyme poly(ADP-ribose)polymerase, which was allowed to automodify itself in the presence of DNA (Naegeli et al., 1989). In summary, we have shown that the poly ADP-ribosylation of proteins involves a strictly processive reaction mechanism. In order to study the consequences of this mechanism on DNA-protein reactions we set up a reconstituted in vitro system involving highly purified components.

478

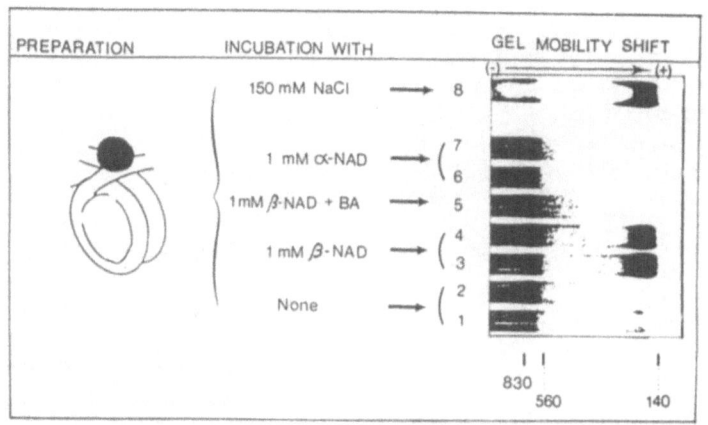

Fig. 1   Release of core DNA from nucleosomal core
         particles following (ADP-ribose)$_n$-modifi-
         cation *in vitro*. Core particles with asso-
         ciated poly (ADP-ribose)polymerase were
         prepared from rat liver chromatin and
         incubated in the absence or presence of
         α-NAD or β-NAD, as indicated in the figure.
         Following incubation for 30 min, the incuba-
         tion mixture was run on a 8% polyacrylamide
         gel at 60 V, and the bands were visualized
         by silver staining. The figures at the
         bottom of the gel indicate the position of
         830, 560 and 140 bp DNA restriction
         fragments.

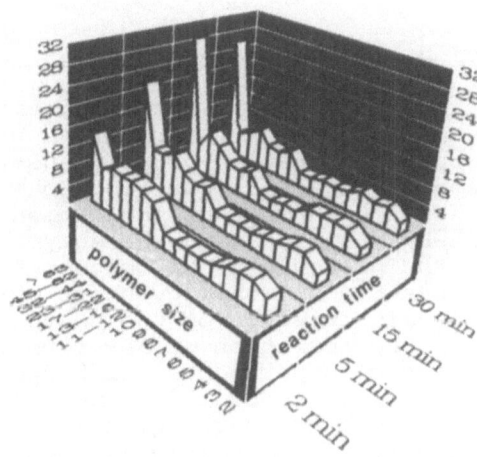

Fig. 2   Results of an experiment, in which [^{32}P]-
         labeled polymers were isolated from core
         particles by boronate chromatography and
         then separated by polyacrylamide gel elec-
         trophoresis as described (Naegeli et al.,
         1989). The relative contribution of each
         size class was determined by quantification
         of the radioactivity in each polymer size
         class.

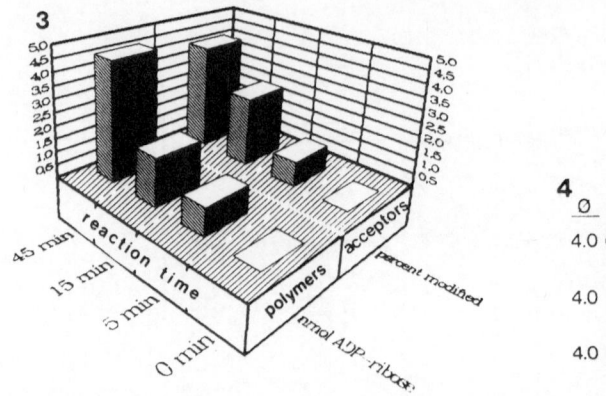

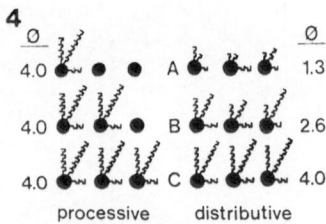

Fig. 3   Overall poly(ADP-ribose) biosynthesis and
         percentage of total proteins modified in
         nucleosomal core particles as a function
         of incubation time.   The reaction conditions
         were as in Fig. 2.

Fig. 4   Scheme illustrating the reaction inter-
         mediates produced in a processive or a
         distributive mode of protein modification
         by poly(ADP-ribose)polymerase.   The letters
         A-C denote three hypothetical stages of the
         reaction, each of them involving the
         production of 12 ADP-ribose residues.   The
         results shown in the present study are in
         agreement with the processivity model shown
         on the left.

In vitro reconstitution of a poly ADP-ribosylation system

     We have reconstituted the following components in vitro:
$5'-[^{32}P]$-end labeled 146 bp core DNA fragments, an
electrophoretically pure preparation of poly(ADP-
ribose)polymerase, and histone H2B, purified to homogeneity by
reverse-phase HPLC.   Fig. 5 shows that incubation of a
saturated histone H2B-DNA complex with the enzyme poly(ADP-
ribose)polymerase in the presence of NAD causes the release of
the DNA fragment as measured by mobility shift gel
electrophoresis.   This dissociation of the DNA-protein complex
was again dependent on the formation of ADP-ribosyl polymers
as shown by various control experiments which specifically
prevented polymer formation.   In addition, we found a linear
dose-response relationship between the quantity of DNA
released and the concentration of polymer formed (data not
shown).   Thus, the DNA binding of a specific protein can be
reversed by the action of poly(ADP-ribose)polymerase.
Furthermore, addition of histone H2B to an incubation with
poly (ADP-ribose) polymerase, which had been automodified in
the presence of core DNA alone, produced identical results
(not shown).   This indicated that ADP-ribosyl polymers
successfully compete for the binding of histone H2B in the
presence of DNA, suggesting a higher binding affinity of H2B
for the polymer.   The following experiments confirmed this
conclusion.   Digestion of protein-bound polymers with the

480

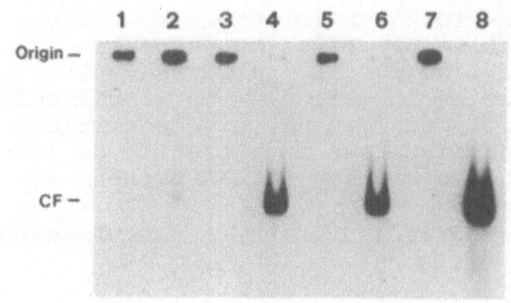

Fig. 5 Release of a 146 bp core DNA fragment for
its association with histone H2B following
incubation of the DNA-histone complex in the
presence of poly(ADP-ribose)polymerase and
100 μM NAD. The incubation times were 0 min
(lane 1), 2 min (lanes 2 & 3), 5 min (lanes
4 & 5), and 10 min (lanes 6 & 7). Benzamide
(10 mM) was present in the incubations run
on lanes 3,5 & 7. On lane 8, 5'-end labeled
146 bp core DNA was run as a marker.

enzyme snake venom phosphodiesterase reestablished the DNA
binding of H2B. Competition experiments, where equimolar
amounts of <u>free</u> polymers were allowed to compete for binding
of histone H2B in the presence of core DNA, confirmed the
preferential binding of H2B to ADP-ribose polymers (data not
shown). In addition, poly(ADP-ribose)polymerase also reversed
the binding of histones H1, H2A, H3 and H4, under similar
conditions. We conclude that the poly ADP-ribosylation
reaction can reverse the binding of histones to DNA templates.

Present status of model

We have found that the mode of posttranslational
poly(ADP-ribose) modification of proteins follows a processive
reaction mode. These results were obtained with two
experimental modes of different complexities, i.e. in the
presence or absence of various other proteins, including known
acceptors of poly(ADP-ribose). Thus, the processive mode of
operation of the enzyme poly(ADP-ribose)polymerase is an
inherent property of the enzyme protein itself and apparently
is not further regulated by other proteins. Another important
aspect of this observation is that this mechanism implies self
termination of polymer elongation, which is also attributable
to the action of the polymerase itself. However, we already
know that other proteins present in our <u>in vitro</u> system do
have an impact on this termination mechanism, generating
different though highly constant polymer size distributions.
We speculate that the distinct polymer size patterns observed
in the presence of various DNA-binding proteins may reflect
an adaptive response of this polymer generating mechanism to
some as yet unrecognized molecular properties of the shuttle
target (e.g. charge distribution within the protein,
hydrophobic vs. hydrophilic domains etc.). This initial
finding and further information amenable in this model system
should greatly enhance our understanding of the complex

polymer patterns found in intact cells following DNA damage, and in other active processes on chromatin. Another important aspect of these findings is that we have for the first time been able to define in molecular terms the complex pattern of protein-bound polymers, which is responsible for a specific function of the polymerase in a clearly defined in vitro system, i.e. the modulation of DNA-histone interactions.

The Poly ADP-ribosylation system: A protein shuttle mechanism in chromatin?

The starting point for our studies on the biological function of poly ADP-ribosylation has been the observation that poly ADP-ribosylation is involved in several chromatin functions (Althaus et al., 1982a; Loetscher et al., 1987), including the repair of various types of DNA damage (Althaus et al., 1982b). Subsequent analyses showed that the nucleosomal unfolding of damaged DNA domains is deficient in poly(ADP-ribose)-depleted cells (Mathis and Althaus, 1986; Mathis and Althaus, 1989). This deficiency was coupled with the lack of excision of bulky DNA adducts. This suggested an involvement of the poly(ADP-ribose) generating system in nucleosomal unfolding, and indirectly, in DNA excision repair. These studies also revealed that the unfolding process generates DNA domains which are indistinguishable from linker DNA with respect to their accessibility to chemical or enzymatic probes (Mathis and Althaus, 1986). Whether this unfolding requires a complete stripping of proteins from DNA, is currently not known. However, it is likely that the increased accessibility of these domains in vivo reflects a reduced binding of associated proteins. In accordance with this concept, we observed a significant reduction of histone binding to nucleosomal core DNA following in vitro ADP-ribosylation of nucleosomal core particles (Fig. 1). In fact, the poly ADP-ribosylation of nucleosomal core particles reduced the separating forces required to release core DNA from histones by a factor of 2.5, an effect which is likely to be underestimated in this model system. Likewise, this phenomenon could be reproduced with electrophoretically pure preparations of histones and poly(ADP-ribose)polymerase in the presence of core DNA and NAD$^+$. In addition, when histones were given the choice to associate either with DNA or preexisting polymerase-bound ADP-ribosyl polymers, they exhibited a clear preference for the polymers. This binding of histones to polymers was saturable (data not shown). Taken together, these results suggest that poly(ADP-ribose)polymerase may act as a histone shuttle mechanism in chromatin, in which the catabolic counterpart, poly(ADP-ribose)glycohydrolase could assume the role of reestablishing DNA-binding of histones. This possibility can now be tested with purified glycohydrolase enzyme. In view of our findings on the unfolding of chromatin domains in DNA excision repair in vivo (Mathis and Althaus, 1986; Mathis and Althaus, 1989), the rapid turnover of ADP-ribosyl residues on chromatin proteins, which varies with the level of DNA damage (Alvarez-Gonzalez and Althaus, 1989), could reflect adaptation of the shuttle mechanism of damaged templates to the level of DNA repair activity.

The concept of poly ADP-ribosylation as part of a protein shuttle mechanism raises a number of questions. For example, it will be very important to understand the specificity, capacity and potency of ADP-ribosyl polymers in reducing DNA binding of a larger spectrum of proteins, which may also act on DNA in DNA excision repair or in other active processes on chromatin such as replication and transcription. For example, it will be important to study the interaction of DNA repair enzymes with their substrate in the presence of histones and poly(ADP-ribose)polymerase. Furthermore, it will be sensible to complement the present in vitro system with purified poly(ADP-ribose)glycohydrolase and study its cooperation with the polymerase and the consequences of glycohydrolase activity on the size distribution of polymers under turnover conditions. Such studies are now in progress in our laboratory.

ACKNOWLEDGEMENTS

This study was supported by grant No. 3.161.0.88 from the Swiss National Foundation for Scientific Research, awarded to F.R.A.

REFERENCES

Althaus, F.R. and Richter, C., 1987, ADP-ribosylation of proteins: Enzymology and biological significance, Mol. Biol. Biochem. Biophys., 43:1-25.

Althaus, F.R., Lawrence, S.D., He, Y.Z., Sattler, G.L., Tsukada, Y. and Pitot, H.C., 1982, Effects of altered (ADP-ribose)$_n$-metabolism on expression of fetal functions by adult hepatocytes, Nature, 300:366-368.

Althaus, F.R., Lawrence, S.D., Sattler, G.L. and Pitot, H.C., 1982, ADP-ribosyltransferase activity in cultured hepatocytes: Interactions with DNA repair, J. Biol. Chem., 257:5528-5535.

Alvarez-Gonzalez, R. and Althaus, F.R., 1989, Poly(ADP-ribose) catabolism in mammalian cells exposed to DNA damaging agents, Mutat. Res., in press.

Jacobson, M.K., Payne, D.M., Alvarez-Gonzalez, R., Juarez-Salinas, H., Sims, J.L. and Jacobson, E.L., 1984, Determination of in vivo levels of polymeric and monomeric ADP-ribose by fluorescence methods, Meth. Enzymol., 106:483-494.

Leduc, Y., de Murcia, G., Lamarre, D. and Poirier, G.G., 1986, Visualisation of poly(ADP-ribose)synthetase associated with polynucleosomes by immunoelectron microscopy, Biochim. Biophys. Acta, 375:243-255.

Loetscher, P., Alvarez-Gonzalez, R. and Althaus, F.R., 1987, Poly(ADP-ribose) may signal changing metabolic conditions to the chromatin of mammalian cells, Proc. Natl. Acad. Sci. USA, 84:1286-1289.

Mathis, G. and Althaus, F.R., 1986, Periodic changes of chromatin organization associated with rearrangement of repair patches accompany DNA excision repair of mammalian cells, J. Biol. Chem., 261:5758-5765.

Mathis, G. and Althaus, F.R., 1987, Release of core DNA from nucleosomal core particles following (ADP-ribose)$_n$-modification in vitro, Biochem. Biophys. Res. Commun., 143:1049-1054.

Mathis, G. and Althaus, F.R., 1989, In: Niacin Nutrition, ADP-ribosylation, and Cancer (Jacobson, M.D. and Jacobson, E.L., eds.) Springer Verlag, pp. 151-157.

McLick, J., Hakam, A., Bauer, P.I., Kun, E., Zacharias, D. and Glusker, J., 1987, Benzamide-DNA interactions: Deductions from binding, enzyme kinetics and from X-ray structural analysis of a 9-ethyladenine-benzamide adduct, <u>Biochim. Biophys. Acta.</u> 909:71-83.

Naegeli, H., Loetscher, P. and Althaus, F.R., 1989, <u>J. Biol. Chem.</u>, 264:14382-14385.

CONSIDERATIONS ON THE MECHANISMS OF UV-INDUCED MUTAGENESIS

Irena Pietrzykowska and Magdalena Felczak

Institute of Biochemistry and Biophysics, Pol.
Acad. Sci. ul. Rakowiecka 36, 02-532 Warszawa
Poland

INTRODUCTION

Mutagenesis induced by UV-light and most chemical mutagenisis requires damage to DNA (which is a signal for induction of SOS functions and a target for mutation), induction of SOS functions controlled by lexA and recA genes, as well as UmuDC and RecA proteins. RecA in its activated form is inolved in derepression of the umuDC gene and in the activation of UmuD protein by proteolytic cleavege between cys24 and gly25. RecA has a third, as yet unknown, role in mutagenesis (Nohmi et al., 1988).

However, the mechanism of mutagenesis is not well understood. Most common hypotheses assume that the mutagenic process involves replication past the lesion that otherwsie would block DNA synthesis. Studies in vitro have shown that a lesion in DNA constitutes a block to replication. In vivo, it is commonly believed, replication of damaged DNA requires RecA and UmuDC proteins, which modify the replication complex in a way that allows replication, but at the cost of fidelity (Marsch and Walker, 1986). Recently Bridges et al. (1985), postulated that activated RecA protein is involved in misincorporation at the lesion, and UmuDC proteins in elongation past the lesion, of replicated UV-irradiated DNA.

We have studied the effect of a mutation in the umuC gene on DNA replcation in UV-irradiated excision repair-deficient E. Coli cells.

EFFECT OF A MUTATION IN umuC GENE ON THE COURSE OF POST-UV DNA REPLICATION

Fig. 1 shows the incorporation of H^3-thymine into DNA of E. coli TK603 thyAuvrA and TK610 thyAuvrAumuC cells irradiated with a dose of 5 J/m^2 at 254 nm during postirradiation incubation. The cells were preincubated with radioactive thymine for 3 generations before irradiation. It may be seen

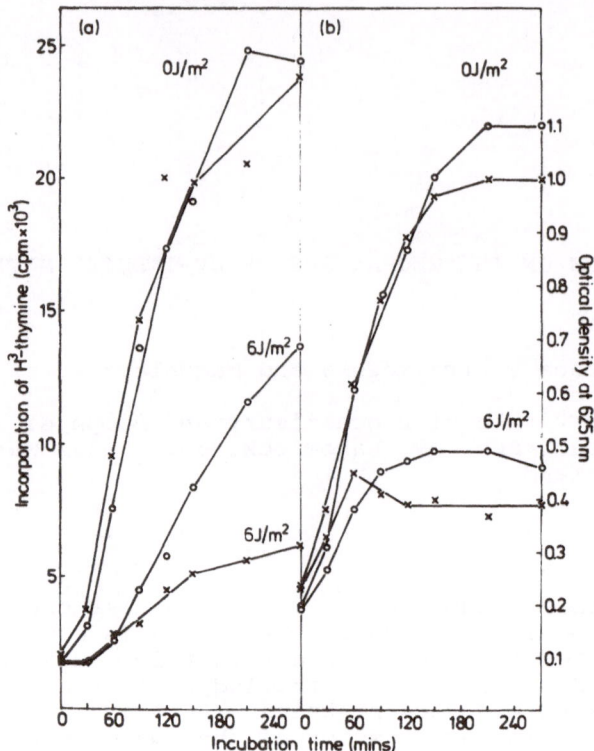

Fig. 1   Effect of umuC mutation on DNA replication in
         UV-irradiated cells.   Incorporation of H³-
         thymine into DNA of irradiated and non-
         irradiated E. coli TK603 umu⁺ (o) and TK610
         umuC (x) during incubation in TN medium (a),
         and the growth curves (b).

that irradiated cells of both strains show similar inhibition
of replication for 30 mins after irradiation, followed by
resumption of replication at a similar rate in both strains.
However, in contrast to the wild type strain in which
incorporation of thymine increases continuously, the rate of
incorporation in umuC cells at 60-90 mins becomes much slower
and at about 150 mins is almost completely inhibited.   Fig. 1
also shows growth curves, measured as optical density of
cultures, indicating that the number of cells for both strains
is similar.   In irradiated wild type cells there is a 9-fold
increase of incorporated thymine and in umuC cells only 3.5-
fold, which corresponds to more than 3 cycles of replication
for wild type and less than 2 for the umuC strain.   It has
been shown by Kato (1977) that, in UV-irradiated umuC cells,
the DNA has a molecular weight similar to that of wild type
cells after 45 mins incubation, indicating that there is no
essential difference in the DNA replicated in UV-irradiated
umuC cells.   Studies on the effect of a umuC mutation on
stable DNA replication showed that the mutation does not
eliminate, but significantly reduces, the level of stable DNA
replication.   Most probably this is a reflection of the effect
of the umC mutation on late step(s) of replication observed in
UV-irradiated cells, but not due to direct involvement of UmuC
protein in stable DNA replication.

The results lead to the following conclusions: (1) UV-irradiation inhibits DNA replication for 30 mins and is then resumed. (2) Resumption is not inhibited in the umuC mutant, indicating that the product of this gene is not involved in resumption of replication. This has also been observed by others (Khidir et al., 1985; Witkin et al., 1987). (3) The umuC gene product is, however, required for later step(s) of replication of UV-irradiated DNA.

ABSENCE OF ANY TIME CORRELATION BETWEEN APPEARANCE OF UV-INDUCED MUTANTS AND THE TIME WHEN umuC AFFECTS DNA REPLICATION

The product of gene umuC is considered to be involved in the mutagenic process since no mutagenesis is induced in umuC mutant. If so, one should expect that mutations are, most likely, formed at the time when umuC mutation affects replication in UV-irradiated cells. We studied the time course of appearance of his[+] revertants, able to grown on the selective medium without histidine, during incubation of UV-irradiated cells in rich liquid medium. It may be seen in Fig. 2 that 30 mins of incubation is sufficient for a maximal level of UV-induced mutagenesis, pointing to the absence of any time correlation between appearance of mutants and the time when umuC affects DNA replication. SEM data are included to show the level of mutagenesis under conditions permissive for DNA replication. More detailed experiments have shown that 15-20 mins incubation in rich medium is sufficient for formation of mutations, but not for resumption of replication, which requires at least 30 mins incubation (Fig. 3). This suggests that: (1) mutagenesis requires a lower level of

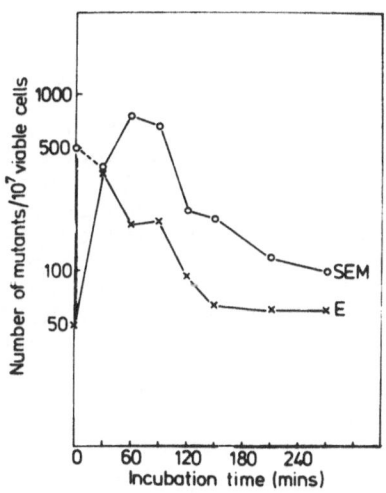

Fig. 2   Time-course of appearance UV-induced his[+] revertants during incubation in TN medium. TK603 uvrA cells after irradiation were incubated in TN medium and at different times were washed and plated on selective medium E without histidine (E) or E medium supplemented with 0.2 percent of TN medium (SEM).

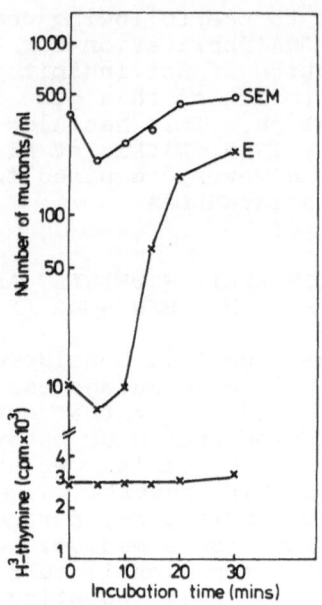

Fig. 3 Appearance of UV-induced his⁺ revertants
and incorporation of H³-thymine during 30
mins incubation in TN medium after irradiation; E. coli TK603 uvrA. Experimental
conditions as in Fig. 2.

induction of SOS functions than resumption of DNA replication
and (2) mutations may be formed before resumption of
replication.

MUTAGENESIS AND POST-UV REPLICATION IN A recF MUTANT

     To clarify the role of DNA replication in UV-induced
mutagenesis, we used a recF mutant, known to be as efficient
in UV-induced mutagenesis as the wild type strain (Kato et
al., 1977), but unable to perform DNA repair dependent on SOS
functions e.g. postreplication repair and W-reactivation
(Thoms and Wackernagel, 1987). recF mutant was presumed to be
defective in induction of RecA protein, but there are
conflicting data regarding induction of RecA in recF cells.
These characteristics of the recF mutant appear to be useful
for determining the role of replication in UV-induced
mutagenesis. The fact that a mutation in the recF gene
inhibits W-reactivation and postreplication repair suggests
that DNA replication in UV-irradiated cells might be
inhibited. Fig. 4 shows that after a UV-dose of 6 J/m², DNA
replication in the wild type cells is resumed after 30 mins,
but not in the recF mutant. The latter shows prolonged
inhibition of replication after irradiation. This suggests
that the recF gene product may be involved in the resumption
of DNA replication inhibited by irradation. The prolonged
inhibition of replication would explain the deficiency in
postreplication repair of the recF mutant.

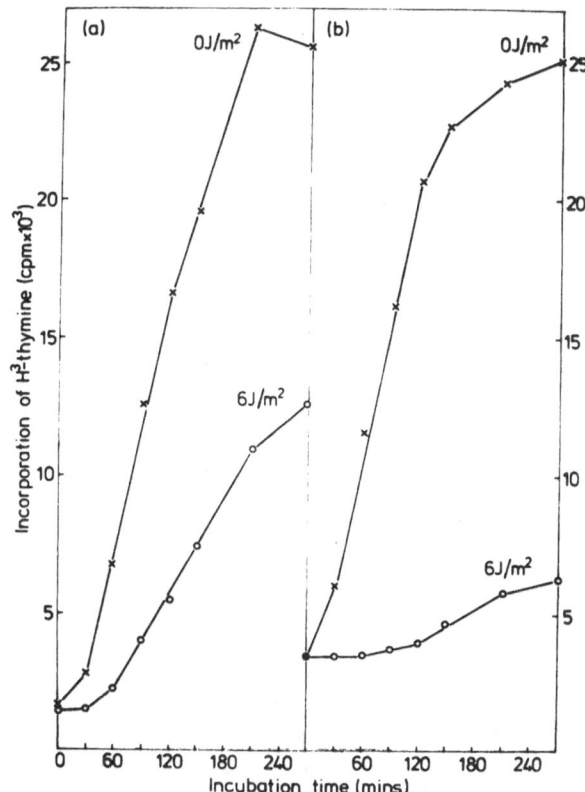

Fig. 4   Effect of recF mutation on DNA replication
in UV-irradiated cells. Incorporation of
H³-thymine into E.coli TK603 uvrA (a), and
MV1178 uvrArecF (b) cells during incubation
in TN medium.

     The time course of appearance of UV-induced his⁺
revertants in the recF mutant during post-UV incubation in
rich medium, and incorporation of H³-thymine into the DNA of
irradiated cells, are shown in Fig. 5.   It may be seen that
most revertants capable of growth on selective medium without
histidine appear after 30 mins incubation, indicating that
formation of mutations precedes resumption of DNA replication.
Control experiments (now shown) indicated that preincubation
of irradiated recF cells in rich medium for 60 mins, which is
sufficient for maximal mutagenesis, is not sufficient for
resumption of DNA replication and its continuation after
transfer of the cells into selective medium without histidine.
The results indicate that UV-induced mutations may be formed
before resumption of DNA replication, suggesting a pathway for
mutagenesis independent of post-UV DNA replication.

UV-INDUCED MUTAGENESIS IN λsus08 PHAGE UNDER NONPERMISSIVE
CONDITIONS FOR DNA REPLICATION

     To test the hypothesis that mutations may be formed
before resumption of DNA replication, we studied UV-induced

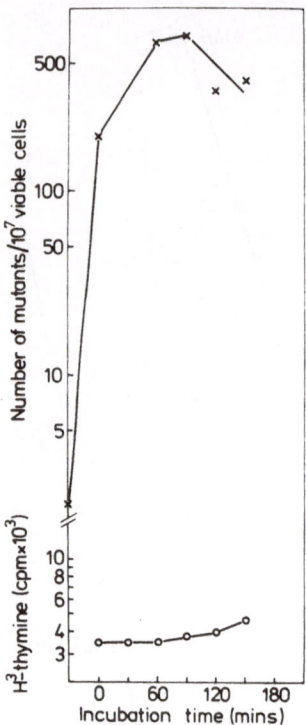

Fig. 5 Time-course of DNA replication (o), and
appearance of his⁺ revertants (x) during
incubation of UV-irradiated E. coli MV1178
uvrArecF cells in TN medium. Experimental
conditions as in Fig. 2.

reversion of the λsus08 amber mutation to sus⁺ in E. Coli 594
su host cells, unable to support growth of the sus08 mutant,
due to inability to replicate its DNA. Mutation in the sus0
gene prevents initiation of phage DNA replication in su⁻ host
(Ogawa and Tomizawa, 1968). In su⁻ cells UV-induced reversion
of λsus08 could be observed only if the mutation occurred
before initiation of DNA replication. Fig. 6 shows that
λsus08 phage is efficiently reverted to sus⁺ in the E. coli
594su⁻ strain nonpermissive for phage DNA replication. The
frequency of reversion is dependent on the UV-dose given to
the phage. Preirradiation of the host before infection
increases the mutation frequncy, indicating that some SOS
functions induced in the host stimulate the mutagenic process.
On the other hand no revertants are observed if only host
cells are irradiated, indicating that nontargeted mutagenesis,
known to be dependent on DNA replication, does not occur in
su⁻ cells. This indicates that the DNA of sus08 phage is not
replicated in su⁻ host, even by SOS-modified replacting
complex, and suggests that protein 0 is also essential for
initiation of phage DNA replication in the host with the
induced SOS system. Further support for the absence of phage
λsus08 DNA replication in su⁻ host is provided by the
observation that a mutation in the dnaQ gene, inactivating
proofreading activity of DNA polymerase holoenzyme, introduced
into E. Coli 594 su⁻ has no mutator effect on UV-induced and

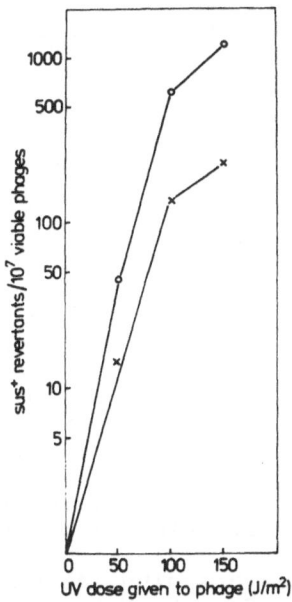

Fig. 6   UV-induced reversion of amber mutation in
          λcl7 sus08 phage as a function of UV-dose
          given to phage in E. coli 594 su⁻ host; non-
          irradiated host cells (x) or irradiated with
          UV-dose of 12 J/m² (o).

spontaneous mutagenesis of λsus08 phage in a su⁻ host. By
contrast, in su⁺ cells, permissive for DNA replication, a high
mutator effect of dnaQ mutation on λsus80 phage mutagenesis is
observed.

     The possiblity that UvrABC-excinuclease might initiate
DNA replication of UV-irradiated λsus08 phage was excluded by
the observation that the efficiency of UV-induced mutagenesis
in λsus08 phage in su⁻ host is similar to that observed in its
uvrA counterpart.

DISCUSSION

     The results indicate that DNA replication in UV-
irradiated cells is inhibited for 30 mins and subsequent
resumption seems to require the recF, but not umuC, gene
product. The umuC gene product is apparently involved in the
later step(s) of replication of damaged DNA since, in this
mutant the rate of DNA replication at 90-270 mins is
remarkably reduced. However, there is no correlation between
the time of appearance of mutants and the time at which a umuC
mutation affects replication of UV-DNA. A significant
proportion of mutations seems to be formed before resumption
of replication as is seen from comparison of mutant numbers on
E and SEM selective medium (Fig. 2 and 3).

     The foregoing poses several questions regarding the
mechanism of mutagenesis.

What is the role of umuC gene product in mutagenesis?

umuC gene product is believed to be directly involved in mutagenesis and, according to Bridges et al. (1985), UmuC protein is involved in elongation of the daughter strand past the lesion during replication of irradiated DNA. This proposal is however inconsistent with the fact that in UV-irradiated umuC cells, resumption of replication is not inhibited and, as observed by Kato (1977), DNA after 45-60 mins is similar to that in umu$^+$ cells. Since a umuC mutation affects later step(s) of replication of damaged DNA, at a time when mutations have already formed, the role of UmuC protein appears to be indirect e.g. in metabolism of damaged DNA rather than direct involvment in the mutagenic process. One possible function of UmuC protein might be involvement in stable DNA replication in UV-irradiated cells. This is supported by the finding of Woodgate and Bridges (1985) that non-mutability of a umuC mutant may be partly reversed by delayed photoreactivation, which by removing pyrimidine photodimers relieves replication of damaged DNA.

In relation to the foregoing, it seems that the functions of the activated forms of UmuD (Nohmi et al., 1988) and UmuC proteins may be separate. The UmuD protein in contrast to UmuC, could be involved directly in a mutagenic process, as suggested by experiments described by Nohmi et al. (1988). The effect of a umuD mutation on post-UV DNA replications requires examination.

The finding that a significant proportion of UV-induced mutations in excision repair-deficient E. coli appears to be formed before resumption of UV-DNA replication, supported further by the experiments with λsus08 phage, raises the question of the relation between replication of damaged DNA and formation of mutations, and calls for a revision of the commonly accepted model of mutagenesis.

Apparently there is more than one mechanism by which UV-induced mutations may be formed. One operates during SOS-dependent replication of damaged DNA and misincorporates bases opposite a lesion or opposite non-damaged sites. This system would be responsible for most transitions and frameshifts, which may be targeted or untargeted. The second mechanism which operates before resumption of replication of damaged DNA, misinserts a base at a special site in nonreplicating DNA, by some inserting system. This system would be responsible for all or most transversions and a part of transition mutations, and could operate at lesions like AP-sites (apyrimidinic and apurinic) or at 6,4-photodipyrimidines which structurally resemble AP-sites (Franklin et al., 1985).

Base inserting activity at AP-sites has been reported in E. coli crude extracts (Livneh et al., 1979; Kataoka and Sekiguchi, 1982) and in a partly purified protein isolated from mammalian cells (Deutsch and Linn 1979), but has hitherto aroused little interest. Possible involvement of insertase activity in the mechanism of mutagenesis would explain the observed formation of mutations in non-replicating DNA, as well as the mechanism of formation of all types of transversion mutations, difficult to explain by a mechanism involving replication of damaged DNA.

It is interesting to note that formation of transversions induced by UV-light and some other mutagens is dependent on the umuDC gene (Matern et al., 1985), possibly mainly on the umuD gene product. It has been observed that a change in a cys-gly site cleavable by RecA in MucA, an analog of UmuD, leads to a change in specificity of mutagenesis (Marsh and Walker, 1987). The presence of pKM101 plasmid carrying the mucAB gene enhances the proportion of transversion mutations even in umu⁺ cells (Eisenstadt et al., 1989).

The existence of mechanisms which replace unusual, but physiologically important, bases from RNA and DNA, e.g. very short patch mismatch repair (Lieb et al., 1987) suggests that an analogous mechanism in mutagenesis may exist. AP-sites appear to be very common lesions in DNA, both spontaneous and induced by a number of mutagens including UV-light, and could be a candidate for operation of an inserting system in mutagenesis. The existance of numerous DNA-glycosylases (Lindahl, 1979) also suggests that insertase may exist and play an important role in repair of AP-sites and in mutagenesis.

REFERENCES

Bridges, B.A and Woodgate, R., 1985, Two step model of bacterial UV mutagenesis, Mutat. Res., 150:133.
Deutsch, W.A., and Linn, S., 1979, DNA binding activity from cultured human fibroblasts that is specific for partially depurinated DNA and that inserts purines into apurinic sites, Proc. Natl. Acad. Sci. USA., 76:141.
Eisenstadt, E., Miller, J.K., Kahng, L-S. and Barnes, M., 1989, Influence of uvrB and pKM101 on the spectrum of spontaneous, UV- and y-ray-induced base substitution that revert hisG 46 in Salmonella typhimurium, Mutat. Res., 210:113.
Franklin, W.A., Doetsch, P. and Heseltine, W.A., 1985, Nucl. Acid Res., 13:5317.
Kataoka, T. and Sekiguchi, M., 1982, Are purine bases enzymatically inserted into depurinated DNA in Escherichia coli? J. Biochem., 92:971.
Kato, T., 1977, Effects of chloramphenicol and Caffeine on postreplication repair in uvrA⁻umuC⁻ and uvrA⁻recF⁻, Mol. Gen. Genet., 156:115.
Kato, T., Rothman, R.H. and Clark, A.J., 1977, Analysis of the recombination repair in mutagenesis of E. coli by UV irradiation, Genetics, 87:1.
Khidhir, M.A., Casaregola, S. and Holland, B., 1985, Mechanism of transient inhibition of DNA synthesis in UV-irradiated E. coli: Inhibition is independent of recA whilst recovery requires RecA protein itself and additional, inducible SOS function, Mol. Gen. Genet., 199:133.
Lieb, M., Allen, E. and Read, D., 1987, Genetics, 114:1041.
Lindahl, T., 1979, DNA-glycosylases, endonucleases for apurinic/apyrimidinic sites, and base excision repair, Progr. Nucl. Acid. Res. Mol. Biol., 22:135.
Livneh, Z., Elad,D. and Sperling, J., 1979, Enzymatic insertion of purine bases into depurinated DNA in vitro, Proc. Natl. Acad. Sci. USA., 76:1089
Marsch, L. and Walker, G.C., 1986, Mutagenic DNA repair in bacteria: The role of UmuDC and MucAB, in: "Mechanisms of

DNA damage repair", M.G. Simic, L. Grossman and A.C. Upton, eds., Plenum Press, New York, London, p.273.

Marsh, L. and Walker, G.C., 1987, New phenotypes associated with mucAB: Alteration of a MucA sequence homologous to the LexA cleavage site, J. Bacteriol, 169:1818.

Matern, I.E., Olthoff-Smith, F.P., Jacobs-Meijsing, B.L.M., Enge-Walk, B.E., Pouwels, P.H. and Lohman, P.H.N., 1985, Mutat. Res., 148:35.

Nohmi, T., Batista, J.R., Dodson, L.A. and Walker, G.C., 1988, RecA-mediated cleavage activates UmuD for mutagenesis: Mechanistic relationship between transcriptional derepression and postransational activation, Proc. Natl. Acad. Sci. USA, 85:1816.

Ogawa, T. and Tomizawa, J., 1968, Replication of DNA of lambda phage deffective in early functions, J. Mol. Biol., 38:217.

Thoms, B. and Wackernagel W., 1987, Regulatory role of recF in the SOS response of E. coli: Impaired induction of SOS genes by UV irradiation and Nalidixic acid in recF mutant, J. Bacteriol., 169:1736.

Witkin, E.M., Roegner-Maniscalco, V., Sweasy, J.B. and McCall, J.O., 1987, Recovery from UV light-induced inhibition of DNA synthesis requires umuDC gene product in recA718 mutant strain but not in recA$^+$ strain of E. coli, Proc. Natl. Acad. Sci. USA., 84:6805.

SURVIVAL, DIMER INDUCTION AND REPAIR IN HUMAN KERATINOCYTES
AND MELANOCYTES AFTER UV IRRADIATION

A.A. Schothorst[1], K.C. Noz[1], L. Evers[1] A.R. Filon[2]
and A.A. van Zeeland[2]

[1]Department of Dermatology, University Hospital
[2]Department of Radiation Genetics and Chemical
Mutagenesis. State University, Leiden
The Netherlands

INTRODUCTION

Changing habits of a large part of the population, such
as more outdoor recreation and the fashion to get tanned, are
responsible for the increasing exposure to ultraviolet (UV)
radiation, especially in the developed countries. The use of
UV-radiation in UV-B phototherapy add to this effect.
Moreover an increased UV-exposure has to be feared because of
a decrease of the ozone layer (NAS 1984, Moan et al., 1989).

By means of several models, attempts have been made to
predict an increase in the incidence of non-melanoma skin
cancer. (Fears et al., 1977; van der Leun, 1984; Slaper et
al., 1986).

Essential factors in these calculations are the UV doses
received by man as estimated by Schothorst et al. (1985) and
Larkö et al. (1983). These UV doses are estimated according
to a weighting function because the carinogenicity of UV
irradiation varies with different wavelengths. An action
spectrum for DNA damage in human skin cells can be used as the
weighting function.

Therefore, we have estimated an action spectrum for dimer
induction in the DNA of human keratinocytes as well as melanocytes, cell types involved respectively in basal and squamous
cell-carcinoma and in melanoma. It is relevant to compare
several effects of UV light such as killing, induction of
dimers and its repair in both cell types since the
relationship between received UV dose and the frequency of
melanoma in a population is still questionable. The values
found for the induction of endonuclease sensitive sites (ESS)
may be useful in models for the risk estimations of skin
cancer and may help to unravel the relationship between UV
exposure and the induction of melanoma.

MATERIALS AND METHODS

Isolation and culture of human epidermal keratinocytes and melanocytes

Keratinocytes

Isolation of keratinocytes from human skin was performed by a modification of the method of Liu et al. (1978). Foreskin originating from children aged between 4 months and 6 years old were cut into small pieces, freed from connective tissue and washed in PBS.

This procedure was followed by an incubation overnight at 4°C in trypsin solution (0.3% trypsin, 0.15 MNaCl, 0.04% KCL, 0.1% glucose pH 7.6). After another 30 min of incubation at 37°C the skin fragments were transferred to growth medium which was a 3:1 mixture of Dulbecco modified Eagle medium (DMEM) and Ham's F-12 (supplemented with 10 fetal calfserum FCS) and the epidermis and dermis were separated. After gentle agitation, both skin layers were discarded and the epidermal cells remaining in the medium were further cultured according to Rheinwald and Green (1975).

For the low-calcium cultures a medium was used composed of calcium free and phenol-red free DMEM mixed with phenol-red free Hams F-12 medium (3:1) supplemented with 10% chelex-treated fetal calfserum (Hennings et al., 1980). The final calcium concentration was 0.06 mM as determined by flame photometry.

Melanocytes

Human foreskin originating from Caucasian children was treated as described at cultivation of keratinocytes. The obtained cell suspension was seeded in Petri dishes and cultured according to Eisinger and Marko, (1982).

The growth medium contained Hams F10, 85 nM 12-0-tetradecanoyl-phorbol-13-acetate (TBA); 2.5 nM cholera toxin and 0.1 mM Isobutyl-methyl-xanthine (IBMX); 5% fetal calfserum; (HyClone); 5000IU/ml streptomycin/penicillin. The medium was free of phenol red.

The cell growth was evaluated by phase contrast microscopy. To cultures contaminated with fibroblasts geneticin (100μgr/ml) was added to the growth medium for 48 to 72 hrs. Cells were cultured at 37° C in a 5% CO_2 atmosphere. The growth medium was refreshed twice a week.

After about 14 days the primary culture was trypsinized (0.01%) and seeded (1:1). Cultured cells were used until passage 6.

Determination of cell survival

Exponentially growing keratinocytes (passage 1) were trypsinized and seeded; 1 x 10⁶ per cm ϕ dish, in the presence of 1 x 10⁶ lethally irradiation 3T3 feedercells.

Exponentially growing malenocytes of passage 3 till 6

were seeded at a density of 250,000 per 6 cm φ dish.

After 1 week the growth medium of both cell kinds was replaced by PBS with 0.1% glucose and the cells were irradiated or placed in darkness.

Immediately after irradiation or dark period cells were trypsinized and seeded at densities estimated to give 100-150 colonies per 9 cm φ Petri-dish. Lethally irradiated 3T3 feedercells (1.5 x 10^6 per 9 cm φ dish were added to the seeded keratinocytes, and 250,000 feeder cells were added to the melanocytes in a 9 cm φ dish. Cultures were refed once after 7 days. After 14 days colonies were stained and counted.

Survival was calculated as the ratio of the cloning efficiency (C.E.) of irradiated cells to the CE of non irradiated cells. The CE keratinocytes kept in the darkness varied between 1 and 2%, for melanocytes this figure varied from 8 to 20%.

Quantification of pyrimidine dimers in DNA

Pyrimidine dimers were quantified by determining the number of DNA sites sensitive to T4 endonuclease V as described by van Zeeland et al. (1981). Briefly, exponentially growing keratinocytes were labelled for 48 hr with medium containing 2 μCi/ml [^{3}H] thymidine (Amersham).

Melanocytes, exponentially growing, were labelled for 6 days with medium containing 1μCi/ml [^{14}C] thymidine (Amersham). [^{14}C] was used to avoid toxic effects of the label.

The medium containing the label was replaced by medium without labelled thymidine 24 hr before irradiation. Before the irradiation this medium was replaced by PBS with 0.1% glucose. Following irradiation the cells were either trypsinized immediately or incubated for various repair periods. After trypsinisation the cells were permeabilised under high salt conditions, (4M), treated with T4 endonuclease V and lysed directly on top of an alkaline sucrose gradient.

In some control experiments no endonuclease V was used; these samples were centrifuged at a reduced speed for a shorter time. After centrifugation, the number average molecular weight (Mn) was calculated from the radioactivity profiles of the gradients and the number of endonuclease sensitive sites (ESS) per Dalton was calculated by the following formula:

$$\frac{1}{Mn.treated} - \frac{1}{Mn.untreated}$$

Irradiation and dosimetry

UV irradiation was provided by a 1000 W Mercury-Xenon arc lamp equipped with narrow band interference filters of 254, 297 and 302 nm (Oriel Corporation, Stamford, Connecticut, USA). Dose rates for these wavelengths were respectively 0.1,

0.97 and 1.7 W/m$_2$. For irradiation with 312 nm we used three
40 W Philips TL-01 lamps. (Philips Nederland BV., Eindhoven,
The Netherlands) emitting the majority (i.e. 51%) of their
radiant energy at a peak around 312 nm. The dose rate of this
UV source was 8.2 W/m^2.

The emission spectrum of this lamp and further details of
the used sources and filters and dosimetry are described by
Enninga et al. (1986).

Statistical analysis of data

The linear regression line of the exposure-response
curves for dimer induction at a certain wavelength was
calculated in each experiment.

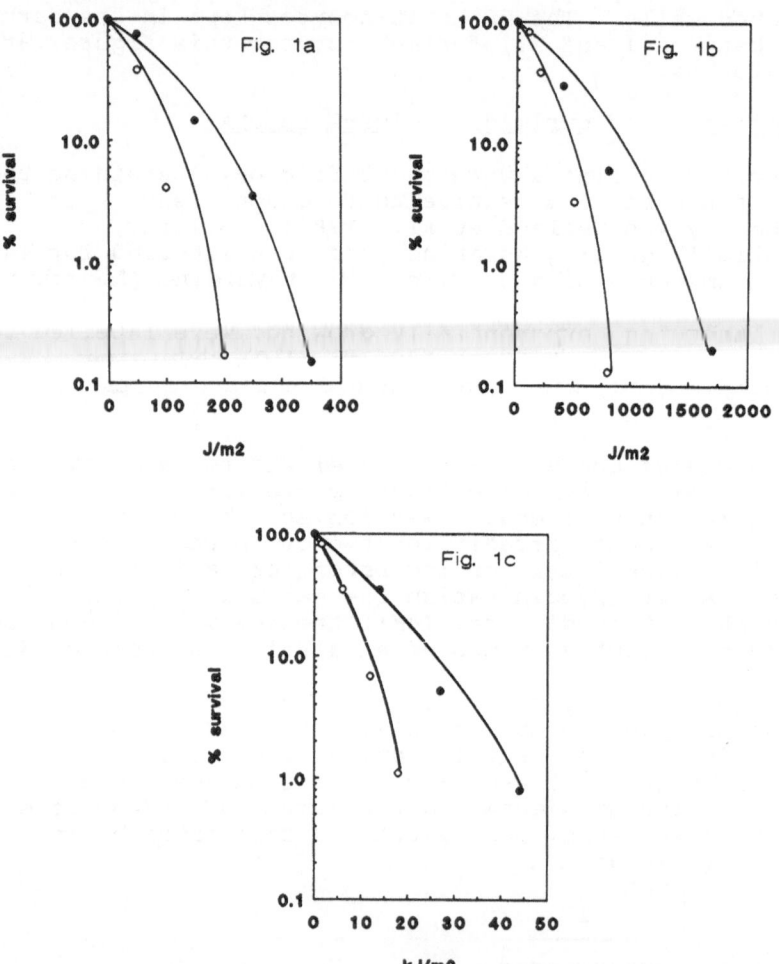

Fig. 1   Survival of cultured human keratinocytes
and melanocytes after irradiation with monochromatic UV-light of a) 297nm, b) 302nm or
c) with a Philips TL-01 lamp (predominantly
312nm). Filled circles - melanocytes; Open
circles - keratinocytes.

RESULTS

The survival of keratinocytes and melanocytes after UV
irradiation was assessed by means of the colony forming
ability.   The survival curves of both cell types after
irradiation with UV of 297, 302 and 312 nm are presented
respectively in Fig. 1a, 1b and 1c.   For longer wavelengths
higher UV-doses are needed to obtain the same extent of cell-
killing.

Induction of UV-endonuclease sensitive sites (E.S.S.) in
keratinocytes and melanocytes by UV of four different
wavelengths is presented in Fig. 2a, b, c and d.

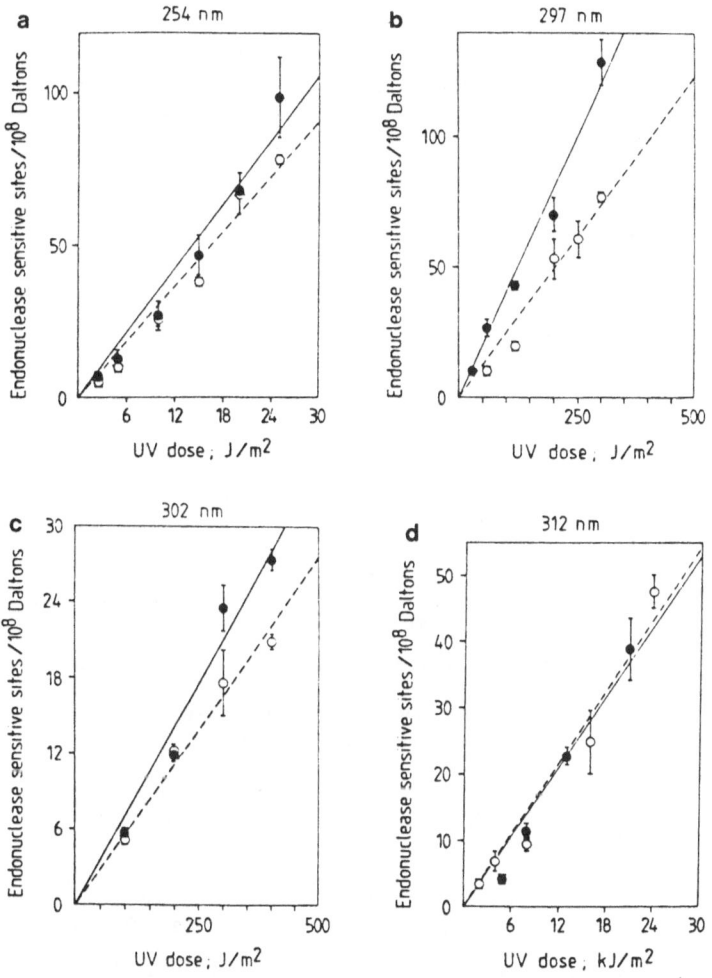

Fig. 2   Induction of endonuclease sensitive sites
in cultured human keratinocytes and melanocytes after irradiation with monochromatic
UV-light of a) 254nm, b) 297nm, c) 302nm, or
d) with a Philips TL-01 lamp (predominantly
312nm).   The points represent mean values
obtained from at least 2 independent
experiments.   Filled circles - melanocytes;
Open circles - keratinocytes.

The relationship between survival and the induction of ESS/10^8 Daltons is presented for three cell types in Fig. 3. In keratinocytes UV of 312 nm exhibits a relatively strong killing effect but a lower induction of ESS in comparison to melanocytes.

The removal of UV induced pyrimidine dimers in keratinocytes as well as in melanocytes is shown in Fig. 4. These results demonstrate that both cell types excise their pyrimidine dimers with approximately the same rate over a period of 24 hr post-irradiation incubation.

The velocity of repair of UV-induced dimers in several human cell types are presented in Table 1.

We calculated the relative sensitivity for the induction of ESS in both investigated cell types; the values found at 254 nm were set as 1. (Fig. 5). Melanocytes show about the same sensitivity for the induction of ESS than keratinocytes do.

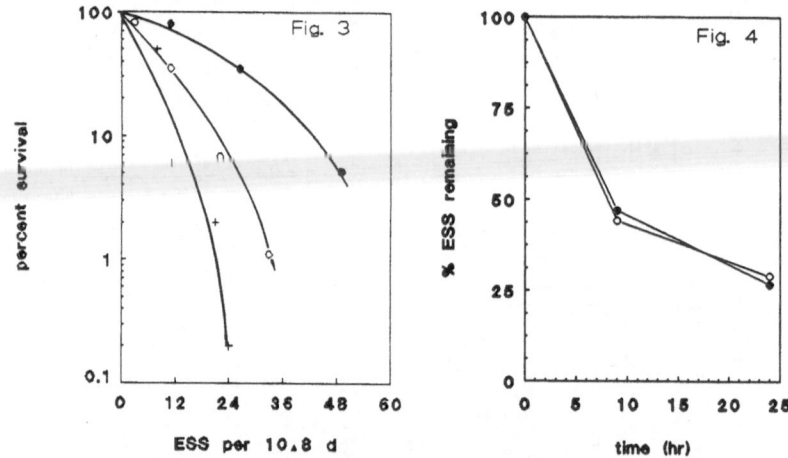

Fig. 3   Relationship between the induction of ESS and the survival after irradiation with UV of 312 nm.   + - Human fibroblasts; (Enninga et al., 1986) Open circles - Human keratinocytes;   Filled circles - Human melanocytes.

Fig. 4   Comparison of the removal of UV-(302 nm) induced pyrimidine dimers from DNA of cultured human diploid keratinocytes and melanocytes.   Keratinocytes were irradiated with 200 J/m^2, melanocytes with 250 J/m^2 while in PBS; 0.1% glucose.   Cultures were provided with the appropriate growth medium and maintained at 37°C until dimer determinations were made at the indicated times. The line with the open circles holds for data obtained from keratinocytes, the line with filled circles is based on the melanocytes results.

500

Table 1 Efficiency of repair UV induced pyrimidine
dimers in cultured human cells. Repair
time 16 h.

Cell type	% repair	Reference	
Keratinocyte	20-60	Taichman et al.	(1979)
Fibroblast	20-80	Taichman et al.	(1979)
Fibroblast	45	Chalmers et al.	(1976)
Hela G B1	20	Chalmers et al.	(1976)
Melanoma MM96	25	Chalmers et al.	(1976)
Melanoma MM200	35	Chalmers et al.	(1976)
Fibroblast	70	Mitchell et al.	(1985)
Fibroblast	50	Enninga	(1985)
Keratinocyte	75	Reusch et al.	(1988)
Keratinocyte	60	this report	
Melanocyte	60	this report	

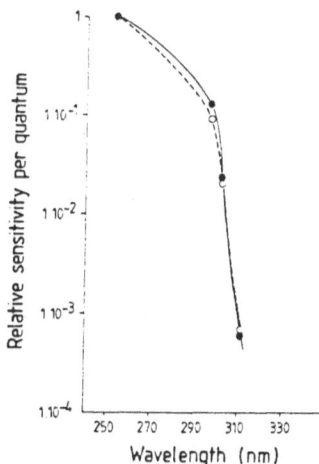

Fig. 5 UV action spectra for dimer induction. All
points are related to the value of 1.00 at
297 nm. Filled circles - dimer induction in
melanocytes; Open circles - dimer induction
in keratinocytes.

DISCUSSION

Comparison of experimental results obtained with tests
using UV-radiation always has its limitations because the
light sources, UV-dose meters, and experimental procedures are
mostly different. We eliminated these restrictions by working
in the same experimental set-up. A very important adjustment
is the culturing method of the keratinocytes since these cells
undergo differentiation and keratinization. We chose medium
with a low Ca^{++} level because keratinocytes grown in a
monolayer do not encounter a shielding effect as occurs in
multilayered differentiating keratinocytes cultures grown at a
normal Ca^{++} level. (Taichman and Setlow 1979).

Estimation of the survival of keratinocytes and
melanocytes after UV irradiation by the assessment of the
colony forming ability is quite well possible. We observed
the necessity for the presence of lethally irradiated 3T3
feedercells for colony formation of both cell types.

The dose-response curves (Fig. 1a, b and c) demonstrate a
lower sensitivity of melanocytes to UVB with regard to
keratinocytes. An explanation of this phenomenon could be the
presence of melanin in melanocytes. However, in cultured
melanoma cells with different melanin contents no difference
in survival after UV irradiation could be assessed by Chalmers
et al. (1976).

The results as shown in Fig. 2a-2d demonstrated that
melanocytes accumulate less pyrimidine dimers after exposure
to UV than keratinocytes. A wavelength-depent difference in
the sensitivity for dimer induction in melanin pigmented and
non-melanin containing mammalian cells is described (London et
al., 1976; Hill and Setlow, 1982).   They compared the number
of dimers induced after UV irradiation of pigmented mouse
cells to non-pigmented carcinoma cells. Both groups found
lower levels of dimers in certain malanin containing cell
lines for UV of 289 nm or less. An explanation may be that
melanins have a higher absorption in the lower end of the
spectrum (Das et al., 1976; Crippa et al., 1978) and play a
role in shielding mammalian cells from UV of wavelengths below
289 nm.

However, using other pigmented mouse cell lines, Hill and
Setlow (1982), as well as Chalmers et al. (1976), could not
find the mentioned differences in sensitivity for UV of 289
nm. This may be caused by the differences in conformation of
melanins in different melanoma cell lines and the tremendous
variation in melanin content of the individual melanoma cells
(Hill and Setlow, 1982).

The possibility that the high number of ESS found in
melanocytes after irradiation with UV of 312 nm is caused by
preexisting breaks is excluded by our control experiments.
These experiments carried out without the addition of
endonuclease V to the dimer essay, demonstrated that the
number of single strand breaks after exposure to 312 nm
wavelength UV and non irradiated melanocytes was nil.

The relatively high number of ESS induced by 312 nm UV in
melanocytes suggests a relatively high number of induced ESS
by UV of longer wavelengths including the UVA range. This
would be of great importance since wavelengths above 302 nm
contribute to a high degree in sunlight at different
geographical latitudes.

Fig. 3 shows clear differences in lethality per dimer at
312 nm for melanocytes, keratinocytes and fibroblasts.
Comparison of dimer and survival data on the basis of
equivalent UV doses yields information about the selective
importance of dimer induction for the killing effect and about
the possible contribution of other damage.

Besides the action spectrum of dimer induction the
removal of UV induced dimers is also of importance. As is

shown in Fig. 4 the repair of ESS in keratinocytes and melanocytes does not differ significantly.

Other authors, Taichman et al. (1979), Chalmers et al. (1976), Mitchell et a.. (1985), Enninga (1985) and Reusch et al. (1988) (Table 1) have also found a recovery of UV induced dimers of the same order of magnitude after 16 hours post irradiation incubation.

Some points need more attention. First Niggly et al. (1983) have made the remark that the in vitro repair percentage is UV-dose dependent while wavelength is less important. Therefore, our experiments were restricted to the 302 nm waveband and we used UV doses of 200 J/m^2 and 250 J/m^2 respectively for keratinocytes and melanocytes. These doses correspond with about 60% surviving cells. UV survival parameters for human melanocytes and keratinocytes have been reported earlier in summary (Schothorst et al., 1988). Secondly, Taichman and Setlow (1979) supposed that the decreased DNA repair which they found in keratinocytes was caused by differentiation of their cultured cells. In our experiments this differentiation was excluded and indeed no increased recovery after irradiation was found.

Comparison of the repair rate found in melanocytes with that found in other UV irradiated pigmented cells is very limited because repair studies carried out with pigmented rodent cells (Hill and Setlow, 1982) are hard to compare because these cells have a limited overall repair, although actively transcribed genes are efficiently repaired (Bohr et al., 1985). A repair study with cultivated human melanocytes derived from a XP patient carried out by Kraemer et al. (1989), demonstrated a decreased repair compared to values obtained from normal human fibroblasts. This decreased repair velocity is probably caused by a genetic defect as reported earlier for other cell types of XP patients (Cleaver, 1968), and not by inhibition of destruction of excision repair enzymes by, for example, photochemical reactions of melanins. Also in our repair studies with melanocytes we did not find any indication of such photochemical processes. Because the two cell types in this study are comparable concerning determination of survival as well as the rate of removal of UV induced ESS a comparison of their relative sensitivity for UV inducable ESS is relevant (Fig. 5).

The presented action spectra might be useful for further investigations of the relationship between UV exposure and the incidence of different types of skin cancer in man.

REFERENCES

Bohr, V.A., Smith, C.A., Okumoto, D.S. and Hanawalt, P.C., 1985, DNA repair in an active gene; Removal of pyrimidine dimers from the BHFR gene of CHO cells is much more efficient than in the genome overall, Cell, 40:359-369.
Chalmers, A.H., Lavin, M., Atisoontornkul, J., Mansbridge and Kidson, C., 1976, Resistance of human melanoma cells to ultraviolet radiation, Cancer Research, 36:1930-1934.
Cleaver, J.E., 1968, Defective repair replication of DNA in xeroderma pigmentosum, Nature, 218:652-656.

Crippa, P.R., Christofoletti, V. and Romeo, N., 1978, A band model for melanin dedured from optical absorption and photoconductivity experiments, <u>Biochim. Biophys. Acta.</u>, 538:164-170.

Das, K.C., Abramson, M.B. and Katzman, R., 1976, A new chemical method for quantifying melanin, <u>J. Neurochem.</u>, 26: p. 695-699.

Eisinger, M. and Marko, O., 1982, Selective proliferation of normal human melanocytes <u>in vitro</u> in the presence of phorbol ester and cholera toxin, <u>Proc. Natl. Acad. Sci. USA</u>, 79:2018-2022.

Enninga, I.C., Groenendijk, R.T.L., Filon, A.R., Van Zeeland, A.A. and Simons, J.W.I.M., 1986, The wavelength dependence of UV induced pyrimidine dimer formation, cell killing and mutation induction in human diploid skin fibroblasts, <u>Carcinogenesis.</u>, 7:1829-1836.

Enninga, I.C., 1985, Personal communication.

Fears, T.R. and Schneiderman, M.A., 1977, Mathematical models of age and ultraviolet effects on the incidence of skin cancer among whites in the United States, <u>Am. J. Epidermiol.</u>, 105:420-427.

Hennings, H.D., Michael, D. and Cheng, C., et al., 1980, Calcium regulation of growth and differentiation of mouse epidermal cells in culture, <u>Cell</u>, 19:245-254.

Hill, H.Z. and Setlow, R.B., 1982, Comparative action spectra for pyrimidine dimer formation in Cloudman S91 mouse melanoma and EMT 6 mouse mammary carcinoma cells, <u>Photochem. and Photobiol.</u>, 36:681-684.

Kraemer, R.H., Herlyn, M., Yuspa, S.H., Clark, W.H., Townsend, G.K., Neises, G.R. and Hearing, V.J., 1989, Reduced DNA repair in cultured melanocytes and nevus cells from a patient with xeroderma pigmentosum, <u>Arch. Dermatol.</u>, 125:263-268.

Larkö, O., Diffey, B.L. and Tate, T.K., 1983, Natural UVB radiation received by people with outdoor, indoor, and mixed occupations and UVB treatment of psoriasis, <u>Clin. Exp. Dermatol.</u>, 8:279-285.

Van der Leun, J.C., 1984, UV Carcinogenesis, <u>Photochem. and Photobiol.</u>, 39:861-868.

Liu, S.C. and Karasek, M., 1978, Isolation and growth of adult human epidermal keratinocytes in cell culture, <u>J. Invest. Dermatol.</u>, 71:157-162.

London, D.A., Carter, D.M. and Condit, E.S., 1976, Effect of pigment on photomediated production of thymine dimers in cultured melanoma cells, <u>J. Invest. Dermatol.</u>, 67:261-264.

Mitchell, D.L., Haipek, C.A. and Clarkson, J.M., 1985, 6-4 photoproducts are removed from the DNA of UV-irradiated mammalian cells more efficiently than cyclobutane pyrimidine dimers, <u>Mutation Research</u>, 143:109-112.

Moan, J., Dahlback, A., Larsen, S., Hendriksen, T. and Stammes, K., 1989, Ozone depletion and its consequences for the fluence of carcinogenic sunlight, <u>Cancer Research</u>, 49:4247-4250.

National Academy of Sciences (NAS), 1984, Causes and effects of changes in Stratospheric Ozone: Update 1983. Washington, D.C.: National Academy Press.

Niggli, H.J. and Cerutti, P.A., 1983, Cyclobutane-type pyrimidine photodimer formation and excision in human skin fibroblasts after irradiation with 313 nm ultraviolet light, <u>Biochemistry</u>, 22:1390-1395.

Reusch, M.K., Maeger, K., Leadon, S.A. and Hanawalt, P.C., 1988, Comparative removal of pyrimidine dimers from human epidermal keratinocytes in vivo and in vitro, J. Invest. Dermatol., 91:349-352.

Rheinwald, J.G. and Green, H., 1975, Serial cultivation of strains of human epidermal keratinocytes: The formation of keratinizing colonies from single cells, Cell, 6:331-344.

Schothorst, A.A., Bruynzeel, F. and Evers, L., 1988, A comparison between ultraviolet -B induced damage in cultivated human keratinocytes and melanocytes, J. Invest. Dermatol., 91 p. 384.

Schothorst, A.A., Slaper, H., Schouten, R. and Suurmond, D., 1985, UVB doses in maintenance psoriasis phototherapy versus solar UVB exposure, Photodermatology., 2:213-220.

Slaper, H., Schothorst, A.A. and Van der Leun, J.C., 1986, Risk evaluation of UVB phototherapy for psoriasis: Comparison of calculated risk for UVB therapy and observed risk in PUVA-treated patients, Photodermatology, 3:271-283.

Taichman, L.B. and Setlow, R.B., 1979, Repair of ultraviolet light damage to the DNA of cultured human epidermal keratinocytes and fibroblasts, J. Invest. Dermatol., 73:217-219.

Zeeland van, A.A., Smith, C.A. and Hanawalt, P.C., 1981, Sensitive determination of pyrimidine dimers in DNA of UV-irradiated mammalian cells Introduction of T4 endonuclease V into frozen and thawed cells, Mutation Res., 82:173-189.

EFFECT OF GENOTOXIC ENVIRONMENT OR TREATMENT ON DNA REPAIR OF UV LIGHT INDUCED DAMAGE OF LYMPHOCYTES

István Vincze[1], Larissa Gueth[1]
and György J. Köteles[2]

[1]Nat. Inst. of Hygiene, P.O.Box 64 H-1966 Budapest
Hungary
[2]F. Joliot-Curie Nat. Res.Inst. for Radiobiol. and
Radiohygiene, P.O.Box 101 H-1775 Budapest, Hungary

INTRODUCTION

In the last ten years following the description of the inducible repair processes in prokaryotic organisms, several experiments have demonstrated the adaptive changes of DNA repair in mammalian cells (Bockstahler and Lytle, 1970; Bockstahler and Lytle, 1971; Lytle and Coppey, 1978; Samson and Schwartz, 1980). In general, not the DNA repair per se, but other parameters such as reactivation of viruses and increasing resistance against different genotoxic effects, i.e. reduction of chromosome aberrations and cytotoxicity were measured in these experiments. The demonstration of this phenomenon in intact organisms is of great importance for the assessment of genetical risks. Pett, (1978), Buckley (1979) and Montesano (1979) and others have shown the enhancement of processes involved in repair of alkylated DNA molecules in the liver of rats following _in vivo_ treatments with genotoxic agents. In these cases, the adaptive response could be accounted for by the increased amount of a single protein (enzyme), the alkyltransferase. In our former work (Gueth et al., 1980) the adaptive change of excision repair in mouse lymphocytes following _in vivo_ gamma irradiation was found. This demonstrated the existence of the adaptive response in a more complex process of DNA repair. In addition to the experimental data, Pero (1982), Tuschl (1983), Benigni (1984), conducted epidemiological investigations to look into the changes of DNA repair processes in lymphocytes of human populations occupationally exposed to a genotoxic environment. However, the reported results are rather inconsistent.

In the work presented here we have examined in animal experiments and epidemiological investigations how the adaptive changes of DNA repair can be induced and how excision repair is related to certain factors during these adaptive changes.

Light in Biology and Medicine, Volume 2 Edited by R.H. Douglas et al.
Plenum Press, New York, 1991

507

MATERIALS AND METHODS

The animal experiments were performed on CCBA F1 male mice weighing 20 g. The _in vivo_ methyl methanesulfonate (Aldrich-Chemie) and 2-acetamido-fluorene (Sigma) treatments were given intraperitoneally in a single dose of 100 mg/kg b.w. and in four doses of 10 mg/kg b.w. on four successive days, respectively. For gamma ray treatment a ^{60}Co source was used (dose rate 0,33 Gy/min) giving whole body irradiation in one session. In the animal experiments lymphocytes were isolated from the spleen of the control and treated animals by squashing the organs in Hanks' medium. In the case of the human studies, the lymphocytes were isolated from blood samples by ficol-uromiro (Böyum, 1968) gradient centrifugation. The cell concentration was 10 x 10^6/ml and 2-5 x 10^6 cells/ml in the animal and human experiments, respectively.

To examine the DNA repair capacity, the cells were exposed in a 1 mm layer to different doses of UV light (0.5-6 J/m^2 and 4 J/m^2 in animal and human studies, respectively). Following irradiation, the cells were incubated for 60 minutes at 37°C in Hanks' medium supplemented with 185 KBq/ml ^{3}H-thymidine (Chemapol Prague, specific activity 2 TBq/mmol) and 5 x 10^{-3}M hyroxyurea (Sigma). The reaction was terminated by perchloric acid, the precipitate was then washed and after hydrolyzing the DNA for 30 minutes at 90°C in 0.5 M perchloric acid, the radioactivity was measured in Aquasol cocktail (New England Nuclear) by liquid scintillation. The DNA content was determined by Burton's method (Burton, 1956). In the case of the human study the DNA repair was expressed by a quotient using the following equation: DNA repair = (I-C)/C where I = the incorporation of the irradiated cells, C= incorporation of the unirradiated cells. The results obtained in this way were much less dependent on the specific activity of dTTP pool of the cells. The DNA repair values measured in the study ranged from 3.1 to 15.2 with an average of 7.2.

The activity of poly(ADP-ribose) polymerase was measured by the method of Altmann et al. (1979) using permeabilized cells and determining the incorporation of ^{3}H-NAD.

The dNTP pool was measured by the method of Hunting et al. (1981) with a slight modification. Briefly, after the extraction of the pool by 70% ethanol the dNTP was measured in the reaction mixture containing 1.4 U/ml of _E. coli_ DNA polymerase I (Sigma), 1.5 mg/ml activated calf thymus DNA (Calbiochem), two unlabeled dNTPs, 555 pmol/ml of each (Sigma), 164 kBq/ml ^{3}H-dTTP in same concentration (Chemapol Prague) or 164 kBq/ml ^{32}P-dATP in same concentration (Izinta Budapest). The incubation was carried out at 37°C for 20 minutes.

Chromosome abberations were analysed by the usual method as described by Almassy et al. (1984).

RESULTS AND DISCUSSION

After the _in vivo_ treatment of mice with genotoxic agents like gamma rays, methyl methanesulfonate (MMS), 2-acetamido-fluorene (AAF) producing different types of DNA damage, the

capacities of excision repair of the lymphocytes were
examined. The cells were isolated from the spleen of the
control or treated animals at different times after the
treatment. The cells were irradiated by different doses of UV
light and the DNA repair was followed by the incorporation of
^{3}H-thymidine. The results were expressed as the percent of
the data of control animals and plotted in Figure 1. The
three different agents changed the excision repair of the
lymphocytes in a different time course. In the case of gamma
irradiation, a marked decrease of DNA repair capacity was
found until the 2nd day. On the 7th day it reached the
control level and then the DNA repair capacity increased up to
the 14th day. The values returned to the control level on the
21st-23rd days. The MMS treatment usually didn't cause a
negative phase - although in some experiments a decrease was
found on the 1st day - the DNA repair capacity of the
lymphocytes increased very rapidly reaching the maximal value
on the 3rd-5th days. The adaptive response was transient and
on the 10th-11th days the values were similar to the control.
In contrast to MMS, following the AAF treatment there was no
remarkable change in the DNA repair capacity of the
lymphocytes for a period of 10-11 days. After this time the
values increased slowly reaching the maximum on the 28th -
30th days. In this case the duration of the adaptive response
was found to be the longest, the DNA repair values returned
to the control level only on the 38th and 43rd days.

The incorporation values at different doses of UV light
gave a saturation like curve which is transformed to a
straight line in double reciprocal plotting (Figures 2 and 3).
The DNA repair capacity was calculated from the intersection
of the straight line on the y axis (reciprocal of the
incorporation values). The UV dose, producing the half
maximal incorporation, in other words the common and apparent
K_M of the consecutive enzyme reactions of the DNA repair was
computed from the intersection of the straight line with the x
axis (reciprocal of UV doses). This value characterizes the
rate limiting step of the enzyme reactions taking part in the
DNA repair. The values found during the adaptive change of
DNA repair provoked by the MMS treatment are shown in Figure 2

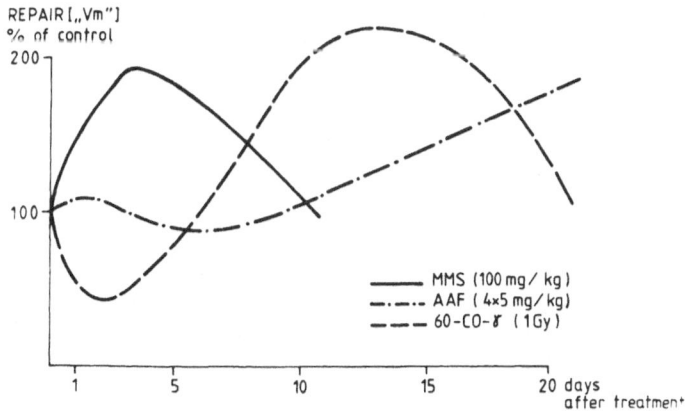

Fig. 1   The time course of the change of DNA
         repair

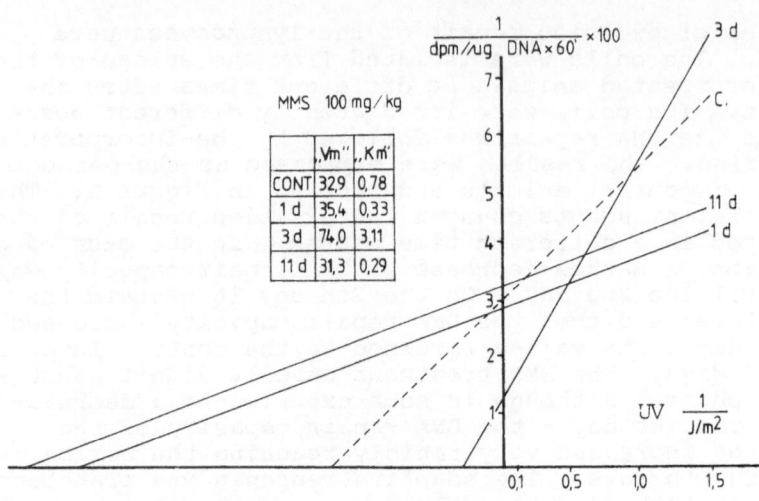

Fig. 2   UV light dose dependency of DNA repair at
         different times following MMS treatment

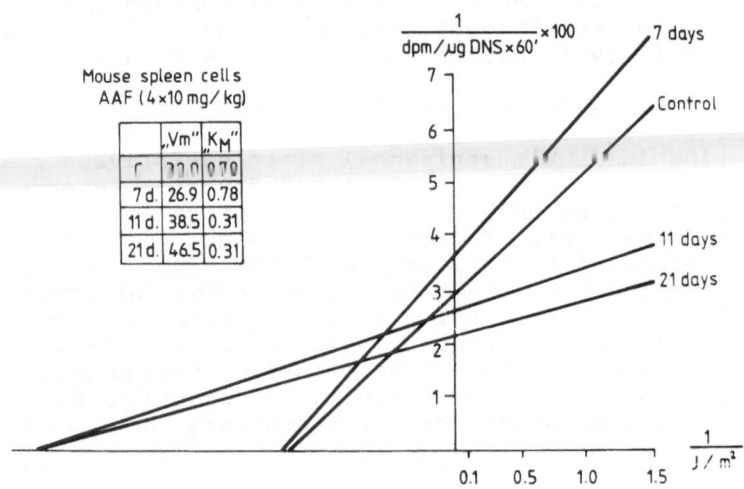

Fig. 3   UV light dose dependency of DNA repair at
         different times following AAF treatment

and the values measured following the AAF treatment are
plotted in Figure 3.   The common and apparent K_M values and
the maximal incorporation, or in other words the DNA repair
capacity, changed during the adaptive response in both cases
and the changes of the two parameters were not coupled
strongly.   The alteration of the K_M values suggests the change
of the rate limiting step of the enzyme reaction, which means
the alteration of the proportion of the different enzymes
during the adaptive response.   During this process the maximum
rate of the DNA repair can remain unchanged or can decrease or
increase depending on the activity of the new rate limiting
enzyme.

Figures 2 and 3 show that all types of these combinations were found in our studies. To verify the hypothesis on the change of the proportion of the enzymes during the adaptive response, the correlation between adaptive change of DNA repair and the <u>in situ</u> measurable enzyme activities were studied. There was no correlation between DNA repair capacity and the endonuclease activity measured by the single strand nick formation after UV light irradiation in the presence of arabinofuranosylcytosine (data not shown). The role of the poly(ADP-ribose) polymerase (PAR) in DNA repair was demonstrated by several works (Durkacz et al., 1980; Miwa et al., 1981; Shall, 1984). More recently Vijayalaxmi and Burkart (1989), found that the in <u>vitro</u> adaptive response of human lymphocytes was inhibited by 3-aminobenzamide, an inhibitor of PAR. The activity of the poly(ADP-ribose) polymerase measured by the incorporation of ^{3}H-NAD showed a linear correlation (r = 0.778) between the adaptive change of DNA repair capacity and the change of the PAR, irrespective of the agent inducing the adaptive response. The data are plotted as a proportion of control value (Figure 4).

The DNA repair measured by the incorporation of ^{3}H-thymidine more or less depends on the <u>in situ</u> activity of DNA polymerase (taking part in the repair). The activity of this enzyme might be regulated by the dNTP pool, therefore, the correlation between the repair incorporation and the dNTP pool was studied. Different, but in all cases acceptable correlations were found with the correlation coefficients of 0.81, 0.67, 0.76, 0.73 in the case of dTTP, dCTP, dATP, and dGTP respectively. The relationships were significant (p smaller than 0.05). The function describing the correlation between the dATP, dGTP and the DNA repair was different from that found in the case of dTTP, dCTP (data not shown).

Previous studies measuring the DNA repair of the lymphocytes in different human populations occupationally exposed to genotoxic environments gave diverse conclusions about the adaptive response of the DNA repair. Since

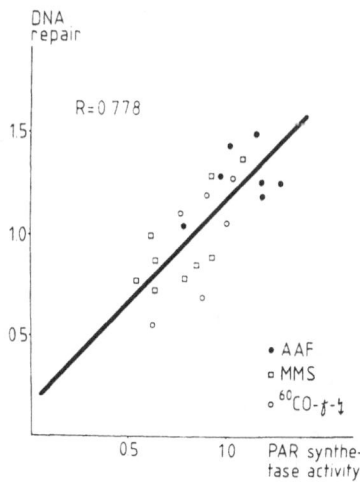

Fig. 4   Adaptive response of DNA repair and change of PAR

511

different methods were used for these studies and the size of populations examined was limited, we studied three populations occupationally exposed to different genotoxic agents using our method.

One of the populations consisted of workers in the rubber industry. It is generally accepted that the main genotoxic agents are, in this case, nitrosoalkyl compounds (Spigelhalder and Preussman, 1982).

The second occupationally exposed population consisted of uranium miners exposed to alpha rays emitting from ^{222}Radon. The third group of workers was selected from aluminium foundries where a relatively high concentration of benso(a)pyrene like compounds can be found. Individual dosimetry was impossible in these cases. Therefore, the exposure was calculated for the uranium miners from the periodically determined ^{222}Radon concentration in the working environment and expressed as WLM accumulated during the working life. In the other two groups the exposure was expressed in working years. The correlation between DNA repair of blood lymphocytes and exposure was investigated by a 3 x 3 contingency table and a Chi^2 test. Samples were divided into three groups according to the exposure. The first row contained the control samples collected from blood donors or newly employed workers. The second row included persons with moderated exposure i.e. with WLM smaller than 400 or less than 5 working years. The third row consisted of highly exposed workers i.e. with WLM higher than 400 or with more than 5 working years. The three coloumns represented three different classes of DNA repair capacity (expressed in quotient, see material and method). The first contained the low values between 0 and 4.6, the second the medium values between 4.6 and 9.0 and the third column contained the high values over 9.0. In all three populations a significant correlation was found according to the Chi^2 test between exposure and DNA repair, the p values were 0.027, 0.0002, 0.003 for the uranium miners, workers of the rubber industry and workers from the aluminium smeltery, respectively. The distribution of data in all the contingency tables was very similar to the table containing the data of the uranium miners (Table 1). In all populations the following phenomena were demonstrated: the frequency of cases with low repair capacity increased with the increasing exposure; the incidence of persons with high repair capacity was significantly higher in the exposed population but was not closely related to the extent of exposure.

We did not find significant correlation between DNA repair and other factors such as age, smoking habits, chronic diseases or medicine consumption either in control or exposed populations.

In the human epidemiological studies and in the animal experiments the gentoxic influences (treatment and environment) were different, as in the animal experiments single and medium doses were given while the human populations were chronically exposed to low doses. Moreover, the intensity of the genotoxic treatment was well known in the animal experiments but it was only estimated in the epidemiological studies. Despite the above specified differences, the _in vivo_ provoked adaptive response of the DNA

Table 1 Correlation between the exposure of the working environment and the DNA repair of uranium miners.

EXPOSURE		REPAIR			SUM
		0 - 4.6	4.6 - 9.5	9.5 -	
CONTROL	N	7	75	10	92
	%	7.6	81.5	10.9	100
100-399 WLM	N	9	54	15	78
	%	11.5	69.3	19.2	100
400-600 WLM	N	15	41	17	73
	%	20.6	56.1	23.3	100

repair of human lymphocytes was justifiable. The presence of this adaptive response was indicated by the increased incidence of the high repair values in the exposed population. It seems to be possible that the adaptive response found in human studies depends mainly on the doses of the last few days, because the duration of this response is not very long according to the animal experiments. This might be the reason why no correlation was found between the incidence of high repair values of the exposed population and the extent of exposure. The increasing incidence of low repair values with increasing exposure is not understandable on the basis of animal experiments. It can be supposed that the regulation of repair is disturbed by the chronic genotoxic effect.

In a conclusion of the epidemiological study on the alpha radiation induced chromosome aberration, Pohl-Ruhling (1979) supposed an adaptive change of DNA repair on the basis of an unexpected correlation. In some in vitro experiments, adaptive responses of the lymphocytes or, in other words, decreased chromosome aberration following low doses of genotoxic treatment was reported by Oliveri et al. (1984) and Shadley and Wolff (1987).

According to these results it was presumed that the adaptive change of DNA repair induced by the genotoxic environment caused a decrease in chromosome aberrations.

To study this hypothesis the chromosome aberrations per 100 WLM were calculated using the values of the last 5 years exposure. The values were divided into three groups: with 0-10, 10-22 and over 22. These groups represented the three rows of the 3 x 3 contingency table and furthermore they were separated into three columns according to the DNA repair values using the same classes presented in the 1st table. Distribution of the values was not random in Table 2. The calculated values of Chi^2 and p were 10.33 and 0.035, respectively, indicating a significant correlation between DNA repair and chromosome aberration per unit exposure. However,

Table 2 Correlation between the chromosome aberration
per exposure in the last 5 years and the DNA
repair of uranium miners

Aberration		REPAIR			SUM
100 WLM		0 - 4.6	4.6 - 9.5	9.5 -	
0-10	N	4	24	1	29
	%	13.8	82.8	3.4	100
10-22	N	3	27	9	39
	%	7.6	69.3	23.1	100
22-	N	15	39	7	61
	%	24.6	63.9	11.5	100

the distribution of the values in the second row differed
markedly from that expected in case of a linear relationship.
One explanation for this phenomenon might be that the
chromosome aberrations correlate well with the cumulated doses
but the adaptive change of the repair depends only on the
doses cumulated shortly before the examination.

REFERENCES

Almássy, S., Kanyár, B., Köteles, G.J., Bank, J., Krommer, I.
and Soós, K., 1984, Cytogenetic effects of radon-
daughters., in: "Proc. Symp. on Occupational Hygiene in
Mines", Pécs, Hungary.
Altmann, H., Dolejs, E., Topaloglou, A., Soóki-Tóth, A., 1979,
Faktoren die die DNA Reparatur in "spacer und core DNA"
von Chromatin menschlicher Zellen beinflussen, Studia
Biophysica, 76:195.
Benigni, R., Calcagnile, A., Fabri, G., Giuliani, A.,
Leopardi, P. and Paoletti, A., 1984, Biological
monitoring of workers in rubber industry, Mutation Res.,
138:468.
Bockstahler, L.E. and Lytle, C.D., 1970, Ultraviolet light
enhanced reactivation of a mammalian virus, Biochem. and
Biophys. Res. Com., 41:184.
Bockstahler, L.E. and Lytle, C.D., 1971, X-ray enhanced
reactivation of ultraviolet-irradiated human virus, J.
Virology., 8:601.
Böyum, A., 1968, Isolation of mononuclear cells and
granulocytes from human blood, Scand. J. Clin. Lab.
Invest., 21:77.
Buckley, J.D. and O'Connor P.J. and Craig, A.W., 1979,
Pretreatment with acetylaminofluorene enhances the repair
of 0^6-methylguanine in DNA, Nature, 281:403.
Burton, KI., 1956, A study of the conditions and mechanism of
diphenylamine reaction for colorimetric estimation of
deoxyribonucleic acid, Biochem. J., 62:315.
Durkacz, B.W., Omidiji, O., Gray, D.A. and Shall, S., 1980,

(ADP-ribose) participates in DNA excision repair, <u>Nature</u>, 283:593.

Gueth, L., Szabo, L. and Vincze, I., 1980, Study of UV-radiation-induced DNA repair in mouse spleen cells after total-body ^{60}Co- irradiation, <u>Radiobiologiia.</u>, 20:508 (in Russian).

Hunting, D. and Henderson, J.F., 1981, Determination of deoxyribonucleoside triphosphates using DNA polymerase: a critical evaluation, <u>Can. J. Biochem.</u>, 59:723.

Lytle, C.D. and Copper, J., 1978, Enhanced survival of ultraviolet irradiated herpes simplex virus in carcinogen-pretreated cells, <u>Nature.</u>, 272:60.

Montesano, R., Bresil, H. and Margison, G.P., 1979, Increased excision of 0^6-methylguanine from rat liver DNA after chronic administration of dimethylnitrosamine, <u>Cancer Res.</u>, 40:452.

Oliveri, G., Bodycote, J. and Wolff, S., 1984, Adaptive response of human lymphocytes to low concentrations of radioactive thymidine, <u>Science.</u>, 223:594.

Pegg, A.E., 1978, Effect of pretreatment with other dialkylnitrosamines on excision from hepatic DNA of 0^6-methylguanine produced by dimethylnitrosamine, <u>Chem. Biol. Interact.</u>, 22:109.

Pero, R.W., Bryngellson, T., Högsted, B. and Akesson, B., 1982, Occupational and <u>in vitro</u> exposure to styrene assessed by unscheduled DNA synthesis in resting human lymphocytes, <u>Carcinogenesis.</u>, 3:681.

Pohl-Rüling, J., Fischer, P., 1979, The dose-effect relationship of chromosome aberrations to alpha and gamma irradiation in population subjected to an increased burden of natural radioactivity, <u>Radiation Res.</u>, 80:61.

Samson, L. and Schwartz, J.L., 1970 Evidence for an adaptive DNA repair pathway in CHO and human skin fibroblast cell lines, <u>Nature.</u>, 287:861.

Shadley, J.D. and Wolff, S., 1987, Very low doses of X-rays can cause human lymphocytes to become less susceptible to ionizing radiation, <u>Mutagenesis.</u>, 2:95.

Shall, S., 1984, ADP-ribose in DNA repair: a new component of DNA excision repair, <u>Adv. Radiat. Biol.</u>, 11:1.

Spigelhalder, B. and Preussmann, R., 1982, Nitrosamines and rubber, <u>in</u>: "N-Nitroso Compound: Occurrence and Biological Effects", H. Barts, I.K. O'Neil, M. Castegnaro and M. Okada, eds., (IARC Scientific publications No. 41), International Agency for Research Cancer, Lyon.

Tuschl, H., Kovac, R., Altman, H., 1983, UDS and SCE in lymphocytes of persons occupationally exposed to low levels of ionizing radiations, <u>Health Phys.</u>, 45:1.

Vijayalaxami and Burkart W., 1989, Effect of 3-aminobenzamide on chromosome damage in human blood lymphocytes adapted to bleomycin, <u>Mutagenesis.</u>, 4:187.

PICOSECOND LASER UV INACTIVATION OF λ BACTERIOPHAGE AND VARIOUS <u>Escherichia coli</u> STRAINS

David N. Nikogosyan

Institute of Spectroscopy, USSR Academy of
Sciences, Troitzk, Moscow Region, 142092, USSR

INTRODUCTION

This report is a short review of the results obtained in the last decade at the Institute of Spectroscopy, USSR Academy of Sciences, and at the Insitute of Molecular Biology, USSR Academy of Sciences, on inactivation of bacterial viruses and bacteria by high-intensity laser UV pulses of picosecond duration ($I=10^{10} \div 10^{14}$ W/m^2, λ =266 nm, τ_p=23 or 29 ps, Δf=1 pps).

The following two factors are essential for the correct performance of laser photobiological experiments. Firstly, it is necessary that the laser beam should cover every viral particle in the test cell, no matter what the irradiation intensity. The standard rectangular spectrophotometric quartz cells utilised by us in the first experiments (Belogurov et al. 1980; Gurzadyan et al. 1981) are disadvantageous in that they always have some of the preparation left on their walls which is not inactivated by the laser (especially in the case of high laser intensities where the laser beam cross sectional area is small). In the early works (Belogurov et al. 1980; Gurzadyan et al. 1981), this led to distortions in the survival curves that were explained by the presence of a UV-resistant fraction in the population. Secondly, to know exactly the actual intensity and the dose of radiation reaching inside the test cell, one should take account of the spatial distribution of local laser intensity across the beam.

Both factors were taken into consideration by Nikogosyan et al, (1986) in studying the picosecond laser UV inactivation of the λ_{11} bacteriophage and <u>E.coli</u> bacteria. They have shown that, as the irradiation intensity increases up to $I=2 \times 10^{11}$ W/m^2, the inactivation cross-section does not change at all and then, starting from $I> (3-7) \times 10^{11}$ W/m^2, it begins to drop sharply, i.e. the UV resistance of the λ phage starts to increase. As the picosecond UV irradiation intensity is raised from 2×10^{11} W/M^2 to 2×10^{13} W/m^2, the UV resistance of the λ phage increases by around 67 times when it is plated onto the strains AB 2480 and AB 1886, and by 45 times when plated onto the strain AB 2463, and 33 times when plated onto the strain AB 1157.

According to Setlow and Setlow (1962), the inactivation of bacterial viruses in the case of low-intensity continuous UV irradiation is mainly due to the induction of single-quantum photolesions (pyrimidine dimers) in the viral DNA. The results obtained by us (Gurzadyan et al. 1981, Nikogosyan et al. 1986) show that the change-over to high-intensity picosecond UV irradiation reduces the quantum yield of lethal single-quantum photolesions in the phage DNA.

It should be noted that as one changes over to UV irradiation with $I>3-7x10^{11}$ W/m^2, the relative contribution of single-quantum DNA photolesions to the total lethal photoproduct in the UV-irradiated λ phage is observed to decrease. Table I lists data on the relative contribution of photolesions repaired by a given repair system of the host cell to the total lethal photoproduct produced in the phage in the course of its picosecond UV irradiation. As the mutant strain E.coli K12 AB 1886 is devoid of uvr ABC endonuclease capable of excising pyrimidine dimers, the reduction of the proportion of lesions repaired by the uvr system from 79 to 54% is indicative of a decrease in the relative contribution of single-quantum photolesions (cyclobutyl pyrimidine dimers included) to the total lethal photoproduct. It can be seen from Table I that as the picosecond UV irradiation intensity is increased, the proportion of photolesions repaired by the rec system decreases from 40 to 18%. This fact also bears witness to a decrease in the relative contribution of single-quantum photolesions (cyclobutyl pyrimidine dimers) to the total lethal photoproduct. Nevertheless, even at the intensity of $I=2 \times 10^{13}$ W/m^2, the contribution of single-quantum photolesions to the lethal effect remains considerable.

The conclusion that the relative contribution of single-quantum photolesions to the total lethal photoproduct decreases is borne out by the results of the experiments on the photoreactivation of the UV-irradiated λ phage (Belogurov et al. 1980; Gurzadyan et al. 1982). It was shown by Gurzadyan et al. (1982) that whereas the contribution of

Table 1   Fraction (%) of photolesions of the λ_{11} bacteriophage repaired by different host-cell repair system relative to the total lethal photoproduct at high-intensity picosecond UV irradiation (Nikogosyan et al. 1986).

Repair system	Irradiation intensity (W/m^2)		
	$7x10^{10}$	$2x10^{12}$	$2x10^{13}$
uvr	79	65	54
rec	40	27	18
uvr+rec	81	75	67

pyrimidine dimers to the lethal effect amounts to 85% of $I<10^{11}$ W/m^2, it rapidly diminishes as the irradiation intensity is increased in excess of 10^{11} W/m^2, coming down to around 10% at $I=4 \times 10^{13}$ W/m^2.

The same principal results were observed with picosecond laser UV irradiation of plasmid pBR 322 DNA (Gurzadyan et al., 1981, 1982). A decrease in inactivation cross section together with the decrease of the photoreactivation sector value was observed to take place simultaneously with a decline in the picosecond UV irradiation intensity. This suggests that the contribution of protein to the picosecond laser UV inactivation of the λphage is insignificant.

Similar results were obtained in the case of picosecond UV inactivation of E.coli (Nikogosyan et al. 1986). It was shown that increasing the picosecond UV laser intensity from 2×10^{10} to 2×10^{13} W/m^2 enhances the UV resistance of AB 2463 strain by 476 times, that of AB 1886 strain by 411 times and that of the AB 1157 strain by 13 times. Table 2 lists data on the relative contribution of photolesions repaired by a given repair system to the total lethal product produced in E.coli in the course of its high-intensity picosecond UV irradiation. At an intensity of 2×10^{10} W/m^2, the single-quantum lesions (cyclobutyl pyrimidine dimers included) repaired by the uvr system amount to 97% of all photolesions induced in E.coli. It also follows from Table II that as the intensity of picosecond-pulse UV irradiation is increased, the proportion of photolesions repaired by uvr system is reduced from 97 to 14%. At the same time, the proportion of photolesions repaired by the rec system decreases from 98 to 41%. Thus, in the case of picosecond UV irradiation of E.coli at $I=2 \times 10^{13}$ W/m^2, single-quantum products are no longer the main lethal photolesions, this place being filled by photoproducts of two-quantum origin.

In experiments with model compounds we tried to demonstrate directly a decrease in the quantum yield of single-quantum dimerization reaction with an increase in the intensity of picosecond laser UV irradiation. In irradiating ribonucleoside Urd (Khoroshilova and Nikogosyan 1990), dinucleoside-monophosphate dTpdT (Esenaliev et al. 1987), oligonucleotide poly (dT) (Zavilgelsky et al. 1984) the increase of irradiation intensity, starting from $I=3 \times 10^{12}$ W/m^2

Table 2 Fraction (%) of E.coli photolesions repaired by the given E.coli repair system relative to the total lethal photoproduct at high-intensity picosecond UV irradiation (Nikogosyan et al. 1986)

Repair system	Irradiation intensity (W/m^2)		
	2×10^{10}	4×10^{12}	2×10^{13}
uvr	97	35	14
rec	98	53	41

led to a decrease in dimerization quantum yield by about two times. On the other hand, at picosecond laser UV irradiation of the pAT48 plasmid DNA the quantum yield of dimerization reaction decreased from the maximum value Φ_{dim} 8×10^3 at $I=10^{11}$ W/m^2 to $\Phi_{dim}= 1 \times 10^{-4}$ at $I=3 \times 10^{13}$ W/m^2, that is about 80 times (Esenaliev et al. 1987). This latter fact agrees well with the results of the above-mentioned photobiological experiments (in the case of λ phage plated onto the AB 1886 uvrA$^-$ strain, defective with respect to excision repair).

It should be emphasized that in all these cases at picosecond UV irradition two-step excitation of chromophores (bases) is realised via the $S_0 \rightarrow S_1 \rightarrow S_N$ channel (Zavilgelsky et al. 1984). The observed difference between the degrees of Φ_{dim} decrease, as shown by Khoroshilova and Nikogosyan (1989), is connected to the difference in the dimerization mechanisms. In Urd, dTpdT, poly (dT) dimers are formed from the T_1 triplet level which at high excitation intensities is populated, in addition to the ordinary channel $S_0 \rightarrow S_1 \rightarrow T_1$, due to the intersystem transition $S_N \rightarrow T_N$ and radiationless relaxation $T_N \rightarrow T_1$. Therefore, the dimerization quantum yield drops by only two times. In DNA dimers are formed from the S_1 singlet level, which is not populated as a result of the $S_N \rightarrow S_1$ transition. According to Khoroshilova and Nikogosyan (1989), in this case mainly $S_N \rightarrow S_0$ relaxation takes place.

Such interpretation agrees with the conformation properties of these molecules. Indeed, it is known that definite mutual orientation of two adjacent pyrimidine bases is required to form dimers. In Urd, dTpdT, poly (dT) such orientation occurs with higher probability at excitation to the long-lived triplet state T_1. On the other hand, in DNA, pyrimidine bases are densely packed and oriented steadily in a favourable way. Therefore, when DNA is irradiated, almost all dimers are formed during excitation to the short-lived singlet state S_1.

It should be noted that the threshold intensity value, from which the Φ_{dim} value starts to fall is much lower in DNA than in Urd, dTpdT or poly (dT). This is due to cooperative effects in the two-step UV excitation of DNA (Nikogosyan et al. 1985).

Thus, the effect of UV resistance increase at high-intensity picosecond UV irradiation of biological objects is explained by a drop of the quantum yield of single-quantum photoreaction of pyrimidine dimer formation in DNA.

REFERENCES

Belogurov, A.A., Zavilgelsky, G.B., Angelov, D.A., Krykov, P.G., Letokhov, V.S. and Nikogosyan, D.N., 1980, Action of ultrashort laser pulse on λ bacteriophage, Stud. Biophys. 80:45

Esenaliev, R.O., Panyutin, I.G., Oraevsky, A.A., Nikogosyan, D.N., and Zavilgelsky, G.B., 1989, Quantum yields of single-quantum and two-quantum photochemical reactions in dinucleoside-monophosphate dTpdT and DNA under high intensity picosecond UV irradiation, Dokl. Akad. Nauk SSSR, 293:232.

Gurzadyan, G.G., Nikogosyan, D.N., Kryukov, P.G., Letokhov,
    V.S., Balmukhanov, T.S., Belogurov, A.A., and
    Zavilgelsky, G.B., 1981, Mechanism of high power laser UV
    inactivation of viruses and bacterial plasmids,
    Photochem. Photobiol., 33:835.
Gurzadyan, G.G., Nikogosyan, D.N., Balmukhanov, T.S., and
    Zavilgelsky, G.B., 1982, The study of formation of
    single-strand break in the DNA chain under picosecond
    laser UV irradiation, Photobiochem. Photobiophys. 4:87.
Khoroshilova, E.V., and Nikogosyan, D.N., 1990, Photochemistry
    of uridine on high intensity laser UV irradiation, J:
    Photochem. Photobiol. B. Biol., 5:413.
Nikogosyan, D.N., Oraevsky, A.A., Letokhov, V.S., Arbieva,
    Z.Kh., and Dobrov, E.N., 1985, Two-step picosecond UV
    excitation of polynucleotides and energy transfer, Chem.
    Phys., 97:31.
Nikogosyan, D.N., Oraevsky, A.A., and Zavilgelsky, G.B., 1986,
    Picosecond laser UV inactivation of λbacteriophage and
    various Escherichia Coli strains, Photobiochem.
    Photobiophys.
Setlow, R.B., and Setlow, J.K., 1962, Evidence that
    ultraviolet-induced thymine dimers in DNA cause
    biological damage, Proc. Natl. Acad. Sci. USA, 48:1250.
Zavilgelsky, G.B., Gurzadyan, G.G., and Nikogosyan, D.N.,
    1984, Pyrimidine dimers, single-strand breaks and
    crosslinks induced in DNA by powerful laser UV
    irradiation, Photobiochem. Photobiophys., 8:175.

PHOTORECEPTION IN UNICELLULAR FLAGELLATES: BIOELECTRIC PHENOMENA IN PHOTOTAXIS

Oleg A. Sineshchekov

Biology Department, Moscow State University
Moscow, USSR

INTRODUCTION

Light control of motile behaviour is widely spread in microorganisms from bacteria to ciliates (Nultsch and Hader, 1988). Photomovements play a crucial role in cell survival and development and represent the fast and obvious examples of primitive photosensory responses (Foster and Smyth, 1980; Lenci et al., 1984). Photosynthetic flagellates possess the highly specialized light reception mechanism, which enables them to sense the spatial distribution of illumination and the direction to the light source. The idea that specialized phototaxis could be considered as a primitive analog of vision has been strongly supported in recent years. The photoreceptor pigment in Clamydomonas was shown to be a retinal-protein, probably similar to the bovine rhodopsin, (Foster et al., 1984; Hegemann et al., 1988) and the photoelectric responses, which are known to be involved in vision, were found in its close relative Haematococcus (Litvin et al., 1978; Sineshchekov, 1978; Sineshchekov et al., 1978). The aim of this presentation is to summarize briefly our data on the key role of the photoelectric phenomena in the light stimulus transduction chain in phototaxis and to describe some features of this mechanism, revealed by photoelectric measurements in a single cell.

The advance in investigation of phototactic electrical responses was achieved by the development of the measuring suction micropipette technique (Sineshchekov, 1978). In this method the cell is sucked into the tip of micropipette, causing two parts of cell surface (those inside and outside the pipette) to become electrically isolated by the glass (see schemes on Fig. 3). If an electrical response is local and takes place only in one of these two parts of membrane, a potential difference appears between the inside and outside of the pipette. The method allows one to measure the photoelectric responses during several hours without damaging the cell and has several other benefits (discussed below).

Light in Biology and Medicine. Volume 2 Edited by R.H. Douglas *et al.*
Plenum Press, New York. 1991

523

For quantitative determination of the kinetics of photomotile responses we used the continuous monitoring of the beat frequency of the individual flagellum as the cell was fixed on the tip of a micropipette. The light probe was passed through the area of flagellum motion near its base. The signal from a photomultiplier, modulated by the flagellum movement, was then recorded and analyzed (Sineshchekov and Litvin, 1988). Some results obtained by usual single cell or population methods are also taken into account in this presentation.

Light induces in a cell of <u>Haematococcus pluvialis</u> a complex of bioelectrical processes (Litvin et al., 1978; Sineshchekov, 1978, 1988) (Fig. 1). The three main responses can be distinguished in the electrical signal by their different kinetics and features, discussed below. Transient <u>primary potential difference</u> (PPD) appears in the millisecond time range upon the intense flash or continuous light stimulus. It is followed by the permanent <u>late potential difference</u> (LPD), which has lower amplitude and dissipates with a time constant of 15 to 40 ms as the light is switched off. Both these responses depend in a graded fashion on light intensity and above the threshold level become superimposed by an all-or-none <u>regenerative response</u> (RR). Its delay time (but not the amplitude) depends on light intensity.

The spectral sensitivity relates the found electrical responses to phototaxis (Litvin et al., 1978; Sineschchekov, 1978; Sineshchekov and Litvin, 1988). Action spectra of potential generation and photomotile behaviour coincide with each other even in their complex fine structure (Fig. 2), which indicates the high rigidity of the chromophore in the protein moiety. But spectral characteristics of a single pigment could hardly account for at least 6 bands in the spectra. This led us to the conclusion that probably more than one pigment participates in the absorption of phototactically active light in flagellates (Sineshchekov and Litvin, 1982). The involvement of two photoreceptor pigments in phototaxis of <u>Halobacterium halobium</u> has been firmly established in recent years (Spudich and Bogomolni, 1988).

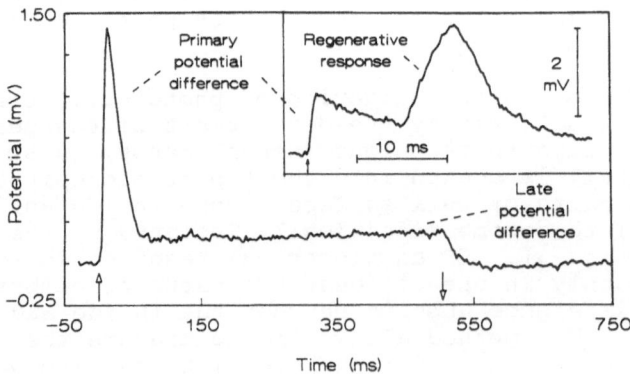

Fig. 1 Three photoelectric responses in phototaxis.

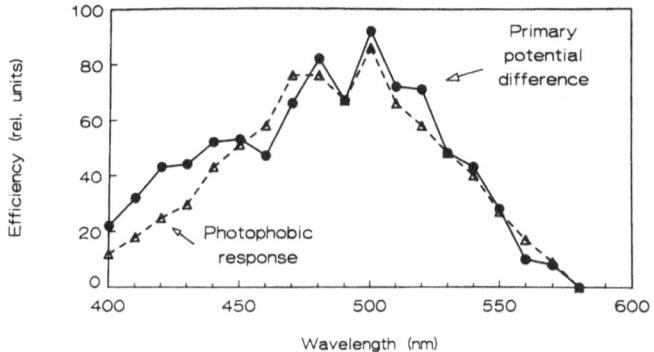

Fig. 2   Spectral dependency of photoelectric and
          photomotile responses.

        Photoinduced electric responses are generated only when
the stigma region of the cell is illuminated (Sineshchekov,
1978; Ristori et al., 1981).   This proves the widely accepted
view that the photoreceptor is located in the part of the
membrane covering the stigma (Foster and Smyth, 1980).
Changing the position of the cell in the tip of the
micropipette it was also possible to determine the
localization of electrical generators responsible for the
electrical responses (Fig. 3).   The sign of PPD (as well as
LPD, data not shown) is changed from positive to negative only
when the stigma region of the cell is transferred during
suction from outside to inside the pipette.   On the other
hand, the sign of RR depends upon whether the flagella are
located outside the pipette (positive) or inside the pipette
(negative response).   It means that the gradual PPD and LPD
are localized in the photoreceptor region and the threshold RR
is related to the flagella, all of the responses being
depolarizing in their initial direction.

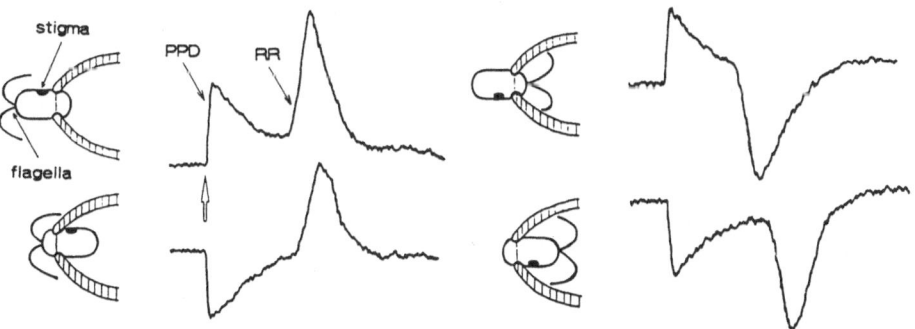

Fig. 3   Dependence of the sign of primary potential
          difference and regenerative response on the
          position of stigma and flagella in the
          pipette.

The RR can be considered as the driving force of the
photophobic behaviour of the cell. This motile response
appears upon an abrupt change in light intensity and prevents
the organism from crossing the light/dark boundary. It is
strongly Ca^{2+}-dependent in <u>Haematococcus</u>, as well as in many
other flagellates (Nultsch and Hader, 1988). The electric RR
is also strictly Ca^{2+}-dependent and disappears upon its
removal by EGTA or by addition of a low concentration of
calcium blockers. Both electric and motile responses are all-
or-none with similar light dependent delay. They have
identical spectral sensitivity and depend in the same way on
red background illumination and other experimental condtions.
Finally, direct microscopic observation reveals that flagella
reorientation takes place only when the RR is registered.
Thus, the massive Ca^{2+} influx during the described RR leads to
photophobic response of flagella.

The quantitative analysis of the motile responses in
individual flagella shows that the behaviour of two flagella
are similar but not identical (Sineshchekov and Litvin, 1988)
(Fig. 4). Moreover, as the stimulus rises in the narrow light
intensity range near the threshold the response of only one
flagellum could be observed. Independently on the position of
the cell in the pipette the flagellum on the sitgma bearing
side (cis-stigma flagellum) is more sensitive to illumination.
This observation indicates that each flagellum behaves like an
independent excitable organelle. Thus, Ca^{2+} channels
responsible for RR mediating the photophobic behaviour are
most probably located in the membrane covering each flagellum.
This is confirmed by the observation that two successive
electric regenerative responses could be registered under
intermediate stimulus intensities, each of them probably
reflecting excitation in individual flagella.

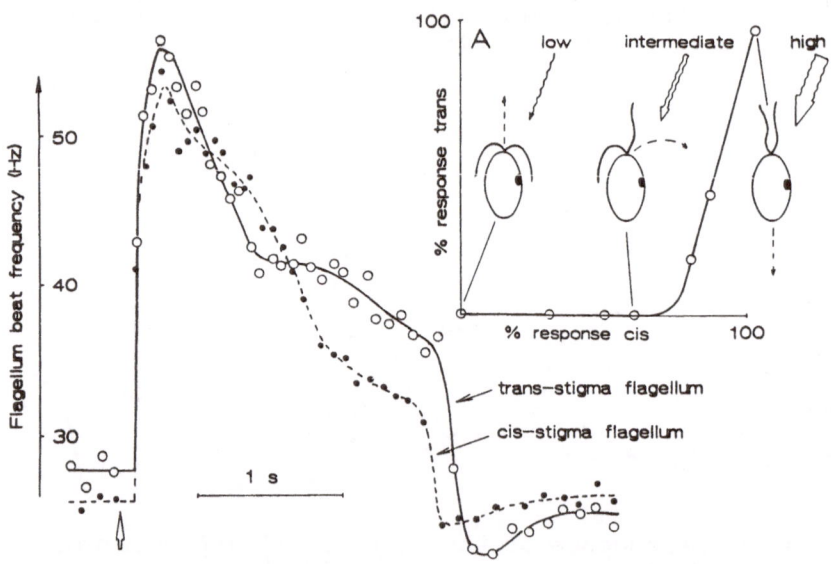

Fig. 4   Photophobic response of individual
         flagellum. A: different light sensitivity
         of cis- and trans-stigma flagella.

The functional _in vivo_ difference between the two
flagella found in these experiments provides the basis for the
second main type of photomotile behaviour - photoorientation,
or true phototaxis. According to a widely accepted hypothesis
a cell uses the spatially sensitive antenna for its phototaxis
(Foster and Smyth, 1980). We have checked this assumption
directly. The measuring pipette with the cell was rotated so
that different parts of the cell surface were exposed to
light. Indeed, the maximum photoinduced potential was
observed when the cell was oriented toward the light source
with its stigma bearing side. Maximum difference between the
electrical signals of two opposite orientations of the cell
was found at the wavelength actively absorbed by stigma and
chloroplast. This confirms the "periodic shading hypothesis"
and the localization of photoreceptors in the membrane
covering the stigma. In freely motile organism the light
absorbed by the photoreceptor should be modulated (due to the
cell rotation) about three times in amplitude and one second
in period. These conditions were simulated for the fixed
cell. Due to the existence of the refractive period of the RR
and low amplitude of light intensity variation, only gradual
LPD was observed under these conditions. The motile responses
of individual flagella were also found to be smooth, gradual
and without changing the type of undulation (Fig. 5). The two
flagella not only show different, but in most cases opposite
motile behaviour. The cis-stigma flagellum shows a positive
photokinesis and the trans-stigma flagellum a negative
photokinesis under the periodic illumination. Thus, the
correction of cell orientation during phototaxis in biflagellated _Haematococcus_ is based on the functional
difference in two morphologically identical flagella and takes
place twice during one rotation cycle.

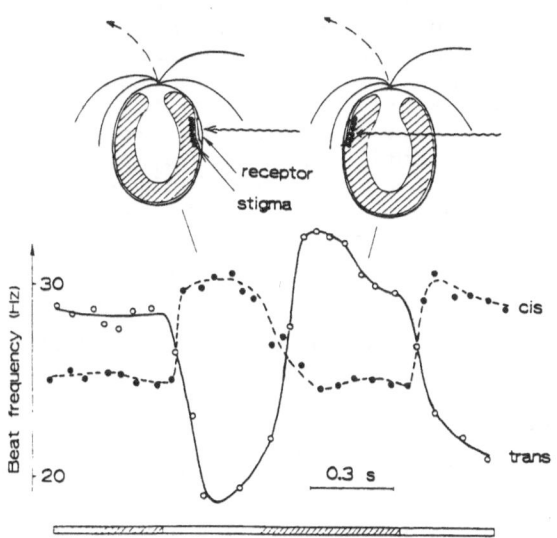

Fig. 5   Opposite photokinetic responses in cis-
          and trans-stigma flagella, which provide
          the correction of cell orientation in
          phototaxis (scheme above).

527

The relation between the gradual electrical events in the stigma region and RR reveal that signal transduction from the photoreceptor to the motor apparatus is based most probably on the electrotonic spread of a potential difference change along the cell membrane (Sineschchekov, 1988). It was found that the integral of light induced potential difference before the RR start is constant for different light intensities. It means that equal amounts of charge should enter the cell to initiate the response. According to our calculation the membrane must be discharged by several tens of millivolts to open the potential dependent Ca^{2+} channels and initiate the threshold electrical response. Under photosynthetically active red background illumination, which is known to hyperpolarize the cell (Sineshchekov et al., 1976), the amplitude of the critical discharge increases more than twice.

The saturation of LPD begins at very low light intensities. Its amplitude rises not more than twice while the intensity of light increases by three orders of magnitude. It means that LPD is limited by secondary (biochemical) events in the signal transduction chain. At an intensitie of 0.1 W/m^2 the amplitude of LPD gives a value for the photoinduced current through the receptor up to 10 unit charge/s. The amount of quanta absorbed at this intensity by the photoreceptor is not more than 10^3 per second (see Foster and Smyth, 1980). Thus, at least 4 orders of magnitude amplification is involved in the photosensory chain of phototaxis. These features allow one to consider the LPD observed in Haematococcus to be the analog of the late receptor potential in vision (Stavenga, 1980).

The ionic nature of the late photoreceptor current remains unknown, but it is most probably not linked to calcium. Neither PPD nor LPD are completely inhibited by the removal of the traces of Ca^{2+} from the medium and by the Ca^{2+} channel blocker ruthenium red, although the effectiveness of these treatments on Ca^{2+} transport is obvious from the disappearance of RR and the motile responses. On the other hand the late receptor potential of phototaxis as well as the photomotile responses are inhibited by 1-cis-diltiazem - the blocker of c-GMP-dependent ion channels, at usually effective concentrations. This fact and the highly efficient signal amplification mentioned above allow the suggestion that the light signal transduction chain in phototaxis involves a biochemical cascade probably similar to that in vision (Kaupp and Koch, 1986).

The question arises whether the high amplitude transient PPD reflects only the fast appearance and decay of ion currents or whether it could originate from charge movements in photoreceptor molecules. In the experiments with laser excitation and improved time resoltuion it was found that PPD is biphasic both in its rise and decay (Fig. 6). The fast component of PPD has no, or less than 30 microseconds, delay and thus most probably precedes the biochemical steps of light signal transduction (Sineshchekov et al., 1990). This assumption is confirmed by the very high level of light saturation of PPD. The data fit with better than 1% accuracy to the exponential function, assuming that the high light intensity saturated component of PPD is limited only by the product of quantum yield and optical cross-section of

photoreceptor molecule (Fig. 7). The experimental data give a
value of approximately 0.8 Å^2 for this product, which is
reasonable for the single retinal-protein molecule. Thus, the
fast component of PPD most probably reflects the charge
movement in pigment molecules and can be considered as the
early receptor potential of phototaxis. The second component
which has a delay of 150 to 400 microseconds could indicate
the beginning of the generation of the late receptor
potential.

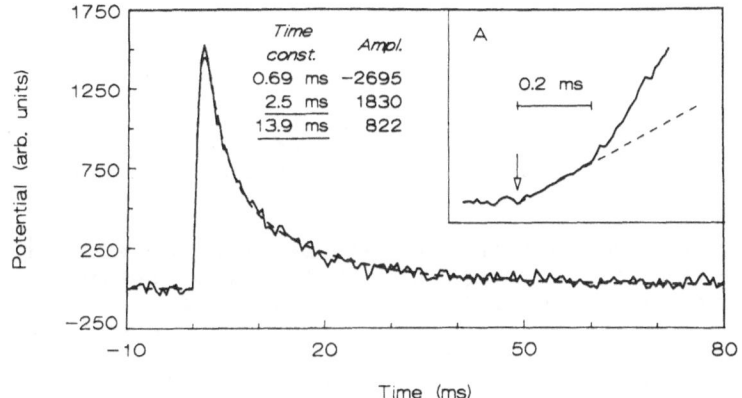

Fig. 6 Computer fit of primary potential
        difference by the sum of two exponentials
        (the rise of the signal was fitted by one
        exponential with negative amplitude). A:
        two components of laser-induced electrical
        signal rise. 10 ns flash is indicated by
        arrow.

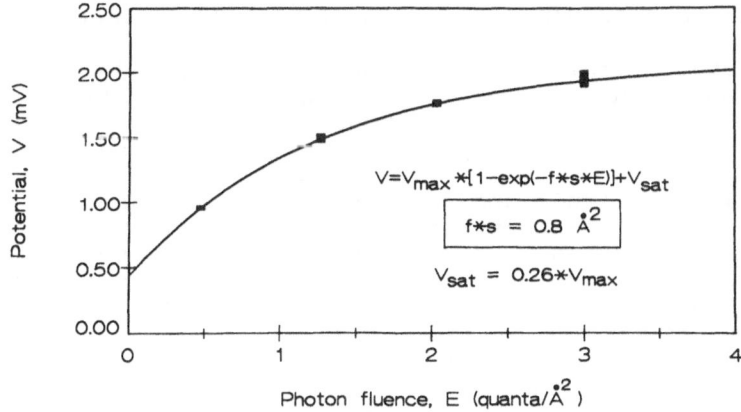

Fig. 7 Computer fit of light intensity saturation
        of primary potential difference by the
        exponential equation. f - quantum yield of
        potential generation, s - optical cross
        section of pigment, V_{sat} amplitude of
        low light intensity saturated component.

The early receptor potential has a refractive period of more than 100 ms after the saturation flash. Since this signal originates directly from pigment molecules, it reveals a limit for the turn-over time of the photoreceptor cycle. The orientation of the chromophore in the membrane was also determined. The maximum signal was found when the plane of polarization of the actinic light was oriented parallel to the receptor surface, indicating that the retinal is oriented in the plane of the membrane (Sineshchekov, 1988). These and above mentioned features of the photoreceptor, revealed by electric measurements, show that this method provides a powerful tool for the investigation of the photoreceptor mechanism in vivo.

CONCLUSION

In conclusion, the following hypothetical cascade of processes of light signal transduction for phototaxis could be suggested:

1. Light absorption: in the cell membrane covering stigma; more than one pigment; high rigidity of retinal chromophores; absorption bands: 420, 445, 480, 500, 520, 545 nm; quantum yeild * cross-section - about 0.8 $Å^2$; turn-over time - several hundreds of milliseconds; dipole transitional moment - in the plane of membrane; spatial sensitivity by shading and pigment orientation.

2. Early receptor potential: positive charge movement into the cell within the photoreceptor pigment molecules; high light intensity saturation; no lag-period; less than 1 ms rise-time.

3. Late receptor potential: in the photoreceptor region of the cell membrane; low light intensity saturation; depolarization; delayed by 150 - 400 microseconds; amplification - about 10^4; non-calcium (?).

4. Electrotonic depolarization of flagella membrane.

5a. If step stimulus: activation of potential-dependent Ca^{2+} channels in membrane surrounding each flagellum; regenerative electric response; massive influx of Ca^{2+} into the flagellum space; photophobic response.

5b. If low amplitude and relatively high frequency modulation of receptor illumination (due to the cell rotation in unilateral light): operation of voltage-dependent divalent cation channels, or the direct modulation of divalent cation fluxes by electrotonically spread late receptor potential; changes in Ca^{2+} (and probably other ions) concentration in flagella; unbalanced or opposite photokinetic responses of cis- and trans-flagellum; orientation corrections; phototaxis.

ACKNOWLEDGEMENTS

I wish to thank Professor F. F. Litvin and Professor L. Keszthelyi for useful discussion, Dr. Cs. Bagyinka for expert help in computer evaluation of data, E. Govorunova for assistance in experimental work and members of the Biology Department of Moscow State University and Institute of Biophysics BRC Hungarian Academy of Sciences for the help and friendly atmosphere.

REFERENCES

Foster, K.W. and Smyth R.D., 1980, Light antennas in phototactic algae. Microbiol. Rev., 44: 572-630.

Foster, K.W., Saranak, L., Patel, N., Zarilli, G., Okabe, M., Kline, T. and Nakanishi, K., 1984, A rhodopsin in the functional photoreceptor for phototaxis in the unicellular eucaryote Chlamydomonas. Nature 311: 756-759.

Hegemann, P., Hegemann, U. and Foster, K.W., 1988, Reversible bleaching of Chlamydomonas rhodopsin in vivo. Photochem. Photobiol., 48: 123-128.

Kaupp, U.B. and Koch, K.-W., 1986, Mechanism of Photoreception in vertebrate vision. Trends Biochem. Sci., 11: 43-47.

Lenci, F., Hader, D.-P. and Colombetti, G., 1984, Photosensory responses in freely motile microorganisms. In: Membranes and Sensory Transduction. G. Colombetti and F. Lenci, eds., Plenum Press, New York.

Litvin, F.F., Sineshchekov, O.A. and Sineshchekov, V.A., 1978, Photoreceptor electric potential in the phototaxis of the alga Haematococcus pluvialis. Nature, 271: 476-478.

Nultsch, W. and Hader, D.-P., 1988, Photomovement in motile microorganisms - 11. Photochem. Photobiol., 47: 837-869.

Ristori, T., Ascoli, C., Banchetti, R., Parrini, P and Petracci, D., 1981, Localization of photoreceptor and active membrane in the green alga Haematococcus pluvialis. Proceedings of the Sixth International Congress on Protozoology, Warsaw, 314.

Sineshchekov, O.A., 1978, The investigation of the photoelectrical processes in phototaxis in green alga. (in Russian). Ph.D Thesis. Moscow State University.

Sineshchekov, O.A., 1988, Phototaxis in microorganisms and its role in photosynthesis regulation. In Phototrophic microorganisms. (I. N. Gogotov, editor). (in Russian). Acad. Sci. USSR, Puschino, 1988. 11-18.

Sineshchekov, O.A, and Litvin, F.F., 1982, Photoregulation of movement in microorganisms. (in Russian). Usp. Sovr. Microbiol., 17: 62-87.

Sineshchekov, O.A. and Litvin, F.F., 1988, The mechanisms of phototaxis in microorganisms. In: "Molecular mechanisms of Biological Action of Optic Radiation. A. B. Rubin, ed., (in Russian). Nauka, Moscow.

Sineshchekov, O.A., Kurella, G.A., Andrianov, V.A. and Litvin, F.F., 1976, Bioelectric phenomena in unicellular flagellated alga, their relation in phototaxis and photosynthesis. (in Russian). Fiziol. Rasteniij 23: 229-236.

Sineshchekov, O.A., Sineshchekov, V.A, and Litvin, F.F., 1978, Photoinduced bioelectrical responses in phototaxis of the unicellular flagellated alga. (in Russian). Dokl. Akad. Nauk. SSSR 239: 471-474.

Sineshchekov, O.A., Litvin, F.F. and Keszthelyi, L., 1990. Two components of photoreceptor potential in phototaxis

of the flagellated green alga <u>Haematococcus pluvialis.</u>
<u>Biophys. J.</u>, 57: 33-39.

Spudich, J.L. and Bogomolni, R.A., 1988, Sensory rhodopsins of
<u>Halobacteria.</u> <u>Ann. Rev. Biophys. Byophis. Chem.</u>, 17:
193-215.

Stavenga, D.G., Short wavelength light in invertebrate visual
sense cell-pigments, potentials and problems. 1980. <u>In</u>:
"The blue light syndrome", H. Senger, ed., Springer,
N.Y., 1980.

PHOTORECEPTORS AND PHOTOMOVEMENTS OF MICROORGANISMS

Francesco Lenci and Francesco Ghetti

C.N.R. Istituto di Biofisica
Via San Lorenzo 26
56127 Pisa (Italy)

INTRODUCTION

Photomotile responses of microorganisms are photobiological processes in which light is used not as a source of energy (photocoupling) but to get information about the surroundings (photosensing). A wide variety of simple biological systems, such as bacteria, unicellular algae and protozoa, are able to detect spatial and temporal variations in the external light field and to react to these environmental stimuli by modifying their motion, usually to get the best illumination conditions for their growth and metabolism (Lenci and Colombetti, 1978; Lenci and Colombetti, 1980; Haeder, 1988; Nultsch and Haeder, 1988; Colombetti and Petracchi, 1989).

The main photobehaviours of microorganisms are photophobic responses, elicited by changes in light fluence rate, and phototaxis, resulting from the detection of light direction (Diehn et al., 1977).

The perception of the different characteristics of photic stimuli (spectral composition, intensity, propogation direction) and the eventual alteration of the ciliary or flagellar beating pattern are connected through morphological structures and functional processes which make up the photosensory transduction chain. In all microorganisms this phenomenon is entirely based on molecular events, started by modifications induced by light in the photoreceptive unit (Song, 1981; Colombetti et al., 1985; Song and Suzuki, 1988).

The identification of the photoreceptor pigments and the study of the primary molecular photoreactions they undergo is, therefore, of crucial importance to setting up a consistent scheme for the fundamental mechanisms of photoreception and subsequent transduction process (Lenci and Colombetti, 1980; Lenci and Ghetti, 1989; and references therein).

Light in Biology and Medicine. Volume 2 Edited by R.H. Douglas *et al.*
Plenum Press, New York, 1991

533

In this paper the main aspects of photoreception processes are discussed using a few case-examples: the polarly biflagellated bacterium <u>Halobacterium halobium</u>, the flagellated green alga <u>Euglena gracilis</u> and the ciliated protozoa <u>Blepharisma japonicum</u> and <u>Stentor coeruleus</u>.

PHOTOMOTILE RESPONSES

As previously mentioned, free-swimming photoresponsive microorganisms show two principal types of photoreactions: photophobic responses and phototaxis (Lenci et al., 1984; Feinleib, 1985).

Following a sudden increase and/or decrease in light fluence rate, the vast majority of microorganisms exhibit respectively step-up and/or step-down photophobic responses, usually consisting in an alteration in the activity of the motor apparatus (flagella, cilia), which, in its turn, causes a change in one or more kinematic parameters. Upon adaptation to the new illumination conditions, cells cease reacting and resume their unperturbed movement. In the case of exceedingly high fluence rates, however, step-up photophobic responses may persist until general damage of the microorganism occurs.

This general definition, of course, does not give an exhaustive picture of the peculiarities of phobic responses and avoidance strategies with which evolutionary adaptation has endowed the various microorganisms. For example, upon increasing light fluence rate, the ciliated protozoa <u>Stentor coeruleus</u> and <u>Blepharisma japonicum</u> suddenly stop, recoil and finally change their swimming direction, whereas they do not show step-down photophobic responses (Song, 1981; Kraml and Marwan, 1983; Passarelli et al., 1984). The flagellated photosynthetic alga <u>Euglena gracilis</u> shows step-up as well as step-down responses: in both cases the cell stops its forward movement and turns on the spot (Colombetti et al., 1982). In the biflagellated bacterium <u>Halobacterium halobium</u>, finally, light modulates the frequency of swimming direction reversal: this can either increase or decrease with respect to the spontaneous reversal frequency, depending on the wavelength of the actinic light (Hildebrand and Schimz, 1985; Spudich and Bogomolni, 1988).

Phototaxis results from the detection of light source direction and is defined as positive or negative according to whether the oriented movement is toward or away from the source. Actually, this photoresponse is elicited through a process of detection of fluence rate modulation, in general assisted by an asymmetry in the photoreceptor apparatus, which allows the cell to sense the anisotropy of the external light field. This directionality of the light-antenna system can be provided by:

a) periodic shading of one photosensitive unit caused, during cell rotation, by an absorbing organelle;

b) existence of two separate photoreceptors in different parts of the cell, possibly with a screening organelle as well;

c)   asymmetric distribution of several photosensitive units along the cellular body.

Obviously this oriented photoresponse implies the presence of a comparator device that analyses two or more output signals from the photoreceptor(s), activated by two or more input light signals detected either in succession or simultaneously (Foster and Smyth, 1980; Feinleib, 1985; and references therein).

Here we only want to mention that whereas Euglena shows both positive and negative phototaxis, Stentor only orients itself away from the light source and Blepharisma does not show any phototactic response at all.

Elaborate devices and mechanisms for locating a light source have been proposed such as, for example, an interference reflector, like a quarter-wave stack, assigned to Chlamydomonas stigma (Foster and Smyth, 1980) or an accommodation process similar to that of crystalline lenses occurring in Euglena photoreceptors (Gualtieri et al., 1988). It should be taken into account that because of the small size of the detection systems of microorganisms and the quite complex distribution of light sources and intensities in any natural environment, extremely precise hair-triggering and high-resolution systems may play a very minor role in nature, in the orientation mechanism, which could efficiently be brought about by coarse trajectory adjustments resulting from trial and error sequences.

PHOTORECEPTORS AND PRIMARY MOLECULAR PHOTOREACTIONS

Halobacterium halobium

Halobacterium light-dependent behaviour has been extensively studied (Hildebrand and Dencher, 1975; Baryshev et al., 1981; Hildebrand and Schimz, 1985; Spudich and Bogomolni, 1988; and references therein). Action spectra for photoinduced reversal of swimming direction led to the suggestion that the photoreceptor pigment PS-565, absorbing in the visible range, is responsible for the step-down response (under PS-565 irradiation the frequency of swimming direction reversal is lower than in unstimulated cells until the microorganism adapts to the new illumination conditions) while the photoreceptor pigment PS-370, absorbing in the near-UV region of the spectrum, is responsible for the step-up response (under PS-373 irradiation, a transient increase in the frequency of swimming direction reversal is observed) (Hildebrand and Dencher, 1975; Spudich and Stoeckenius, 1979).

It has been suggested that the rhodopsin-like all-trans retinal protein bacteriorhodopsin, with absorption maximum of 568 nm, which operates as a light-driven vectorial proton pump to utilize light energy for ATP synthesis (Oesterhelt and Stoeckenius, 1971 and 1973), acts as the antenna pigment in the PS-565 system and that another retinylidene protein, with absorption maximum around 370 nm, acts, together with carotenoids, as an accessory pigment, as does the photosensory pigment in the PS-370 system (Dencher, 1978 and references therein).

More recently substantial progress in understanding
Halobacterium photoreception processes has been made thanks to
the combination of different experimental techniques.
Selection and isolation procedures have provided, for
instance, mutants lacking one or more photopigment systems and
mutants deficient in photomotile and/or chemotactic responses.
These have been of crucial importance in both spectroscopic
and photobehavioural investigations (Spudich and Spudich,
1982; Sundberg et al., 1985).

The role of bacteriorhodopsin (bR) in photosensory
reception was seriously questioned by the finding that bR-
lacking mutants, like the L-33 and ET-15 strains, still
exhibit motile response upon stimulation with 565 nm light
(Traulich et al., 1983; Hildebrand and Schimz, 1983).
Moreover, the significant spectral shifts (about 20 nm)
observed between the maxima of step-down response action
spectra and of bR absorption spectra in bacterial populations
in which retinal has been substituted by different analogs,
were against the identification of PS-565 with bR (Dencher,
1983 and references therein).

Halorhodosin (hR), another rhodopsin-like all-trans
retinal, which also acts as a light-driven chloride pump
(Schobert and Laniy, 1982), seemed to play some role in the
photosensory process of Halobacterium, but the isolation of
fully photoresponsive mutants lacking both bR and hR - Flx
mutants (Bogomolni and Spudich, 1982) - necessitated a
reconsideration of the whole problem.

A series of flash photolysis experiments with microsecond
resolution were performed on membranes of wild-type and Flx's
Halobacterium mutants. The analysis of the decay kinetics of
flash generated transient species led to the identification of
two new all-trans retinal membrane proteins, both of them
responsible for Halobacterium photobehaviour. These sensory
rhodopsins undergo photoinitiated chemical cycles, consisting
of a slow (milliseconds) and a fast (microseconds) phase
(Spudich and Bogomolni, 1984 and 1987, and references
therein).

The first slow-cycling sensory rhodopsin, sR-I, exists in
two spectrally different forms, both active as photosensing
systems. The orange-red absorbing sR-I-587 mediates
"attractant" responses (frequency of swimming direction
reversal lower than in unstimulated samples) and, at the same
time, photoconverts, on a submillisecond scale, to a long-
lived intermediate, sR-I-373, which, upon absorption of a
second photon, triggers the "repellent" response (frequency of
swimming direction reversal higher than in unstimulated
cells). In the dark sR-I-373 thermally decays, in about 0.8
seconds, to sR-I-587.

The second slow-cycling sensory rhodopsin, sR-II-480 or
P480, seems to mediate only avoidance reactions from blue
light (Spudich and Bogomolni, 1987, Takahashi et al., 1985);
it photoconverts to a species absorbing at about 360 nm, which
thermally decays to sR-II-480 within about 0.3 seconds. The
photocycle of this "phoborhodopsin" (Tomioka, 1986) and its
physiological role are at present under investigation
(Takahashi, 1989).

The hypothesis that sR's might only play the role of accessory pigments and that the "main" sensory pigments which trigger photomotile responses have not yet been identified because of their extremely low concentration (Hildebrand and Schimz, 1986), seems to be ruled out by the fact that Phonon-photoresponsive mutants, still able to react to chemical stimuli, do not show any photomotile response and neither sR's nor other photochemically reactive pigments are spectroscopically detectable in their membranes (Sundberg et al., 1985).

Work is, finally, in progress on retinal analog substitutions of sensory rhodopsins (Takahashi et al., 1989a and 1989b).

Euglena gracilis

Even though sensory responses of Euglena to photic, chemical and mechanical stimuli have been widely investigated (Diehn, 1979; Colombetti et al., 1982; Mikolajczyk, 1984; Nultsch and Haeder, 1988; and references therein), the understanding of the primary photoinduced molecular events at the basis of its photomotile responses is still rather poor.

By means of sophisticated microspectrofluorometric techniques, fluorescence emission (Benedetti and Lenci, 1977) and excitation (Ghetti et al., 1985) spectra of the pigments contained in the photoreceptor structure (ParaFlagellar Body) were measured "in vivo". These results are consistent with the hypothesis that the photoreceptor pigments are flavin-type chromophores, most likely embedded in a rigid and/or hyrophobic molecular matrix. In the physiological molecular environment, according to the low value of fluorescence quantum yield (0.005, as compared with 0.25 of free riboflavin solutions), the first excited singlet state of the pigment undergoes de-excitation pathways other than radiative decay. This low value of fluorescence quantum yield of PFB flavins is in agreement with their photoreceptor function, as no photoreceptor pigment is likely to decay from its first excited singlet state mainly through a radiative transition, but rather through reactions which can serve to trigger the first molecular steps of the photosensory transduction chain.

Concerning the primary reactive state of the photopigment that triggers and initiates the transduction process, a few hypotheses have been put forward (Lenci et al., 1984 and references therein), also utilizing model systems like lipid vesicles and lipid bilayers (Colombetti et al., 1982 and references therein). The first excited singlet state seems to be the most probable candidate, because of its higher energy and higher population with respect to the triplet state.

Recently, the PFB of Euglena has been isolated and its morphology investigated by electron microscopy (Gualtieri et al., 1986 and 1988). The structure of the optical absorption spectrum of an isolated PFB, measured by means of a microspectrophotometer, has prompted the suggestion that rhodopsin-like molecules, rather than flavins, can play the role of photoreceptor pigments in Euglena (Gualtieri, et al., 1989), similar to what has been hypothesized for Chlamydomonas (Foster and Saranak, 1989 and references therein). In both cases, however, further experimental evidence is necessary before drawing definitive conclusions.

Stentor coeruleus and Blepharisma japonicum

The ciliated protozoa Stentor and Blepharisma are closely related not only in the morphological structure of their photoreceptor organelles (deeply pigmented granules, immediately beneath the cell surface, arranged so as to form bands near the ciliary rows), but also in the chemical structure of their native photoreceptor chromophores (stentorin and blepharismin), which seem to be quite similar to the natural photosensitizer hypericin. In spite of the pigment similarity, Stentor is blue-green colored while Blepharisma is red colored; this difference is apparently due to a relatively intense fluorescence emission from Blepharisma pigment. Moreover, blepharismin seems to be a much stronger photosensitizer than stentorin. Actually, upon high light intensity irradiation, sensitized photokilling of Blepharisma cells is observed. Both microorganisms exhibit step-up photophobic responses, the lag time of which decreases on increasing the intensity of the light stimulus; in Stentor negative phototaxis has also been observed (Giese, 1973; Song et al., 1980a; Song et al., 1980b; Giese, 1981; Song, 1981; Kraml and Marwan, 1983; and references therein).

In the case of Stentor two different chromo-proteins have been isolated and identified as Stentorin I and Stentorin II; they have identical absorption spectra and similar fluorescence spectra, but different fluorescence quantum yields (q. y. of Stentorin II relative to Stentorin I approximately 0.014) (Walker et al., 1979; Kim et al., 1989).

The action spectra of photoinduced ciliary reversal, of step-up photophobic responses and of negative phototaxis are fully consistent with the hypothesis that one or both stentorins play the role of photoreceptors (Wood, 1976; Song et al., 1980a and 1980b). Moreover, the lack of KI, CsCl or acrylamide quenching effect of the photoreceptor fluorescence (quenching which, on the contrary, is observed with hypericin or with denaturated stentorin) indicates that the chromophore is deeply buried in the hydrophobic core of the protein (Song et al., 1981).

Fluorescence from the photoreceptor pigment in its physiological state (e.g. in living cells) is red-shifted (emission maximum at 660 nm) with respect to that from isolated pigment solutions (emission maximum at 610 nm), even though the fluorescence excitation spectra are identical; this red shifted fluorescence is observed also in hypericin solutions and in stentorin preparations at high pH values. These findings suggest that a fast, highly efficient process occurs in the excited state of stentorin which results in the formation of the anionic species of the chromophore (Walker et al., 1981).

Upon irradiation with light absorbed by stentorin, a light-induced fluorescence quenching of membrane-embedded 9-amino-acridine is observed in suspensions of both living Stentor and liposomes incorporating photoreceptor proteins; this quenching effect can be inhibited by adding protonophores such as CCCP (carbonyl cyanide m-chlorophenil hydrazone) or nigericin. These findings are a clear indication that light generates a pH gradient across the cell membrane (Song, 1981; Song et al., 1981; Walker et al., 1981; Huh, 1987).

Recent spectroscopical and biochemical investigations, providing detailed information on the molecular properties of Stentorins I and II suggest that Stentorin II is the primary photoreceptor, from the excited state of which proton dissociation occurs, whereas Stentorin I may only play the role of an accessory pigment (Song et al., 1989; Kim et al., 1989). Photoreleased protons would generate a transient pH gradient which triggers a Ca^{2+} influx across the plasma membrane, this transient increase in cytoplasmic and/or intraciliary Ca^{2+} concentration is finally responsible for ciliary reversal and therefore for the photophobic response of Stentor (Song, 1981).

In the case of Blepharisma two different action spectra have been proposed for its step-up photophobic response, the first one indicating a blue absorbing pigment as the photoreceptor (Kraml and Marwan, 1983), the other one fully consistent with the hypothesis that blepharismin plays this role and that granules are the organelles where the photosensory transduction process starts (Scevoli et al., 1987). This last conclusion is strengthened by the fact that fluorescence lifetime and quantum yield of cold extracted granules were found to be about half those in acetone extracted pigment solutions, a fact suggesting that in the natural environment the excited pigment finds an alternative decay pathway that may well be the first step of the photosensory transduction process (Scevoli et al., 1987).

Also in the case of Blepharisma, a vectorial light-driven proton translocation process was proposed as being involved in the primary steps of the transduction chain (Passarelli et al., 1984). A series of steady-state fluorescence studies on pigment crude extracts were performed, to check whether the proton transfer process could take place at the level of the photoreceptor. In pure ethanol blepharismin solutions, the emission spectrum showed a single band centered around 600 nm; increasing the basicity of the medium, the appearance of a band centered at about 660 nm was observed, together with a decrease of the fluorescence emission at 600 nm, whereas only minor changes in optical absorption spectra were detected; these emission bands at 600 nm and 660 nm were attributed respectively to the first excited singlet states of the neutral and of the anionic form of blepharismin. The hypothesis was put forward that the charged species resulted from deprotonation of a hydroxyl group of the chromophore in its first excited singlet state (Lenci et al., 1989).

More recently, fluorescence lifetimes, time-integrated and time-gated spectra of blepharismin have been measured in ethanol, aqueous solutions and detergent micelles. A short-living (0.2-0.4 ns) molecular species, emitting at 600 nm, is predominant in aqueous solutions at pH < 11.7, whereas in pure ethanol solutions an intermediate-living species (about 1 ns), still emitting at 600 nm, prevails. Upon increasing OH⁻ concentration, a third, long-living (about 4-6 ns) molecular species, emitting at 660 nm, is formed. This species has been identified as the negatively charged form of the photoreceptor pigment, whereas the short-living and the intermediated-living fluorescence emissions have been attributed respectively to the phenolic and the quinonic neutral forms of blepharismin (Cubeddu et al., 1989).

The fact that the emission at 660 nm is predominant only at pH values above 11.7 seems to indicate that, at physiological pH, the deprotonation of blepharismin in its first excited singlet state occurs with a quite low yield. This implies that, even though in a photosensing phenomenon light acts as a trigger, so that low-efficiency processes may be sufficient to initiate the phototransduction chain, the hypothesis of deprotonation of blepharismin in its first excited singlet state as the primary molecular event in Blepharisma photoreception deserves further investigation.

Work is in progess on the one hand to clarify this point and on the other hand to purify and characterize blepharismin photoreceptor protein (Ghetti et al., 1989).

CONCLUDING REMARKS

Regardless of their chemical nature, all photoreceptor pigments have to detect and transduce light signals, thus initiating the sensory processes which allow the organisms to actively interact with their natural environment. To be efficient phototransducing systems, light antennas will therefore have a high degree of structural organization, for example with photopigment molecules embedded in or associated with membranes (Lenci et al., 1984). Despite their diversity, moverover, all photosensing systems have the same final aim: to induce alterations in a motor apparatus which is substantially identical in all eukaryotes.

REFERENCES

Baryshev, V.A., Glagolev, A.N. and Skulachev, V.P., 1981, Sensing of in phototaxis of Halobacterium halobium, Nature, 292:338-340.
Benedetti, P.A., and Lenci, F., 1977, In vivo microspectrofluorometry of photoreceptor pigments in Euglena gracilis, Photochem. Photobiol., 26:315-318.
Bogomolni, R.A., and Spudich, J.L., 1982, Identification of a third rhodopsin-like pigment in phototactic Halobacterium halobium, Proc. Natl. Acad. Sci. USA, 79:6250-6254.
Colombetti, G., Lenci, F. and Diehn, B., 1982, Responses to photic, chemical and mechanical stimuli, in: "The Biology of Euglena", vol. 3, D.E. Buetow, ed., Academic, New York, pp. 169-195.
Colombetti, G., Lenci, F. and Song, P.-S., eds., 1985, "Sensory perception and transduction in aneural organisms", Plenum, London.
Colombetti, G. and Petracchi, D., 1989, Photoresponse mechanisms in flagellated algae, in: "CRC Critical Reviews in Plant Sciences", B.V. Conger, ed., CRC, Boca Raton, pp. 309-355.
Cubeddu, R., Ghetti, F., Lenci, F., Ramponi,R. and Taroni, P., 1989, Time-gated fluorescence of belpharismin, the

photoreceptor pigment for photomovement of <u>Blepharisma</u>,
   <u>Photochem. Photobiol.</u>, 51:567-573.
Dencher, N., 1978, Light-induced behavioral reactions of
   <u>Halobacterium halobium:</u> evidence for two rhodopsins
   acting as photopigments, <u>in</u>: "Energetics and Structure of
   Halophilic Microorganisms", S.R. Caplan and H. Ginzburg,
   eds., Elsevier, Amsterdam, pp. 67-88.
Dencher, N., 1983, The five retinal-proteins of halobacteria:
   bacteriorhodopsin, halorhodopsin, PS-565, PS-370 and
   slow-cycling rhodopsins, <u>Photochem. Photobiol.</u>, 38:753-
   767.
Diehn, B., 1979, Photic responses and sensory transduction in
   protists, <u>in</u>: "Handbook of Sensory Physiology", vol. 7,
   H. Autrum, ed., Springer, Berlin, pp. 23-68.
Diehn, B., Feinleib, M.E., Haupt, W., Hildebrand, E., Lenci,
   F., and Nultsch, W., 1977, Terminology of behavioral
   responses of motile microorganisms, <u>Photochem.
   Photobiol.</u>, 26:559-560.
Feinleib, M.E., 1985, Behavioral studies of free-swimming
   photoresponsive microorganisms, <u>in</u>: "Sensory Perception
   and Transduction in Aneural Organisms", G. Colombetti, F.
   Lenci and P.-S. Song, eds., Plenum, New York, pp. 119-
   146.
Foster, K.W. and Saranak, J., 1989, The rhodopsin of
   <u>Chlamydomonas reinhardtii</u>, <u>Photochem. Photobiol</u>, 49S:38S.
Foster, K.W., and Smyth R.D., 1980, Light antennas in
   phototactic algae, <u>Microbiol. Rev.</u>, 44:572-630.
Ghetti, F., Colombetti, G., Lenci, F., Campani, E., Polacco,
   E. and Quaglia, M., 1985, Fluorescence of <u>Euglena gracils</u>
   photoreceptor pigment: an 'in vivo'
   microspectrofluorometric study, <u>Photochem. Photobiol.</u>,
   42:29-33.
Ghetti, F., Gioffre', D., Lenci, F., Balestreri, E., Song, P.-
   S. and Kim, I.-H., 1989, Isolation and purification of
   <u>Blepharisma japonicum</u> photoreceptor proteins, <u>Med. Biol.
   Environment</u>, 17:839-843.
Giese, A.C., 1973, "<u>Blepharisma</u>: the biology of a light-
   sensitive protozoan", Stanford University Press,
   Stanford.
Giese, A.C., 1981, The photobiology of <u>Blepharisma</u>, <u>in</u>
   "Photochemical Photobiological Reviews", K.C. Smith, ed.,
   Plenum, New York, pp. 139-180.
Gualtieri, P., Barsanti, L., Passarelli, V., 1989, Absorption
   spectrum of a single isolated Paraflagellar Swelling of
   <u>Euglena gracilis</u>, <u>Biochim. Biophys. Acta</u>, 993:293-296.
Gualtieri, P., Barsanti, L., Passarelli, V., and Verni, F.,
   1988, Morphological investigations of <u>Euglena gracilis</u>
   isolated Paraflagellar Body, <u>Micron and Microsc. Acta</u>,
   19:241-246.
Gualtieri, P., Barsanti, L. and Rosati, G., 1986, Isolation of
   the photoreceptor (paraflagellar body) of the phototactic
   flagellate <u>Euglena gracilis</u>, <u>Arch. Microbiol.</u>, 145:303-
   305.
Haeder, D.P., 1988, Ecological consequences of photomovement
   in microorganisms, <u>J. Photochem. Photobiol. B</u>, 1:385-414.
Hildebrand, E. and Dencher, N., 1975, Two photosystems
   controlling behavioral responses of <u>Halobacterium
   halobium</u>, <u>Nature</u>, 257:46-48.
Hildebrand, E., and Schimz, A., 1983, Photosensory behavior of
   a bacteriorhodopsin-deficent mutant, ET-15, of
   <u>Halobacterium halobium</u>, <u>Photochem. Photobiol.</u>, 37:581-
   584.

Hildebrand, E., and Schimz, A., 1985, Sensory transduction in Halobacterium, in: "Sensory Perception and Transduction in Aneural Organisms", G. Colombetti, F. Lenci and P.-S. Song, eds., Plenum, New York, pp. 93-111.

Hildebrand, E., and Schimz, A., 1986, The rhodopsin-like sensory pigments of halobacteria, TIBS 11, 402.

Huh, J.W., 1987, "Characterization and function of stentorin, the photoreceptor protein in Stentor coeruleus", Ph. D. Thesis, Texas Tech University.

Kim, I.-H., Rhee, J.S., Huh, J.W., Florell, S., Lee, K.W., Song, P.-S., Tamai, N., Yamazaki, T. and Yamazaki, I., 1989, Structure and function of the photoreceptors stentorins in Stentor coeruleus. I. Partial characterization of the photoreceptor organelle and stentorins, Biochim. Biophys. Acta., in press.

Kraml, M., and Marwan, W., 1983, Photomovements responses of the heterotrichous ciliate Blepharisma japonicum, Photochem. Photobiol., 37:313-319.

Lenci, F. and Colombetti, G., 1978, Photobehavior of micro- organisms. A biophysical approach, Ann. Rev. Biophys. Bioeng., 7:341-361.

Lenci, F. and Colombetti, G., eds., 1980, "Photoreception and Sensory Transduction in Aneural Organisms", Plenum, London.

Lenci, F., and Ghetti, F., 1989, Photoreceptor pigments for photomovement of microorganisms: some spectroscopic and related studies, J. Photochem. Photobiol. B, 3:1-16.

Lenci, F., Ghetti, F., Gioffre', D., Passarelli, V., Heelis, P.F., Thomas, B., Phillips, G.O., and Song, P.-S., 1989, Effect of molecular environment on some spectroscopic properties of Blepharisma photoreceptor pigment, J. Photochem. Photobiol. B, 3:449-453.

Lenci, F., Haeder, D.P., and Colombetti, G., 1984, Photosensory responses in freely motile microorganisms, in: "Membranes and Sensory Transduction", G. Colombetti and F. Lenci, eds., Plenum, New York, pp.199-229.

Mikolajczyk, E., 1984, Photophobic responses in Euglenina. 1. Effect of excitation wavelength and external medium on the step-up response of light- and dark-grown Euglena gracilis, Acta Protozool., 23:1-10.

Nultsch, W. and Haeder, D.P., 1988, Photomovement in motile microorganisms. II, Photochem. Photobiol., 47:837-869.

Oesterhelt, D. and Stoeckenius, W., 1971, Rhodopsin-like protein from the purple membrane of Halobacterium halobium, Nature, 233:149-152.

Oesterhelt, D. and Stoeckenius, W., 1973, Functions of a new photoreceptor membrane, Proc. Natl. Acad. Sci. USA, 70:2853-2857.

Passarelli, V., Lenci, F., Colombetti, G., Barone, E. and Nobili, R., 1984, The possible role of H^+ and Ca^{2+} in photobehavior of Blepharisma japonicum, in: "Blue light effects in biological systems", H. Senger, ed., Springer, Berlin, pp. 480-483.

Scevoli, P., Bisi, F., Colombetti, G., Ghetti, F., Lenci, F., and Passarelli, V., 1987, Photomotile responses of Blepharisma japonicum. I: Action spectra determination and time-resolved fluorescence of photoreceptor pigments, J. Photochem. Photobiol B, 1:75-84.

Schobert, B. and Lanyi, J.K., 1982, Halorhodopsin is a light- driven chloride pump, J. Biol. Chem., 257:10306-10313.

Song, P.-S., 1981, Photosensory transduction in Stentor coeruleus and related organisms, Biochim. Biophys. Acta, 639:1-29.

Song, P.-S., 1983, The electronic spectroscopy of photoreceptors (other than rhodopsin), in: "Photochemical photobiological Reviews" 7, K.C. Smith, ed., Plenum, New York, pp. 77-139.

Song, P.-S., Haeder, D.P., and Poff, K.L., 1980a, Step-up photophobic response in the ciliate Stentor coeruleus, Arch. Microbiol., 126:181-186.

Song, P.-S., Haeder, D.P. and Poff, K.L., 1980b, Phototactic orientation by the ciliate Stentor coeruleus, Photochem. Photobiol., 32:781-786.

Song, P.-S., Kim, I.-H., Florell, S., Faure, B., Tamai, N., Yamazaki, T. and Yamazaki, I., 1989, Structure and function of the photoreceptors stentorins in Stentor coeruleus. II. Primary photoprocess and picosecond time-resolved fluorescence, Biochim. Biophys. Acta, in press.

Song, P.-S. and Suzuki, S., 1988, Properties and evolution of photoreceptors, in: "Photoreceptor Evolution and Function", M.G. Holmes ed., Academic Press, London.

Song, P.-S., Walker, E.B., Auerbach, R.A. and Robinson, G.W., 1981, Proton release from Stentor photoreceptors in the excited states, Biophys. J., 35:551-555.

Spudich, E.N. and Spudich, J.L., 1982, Control of transmembrane ion fluxes to select halorhodopsin-deficient and other energy transduction mutants of Halobacterium halobium, Proc. Natl. Acad. Sci. USA, 79:4308-4312.

Spudich, J.L. and Bogomolni, R.A., 1984, Mechanism of color discrimination by a bacterial sensory rhodopsin, Nature, 312:509-513.

Spudich, J.L. and Bogomolni, R.A., 1987, Sensory rhodopsins in Halobacterium Halobium, in "Biophysical Studies of Retinal Proteins", T.G. Ebrey, H. Frauenfelder, B. Honig and K. Nakanishi, eds., University of Illinois, Urbana, pp. 24-30.

Spudich, J.L. and Bogomolni, R.A., 1988, Sensory rhodopsins of halobacteria, Ann. Rev. Biophys. Biophys. Chem., 17:193-215.

Spudich, J.L. and Stoeckenius, W., 1979, Photosensory and chemosensory behavior of Halobacterium halobium, Photobiochem. Photobiophys., 1:43-53.

Sundberg, S.A., Bogomolni, R.A., and Spudich, J.L., 1985, Selection and properties of phototaxis-deficient mutants of Halobacterium halobium, J. Bacteriol., 164:282-287.

Takahashi, T., 1989, Phoborhodopsin - A blue light photoreceptor in Halobacteria, Photochem. Photobiol., 49S:37S.

Takahashi, T., Spudich, E.N. and Spudich, J.L., 1989a, Retinal reconstitution study of two phototaxis receptor mutants of Halobacterium halobium, Bioph. J., 55:387a.

Takahashi, T., Tomioka, H., Kamo, N. and Kobatake, Y., 1985, A photosystem other than PS370 also mediates the negative phototaxis of Halobacterium halobium, FEMS Microbiol. Lett., 28:161-164.

Takahashi, T., Yan, R., Rao, B., Nakanishi, K., and Spudich, J.L., 1989b, Retinal analog reconstitution of bacterial sensory rhodopsin I, Bioph. J., 55:383a.

Tomioka, H. Takahashi, T., Kamo, N. and Kobatake, Y., 1986, Flash spectrophotometric identification of a fourth rhodopsin-like pigment in <u>Halobacterium halobium</u>, <u>Biochem. Biophys. Res. Commun.</u>, 139:389-395.

Traulich, B., Hildebrand, E., Schimz, A., Wagner, G. and Lanyi, J.K., 1983, Halorhodopsin and photosensory behavior in <u>Halobacterium halobium</u> mutant strain L-22, <u>Photochem. Photobiol.</u>, 37:577-579.

Walker, E.B., Lee T.-Y and Song, P.-S, 1979, Spectroscopic characterization of the <u>Stentor</u> photoreceptor, <u>Biochim. Biophys. Acta.</u>, 587:129-144.

Walker, E.B., Yoon, M. and Song, P.-S., 1981, The pH dependece of photosensory responses in <u>Stentor coeruleus</u> and model system, <u>Biochim. Biophys. Acta.</u>, 634:289-308.

Wood, D.C., 1976, Action spectrum and electrophysiological responses correlated with the photophobic response of <u>Stentor coeruleus</u>, <u>Photochem. Photobiol.</u>, 24:261-266.

RETINAL ISOMERS IN PHOTO- AND ELECTROINDUCED RHODOPSIN
INTERMEDIATES

Tatyana B. Protasova, Irina B. Fedorovich
and Michail A. Ostrovsky

Institute of Chemical Physics of the USSR Academy
of Science, ul Kosygina 4, Moscow 117977, USSR

INTRODUCTION

This paper investigates the contribution of retinal
chromophore isomeric state and protein conformational change
on the absorption spectra of rhodopsin and its photo- and
electroinduced intermediates.

A quantitative comparative analysis of retinal isomeric
content and absorption spectra of intermediates has been
carried out. For investigation of isomeric composition the
HPLC technique has been applied with two modifications: one
for free retinals, the other for retinaloximes (Groenendijk et
al., 1980; Suzuki et al., 1986). The latter modification was
used because all of the spectral data have been obtained in
the presence of hydroxylamine.

We have examined rhodopsin photolysis and
photoregeneration at different stages, using the low
temperature technique.

Prior to irradiation the experimental material (Fig. 1 -
A1) contained only 3% of all-trans retinal with almost all of
the remainder being 11-cis retinal (Fig. 1 - B1). In
accordance with the classical picture (Yoshizawa and Kito,
1958) rhodopsin irradiation by blue light (437 nm) at 77°K
resulted in a bathochromic shift of its absorption spectrum
maximum (Fig. 1 - A3). As a result of such illumination a
photosteady state mixture composed of (11-cis) rhodopsin, (9-
cis) isorhodopsin and (all-trans) bathorhodopsin was produced
(Fig. 1 - B2). Fig. 1 - A5 shows the result of bleaching this
photosteady state mixture with yellow light (579 nm) i.e. the
process of photoregeneration. In this case a hypsochromic
shift of absorption spectrum maximum took place and the HPLC
analysis showed a mixture of 11-cis and 9-cis and only 5% of
all-trans retinals (Fig. 1 - B3). This means that during such
illumination practically complete photoregeneration occurred.

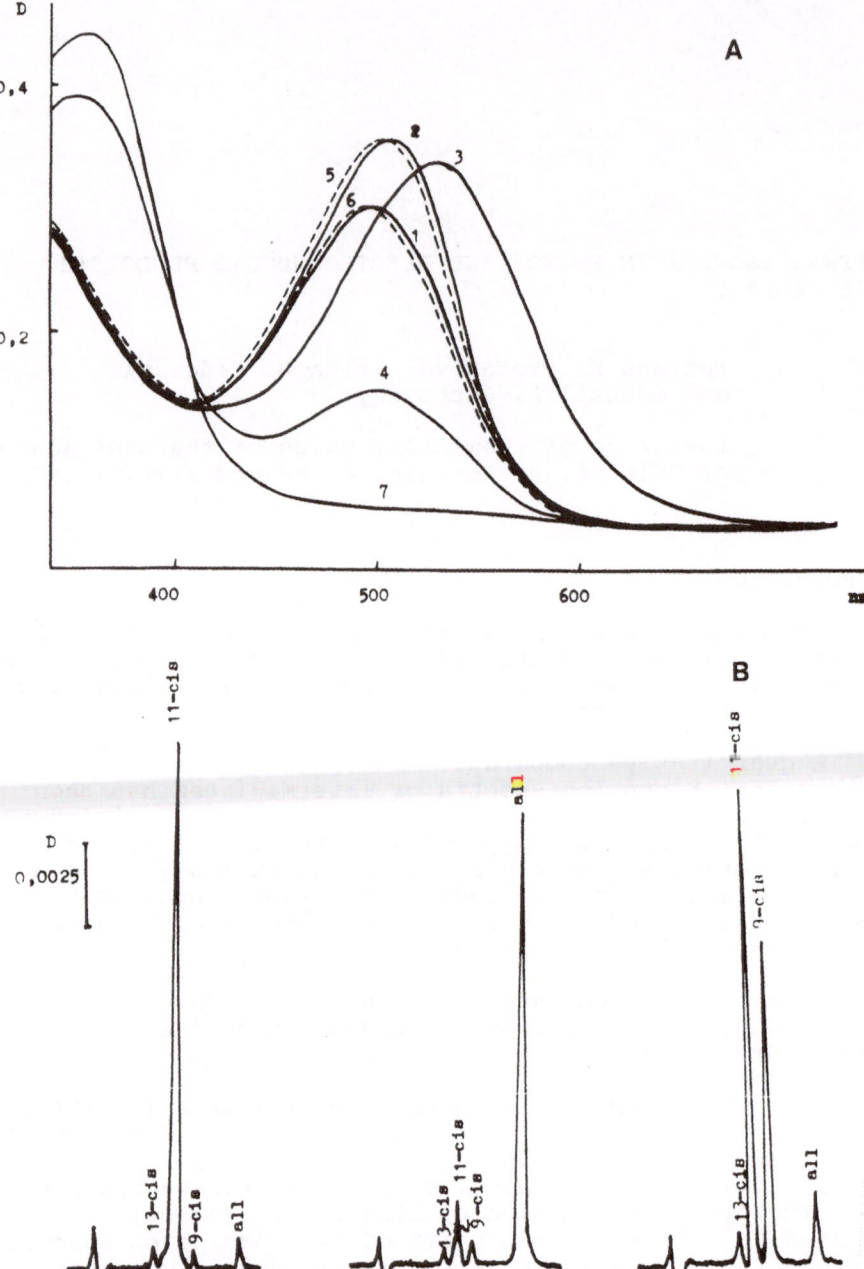

Fig. 1   Photoconversion of frog rhodopsin digitonin
extract at 77°K.   A)   Absorption spectra:
1. at 293°K, 2. at 77°K, 3. illuminated for
10 minutes at 437 nm, 4. warmed to 293°K,
5. product (3) illuminated for 10 minutes
at 579 nm at 77°K; 6. warmed to 293°K;
7. full bleached at 293°K.   B)   HPLC data:
1. rhodopsin; 2. illuminated at 437 nm,
3. illuminated at 579 nm.

It is necessary to mention that the absorption maximum of
this photoregenerated mixture depends on the duration of
yellow light illumination: the higher the light dose the
larger the hypsochromic shift of the absorption maximum. The
reason for this phenomenon is seen from Table 1, which shows
that longer irradiation favours 9-cis retinal isomer
accumulation.

The same conclusion may be drawn from our old results,
(Ostrovsky et al., 1974). The bleaching of cattle rhodopsin
digitonin extract by short laser flash (20 ns, 530 nm) at
room temperature resulted in the appearance of a mixture
containing rhodopsin, isorhodopsin and bleached product. Fig.
2 shows that increasing flash power resulted in an increase in
the amount of isorhodopsin and a decrease in rhodopsin.

Table 1 The dependence of retinal isomer content
on the duration of illumination with yellow
light (579 nm) at 77°K.

N	t(sec)	11-cis(%)	9-cis(%)	all- trans(%)	13-cis(%)
1	360	61	34	3	2
2	420	56	38	4	2
3	480	50	44	4	2
4*	1800	8	82	4	6

*Maeda el al., 1979

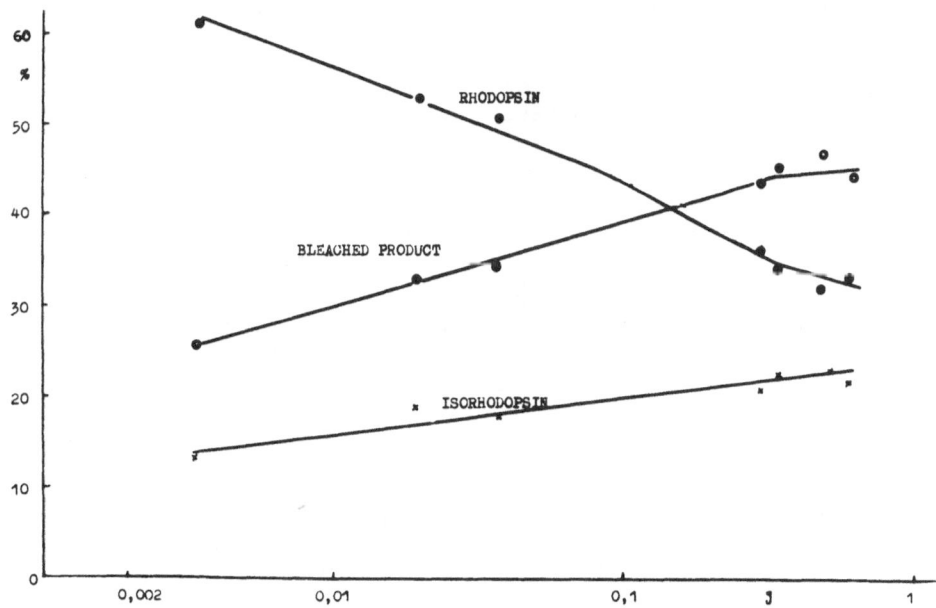

Fig. 2 Composition of cattle rhodopsin mixture
obtained by laser flash (530 nm, 20 nsec)
at room temperature.

Thus blue light favours isomeric conversion of cis to all-trans and yellow favours all-trans to cis conversion. It should be mentioned that the 9-cis chromophore accumulates independently of the light dose.

Table 2 gives a quantitative comparison of spectral (Δ D_{500}) and HPLC data for intermediates obtained at 77°K. Native rhodopsin, i.e. original as well as the mixture photoregenerated by yellow light (579 nm) showed a good correlation between HPLC and spectral data.

A different picture was observed in relation to the photosteady state mixture obtained under the same experimental conditions except using blue light illumination. According to the spectral data, 40% of the rhodopsin and isorhodopsin remained stable after defreezing. But according to the HPLC data the amounts of 11-cis and 9-cis isomers were only about 20% (free retinals) or about 30% (retinaloxims) and the remainder was all-trans. This result means that the product, stable at room temperature (Fig. 1 - A4) with maximum near 500 nm, contained not only cis but also trans isomers.

This result is rather difficult to explain. It is possible to surmise that the photosteady state mixture obtained as a result of blue light irradiation at 77°K also contained a specific intermediate which had a chromophore in some changed (perhaps transoid) form but its protein conformations are similar to those of rhodopsin, which accounts for its stability at room temperature. But denaturation of opsin during the preparation of retinal samples for HPLC led to the liberation of retinal in the all-trans form (Protasova et al., 1989). A good fit between spectral (D_{500}) and HPLC data in original and product produced by yellow light gives a basis for thinking that these data are not the result of experimental error.

Table 2   Isomeric state of retinals in frog rhodopsin photointermediates at 77°K.

	11-cis	9-cis	all-trans	13-cis
Native rhodopsin				
Spectral data	100	–	–	–
HPLC of retinaloximes	96.0±0.1	–	3.0±0.1	1.0±0.1
HPLC of retinals	94.5±0.5	–	3.5±0.5	1.1±0.1
Irradiated at 437 nm				
Spectral data	40.4 ± 1.3		59.6 ± 1.6	
HPLC of retinaloximes	26.2±1.0	4.2±0.6	67.8±1.4	1.8±0.3
HPLC of retinals	16.7±1.1	4.3±1.8	77.7±2.3	1.3±0.4
Irradiated at 579 nm				
Spectral data	98.0 ± 1.0		2.0 ± 0.2	
HPLC of retinaloximes	94.3 ± 0.4		3.7±0.4	1.9±0.1
HPLC of retinals	94.0 ± 0.6		5.0±1.6	1.0±0.1

Many years ago we investigated the reversible phototransformation of rhodopsin at a temperature favourable to the
formation and persistence metarhodopsin I (Krongaus et al.,
1975a, 1975b, 1975c). Blue light illumination of frog
metarhodopsin I at 251°K led to a product which was
practically isochromic to the original rhodopsin photosteady
state mixture (Fig. 3). Warming the mixture to room
temperature, however, resulted in bleaching about 2/3 of it.
We named this thermally unstable product pseudorhodopsin,
since it was practically isochromaic to rhodopsin.

The present HPLC data showed that our "pseudorhodopsin"
contains all-trans retinal (Table 3).

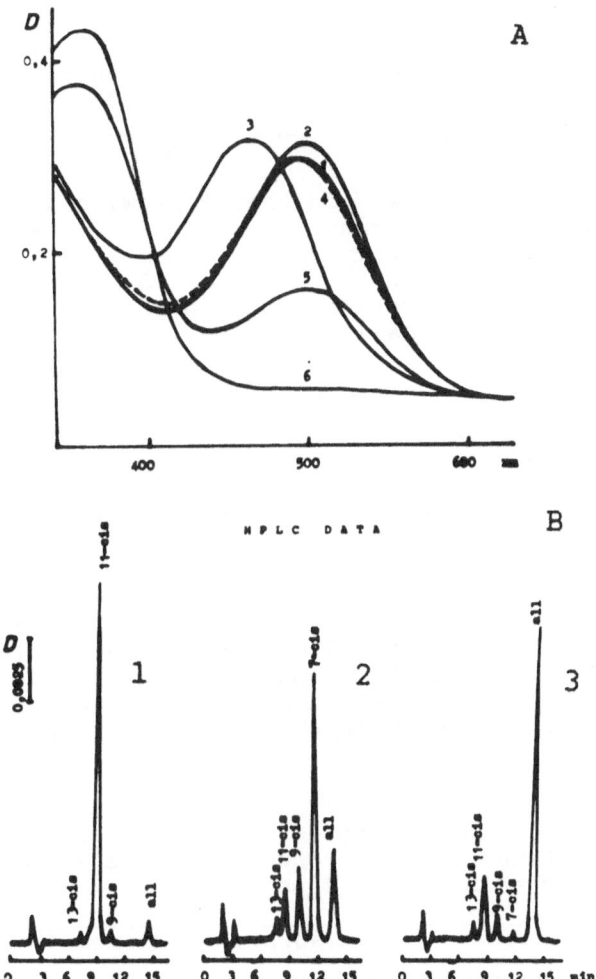

Fig. 3   Photoconversion of frog digitotin extract
         at 251°K. A) Absorption spectra: 1. at
         293°K, 2. at 251°K, 3. illuminated at 579 nm
         at 251°K, 4. illuminated at 437 nm at 251°K,
         5. warmed to 293°K, 6. full bleached at
         293°K. B) HPLC data: 1. rhodopsin,
         2. illuminated at 579 nm, 3. illuminated at
         437 nm.

Table 3   Isomeric composition in % of the retinals
in the product of frog rhodopsin photoconversion at 251°K by blue light
irradiation

	11-cis	9-cis	all-trans	13-cis
Native rhodopsin				
Spectral data	100	-	-	-
HPLC of retinaloximes	96.5±0.7	1.2±0.2	1.7±0.5	0.7±0.1
HPLC of retinals	96.0±0.5	1.2±0.2	2.2±0.4	1.0±0.1
Irradiated at 437 nm				
Spectral data	43.0 ± 1.4		57.0 ± 1.4	
HPLC of retinaloximes	24.5±0.7	9.3±0.1	62.5±0.7	4.7±1.0
HPLC of retinals	16.7±0.6	9.0±1.0	70.3±0.8	4.0±1.0

It is necessary to mention that following irradiation by
blue light the absorption spectrum maximum of the photosteady
state mixture depends on temperature. Table 4 shows that the
maximum shifted from 525 nm at 77°K to 490 nm at 253°K. But
the amount of all-trans retinal in such photosteady state
mixtures remained practically the same (see also Table 2 and
3).

What this means is that progressive changes in protein
conformation of intermediates determine the position of the
absorption spectrum maximum. Thus it is difficult to speak
about fixed positions for absorption spectrum maxima for such
intermediates as lumirhodopsin or metarhodopsin I. The
illumination of frog rhodopsin digitonin extract by yellow
light at temperatures higher than 233°K resulted in a
hypsochromic shift of the absorption spectrum maximum up to
450 nm (Fig. 3 - A3). HPLC analysis showed that in the course
of such a hypsochromic shift 7-cis retinal accumulated. The
appearance of the 7-cis isomer during isomerization of the
all-trans retinal means that at stages such as lumirhodopsin
and metarhodopsin I the chromophore binding site becomes
enlarged.

Table 4   The temperature dependence of photosteady
state mixture produced by blue light
irradiation in frog rhodopsin

T°,K	Absorption maximum (nm)
77	525
188	508
223	505
243	498
248	495
253	490

The role of the protein conformation state in determining rhodopsin intermediate absorption maximum position was also demonstrated in the next set of experiments. We investigated the influence of an external electric field on spectral properties of rhodopsin in dry films containing photoreceptor membranes (for more detail see : Maisel et al., 1984; Protasova et al., 1987).

Application of the electric field (1400-1600V) resulted in a shift of absorption maximum 6-8 nm toward longer wavelengths (Fig. 4). The field strength across the film was calculated as for a plane capacitor on the assumption of sample homogeneity. It was about 10-100 mV per membrane thickness. The differential absorption spectrum of the product formed in the dark was similar to light produced bathorhodopsin. This shift of the maximum coincided with the conversion of 10-15% of rhodopsin into the bathoproduct. But in this case we failed to obtain any isomerization of the chromophore (Table 5).

Thus, in this case, electroinduced conformation changes of protein occurred without any chromophore isomerization, leading to spectral changes similar to photoinduced ones.

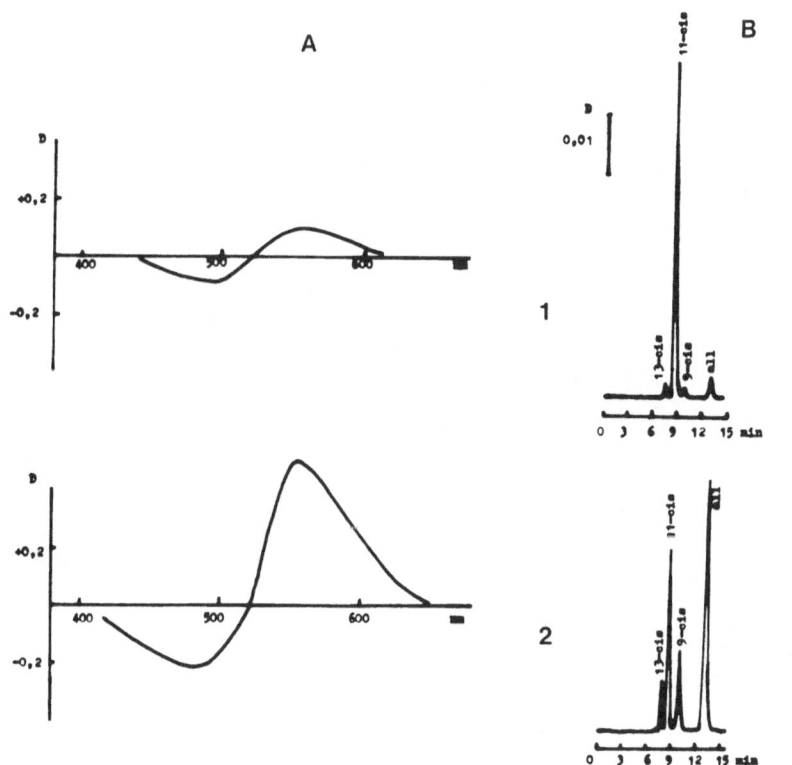

Fig. 4 The action of an electric field on the photo-
        receptor membrane dry film. A) Differential
        spectra of 1) electroinduced bathoproduct
        and7 2) photoinduced bathoproduct. B) HPLC
        data of 1) rhodopsin after voltage
        application and 2) after illumination.

Table 5   Isomeric composition of retinal in electroinduced modification of rhodopsin (Rh) and
bacteriorhodopsin (BRh).

Sample	11-cis	9-cis	all-trans	13-cis
Dark Rh	93.6±0.6	1.5±0.2	3.2±0.4	1.5±0.1
Rh+Ef(3h.)	93.0±0.4	1.5±0.2	3.9±0.4	1.5±0.1
Rh+Ef+dark 6 days	93.6±0.4	1.6±0.1	3.2±0.2	1.5±0.1
Rh+437nm 10 min	35.2±1.1	15.5±1.3	48.3±1.0	1.0±0.1
Rh+437nm 10 min+Ef	34.5±2.1	15.7±0.4	48.8±1.9	0.7±0.2
Rh+Ef+437 nm 10 min	33.5±0.5	16.0±0.2	48.9±1.8	0.6±0.2
BRh			60.0±0.2	40.0±0.1
BRh+Ef			60.0±0.1	40.0±0.1

Electric field (Ef) was 10^6 V/m

Similar electroinduced changes in absorption spectra have
been obtained on purple membrane containing dry films
(Borisevitch et al., 1979). An externally applied electric
field also has an influence on the characteristics of the
bacteriorhodopsin photocycle (Groma et al., 1988; Maximychev
et al., 1988).

Fig. 5 shows that switching on the field caused a very
slow formation of bacteriorhodopsin bathoproduct and switching
off resulted in restoring the spectrum to its original
position. HPLC data show that with bacteriorhodopsin as well
as with rhodopsin we failed to obtain any isomerization of the
chromophore (Table 5).

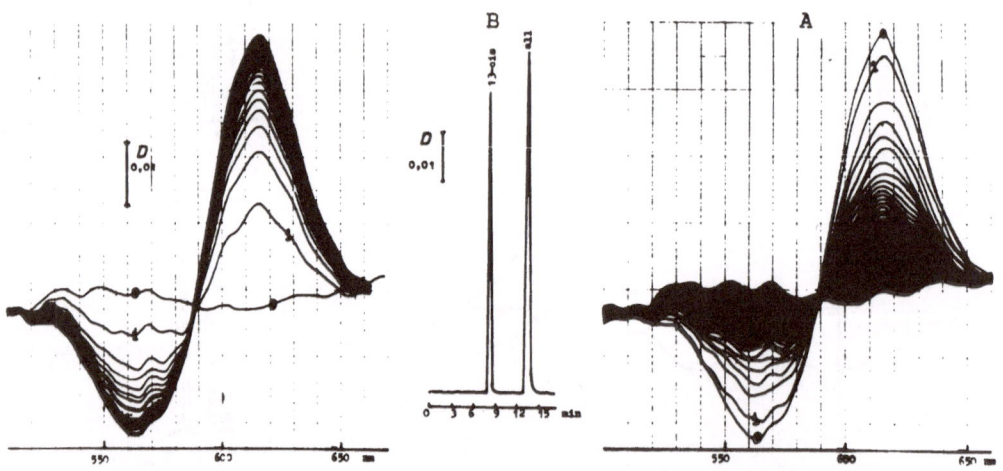

Fig. 5   The action of an electric field on the purple
membrane dry film. A) Differential spectra
during electric field action (10^6 V/m, 30 min,
left) and after switching off the electric
field (about 4 h, right). B) HPLC data.
The isomeric patterns in both cases are the
same.

Switching off the electric field applied to the photoreceptor membrane dry film resulted in dark bleaching of rhodopsin but not a return of the absorption spectrum to its original state as we have observed on the film containing bacteriorhodopsin. The dark bleaching of rhodopsin in a dry film was very slow, but at the end intermediates spectrally similar to metarhodopsin I and metarhodopsin II were produced (Fig. 6). This bleaching also took place without isomerization of the chromophore (Table 5).

Switching on the electric field for a second time after completion of dark bleaching led to spectral photoregeneration. In this case a product spectrally identical to rhodopsin appeared (Fig. 7 - A) but this product was unstable at room temperature. After the electric field was switched off the electroregenerated product bleached away very slowly.

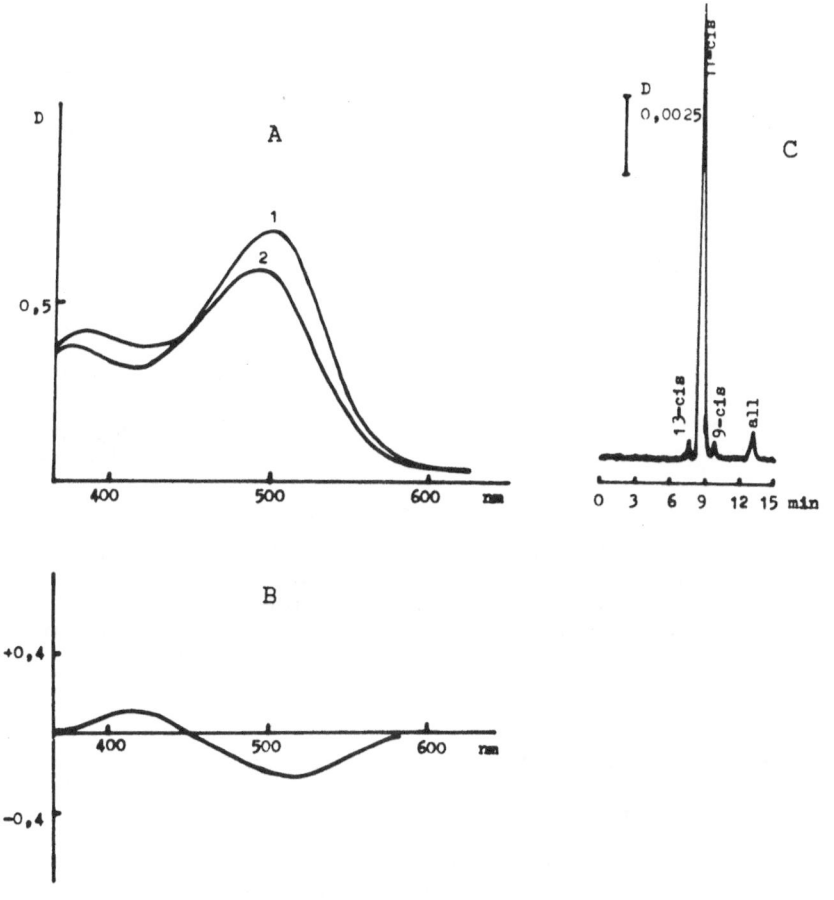

Fig. 6 Dark bleaching of rhodopsin after switching off the electric field. A) Absorption spectra of: 1) Frog rhodopsin in dry film, 2) after 2 days of electric field action. B) Differential spectrum: spectrum (2) minus spectrum (1). C) HPLC data of retinal isomers after electroinduced changes in rhodopsin.

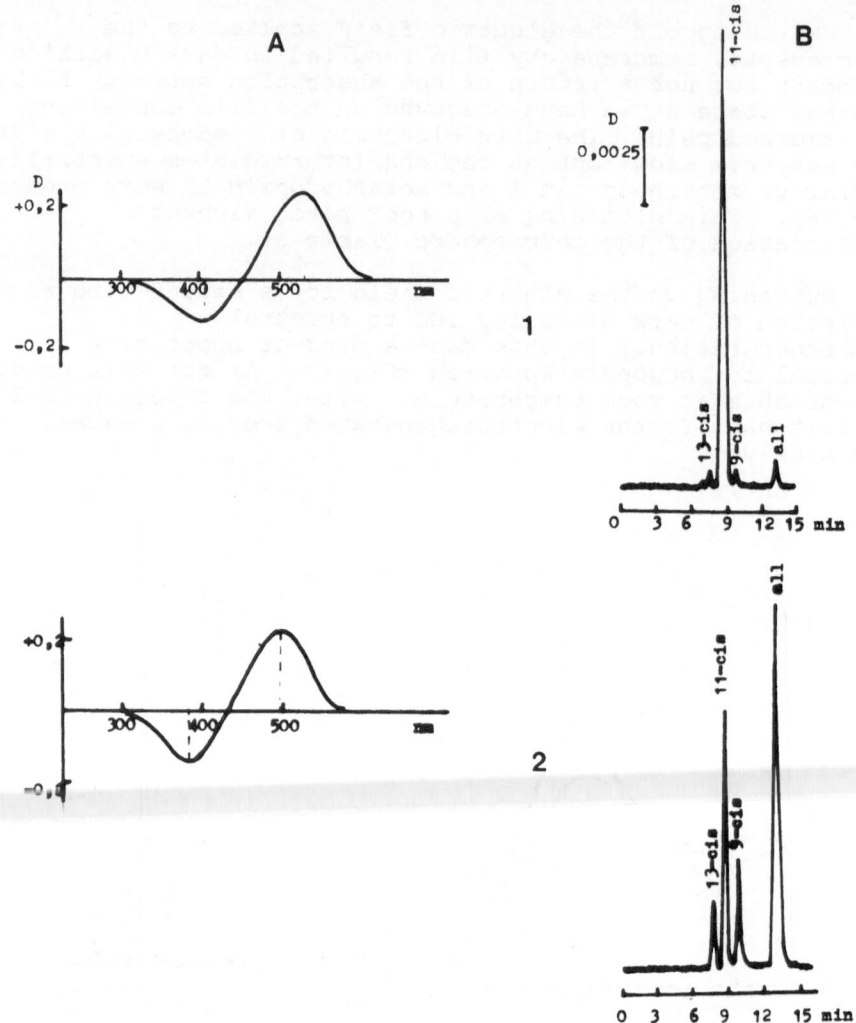

Fig. 7 Electroinduced regeneration of rhodopsin in
photoreceptor membrane dry film. A)
Differential spectra B) HPLC data. The
electric field was applied to (1) electro-
bleached rhodopsin and (2) photobleached
rhodopsin.

Electroregeneration was also seen on the field being
applied to dry photoreceptor membrane films previously
irradiated by light. In this case the product of
electroregeneration had a maximum at about 465 nm (Fig. 7
A(2)). It would appear that this product was similar to the
product of photoregeneration from photo-metarhodopsin II, but
in contrast to photo P_{465} contained all-trans retinal (Table
5).

So changing the conformational state of the protein part
under the influence of an electric field was sufficient for
the appearance of electrointermediates, which were spectrally
similar to photointermediates but without chromophore
isomerisation.

554

Thus the different electronic conformational states of a given pigment can show a great variety of spectral intermediates unconnected to the isomerization of a chromophore.

The further investigation of molecular dynamics of visual pigments will give us more detailed information about the nature of these states.

REFERENCES

Borisevitch, G.P., Lukashev, E.P., Kononenko, A.A. and Rubin, A.B., 1979, Bacteriorhodopsin (BR$_{570}$) bathochoromic band shift in an external electric field, Biochim. et Biophys. Acta, 546:171.
Groenendijk, G.W.T., De Grip, W.J., Daemen, M., 1980, Quantative determination of retinals with complete retention of their geometric configuration, Biochim. Biophys. Acta, 617:430.
Groma, G.I., Varo, G., Keszthelyi, L., 1988, Effect of the electric field on the photocycle of bacteriorhodopsin, Proc. of the Yamada Conf. XXXI:97.
Krongaus, V.A., Shifrina, R., Fedorovich, I.B. and Ostrovsky, M.A., 1975a, Photochromism of visual pigments. Formation of isochromic products at the reversible transformation of frog rhodopsin, Biofisica, 20:219.
Krongaus, V.A., Shifrina, R., Fedorovich, I.B. and Ostrovsky, M.A., 1975b, Photochromism of visual pigments. Kinetics of phototransformation of frog rhodopsin, Biofisica, 20:419.
Krongaus, V.A., Shifrina, R., Fedorovich, I.B. and Ostrovsky, M.A., 1975c, Photochromism of visual pigments. Comparative study of phototransformation of bovine and frog rhodopsin, Biofisica, 20:426.
Maeda, A., Schichida, Y. and Yoshizawa, T., 1979, Formation of 7-cis and 13-cis retinal pigments by irradiating squid rhodopsin, Biochimestry, 18:1449.
Maisel, S.M., Fedorovich, I.B., Borisevich, G.P., Ostrovsky, M.A. and Rubin, A.B., 1984, Electrochromic reactions of rhodopsin, Dokl. Acad. Nauk USSR, 277:723.
Maximychev, A.B. and Kononenko, A.A., 1988, Photochromic, electrochromic and photoelectric properties of bacteriorhodopsin oriented films. Possible applied aspects, Fizichescaj chimia, 62:2753.
Ostrovsky, M.A., Fedorovich, I.B., Darmanyan, A.P., Williams, T.P. and Kuxmin, V.A., 1974, Visual pigment regeneration by flash in Abstracs II Soviet conference on photochemistry, Suchumi, p. 253.
Protasova, T.B., Fedorovich, I.B., Borisevich, G.P., Ostrovsky, M.A. and Rubin, A.B., 1987, Photo- and electroinduced changes of rhodopsin in photoreceptor membrane dry films. Biol. membranes. 4:695.
Protasova, T.B., Tarasov, V.F. and Fedorovich, I.B., 1989, The retinal isomeric composition in the rhodopsin illuminated at 77K, Sensory Systems, 3:12.
Suzuki, T., Fujita, Y., Noda, Y. and Miyata, S., 1986, A simple procedure for the native chromophore of visual pigments: the formaldehyde method, Vision Res., 26:425.
Yoshizawa, T. and Kito, Y., 1958, Chemistry of rhodopsin cycle, Nature, 182:1604.

ARCHAERHODOPSINS: NOVEL BACTERIAL PROTON PUMPS

Yasuo Mukohata, Yasuo Sugiyama and Koichi Uegaki

Department of Biology, Faculty of Science
Nagoya University, Nagoya 464-01 Japan

INTRODUCTION

In 1983 a Japanese survey team drove around the desert in Western Australia to collect new extreme halophiles which may contain new types of retinal proteins. Many bacterial colonies were found on high-salt agar plates where soil from clay pans and water from salt lakes had been placed and incubated. These colonies were coloured yellow, orange, vermilion, pink, red and purple and were either translucent or opaque (Mukohata et al. 1988).

These Australian bacteria were rod shaped, around 3-7 μm in length, and grew only aerobically in 3 - 4 M NaCl media with peptone as a nutrient. The bacteria were resistant against 100 μg/ml penicilline G in culture media. The membrane lipids were not saponifiable and were thus postulated to be ether lipids. The GC content of major DNA was about 65% by hyperchromic melting point and bouyant density measurements. Such features (Mukohata et al. 1988) are characteristics of archaebacteria and these bacteria were therefore tentatively and successively classified into Halobacterium sp. aus-1, aus-2 and so on.

The intracellular ATP level of these bacteria was maintained by respiration and thus under nitrogen it decreased to almost nil. The initial aerobic ATP level was restored not only by aeration but by actinic illumination. This suggested that photophosphorylation as found with bacteriorhodopsin (bR) (Danon and Stoeckenius, 1974) or halorhodopsin (hR) (Mukohata et al. 1980) also occurred in these bacteria. That is, these bacteria would contain such a light-energy transducer. ATP was also synthesized by a pH jump (outside acidic) in the substrate-stuffed cell envelope vesicles as well as those of H. halobium (Mukohata et al, 1986). The ATP formation was abolished by 1 mM triphenylmethylphosphonium cation (TPMP$^+$; lipophilic cation uncoupler).

The polyclonal antibody raised against the H. halobium ATPase/synthase (Mukohata et al, 1987) identified its characteristic subunits (86k and 64 kDa on SDS-PAGE; electrophoresis on polyacrylamide gel in the presence of sodim dodecylsulfate; Nanba and Mukohata, 1987) in the membrane proteins of H. sp. aus-1 and aus-2 by Western blotting. These results suggested that H. sp. aus-1 and aus-2 carry a H^+-translocating archae-ATP-synthase (Mukohata and Yoshida, 1987) (A-type ATPase), which is specific to archaebacteria (Mukohata et al, 1987).

The pH of suspensions of envelope vesicles of both aus-1 and aus-2 cells in 3 M KCl (instead of NaCl in order to avoid complications due to the Na^+/H^+ antiport; Lanyi and MacDonald, 1976) decreased under actinic light together with the rise of negative membrane potential which increased monotonously to about -100 mV. Carbonylcyanide m-chlorophenylhydrazone (CCCP; protonophore) abolished both the pH decrease and the membrane potential increase. $TPMP^+$ and valinomycin (in a KCl suspension) intensified the pH decrease possibly by nullifying the potential. These results suggested that a light-driven electrogenic proton-extruding pump functions on the membranes. The action spectrum for proton pumping also suggested that the functioning pigment is similar to bacteriorhodopsin (Mukohata et al. 1988).

Fractions of claret-coloured membrane-1 and -2 were separated from the lysates of the aus-1 and aus-2 cells, respectively, by sucrose density gradient centrifugation at 35 - 30 w/w % (purple membrane at 40 %). When these claret membranes were incorporated into asolectin liposomes, the proteoliposomes showed CCCP-sensitive proton uptake in the light as well as those of purple membrane did. When the claret membranes were bleached by NH_2OH-light treatment a spectrum of bacterioruberin, a characteristic isoprenoid pigment of halobacteria, was left unbleached (Mukohata et al., 1988). The NH_2OH-bleaching, which cleaves a retinal Schiff base in bacteriorhodopsin (Oesterhelt and Stoeckenius, 1974), gave the difference spectra on both claret membranes. The difference spectra resembled the absorption spectrum of bacteriorhodopsin in both cases. The bleached pigments were converted back to the original ones by adding all-trans retinal. Restoration of the visible absorption spectra were parallelled by the recovery of their proton pumping activity. These observations suggested that the claret-coloured membrane-1 and -2 contain a retinal-protein or retinal-proteins which pump protons in the light, as bacteriorhodopsin does. SDS-PAGE of the claret-colored membranes suggested that the pigment proteins are almost only one protein component in individual membranes. The claret membrane-1 contained bacterioruberin and retinal in about 1:1 mole ratio. This high content of bacterioruberin makes the membrane claret instead of purple. Because this bacterioruberin shows circular dichroism (Sugiyama et al, to be published), the isoprenoid may be present in the membrane with a strong interaction to a protein moiety.

By laser flash photolysis (Mukohata et al, to be published), the pigment protein in the claret membranes were transiently bleached, and spontaneously recovered within about 10 msec. The photochemical intermediate with absorption maximum at around 410 nm appeared in both claret membrane-1 and -2 and showed the longest life time similar to the M intermediate of bacteriorhodopsin. So far no appreciable difference in the photocycle has been observed, nor has energy transfer from bacterioruberin to the pigment protein been detected. However, bacterioruberin seemed to accelerate the photocycle to some extent.

Two individual pigment protiens were solubilised from their claret-colored membranes and isolated by column chromatography by a similar method to that used for halorhodopsin isolation (Sugiyama and Mukohata 1984). Their absorption spectra, with the maximum at around 560 nm, were very similar to that of bacteriorhodopsin. The proteins ran on SDS-PAGE gel at a rate very similar to bacteriorhodopsin, suggesting their molecular sizes are close to that of bacteriorhodopsin. However, the pigment proteins in the claret-membrane-1 and -2 and those proteins isolated from the membranes differed in susceptibility to protease digestion. The peptide mapping clearly showed that the pigment proteins from claret membrane-1 and -2 differed not only from each other but also from bacteriorhodopsin (Mukohata et al, 1988). In addition, the N-terminus of the aus-1 pigment protein was analysed to be Thr instead of pyro-Glu of bacteriorhodopsin (Gerber at al. 1979). Therefore, these newly found retinal proteins were clearly the light-driven proton pumps, and we named them as archaerhoropsin (aR) (Mukohata et al, 1988) and archaerhodopsin-II (aR-2).

In order to clarify the nature of these new proton pumps their amino acid sequences were determined. The gene coding for aR has been cloned and the DNA sequence analysed (Sugiyama et al, 1989). The gene coding for aR-2 was also cloned and analysed (Uegaki et al, 1991). Here we compare three proton pumps which are presently available together with halorhodopsin (hR) in their primary structures and will discuss the structures in relation to function.

MATERIALS AND METHODS

Halobacterium sp. aus-1 and aus-2 were grown as described previously (Matsuno-Yagi and Mukohata, 1977). Apoproteins of aR and aR-2 were purified by column chromatography from the corresponding claret-coloured membranes after being solubilised with 1 % sodium dodecyl sulfate and digested by lysylendopeptidase (aR) or proteinase K (aR-2). The peptide fragments were separated by reversed phase columns and sequenced on a gas phase sequencer. Oligonucleotide probes were synthesised on the basis of these partial sequence data.

Total DNA was prepared from H. sp. aus-1 and aus-2 as described elswhere (Betlach et al. 1983). The DNA was digested with Sau3AI (aR DNA) or SmaI (aR-2 DNA) and ligated into BamHI (aR DNA) or SmaI (aR-2 DNA) site of pUC18. Recombinant DNA was introduced into E.coli JM83 and ampicillin resistant colonies were screened by filter hybridisation. Plasmid DNA was prepared and sequenced from both directions by the dideoxy chain termination method. Details of these sequence processes are described separately (Sugiyama et al, 1989, Uegaki et al, 1991). Restriction enzymes were purchased

from Toyobo and other chemicals (highest grade avaialble) from
Nakarai Tesgue Co.

RESULTS AND DISCUSSION

The genes coding for aR (Sugiyama et al, 1989) and aR-2
(Uegaki et al, 1991) were isolated and the amino acid
sequences were deduced from the corresponding DNA sequences.

In Table 1 the amino acid compositions of the four ion
pumps are compared. The Cys residue is absent in bR and aR-2
but two are present in hR (Cys-149 and -174; the number of
amino acid is given as shown in Table 2) and one in AR (Cys-
174), which is conserved at one of the two Cys in hR. It was
reported that there is one (actually two) Cys residue in hR
which can be modified by mercuric compounds and can modulate
photochemical activity (Ariki et al, 1984). Since the
photochemical and pump activiities of aR were not inhibited by
mercuric compounds, the Cys residue which affects the
photochemical activity may be Cys-149 in the helix D or one
may speculate a specific role in Cl⁻ pumping for Cys-149. His
residue is missing in bR and aR-2 but present in hR (His-99)
and aR (His-159). These His residues would not have any
importance in structure and function of these ion pumps.

Table 2 compares the amino acid sequences of four retinal
protein ion pumps which are currently available. In Table 2
the amino acids involved in the transmembrane helices are also
indicated based on the hydropathy profiles of the proton pumps
plotted with the same amino acid span of 7. In Table 3 the
percentage homologies in amino acid sequences between ion
pumps are computed. The homology between aR and aR-2 is as
high as 85% and that between two aR's and bR is 50 - 60 %.
The homology between new and old proton pumps and the

Table 1. The amino acid composition of three proton
          pumps (aR, aR-2 and bR) and a chloride
          pump (hR)

	aR	aR-2	bR	hR		aR	aR-2	bR	hR
Gly	26	24	25	21	Ser	12	11	13	20
Ala	29	32	29	35	Thr	23	24	18	17
Val	22	20	21	30	Asn	1	2	3	4
Leu	40	40	36	37	Gln	1	1	4	3
Ile	18	17	15	14	Cys	1	0	0	2
Met	6	4	9	10					
Phe	11	13	13	10	Lys	5	5	7	1
Trp	7	7	8	10	His	1	0	0	1
Pro	8	7	11	6	Arg	9	11	7	11
Asp	10	10	9	6	Tyr	10	11	11	7
Glu	12	12	9	6					
					Total	252	251	248	251

Table 2. The primary amino acid sequences of archaerhodopsin (aR), archaerhodopsin-II (aR-2), bacteriorhodopsin (bR) and halorhodopsin (hR).

```
       1          11      21        31      41
aR     ------------MDPIALTAAVGADLLGDGRPETLWLGIGTLLMLIGTFY
aR-2              MDPIAL--QAGFDLLNDGRPETLWLGIGTLLMLIGTFY
bR               MLELLPTAV--EGVSQAQITGRPEWIWLALGTALMGLGTLY
hR     MSITSVPGVVDAGVLGAQ--SAAAVRENALLSSSLWVNV--ALAGIAILV
                                        ------ helix A -
       51     61       71        81       91
aR     FIVKGWGVTDKEAREYYSITILVPGIASAAYLSMFFGIGLTEVQVGS---
aR-2   FIARGWGVTDKEAREYYAITILVPGIASAAYLAMFFGIGVTEVELASG--
bR     FLVKGMGVSDPDAKKFYAITTLVPAIAFTMYLSMLLGYGLTMVPFGG---
hR     FVYMGRTIRPGRPRLIWGATLMIPLVSISSYLGLLSGLTVGMIEMPAGHA
       ------           ------ helix B -----------
       101    111      121       131      141
aR     ---EMLDIYYARYADWLFTTPLLLLDLALLAKVDRVSIGTLVGVDALMIV
aR-2   ---TVLDIYYARYADWLFTTPLLLLDLALLAKVDRVTIGTLIGVDALMIV
bR     ---EQNPIYWARYADWLFTTPLLLLDLALLVDADQGTILALVGADGIMIG
hR     LAGEMVRSQWGRYLTWALSTPMILLALGLLADVDLGSLFTVIAADIGMCV
                ------ helix C ----     --- helix D -
       151     161      171       181      191
aR     TGLVGALSHTPLARYTWWLFSTICMIVVLYFLA---TSLR-AAAKERGPE
aR-2   TGLIGALSKTPLARYTWWLFSTIAFLFVLYYLL---TSLRSAAAK-RSEE
bR     TGLVGALTKVYSYRFVWWAISTAAMLYILYVLFFGFTS-K-AESM-R-PE
hR     TGLAAAMTTSALL-FRWAFYAIMC-AFFVVVLSALVTDW--AASA-SSAG
       -------           ------ helix E --------
       201      211     221       231      241
aR     VASTFNTLTALVLVLWTAYPILWIIGTEGAGVV-GLGIETLLFMVLDVTA
aR-2   VRSTFNTLTALVAVLWTAYPILWIVGTEGAGVV-GLGIETLAFMVLDVTA
bR     VASTFKVLRNVTVVLWSAYPVVWLIGSEGAGIVP-LNIETLLFMVLDVSA
hR     TAEIFDTLRVLTVVLWLGYPIVWAVGVEGLALVQSVGVTSWAYSVLDVFA
           ----- helix F ------      -------- helix G -
       251      261     271       281
aR     KVGFGFILLRSRAILGDTEAPEPSAGAEASAAD
aR-2   KVGFGFVLLRSRAILGETEAPEPSAGADASAAD
bR     KVGFGLILLRSRAIFGEAEAPEPSAG-DGAAATSD
hR     KYVFAFILLRWVA---NNERTVAVAG-QTLGTMSSDD
       ---------
```

Those residues conserved in all ion pumps are marked by a double line and those conserved in the top three (proton pumps) by a single line. The amino acids are numbered from the longest N-terminus to the longest C terminus and this numbering is used in the text. The positions of (possible) cleavage by processing are shown by arrow heads. The amino acids involved in the transmembrane helices are estimated by hydropathy plot (Kyte and Doolittle, 1982) and indicated by A to G from the N Terminus.

chloride pump is 32 %. Although there is 20% mutation between aR's and 40% between bR and aR's, the hydropathy profiles of aR's suggest a seven-helix structure (cf. Table 2) as reported for bR (Kyte and Doolittle, 1982). This means that mutation took place mostly by conservative substitution of amino acids or at the margins and loops of the molecules. This is true especially on helix C whose amino acids are 100% conserved in all three proton pumps (Table 2). Most of the amino acids on helix G are also conserved. These two helices would be specially important for the transmembrane architecture of every proton pump molecule. The amino acid residues conserved in helix F are commonly found in all ion pumps. These residues and/or the transmembrane architecture which holds these residues at fixed distances may play an important role (or roles) in the ion pumps. It should be noted that helices C, F and G compose a trimer structure facing each other with a central space in which retinal is located.

Lys residue (Lys-251) on helix G, to which retinal forms a Schiff base, is conserved in all pumps. Although additional Lys residues are found in the proton pumps, none of them in the helical region is common to all three proton pumps. Even from this point of view, Lys residues may be hardly involved in proton pumping. The Arg-112 on helix C is conserved in all pumps as the possible positive charge contribution to the opsin shift (Nakanishi et al 1980). When compared only with bR, hR seemed to have more Arg residues and thus it is suggested that these, but not all, Arg residues are part of the Cl⁻ pump machinery (Dlanck and Oesterhelt, 1987) in analogy to the carboxyl cascade machinery of bR proton pump (Nagle and Mille, 1981). However, the number of Arg residues in aR-2 and hR are the same and these are located near or in the hydrophilic region as judged from hydropathy profiles (cf. Table 2). So, although some of these Arg residues would be the Cl⁻ binding sites, their direct involvement in Cl⁻ pumping is still ambiguous unless one assumes more helical turns than expected from the hydropathy plots.

One may pay attention to the sum of acidic amino acids in hR, which is two third of bR and almost one half of aR and aR-2. This enrichment of Asp and Glu residues in the proton pumps, of course relates to the carboxyl cascade in proton pumping. Along the transmembrane helices of the proton pumps, two acidic residues, Asp-115 and Asp-126 on helix C and Glu-239 on the helix G, are conserved in proton pumps but missing in hR. This implies that these acidic residues are essential for proton pumping, as many other data by site-directed mutagenesis (Mogi et al., 1988) and FTIR (Englehard et al., 1985) have suggested. These negative charges of Asp-145 on helix D, Glu-228 on helix F and Asp-247 on helix G are conserved in all ion pumps, suggesting their (possibly two out of three residues) involvement in the opsin shift. Other negative charges (Glu-33, Asp-60, Glu-200, -271 and -274 in proton pumps and Asp-134 in all ion pumps) are located near the edge of the helices or on the loops between helices or on the margins of the C and N termina. The role(s) of these charges and hydrophilic residues near both ends (edges) of helices facing hydrophilic milieu may be to hold individual helices at adequate positions. The roles of the long and short loops and margins has not been clarified although several amino acides are commonly conserved there. In visual rhodopsins some loops are important for signal transduction.

Among Tyr residues in the helix structure, Tyr-50 on helix A and Tyr-179 on helix E are conserved among the proton pumps, while Tyr-81, - 113 and -219 are conserved in all ion pumps. Tyr-50 and Tyr -179 would be involved in the proton pumping mechanism. Tyr-219 was also reported to be involved in the pumping mechanism (Braiman et al 1988).

One may also pay attention to the Thr/Ser residues in the proton pumps; Thr-41 and -48 on helix A, Ser-171 and Thr-172 on helix E, Ser/Thr-217 and -227 on helix F and Ser/Thr-249 on helix G, which are all conserved only in the proton pumps and not present in hR in their vicinities. They may serve as mediators for proton pumping.

Homologies between these ion pumps are shown in Table 3. Between aR's the homology is computed to be 85% while it is about 60% between bR and either one of aR's. Between the hR chloride pump and each one of those proton pumps the homology values are about 32%. This would suggest that the chloride pump and the proton pump diverged relatively long ago from the ancestry rhodopsin on halobacteria, and that bR and aR's were, most likely, diverged when the Gondwanaland was separated from the Australia/Antarctic continent (say 100 million years ago). Within these 100 million years aR's in Australia has mutated to result in two aR's with 85% amino acid homology. If mutation took place at a constant rate, rough estimation suggests that the ancestry rhodopsin may have diverged into hR and proton pumps about 220 million years ago. AR and aR-2 may have diverged 30 million years ago. (If another proton pump with less homology is found in either one of the continents, this value becomes smaller or the assumption of the bR-aR diversion time becomes inadequate. The rate of mutation would be 2 x 10^{-9}/amino acid/year.

Here we describe some charateristics of aR and aR-2. We are cloning the aR-3 gene from $\underline{H}$. sp. aus-3 and the halo-aR gene from $\underline{H}$. sp. aus-4. We have predicted there should be more pumps in other halobacteria (Mukohata et al 1988). Actually not only from our Australian strains but even from the available collection of halobacterial strains, new bacterial rhodopsins are found. As partly shown here, aR and aR-2 give considerable information on the primary structure of

Table 3. Percentage homology between amino acid
          sequences of bacterial rhodopsins

	aR	aR-2	bR	hR
aR		85	60	32
aR-2	85		56	33
bR	60	56		32
hR	32	33	32	

the pumps. Accumulation of this kind of data gives us new
insight not only on the conserved essential amino acid
residues but also on the domain or module (whole or part of
helix or loop or margin) essential for stucture and/or
function of the rhodopsin ion pumps.

SUMMARY

Cells of two strains of extreme halophile collected in
Western Australia are rod shaped, show the characteristic
features of archaebacteria and thus are tentatively classified
into Halobacterium sp. aus-1 and aus-2. These cells contain
retinal proteins which pump out protons under actinic
illumination. The pump proteins are different from each other
and, moreover, clearly distinguished from bacteriorhodopsin of
Halobacterium salinarium (halobium). These two novel proton
pumps of archaebacterial rhodopsin are named
"archaerhodopsins", archaerhodopsin in H. sp. aus-1 and
archaerhodopsin-II in H sp. aus-2.

The genes coding for archaerhodopsin and archaerhodopsin-
II have been isolated and the DNA sequences analysed. From
the amino acid sequences deduced from them the homology
between archaerhodopsin and archaerhodopsin-II is computed to
be 85%, while those between the new and old proton pumps are
about 60%. Homology between one of these proton pumps and
halorhodopsin was about 32%. The hydropathy profiles of the
new proton pumps imply the presence of seven bacteriorhodopsin like transmembrane helices. The amino acid sequence
of helix C is perfectly conserved among three proton-pumping
rhodopsins and that of the retinal-binding helix G is
preserved almost as well. This kind of comparative
biochemistry of rhodopsin ion pumps in relation to
physiochemical functions is discussed.

ACKNOWLEDGEMENTS

This work was supported in part by Grant-in-Aid for
Scientific Research in Priority Areas of "Bioenergetics
(Energetics of Extremophiles)" to Y.M. (#62617003 & #63617003)
from the Ministry of Education, Science and Culture of Japan.

REFERENCES

Ariki, M and Lanyi, J.K, 1984, Evidence for a sulfhydryl group
    near the retinal-binding site of halorhodopsin. J. Biol.
    Chem. 259:3504-3510
Betlach, M.J., Pfeifer, F., Friedman, J., and Boyer, H.W,
    1983, Bacterio-opsin mutants of Halobacterium halobium.
    Proc. Natl. Acad. Sci. USA 80:1416-1420
Blanck A. and Oesterhelt, D., 1987, The halo-opsin gene. II.
    Sequence, primary structure of halorhodopsin and
    comparison with bacteriorhodopsin. EMBO J. 6:265-273
Braiman, M.S. Mogi, T., Marti, T., Stern, L.J., Khorana, H.G.
    and Rothchild, K.J., 1988, Vibrational spectroscopy of
    bacteriorhodopsin mutants: light-driven proton transport
    involves protonation changes of aspartic acid residues
    85, 96 and 212. Biochemistry 27:8516-8520

Danon, A. and Stoeckenius, W., 1974, Photophosphorylation in
Halobacterium halobium. Proc. Natl. Acad. Sci. USA
71:1234-1238

Engelhard, M., Gerwert, K., Hess, B., Kreutz, W. and Siebert,
F., 1985, Light-driven protonation changes of internal
aspartic acides of bacteriorhodopsin: An investigation
by static and time-resolved infrared difference
spectrosocpy using $[4-^{13}C]$ aspartic acid labelled purple
membrane. Biochemistry 24:400-407

Gerber, G.E., Anderegg, R.J., Herligy, W.C., Gray, C.P.,
Biemann, K. and Khorana, H.G., 1979, Partial primary
structure of bacteriorhodopsin: sequencing methods for
membrane proteins. Proc. Natl. Acad. Sci. USA 76:227-231

Kyte, J. and Doolittle, R.F., 1982, A simple method for
displaying the hydrophobic character of a protein. J.
Mol. Biol. 157:105-132

Lanyi, J.K. and MacDonald, R.E., 1976, Existence of
electrogenic hydrogen ion/sodium ion antiport in
Halobacterium halobium cell envelope vesicles.
Biochemistry 15:4608-4614

Matsuno-Yagi, A. and Mukohata, Y., 1977, Two possible roles
of bacteriorhodopsin; a comparative study of strains of
Halobacterium halobium differing in pigmentation.
Biochem. Biophys. Res. Commun. 78:237-243

Mogi, T., Stern, L.J., Marti, T., Chao, B.H. and Khorana,
H.G., 1988, Aspartic acid substitutions affect proton
translocation by bacteriorhodopsin. Proc. Natl. Acad.
Sci. USA 85:4148-4152

Mukohata, Y., Matsuno-Yagi, A. and Kaji, Y., 1980, Light-
induced proton uptake and ATP synthesis by
bacteriorhodopsin-depleted Halobacterium, in "Saline
Environment" (H. Morishita and M. Masui eds), pp. 31-37,
Business Ctr. Acad. Soc. Japan, Tokyo

Mukohata, Y., Isoyama, M. and Fuke, A., 1986, ATP synthesis
in cell envelope vesicles of Halobacterium halobium
driven by membrane potential and/or base-acid transition.
J. Biochem. 99:1-8

Mukohata, Y. and Yoshida, M., 1987, The H^+-translocating ATP
synthase in Halobacterium halobium differs from F_oF_1-
ATPase/synthase. J. Biochem. 102:797-802

Mukohata, Y., Ihara, K., Yoshida, M., Konishi, J., Sugiyama,
Y. and Yoshida, M., 1987, The halobacterial H^+-
translocating ATP Synthase relates to the eukaryotic
anion-sensitive H^+-ATPase. Arch. Biochem. Biophys.
259:650-653

Mukohata, Y., Sugiyama, Y., Ihara, K. and Yoshida, M., 1988,
An Australian halobacterium contains a novel proton pump
retinal protein: Archaerhodopsin. Biochem. Biophys. Res.
Commun. 151:1339-1345

Nagle, J.F. and Mille, M., 1981, Molecular models of proton
pumps. J.Chem. Phys. 74:1367-1372

Nakanishi, K., Balogh-Nair, V., Arnaboldi, M., Tsujimoto, K.
and Honig, B., 1980, An external point-charge model for
bacteriorhodopsin to account for its purple colour. J.
Am.Chem. Soc. 102:7945-7947

Nanba, T. and Mukohata, Y., 1987, A membrane-bound ATPase
from Halobacterium halobium: Purification and
characterisation. J. Biochem. 102:591-598

Oesterhelt, D. and Stoeckenius, W., 1971, Rhodopsin-like
protein from the purple membrane of Halobacterium
halobium. Nature new Biol. 233:149-152

Sugiyama, Y. and Mukohata, Y., 1984, Isolation and
     characterisation of halorhodopsin from <u>Halobacterium
     halobium</u>. <u>J. Biochem.</u> 96:413-420
Sugiyama, Y., Maeda, M., Futai, M. and Mukohata, Y., 1989,
     Isolation of a gene that encodes a new retinal protein,
     archaerhodopsin, from <u>Halobacterium</u> sp. aus-1, <u>J. Biol.
     Chem</u>. 264:20859-20862
Uegaki, K., Sugiyama, Y. and Mukohata, Y., 1991,
     Archaerhodopsin-2, from <u>Halobacterium</u> sp. aus-2 further
     reveals essential amino acid residues for light-driven
     proton pumps, <u>Arch. Biochem. Biophys</u>., 286:107-110

INDEX